Computational Statistics in the Earth Sciences

Based on a course taught by the author, this book combines the theoretical underpinnings of statistics with the practical analysis of earth sciences data using MATLAB. The book is organized to introduce the underlying concepts and then extends these to data, covering methods that are most applicable to earth sciences. Topics include classical parametric estimation and hypothesis testing and more advanced nonparametric and resampling estimators. The method of least squares is explored in detail. Multivariate data analysis, not often encountered in introductory texts, is presented later in the book, and compositional data are treated at the end. Data sets and bespoke MATLAB scripts used in the book are available online, as well as additional data sets and suggested questions for use by instructors. Aimed at entering graduate students and practicing researchers in the earth and ocean sciences, this book is ideal for those who want to learn how to analyze data using MATLAB in a statistically rigorous manner.

Alan D. Chave is a senior scientist at Woods Hole Oceanographic Institution and holds the Walter A. and Hope Noyes Smith Chair for Excellence in Oceanography. He has also been a Chartered Statistician (United Kingdom) since 2003 and has taught a graduate-level course in statistics in the MIT/WHOI Joint Program for 20 years. For over 30 years he has conducted research using the magnetotelluric method, primarily in the oceans, and electromagnetic measurements to define the barotropic water velocity. Dr. Chave has also designed instrumentation for optical and chemical measurements in the ocean and has played a leadership role in developing long-term ocean observatories worldwide. He has been an editor of the *Journal of Geophysical Research* and editor-in-chief of *Reviews of Geophysics* and is the co-author of *The Magnetotelluric Method* (Cambridge University Press, 2012).

Computational Statistics in the Earth Sciences

With Applications in MATLAB

ALAN D. CHAVE
Woods Hole Oceanographic Institution, Woods Hole, Massachusetts

CAMBRIDGE
UNIVERSITY PRESS

University Printing House, Cambridge CB2 8BS, United Kingdom

One Liberty Plaza, 20th Floor, New York, NY 10006, USA

477 Williamstown Road, Port Melbourne, VIC 3207, Australia

314–321, 3rd Floor, Plot 3, Splendor Forum, Jasola District Centre, New Delhi – 110025, India

79 Anson Road, #06–04/06, Singapore 079906

Cambridge University Press is part of the University of Cambridge.

It furthers the University's mission by disseminating knowledge in the pursuit of education, learning, and research at the highest international levels of excellence.

www.cambridge.org
Information on this title: www.cambridge.org/9781107096004
DOI: 10.1017/9781316156100

First published 2017

Printed in the United Kingdom by TJ International Ltd. Padstow Cornwall

A catalogue record for this publication is available from the British Library.

Library of Congress Cataloging-in-Publication Data
Names: Chave, Alan Dana.
Title: Computational statistics in the earth sciences : with applications in MATLAB / Alan D. Chave, Woods Hole Oceanographic Institution, Woods Hole, Massachusetts.
Description: Cambridge : Cambridge University Press, 2017. | Includes bibliographical references and index.
Identifiers: LCCN 2017009160 | ISBN 9781107096004 (hardback)
Subjects: LCSH: Earth sciences–Statistical methods. | Earth sciences–Statistical methods–Data processing. | Mathematical statistics–Data processing. | MATLAB.
Classification: LCC QE26.3 .C43 2017 | DDC 519.50285–dc23 LC record available at https://lccn.loc.gov/2017009160

ISBN 978-1-107-09600-4 Hardback

Additional resources for this publication at www.cambridge.org/chave.

Contents

Preface *page* xi

1 Probability Concepts 1
1.1 Fundamental Concepts and Motivating Example 1
1.2 Probability Philosophies 4
1.3 Set Theory 5
1.4 Definition of Probability 8
1.5 Finite Sample Spaces 10
1.5.1 Simple Sample Spaces 10
1.5.2 Permutations 11
1.5.3 Combinations 12
1.6 Probability of a Union of Events 13
1.7 Conditional Probability 14
1.8 Independent Events 15
1.9 Bayes' Theorem 16

2 Statistical Concepts 18
2.1 Overview 18
2.2 The Probability Density Function 18
2.2.1 Discrete Distributions 18
2.2.2 Continuous Distributions 20
2.2.3 Mixed Distributions 22
2.3 The Cumulative Distribution and Quantile Functions 23
2.4 The Characteristic Function 25
2.5 Bivariate Distributions 27
2.6 Independent and Exchangeable Random Variables 28
2.7 Conditional Probability Distributions 30
2.8 Functions of a Random Variable 32
2.9 Functions of Two or More Random Variables 34
2.10 Measures of Location 36
2.11 Measures of Dispersion 39
2.12 Measures of Shape 41
2.13 Measures of Direction 42
2.14 Measures of Association 44
2.15 Conditional Expected Value and Variance 44

2.16 Probability Inequalities 45
2.17 Convergence of Random Variables 46

3 Statistical Distributions 48
3.1 Overview 48
3.2 MATLAB Support for Distributions 48
3.3 Discrete Distributions 49
3.3.1 Bernoulli Distribution 49
3.3.2 Binomial Distribution 50
3.3.3 Negative Binomial Distribution 53
3.3.4 Multinomial Distribution 54
3.3.5 Hypergeometric Distribution 56
3.3.6 Poisson Distribution 59
3.4 Continuous Distributions 62
3.4.1 Normal or Gaussian Distribution 62
3.4.2 Stable Distributions 65
3.4.3 Rayleigh Distribution 67
3.4.4 Lognormal Distribution 69
3.4.5 Gamma Distribution 70
3.4.6 Exponential Distribution 72
3.4.7 Weibull Distribution 74
3.4.8 Beta Distribution 76
3.4.9 Generalized Extreme Value Distribution 78
3.4.10 Bivariate Gaussian Distribution 80
3.4.11 Directional Distributions 81

4 Characterization of Data 86
4.1 Overview 86
4.2 Estimators of Location 86
4.3 Estimators of Dispersion 91
4.4 Estimators of Shape 93
4.5 Estimators of Direction 95
4.6 Estimators of Association 100
4.7 Limit Theorems 101
4.7.1 The Laws of Large Numbers 101
4.7.2 Classic Central Limit Theorems 102
4.7.3 Other Central Limit Theorems 104
4.7.4 The Delta Method 105
4.8 Exploratory Data Analysis Tools 106
4.8.1 The Probability Integral Transform 106
4.8.2 The Histogram and Empirical CDF 107
4.8.3 Kernel Density Estimators 111
4.8.4 The Percent-Percent and Quantile-Quantile Plots 115
4.8.5 Simulation 120

4.9 Sampling Distributions 122
4.9.1 Chi Square Distributions 122
4.9.2 Student's t Distributions 125
4.9.3 The F Distributions 128
4.9.4 The Correlation Coefficient 130
4.10 Distributions for Order Statistics 135
4.10.1 Distribution of a Single Order Statistic 135
4.10.2 Distribution of the Sample Median 137
4.10.3 Joint Distribution of a Pair of Order Statistics 138
4.10.4 Distribution of the Interquartile Range 138
4.11 Joint Distribution of the Sample Mean and Sample Variance 140

5 Point, Interval, and Ratio Estimators 142
5.1 Overview 142
5.2 Optimal Estimators 142
5.2.1 Consistency 142
5.2.2 Unbiased Estimators 143
5.2.3 Efficiency and the Cramér-Rao Lower Bound 144
5.2.4 Robustness 147
5.2.5 Sufficient Statistics 148
5.2.6 Statistical Decision Theory 152
5.3 Point Estimation: Method of Moments 154
5.4 Point Estimation: Maximum Likelihood Estimator 155
5.5 Interval Estimation: Confidence and Tolerance Intervals 160
5.6 Ratio Estimators 166

6 Hypothesis Testing 169
6.1 Introduction 169
6.2 Theory of Hypothesis Tests I 171
6.3 Parametric Hypothesis Tests 177
6.3.1 The z Test 177
6.3.2 The t Tests 178
6.3.3 The χ^2 Test 186
6.3.4 The F Test 188
6.3.5 Bartlett's M Test for Homogeneity of Variance 189
6.3.6 The Correlation Coefficient 190
6.3.7 Analysis of Variance 192
6.3.8 Sample Size and Power 194
6.4 Hypothesis Tests and Confidence Intervals 195
6.5 Theory of Hypothesis Tests II 196
6.5.1 Likelihood Ratio Tests for Simple Hypotheses 197
6.5.2 Uniformly Most Powerful Tests 198
6.5.3 Likelihood Ratio Tests for Composite Hypotheses 200

6.5.4 The Wald Test 207
6.5.5 The Score Test 208
6.6 Multiple Hypothesis Tests 210

7 Nonparametric Methods 214
7.1 Overview 214
7.2 Goodness-of-Fit Tests 214
7.2.1 Likelihood Ratio Test for the Multinomial Distribution 214
7.2.2 Pearson's χ^2 Test for Goodness-of-Fit 219
7.2.3 Kolmogorov-Smirnov Test 222
7.2.4 Cramér–von Mises Tests 228
7.2.5 Jarque-Bera Test 230
7.3 Tests Based on Ranks 231
7.3.1 Properties of Ranks 231
7.3.2 Sign Test 232
7.3.3 Signed Rank Test 235
7.3.4 Rank Sum Test 237
7.3.5 Ansari-Bradley Test 240
7.3.6 Spearman Rank Correlation Test 241
7.3.7 Kendall's Tau 242
7.3.8 Nonparametric ANOVA 243
7.4 Meta-analysis 245

8 Resampling Methods 247
8.1 Overview 247
8.2 The Bootstrap 247
8.2.1 The Bootstrap Distribution 247
8.2.2 Bootstrap Parameter Estimation 251
8.2.3 Bootstrap Confidence Intervals 255
8.2.4 Bootstrap Hypothesis Tests 259
8.2.5 Bias Correction for Goodness-of-Fit Tests 265
8.3 Permutation Tests 267
8.3.1 Principles 267
8.3.2 One-Sample Test for a Location Parameter 268
8.3.3 Two-Sample Test for a Location Parameter 270
8.3.4 Two-Sample Test for Paired Data 274
8.3.5 Two-Sample Test for Dispersion 275
8.4 The Jackknife 277

9 Linear Regression 281
9.1 Motivating Example 281
9.2 Statistical Basis for Linear Regression 283

9.3 Numerical Considerations 286
9.4 Statistical Inference in Linear Regression 289
9.4.1 Analysis of Variance 289
9.4.2 Hypothesis Testing on the Regression Estimates 291
9.4.3 Confidence Intervals 292
9.4.4 The Runs and Durbin-Watson Tests 294
9.5 Linear Regression in Practice 295
9.5.1 Assessing the Results 295
9.5.2 Examples 298
9.6 Robust and Bounded Influence Regression 316
9.6.1 Robust Estimators 317
9.6.2 Bounded Influence Estimators 327
9.7 Advanced Linear Regression 335
9.7.1 Errors in Variables 335
9.7.2 Shrinkage Estimators 336
9.7.3 Logistic Regression 340

10 Multivariate Statistics 344
10.1 Concepts and Notation 344
10.2 The Multivariate Gaussian Distribution 346
10.2.1 Derivation of the Multivariate Gaussian Distribution 346
10.2.2 Properties of the MV Gaussian Distribution 347
10.2.3 The Sample Mean Vector and Sample Covariance Matrix 348
10.2.4 The Complex Multivariate Gaussian Distribution 349
10.3 Hotelling's T^2 Tests 350
10.4 Multivariate Analysis of Variance 354
10.5 Hypothesis Tests on the Covariance Matrix 362
10.5.1 Sphericity Test 363
10.5.2 Comparing Covariance Matrices 364
10.5.3 Test of Independence 365
10.6 Multivariate Regression 366
10.7 Canonical Correlation 371
10.8 Empirical Orthogonal Functions 373
10.8.1 Theory 374
10.8.2 Choosing the Number of Eofs 377
10.8.3 Example 378
10.8.4 Empirical Orthogonal Function Regression 385

11 Compositional Data 391
11.1 Introduction 391
11.2 Statistical Concepts for Compositions 392
11.2.1 Definitions and Principles 392

11.2.2 Compositional Geometry 395
11.2.3 Compositional Transformations 400
11.3 Exploratory Compositional Data Analysis 405
Appendix 11A: MATLAB Functions to Produce Ternary Diagrams 429

References 435
Index 444

Preface

Statistics, and especially its application to data, is an essential tool in the sciences, and the earth and ocean sciences are no exception. However, the study of statistics beyond a very rudimentary level is usually neglected in undergraduate curricula, and hence entering graduate students in the earth and ocean sciences need to acquire a background in statistics early in their tenure. For this reason, a course entitled, "Computational Data Analysis" (MIT 12.714), was devised more than 20 years ago for the Massachusetts Institute of Technology/Woods Hole Oceanographic Institution Joint Program in Oceanography. An abbreviation of this book constitutes the first half of the course, with the remainder being devoted to spectral analysis. The computational tool used in the course and book is MATLAB, which has become nearly ubiquitous in geophysics and oceanography. The emphasis in the course is on analyzing data rather than abstract theory, and given that it is a graduate course, homework constitutes the analysis of data sets using the tools presented in lectures. Representative data sets and questions are available on the Cambridge University Press website at http://www.cambridge.org/chave. In addition, all the exemplar data sets and MATLAB functions used in this book can be found at the same location.

The book constitutes 11 chapters that present introductory statistics at an entering graduate level in a manner that is logical (at least to the author). The material in the first two chapters appears in every book on statistics, constituting the theory of probability and statistical concepts that underlie the remainder of the book. Chapter 1 introduces probability concepts first through MATLAB examples and then describes set theory and the Kolmogorov axioms that constitute the basis for probability theory. Permutation and combination are described, leading to the probability of a union of events, conditional probability, the concept of independence, and finally Bayes' theorem.

Chapter 2 is entitled "Statistical Concepts" and covers the probability density, cumulative distribution, quantile, and characteristic functions for discrete and continuous distributions. These concepts are extended to the bivariate case, and then independence and exchangeability are formally defined. Conditional, joint, and marginal distributions are introduced, and methods to transform distributions as random variables change are described. The chapter then sets out population measures of location, dispersion, shape, direction, and association. It closes by defining conditional expectation and variance that underlie the theory of least squares, and probability inequalities and convergence.

Chapter 3 describes the major discrete and continuous distributions that are encountered in the sciences and contains considerable material that is useful for reference.

Chapter 4 moves from the theoretical basis of statistics into the characterization of data. It first extends the population measures of location, dispersion, shape, direction, and association to sample entities and introduces the MATLAB tools for this purpose. Limit theorems are then described. A set of exploratory data analysis tools is set out, including the histogram, the empirical cumulative distribution function, the kernel density estimator, percent-percent and quantile-quantile plots, and simulation. These are extensively illustrated by examples using real data. The major sampling distributions used throughout the remainder of the book are introduced, and then the distributions of the order statistics are derived, including that for the sample median and interquartile distance. The chapter closes by describing the joint distribution of the sample mean and variance.

Chapter 5 first covers a set of estimator optimality criteria, including consistency, unbiasedness, efficiency, robustness, and sufficiency, which are essential to characterize estimator performance. It then introduces the methods of moments and maximum likelihood, confidence and tolerance intervals, and ratio estimators. All of these topics are characterized through MATLAB examples.

Chapter 6 describes the theory of hypothesis testing, including the concept of a likelihood ratio test and its asymptotic approximation via the Wald and score tests. A set of standard parametric hypothesis tests is introduced and illustrated with data. The concept of statistical power receives considerable emphasis. The chapter closes by setting out the testing of multiple simultaneous hypotheses that is becoming essential in the modern world of big data.

Chapter 7 covers goodness-of-fit and rank-based testing. The multinomial likelihood ratio test is emphasized over its widely used asymptotic approximation, the Pearson χ^2 goodness-of-fit test. The Kolmogorov-Smirnov test is defined, and its use to place confidence bounds on percent-percent and quantile-quantile plots is described. The more powerful Anderson-Darling goodness-of-fit test is also described. A set of standard nonparametric rank hypothesis tests is then introduced that typically lack power over their parametric counterparts but are less sensitive to mixtures of distributions that are common in earth and ocean sciences data. The chapter closes by discussing meta-analysis.

Chapter 8 describes resampling methods that are being used more extensively as computational power increases. The most widely used resampling tool is the bootstrap that is based on sampling with replacement, hence yielding approximate results that are useful in characterizing complicated estimators and confidence intervals for them. A Monte Carlo approach to removing bias from goodness-of-fit tests is then described. Permutation hypothesis testing based on resampling without replacement then receives considerable attention. Permutation methods yield exact tests and, in the author's opinion, are the tool of choice for earth and ocean sciences data.

Chapter 9 sets out the theory of linear regression beginning with examples and then the statistical basis for the realistic case where the predictors are random variables. Numerical considerations receive attention, followed by a set of statistical tools to characterize least squares estimates and quantify their consistency with their statistical basis. Procedures for assessing a linear regression are described and illustrated through examples. Robust and bounded influence extensions that are the linear regression tools of choice in the earth sciences are then defined and illustrated. Errors in variables, shrinkage estimators, and the general linear model close the chapter.

Chapter 10 introduces multivariate statistical methods beginning with the real and complex multivariate Gaussian distributions. It then covers the multivariate extensions of the Student t and F distributions, Hotelling's T^2 and Wilks' Λ, and their application in hypothesis testing, multivariate analysis of variance, and multivariate regression. Canonical correlation is introduced, and the chapter closes with the theory and application of empirical orthogonal functions (often called principal components).

Chapter 11 covers the analysis of compositional data that are highly multivariate but constrained so that the sum of their parts equals a constant for each observation. Such data are very common in the earth sciences; rock composition in weight percent or parts per million is a prominent example. The theory of compositional data has only evolved in the past 30 years and is beginning a migration from mathematical statistics into science domains. A set of MATLAB functions for analyzing compositional data and their presentation on ternary diagrams is provided to facilitate this process. Exploratory analysis of rock composition data provides an exemplar.

The MATLAB version used to produce this book is R2015b. The only changes in functionality that would affect the treatment is the addition of stable distribution objects in R2016a, and a change from **rose** to **polarhistogram** to create a polar histogram, as used in Chapters 4 and 6.

I acknowledge the two co-lecturers for 12.714, first Marcia McNutt and then Tom Herring, for their influence on the material in this book. I also recognize the graduate students at both MIT and WHOI who have suffered through the course over the past two decades.

1 Probability Concepts

1.1 Fundamental Concepts and Motivating Example

Probability is the mathematical description of uncertainty, and many of its abstractions underlie the statistical concepts that are covered in Chapter 2. The material in the first two chapters of this book is very standard, and is covered at an elementary level by DeGroot & Schervish (2011) and at a more advanced level by Wasserman (2004) and Rice (2006).

A core purpose of this section is to illustrate a set of concepts using MATLAB as a tool. An *experiment* is an activity whose result is not known a priori with certainty. An *event* or *outcome* is one element of a collection of results obtained by performing an experiment. A *simple event* is an event that cannot be broken down into some combination of other events. A *composite event* is an event that is not simple. The *sample space* is the collection of all possible outcomes of an experiment. Since the possible outcomes are known in advance, the sample space can be defined before an experiment is performed.

A motivating example can be produced using the MATLAB **unidrnd**(N, M, k) function that gives k realizations of M random draws from the integers ranging between 1 and N. For $M = 5$ and $N = 50$, the outcome of a single experiment is

```
unidrnd(50, 5, 1)
ans =
  41
  46
   7
  46
  32
```

Note that the MATLAB result depends on a random number seed and may differ depending on the version. The random number generator may be initialized with the same value by issuing the "**rng** default" command before the **unidrnd** one, and doing so before the first call will duplicate the results of this section.

If **unidrnd** is called a second time, a different outcome is obtained

```
unidrnd(50, 5, 1)
ans =
   5
  14
  28
  48
  49
```

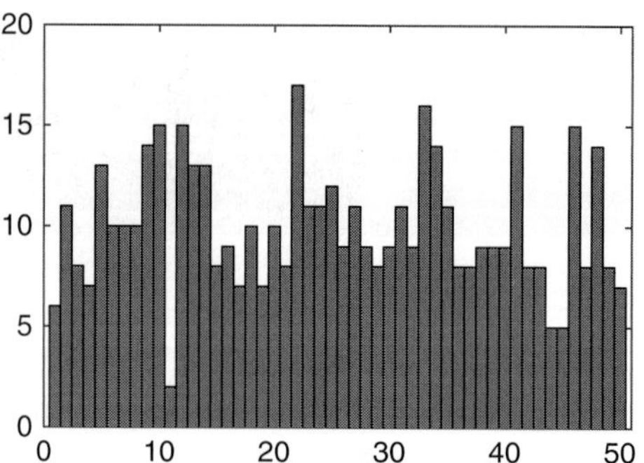

Figure 1.1 Histogram of 100 random draws of five values from the uniformly distributed integers between 1 and 50.

Each time the command is executed, an experiment is performed. The value for one experiment is an event, such as {41, 46, 7, 46, 32}. The sample space is all possible permutations of five integers lying between 1 and 50 and can be written down prior to performing any experiments. As shown in Section 1.5.3, there are 2,118,760 unique elements in the sample space.

A *probability measure* assigns a likelihood in the form of a real number lying between 0 and 1 to each event in a sample space. A probability measure for a given integer lying between 1 and N can be inferred by carrying out a large number of trials and counting the number of occurrences of that integer compared with the number of trials. It is reasonable to expect that any integer lying between 1 and N will occur with an equal probability of $1/N$ for a single trial. This can easily be checked by simulation. A heuristic demonstration obtains from plotting a histogram of 100 draws of five integers ranging from 1 to 50, as shown in Figure 1.1.

```
histogram(reshape(unidrnd(50, 5, 100), 1, 500))
```

The ordinate is the number of times a given value has occurred over 100 draws and 5 variables, whereas the abscissa assigns them to 50 equal interval bins covering the integers between 1 and 50. One would expect intuitively that the ordinate would be about 10 on average, which is approximately true, but with considerable variability. Repeating the experiment with 1000 realizations, whose outcome ought to be about 100

```
histogram(reshape(unidrnd(50, 5, 1000), 1, 5000))
```

The result appears much more uniform (Figure 1.2).

A *random variable* (rv) is a real-valued, measurable function that transforms an event into a real number. Since a random variable depends on an outcome that is not known a priori, the value of a random variable is uncertain until after an experiment has been performed. An example of an rv is the minimum value from a given trial using **unidrnd**. This can be summarized using a histogram (Figure 1.3).

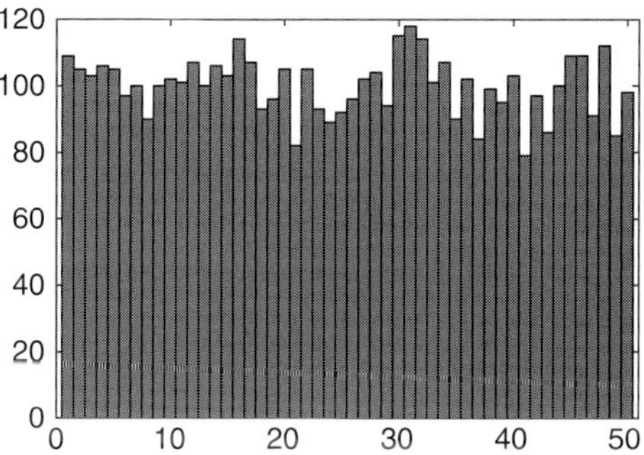

Figure 1.2 Histogram of 1000 random draws of five values from the uniformly distributed integers between 1 and 50.

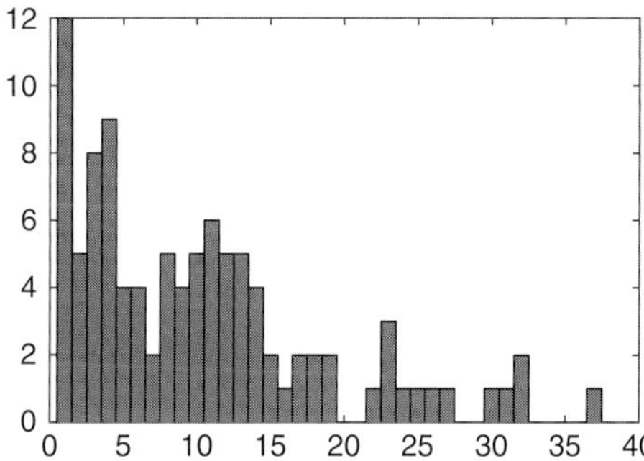

Figure 1.3 Histogram of the minimum value obtained from 100 random draws of five values from the uniformly distributed integers between 1 and 50.

```
histogram(min(unidrnd(50, 5, 100)))
```

Figure 1.4 shows the result from 1000 replications.

```
histogram(min(unidrnd(50, 5, 1000)))
```

Figures 1.3 and 1.4 hint at an important underlying idea: the value of an rv observed many times will exhibit a characteristic pattern with variability. The number of occurrences (i.e., the heights of the histogram bins) for a particular range of the integers (the abscissa) for a sufficiently large number of replications will stabilize around some set of values that defines the *probability distribution*. While the distribution for the integer draws is uniformly distributed, that for the minimum value is not.

The *expectation* or *expected value* of an rv is a single number that gives its average value (in some well-defined sense). If all of the possible values of the rv are not equally likely (as for the minimum value of uniformly distributed integer draws but not for the uniformly distributed integers themselves), it may not make sense to use the simple arithmetic average. Rather, it would be more appropriate to use a weighted average, with the weights given by the probabilities of occurrence. This will be quantified in Chapter 2.

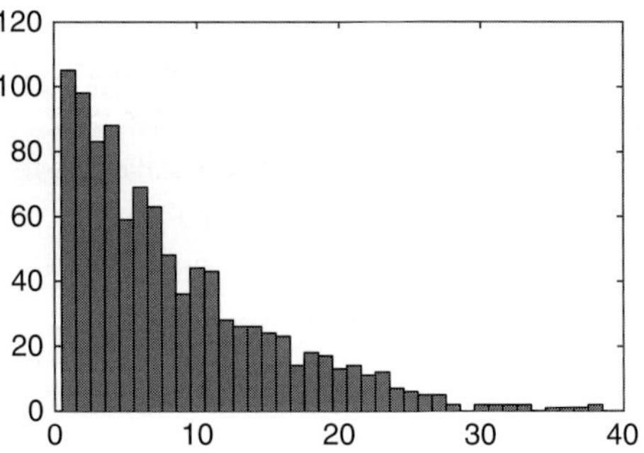

Figure 1.4 Histogram of the minimum value obtained from 1000 random draws of five values from the uniformly distributed integers between 1 and 50.

1.2 Probability Philosophies

The term *probability* enters into daily conversation (e.g., “It probably will rain tomorrow”), although such usage is typically imprecise. In fact, lack of a clear definition exists in formal theory as well as in daily life. There are two main definitions for probability, with their respective schools failing to agree on some basic concepts, and the ensuing conflict sometimes exhibits a quasi-religious dimension.

Frequentist or classical inference is based on three postulates:

1. Probability is the limiting relative frequency of occurrence of an event if the process creating it is repeated many times under identical conditions;
2. Parameters are invariant constants, so probability statements cannot be made about them; and
3. Statistical inference procedures should be designed to yield well-defined large-sample frequency properties.

The frequentist interpretation of probability given by postulate 1 pervades Section 1.1 and is itself not precise. For example, what is meant by “repeated many times” or “under identical conditions”? The remaining two postulates will be examined later in this book.

The Bayesian or subjective interpretation is based on three alternate postulates:

1. Probability statements are equivalent to claims about degree of belief, so the interpretation of probability is based on the judgment of the person making the assignment;
2. Probability statements can be assigned to objects that do not exhibit random variation, such as constant parameters; and
3. Statistical inferences about parameters may be made after defining their probability distributions.

The Bayesian approach has its origins in a paper by Bayes (1763). It is sometimes controversial because of its inherent subjectivity, although it is playing a growing role in applied statistics and machine learning due to the rapid recent increase in computational capability. This book will focus on the frequentist approach. A comprehensive survey of Bayesian data analysis is provided by Gelman et al. (2013).

Despite the controversy about the meaning of probabilities that are assigned to events from experiments, there is agreement that once probabilities are assigned, the mathematical theory of probability gives a methodology to study them. The following sections circumscribe the scope of the theory of probability through defining set theory, the probability axioms and their corollaries, finite sample spaces, conditional probability, and independence.

1.3 Set Theory

A formal mathematical model for probability follows from the theory of sets. All the entities introduced in Section 1.1 can be defined using set theory. Denote the sample space by $\mathcal{S}$. An event or outcome is a part $\mathcal{A}$ of the sample space, written as $\mathcal{A} \in \mathcal{S}$. The symbol $\in$ means "is an element of." Some outcomes in the sample space signify that the event $\mathcal{A}$ occurred, and all other outcomes signify that it did not. Let $\mathcal{B}$ denote an outcome that is not in $\mathcal{S}$. Then $\mathcal{B} \notin \mathcal{S}$, where $\notin$ means "is not an element of."

Let $\mathcal{A} \subset \mathcal{S}$ be an event in the sample space. The symbol $\subset$ means "is a subset of" or "belongs to." Conversely, the symbol $\not\subset$ means "does not belong to." Then $\mathcal{A}$ is contained in another event $\mathcal{B}$ if every outcome in $\mathcal{A}$ also belongs to $\mathcal{B}$ or, in symbols, $\mathcal{A} \subset \mathcal{B}$. Equivalently, if $\mathcal{A} \subset \mathcal{B}$, then $\mathcal{B} \supset \mathcal{A}$, where the symbol $\supset$ means "is contained in."

Some fairly obvious corollaries are

$$\text{If} \quad \mathcal{A} \subset \mathcal{B} \quad \text{and} \quad \mathcal{B} \subset \mathcal{A}, \quad \text{then} \quad \mathcal{A} = \mathcal{B} \tag{1.1}$$

where the last statement indicates that the events $\mathcal{A}$ and $\mathcal{B}$ are identical;

$$\text{If} \quad \mathcal{A} \subset \mathcal{B} \quad \text{and} \quad \mathcal{B} \subset \mathcal{C}, \quad \text{then} \quad \mathcal{A} \subset \mathcal{C} \tag{1.2}$$

The null or empty set $\varnothing$ is the subset of $\mathcal{A}$ that contains no outcomes, so

$$\varnothing \subset \mathcal{A} \subset \mathcal{S} \tag{1.3}$$

Some sets contain a finite number of elements, whereas others may contain an infinite number. In that case, a countably infinite set, in which there is a one-to-one correspondence between the set elements and the natural numbers $\{1, 2, 3, \ldots\}$, must be distinguished from an uncountably infinite set that is neither finite nor countable. Examples of countable sets include the integers, the even integers, the odd integers, and the prime numbers. Examples of uncountable sets include the real numbers and the numbers contained on a specified interval of the real line.

The operations of set theory include union, intersection, and complement. These are equivalent to the logical operations || or "or," && or "and," and ~ or "not" in MATLAB. It is standard practice to use Venn diagrams to visualize these concepts. The Venn diagram for the sample space $\mathcal{S}$ is shown in Figure 1.5.

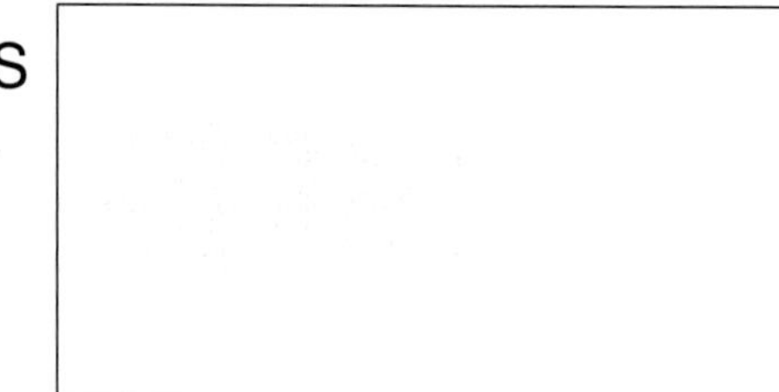

Figure 1.5 Venn diagram of the sample space $\mathcal{S}$.

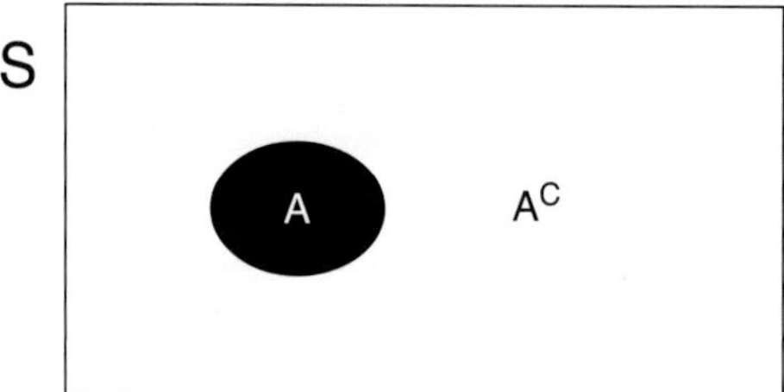

Figure 1.6 Venn diagram for the event $\mathcal{A}$ and its complement.

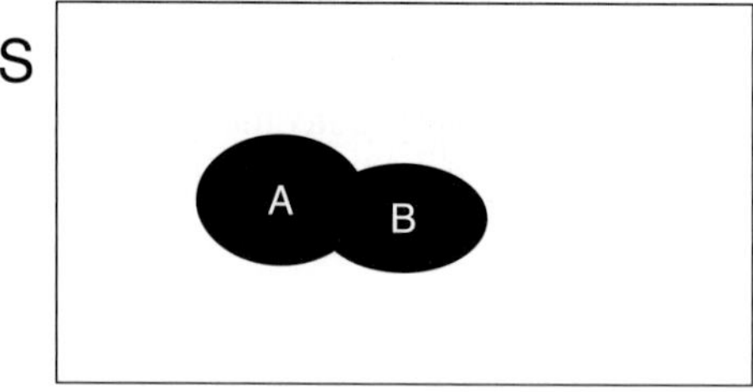

Figure 1.7 Venn diagram illustrating the union of events $\mathcal{A}$ and $\mathcal{B}$.

If $\mathcal{A} \in \mathcal{S}$ is an event in a sample space, then $\mathcal{A}^C$ or "$\mathcal{A}$ complement" or "not $\mathcal{A}$" is the set of possible outcomes in $\mathcal{S}$ that are not in $\mathcal{A}$. The Venn diagram illustrating the concept of complement is shown in Figure 1.6.

Some obvious implications that can all be proved using either logic or Venn diagrams are

$$\left(\mathcal{A}^C\right)^C = \mathcal{A} \tag{1.4}$$

$$\varnothing^C = \mathcal{S} \tag{1.5}$$

$$\mathcal{S}^C = \varnothing \tag{1.6}$$

If $\mathcal{A}$ and $\mathcal{B}$ are events, then the event $\mathcal{A} \cup \mathcal{B}$, or "$\mathcal{A}$ union $\mathcal{B}$" or "$\mathcal{A}$ or $\mathcal{B}$," is the event containing all possible outcomes in $\mathcal{A}$ alone, $\mathcal{B}$ alone, and in both $\mathcal{A}$ and $\mathcal{B}$. The Venn diagram illustrating union is shown in Figure 1.7.

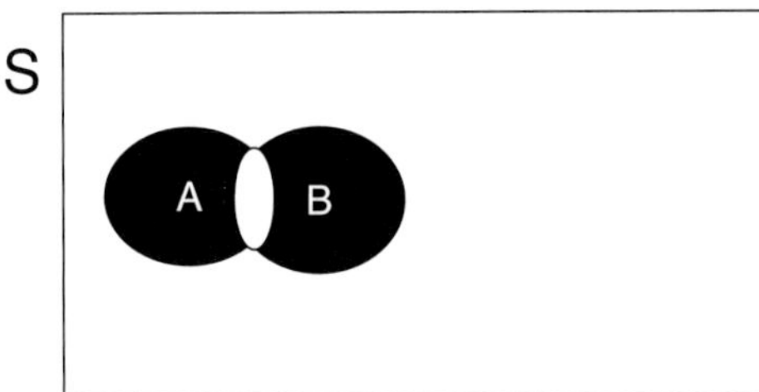

Figure 1.8 Venn diagram illustrating the intersection of two events $\mathcal{A}$ and $\mathcal{B}$. The intersection is the white area between the two events.

Some obvious implications follow that can be derived by drawing Venn diagrams:

$$\mathcal{A} \cup \mathcal{B} = \mathcal{B} \cup \mathcal{A} \tag{1.7}$$

$$\mathcal{A} \cup \varnothing = \mathcal{A} \tag{1.8}$$

$$\mathcal{A} \cup \mathcal{A} = \mathcal{A} \tag{1.9}$$

$$\mathcal{A} \cup \mathcal{S} = \mathcal{S} \tag{1.10}$$

$$\text{If } \mathcal{A} \subset \mathcal{B}, \text{ then } \mathcal{A} \cup \mathcal{B} = \mathcal{B} \tag{1.11}$$

$$\mathcal{A} \cup \mathcal{B} \cup \mathcal{C} = (\mathcal{A} \cup \mathcal{B}) \cup \mathcal{C} = \mathcal{A} \cup (\mathcal{B} \cup \mathcal{C}) \tag{1.12}$$

Equation (1.7) is the commutative property, whereas (1.12) is the associative property. The union of three events may be written

$$\mathcal{A}_1 \cup \mathcal{A}_2 \cup \mathcal{A}_3 = \bigcup_{i=1}^{3} \mathcal{A}_i \tag{1.13}$$

and extends trivially to N events.

If $\mathcal{A}$ and $\mathcal{B}$ are two events, then the event $\mathcal{A} \cap \mathcal{B}$, or "$\mathcal{A}$ intersection $\mathcal{B}$," "$\mathcal{A}$ and $\mathcal{B}$," or $\mathcal{AB}$ (although the last form will be avoided in this book to minimize confusion with multiplication) is the event containing all possible outcomes belonging to both $\mathcal{A}$ and $\mathcal{B}$. The Venn diagram illustrating intersection is shown in Figure 1.8.

Some consequences follow directly and can also be proved using Venn diagrams:

$$\mathcal{A} \cap \mathcal{B} = \mathcal{B} \cap \mathcal{A} \tag{1.14}$$

$$\mathcal{A} \cap \varnothing = \varnothing \tag{1.15}$$

$$\mathcal{A} \cap \mathcal{A} = \mathcal{A} \tag{1.16}$$

$$\mathcal{A} \cap \mathcal{S} = \mathcal{A} \tag{1.17}$$

$$\text{If } \mathcal{A} \subset \mathcal{B}, \text{ then } \mathcal{A} \cap \mathcal{B} = \mathcal{A} \tag{1.18}$$

$$\mathcal{A} \cap \mathcal{B} \cap \mathcal{C} = (\mathcal{A} \cap \mathcal{B}) \cap \mathcal{C} = \mathcal{A} \cap (\mathcal{B} \cap \mathcal{C}) \tag{1.19}$$

Equation (1.14) is the commutative property, whereas (1.19) is the associative property. The intersection of three events may be expressed as

$$\mathcal{A}_1 \cap \mathcal{A}_2 \cap \mathcal{A}_3 = \bigcap_{i=1}^{3} \mathcal{A}_i \tag{1.20}$$

and generalizes trivially to N events.

If $\mathcal{A} \cap \mathcal{B} = \varnothing$, then $\mathcal{A}$ and $\mathcal{B}$ are *mutually exclusive* or *disjoint*. This appears as nonintersecting sets $\mathcal{A}$ and $\mathcal{B}$ on a Venn diagram.

Set theory will conclude with two less intuitive pairs of relations. De Morgan's theorem holds that

$$\begin{aligned} (\mathcal{A} \cup \mathcal{B})^C &= \mathcal{A}^C \cap \mathcal{B}^C \\ (\mathcal{A} \cap \mathcal{B})^C &= \mathcal{A}^C \cup \mathcal{B}^C \end{aligned} \tag{1.21}$$

The distributive properties combine intersection and union:

$$(\mathcal{A} \cup \mathcal{B}) \cap \mathcal{C} = (\mathcal{A} \cap \mathcal{C}) \cup (\mathcal{B} \cap \mathcal{C}) \tag{1.22}$$

$$(\mathcal{A} \cap \mathcal{B}) \cup \mathcal{C} = (\mathcal{A} \cup \mathcal{C}) \cap (\mathcal{B} \cup \mathcal{C}) \tag{1.23}$$

Both of these can be proved using Venn diagrams, and the reader is encouraged to do so.

1.4 Definition of Probability

For each event $\mathcal{A} \in \mathcal{S}$, a number $\Pr(\mathcal{A})$ may be assigned that measures the probability that $\mathcal{A}$ will occur. In order to satisfy the mathematical definition of probability, the number $\Pr(\mathcal{A})$ must satisfy three conditions called the *Kolmogorov axioms*

$$\Pr(\mathcal{A}) \geq 0 \tag{1.24}$$

$$\Pr(\mathcal{S}) = 1 \tag{1.25}$$

and, for every infinite sequence of disjoint events $\{\mathcal{A}_i\}$,

$$\Pr \bigcup_{i=1}^{\infty} \mathcal{A}_i = \sum_{i=1}^{\infty} \Pr(\mathcal{A}_i) \tag{1.26}$$

The mathematical definition of probability holds that on a sample space $\mathcal{S}$, the probability distribution (or probability) for a set of events $\{\mathcal{A}_i\} \in \mathcal{S}$ is the set of numbers $\Pr(\mathcal{A}_i)$ that satisfy the Kolmogorov axioms.

A number of important corollaries may be derived from the Kolmogorov axioms (1.24)–(1.26). Because $\mathcal{A}$ and $\mathcal{A}^C$ are disjoint,

$$\Pr\left(\mathcal{A}^C\right) = 1 - \Pr(\mathcal{A}) \tag{1.27}$$

From corollary 1 and because $\varnothing = \mathcal{S}^C$,

$$\Pr(\varnothing) = 0 \tag{1.28}$$

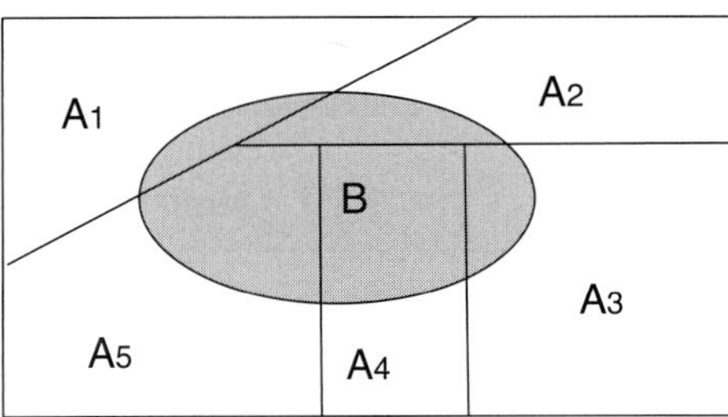

Figure 1.9 The intersection of event $\mathcal{B}$ with a partition $\{\mathcal{A}_i\}$ of a sample space.

Equation (1.26) can be extended to a finite number of events

$$\Pr\left(\bigcup_{i=1}^{N} \mathcal{A}_i\right) = \sum_{i=1}^{N} \Pr(\mathcal{A}_i) \tag{1.29}$$

Since $\mathcal{A}$ can range from $\varnothing$ to $\mathcal{S}$,

$$0 \leq \Pr(\mathcal{A}) \leq 1 \tag{1.30}$$

Finally,

$$\text{If } \mathcal{A} \subset \mathcal{B}, \text{ then } \mathcal{A} = \mathcal{A} \cap \mathcal{B} \text{ and } \Pr(\mathcal{A}) \leq \Pr(\mathcal{B}) \tag{1.31}$$

For any two events $\mathcal{A}$ and $\mathcal{B}$, the *general addition law* is

$$\Pr(\mathcal{A} \cup \mathcal{B}) = \Pr(\mathcal{A}) + \Pr(\mathcal{B}) - \Pr(\mathcal{A} \cap \mathcal{B}) \tag{1.32}$$

and will be used extensively throughout this book. If the two events $\mathcal{A}$ and $\mathcal{B}$ are disjoint, then $\Pr(\mathcal{A} \cap \mathcal{B}) = 0$, and the probability of the union of $\mathcal{A}$ and $\mathcal{B}$ is just the sum of their individual probabilities. *Boole's inequality*

$$\Pr(\mathcal{A} \cup \mathcal{B}) \leq \Pr(\mathcal{A}) + \Pr(\mathcal{B}) \tag{1.33}$$

is a simple corollary of the general addition law. Both the general addition law and Boole's inequality extend to an arbitrary number of events.

Let the set of events $\{\mathcal{A}_i\}$ be disjoint and exhaust $\mathcal{S}$ so that $\{\mathcal{A}_i\}$ fully partitions the sample space. If $\mathcal{B}$ is any other event, the set of events $\{\mathcal{A}_i \cap \mathcal{B}\}$ is also disjoint (see Figure 1.9), and hence

$$\mathcal{B} = \bigcup_{i=1}^{N} (\mathcal{A}_i \cap \mathcal{B}) \tag{1.34}$$

It is not necessary that $\mathcal{B}$ intersect each $\mathcal{A}_i$, as is depicted in Figure 1.9, because $\mathcal{A}_i \cap \mathcal{B} = \varnothing$ in its absence. The *law of total probability* follows directly from corollary 3

$$\Pr(\mathcal{B}) = \sum_{i=1}^{N} \Pr(\mathcal{A}_i \cap \mathcal{B}) \tag{1.35}$$

Both (1.34) and (1.35) hold for $N \to \infty$.

1.5 Finite Sample Spaces

1.5.1 Simple Sample Spaces

Experiments for which there can be only a finite number of possible outcomes are carried out in finite sample spaces such that $\mathcal{S} = \{s_i\}, i = 1, ..., N$. A simple sample space is one for which each s_i has a corresponding probability $p_i = 1/N$. In other words, no single element in the sample space occurs with higher or lower probability than another. In this case, the probability of $\mathcal{A}$ is simply $\text{Nr}(\mathcal{A})/N$, where $\text{Nr}(\mathcal{A})$ is the number of outcomes in $\mathcal{A}$. Computing probability becomes a simple matter of counting.

Example 1.1 A sample of six brands of malt beverage consists of equal parts of distinct brands of stout and ale. Let $\mathcal{A}$ be the event "all stout." There are three possible outcomes in $\mathcal{A}$, each with probability 1/6; hence $\Pr(\mathcal{A}) = 3/6 = 1/2$. Let $\mathcal{A}$ be the event "all stout" and $\mathcal{B}$ be the event "Guinness." Then $\Pr(\mathcal{A} \cap \mathcal{B}) = 1/6$ by counting the intersection of $\mathcal{A}$ and $\mathcal{B}$. However, $\Pr(\mathcal{A} \cup \mathcal{B}) = 3/6 = 1/2$. The general law of addition (1.32) also gives this result because

$$\Pr(\mathcal{A} \cup \mathcal{B}) = \Pr(\mathcal{A}) + \Pr(\mathcal{B}) - \Pr(\mathcal{A} \cap \mathcal{B}) = 3/6 + 1/6 - 1/6 = 1/2$$

Example 1.2 In the game of dice, why does it pay to consistently bet on getting a 6 at least once in four throws of a single die but not on a double 6 at least once in 24 throws of two dice?

Betting on dice was a constant pasttime in the French court of the early 1700s. An observant gambler, the Chevalier de Mere, came up with the preceding rule but lacked an explanation for it. He consulted the mathematician Blaise Pascal, who calculated the probabilities. The probability of not seeing a 6 in a given throw is 5/6. Presuming that successive throws are independent, the probability of not seeing a 6 in four throws is $(5/6)^4$. The probability of seeing a 6 is therefore $1 - (5/6)^4$, or 0.518, which is favorable to the gambler.

One double roll has 36 possible outcomes. The probability of seeing a double 6 once in 24 throws of two dice is $1 - (35/36)^{24} = 0.491$. A gambler would lose on this bet in the long term. It is fair to conclude that the Chevalier was a remarkably observant man to come up with this rule purely based on watching!

Suppose that an experiment is performed in two parts, where the first has M possible outcomes and the second has N possible outcomes independent of the number of outcomes from the first part. Then $\mathcal{S}$ contains MN possible outcomes, where each element is an ordered pair consisting of an outcome from the first and second experiments. For a simple sample space, the counting approach still pertains. This extends trivially to experiments with more than two parts.

Example 1.3 A 16-bit binary word is a sequence of 16 digits that may take a value of either 0 or 1. How many different 16-bit words are there?

Each bit has two possible outcomes (0 or 1). There are 16 bits, so the number of different words is $2 \times 2 \times \ldots \times 2 = 2^{16} = 65{,}536$.

1.5.2 Permutations

A *permutation* is an ordered collection of objects. From a sample of N objects, select one specimen at random and set it aside, then select a second specimen and set it aside, and finally select a third specimen and set it aside. There are $N(N-1)(N-2)$ possible outcomes, or permutations, because the collection was sampled without replacement. The sample space S is some arrangement of three objects, each of which is a permutation.

This is easily generalized to k selections without replacement, yielding for the number of different ordered samples

$$q_{N,k} = N(N-1)\cdots(N-k+1) \tag{1.36}$$

where $q_{N,k}$ is the permutation of N things taken k at a time. In the limit where $k \to N$,

$$q_{N,N} = N! \tag{1.37}$$

meaning that there are $N!$ possible permutations of N objects. Consequently, the permutation of N objects taken k at a time is

$$q_{N,k} = \frac{N!}{(N-k)!} \tag{1.38}$$

where $0! \equiv 1$. In MATLAB, the factorial is given by **factorial**(n). However, even with double-precision numbers, it is only accurate up to $n = 170$, and in practice, it is best to use (1.36) for computation via the **prod** function.

Example 1.4 For a rock collection, suppose that the probability of drawing k samples with different identifiers is wanted. Presuming that $k \le N$, the number of outcomes is the number of data vectors such that all k components are different. Since the first component s_1 can have N possible values, the second component s_2 can have $N-1$ possible values, and so on; this is just the permutation of N objects taken k at a time $q_{N,k}$. The probability that k different rock specimens will be selected is the number of permutations divided by the number of possible k vectors in the N-dimensional sample space, or $q_{N,k}/N^k$. This example mixes sampling with and without replacement, as is appropriate.

Sampling with replacement yields a larger sample space than sampling without replacement. As is intuitively obvious, the distinction between sampling with and without replacement becomes moot as $N \to \infty$.

1.5.3 Combinations

Consider a set of N distinct objects from which k are to be drawn as a subset, where $k \le N$. The number of distinct subsets that may be drawn must be determined. In this instance, ordering of the objects in a subset is irrelevant because no two subsets can consist of the same elements. Such a subset is called a *combination*. Let $c_{N,k}$ denote the number of combinations of N elements taken k at a time. The number of permutations of N elements taken k at a time is $q_{N,k}$, and the number of combinations $c_{N,k}$ must be smaller to correct for duplication. Each combination of k elements has $k!$ permutations, hence $q_{N,k} = k!c_{N,k}$. Consequently,

$$c_{N,k} = \frac{q_{N,k}}{k!} = \frac{N!}{(N-k)!k!} \equiv \binom{N}{k} \tag{1.39}$$

where the last term is the binomial coefficient and is read "N things taken k at a time." In MATLAB, the binomial coefficient is implemented as **nchoosek**(n, k).

Example 1.5 Suppose that the geophysics chair wants to form a committee of 6 out of the 25 staff to constitute a cricket team to challenge the physical oceanography department to a match. How many committee possibilities are there?

The number of groups of six different people who might serve is $c_{25,6} = 177{,}100$. This is much smaller than the number of possible committees $q_{25,6} = 127{,}512{,}000$, ignoring the fact that it is unconventional for a single person to serve in multiple committee slots.

Example 1.6 Suppose that there are 45 glasses of beer on a table, of which 15 are Sam Adams and 30 are Bud Lite. Ten glasses will be consumed at random by a group of graduate student volunteers. What is the probability that all 10 glasses will be Sam Adams?

There are $\binom{45}{10}$ possible combinations, each of which has equal probability because the sampling is random. The number of different combinations in which 10 San Adams can be selected from the pool of 15 is $\binom{15}{10}$. For Bud Lite, it is $\binom{30}{0}$. Consequently, the number of paired combinations with 10 Sam Adams and 0 Bud Lite is

$$\frac{\binom{15}{10}\binom{30}{0}}{\binom{45}{10}} \approx 9.4 \times 10^{-7}$$

The probability for a sample with 3 Sam Adams and 7 Bud Lite is

$$\frac{\binom{15}{3}\binom{30}{7}}{\binom{45}{10}} \approx 0.29$$

This is an example of sampling without replacement because once a specimen is selected and consumed, it is permanently removed from the sample.

1.6 Probability of a Union of Events

For two events $\mathcal{A}_1$ and $\mathcal{A}_2$, the general addition law (1.32) holds that $\Pr(\mathcal{A}_1 \cup \mathcal{A}_2) = \Pr(\mathcal{A}_1) + \Pr(\mathcal{A}_2) - \Pr(\mathcal{A}_1 \cap \mathcal{A}_2)$. Extending the general addition law to three events $\mathcal{A}_1$, $\mathcal{A}_2$, and $\mathcal{A}_3$ gives

$$\begin{aligned}\Pr(\mathcal{A}_1 \cup \mathcal{A}_2 \cup \mathcal{A}_3) &= \Pr(\mathcal{A}_1) + \Pr(\mathcal{A}_2) + \Pr(\mathcal{A}_3) - \Pr(\mathcal{A}_1 \cap \mathcal{A}_2) - \Pr(\mathcal{A}_2 \cap \mathcal{A}_3)\\ &\quad - \Pr(\mathcal{A}_1 \cap \mathcal{A}_3) + \Pr(\mathcal{A}_1 \cap \mathcal{A}_2 \cap \mathcal{A}_3) \end{aligned} \tag{1.40}$$

This can be proved with a Venn diagram and can be extended to N events:

$$\begin{aligned}\Pr\left(\bigcup_{i=1}^{N} \mathcal{A}_i\right) &= \sum_{i=1}^{N} \Pr(\mathcal{A}_i) - \sum_{i<j} \Pr(\mathcal{A}_i \cap \mathcal{A}_j) + \sum_{i<j<k} \Pr(\mathcal{A}_i \cap \mathcal{A}_j \cap \mathcal{A}_k) - \cdots\\ &\quad + (-1)^{N+1}\Pr(\mathcal{A}_1 \cap \cdots \cap \mathcal{A}_N) \end{aligned} \tag{1.41}$$

where $\sum_{i<j} = \sum_{j=1}^{N}\sum_{i=1}^{j-1}$. The terms involving single events are called *marginal probabilities*, whereas those containing intersections of events are called *joint probabilities*. In practice, the latter often are difficult to estimate, and sometimes only an upper bound can be obtained by assuming that the events are disjoint.

Example 1.7 A lazy TA in the rocks for jocks class is faced with a collection of N rock samples that need to be sorted into N bins. Being a geophysicist, and hence convinced that "you've seen one rock, you've seen 'em all," she simply places the specimens in bins at random, figuring that no one will notice until next year when she won't be TA any more. What is the probability that at least one of the rocks will be in the right bin?

Let $\mathcal{A}_i$ be the event that the ith rock is placed in the correct bin. The quantity of interest is the probability that the first rock is in the right bin or the second rock is in the right bin and so on, or the union of all of the possible events

$$p_N = \Pr\left(\bigcup_{i=1}^{N} \mathcal{A}_i\right)$$

It follows that the probability that any particular rock will go into the right bin is $\Pr(\mathcal{A}_i) = 1/N$. Therefore, the first term in (1.41) is $\sum_{i=1}^{N} \Pr(\mathcal{A}_i) = N(1/N) = 1$. Since rock 1 could be placed in any bin and rock 2 in any of the remaining $N - 1$ bins, the probability that both rocks 1 and 2 will be in the correct bin is

$$\Pr(\mathcal{A}_1 \cap \mathcal{A}_2) = \frac{1}{N(N-1)}$$

so the second term in (1.41) is the number of combinations of N things taken two at a time times this probability

$$\sum_{i<j} \Pr(\mathcal{A}_i \cap \mathcal{A}_j) = \frac{\binom{N}{2}}{N(N-1)} = \frac{1}{2!}$$

Applying similar reasoning to the triplet term,

$$\sum_{i<j<k} \Pr(\mathcal{A}_i \cap \mathcal{A}_j \cap \mathcal{A}_k) = \frac{\binom{N}{3}}{N(N-1)(N-2)} = \frac{1}{3!}$$

By induction, the result for N terms is

$$p_N = 1 - \frac{1}{2!} + \frac{1}{3!} - \cdots + \frac{(-1)^{N+1}}{N!}$$

and is a rapidly converging series. In the limit $N \to \infty$, this becomes $1 - 1/e \approx 0.632$. The probability that at least one rock will be in the correct bin is not very large! Note that the probability that all of the rocks are in the correct bins is $1/N!$, which is a very small number for any substantial number of samples.

1.7 Conditional Probability

How does the probability for an event $\mathcal{A}$ change if it is known that $\mathcal{B}$ has already occurred? This new probability is written $\Pr(\mathcal{A}|\mathcal{B})$ and is read "probability of $\mathcal{A}$ given $\mathcal{B}$." It is a conditional probability.

If $\mathcal{B}$ is known to have occurred, then the sample space $\mathcal{S}$ includes $\mathcal{B}$, and the outcome of any experiment must contain $\mathcal{B}$. Consequently, the entity of interest is outcomes in $\mathcal{B}$ that also result in the occurrence of $\mathcal{A}$. This is just $\mathcal{A} \cap \mathcal{B}$. It is natural to define $\Pr(\mathcal{A}|\mathcal{B})$ as the proportion of the total probability $\Pr(\mathcal{B})$ that is represented by $\Pr(\mathcal{A} \cap \mathcal{B})$, or

$$\Pr(\mathcal{A}|\mathcal{B}) = \frac{\Pr(\mathcal{A} \cap \mathcal{B})}{\Pr(\mathcal{B})} \tag{1.42}$$

Equation (1.42) is not defined if $\Pr(\mathcal{B}) = 0$. Therefore, the conditional probability is the joint probability of $\mathcal{A}$ and $\mathcal{B}$ divided by the marginal probability for $\mathcal{B}$. It is also true that

$$\Pr(\mathcal{B}|\mathcal{A}) = \frac{\Pr(\mathcal{A} \cap \mathcal{B})}{\Pr(\mathcal{A})} \tag{1.43}$$

because the intersection of two events is commutative. Equations (1.42) and (1.43) are distinct unless $\Pr(\mathcal{A}) = \Pr(\mathcal{B})$.

Example 1.8 Over the past 200 Ma, the geomagnetic field has been normal and reversed with approximately equal frequency (hence probability). However, during the Cretaceous, there was a 33 Ma interval in which the field was normal. Given that a seamount is normally magnetized, what is the probability that it is Cretaceous in age?

Let the event $\mathcal{A}$ be eruption during the Cretaceous quiet interval and the event $\mathcal{B}$ be normal polarity. It is known that $\Pr(\mathcal{B}) = 0.5$ from observations over 200 Ma. If seamount production and preservation have been uniform over the past 200 Ma, then $\Pr(\mathcal{A}) = 33/200 = 0.17$. It follows that $\Pr(\mathcal{A} \cap \mathcal{B}) = \Pr(\mathcal{A})$ because the entire Cretaceous quiet interval is normal. Then $\Pr(\mathcal{A}|\mathcal{B}) = 0.17/0.5 = 0.34$. The chances are about 33% that a normally magnetized seamount is Cretaceous in age. In reality, this is an underestimate due to variations in the rate of volcanism through time and the subduction of some Cretaceous seafloor.

1.8 Independent Events

Suppose that two events occur independently of one another. Their joint probability is just the product of their marginals, or $\Pr(\mathcal{A} \cap \mathcal{B}) = \Pr(\mathcal{A})\Pr(\mathcal{B})$. Equivalently, if the events are independent, the conditional probability $\Pr(\mathcal{A}|\mathcal{B}) = \Pr(\mathcal{A})$ because the prior occurrence of $\mathcal{B}$ is irrelevant. This yields $\Pr(\mathcal{A} \cap \mathcal{B}) = \Pr(\mathcal{A})\Pr(\mathcal{B})$ from the definition of conditional probability.

Independence of two events is defined by $\Pr(\mathcal{A} \cap \mathcal{B}) = \Pr(\mathcal{A})\Pr(\mathcal{B})$. If independence is known on physical grounds, such as for radioactive decay of an isotope in a sample where it is at a low concentration so that one radioactive decay does not influence another, then $\Pr(\mathcal{A} \cap \mathcal{B})$ can easily be estimated.

Example 1.9 A sample of water from a hydrothermal vent field contains ^{222}Rn and ^{40}K, both of which undergo radioactive decay at known rates. Given this information, plus the concentration of Rn and K in the samples, plus the relative concentration of the two isotopes of Rn and K in earth, the probabilities for decay of ^{222}Rn and ^{40}K in a given unit of time are 0.93 and 0.75. What is the probability of observing either a ^{222}Rn or a ^{40}K decay in the same time interval?

Let the events $\mathcal{A}$ and $\mathcal{B}$ be ^{222}Rn and ^{40}K decay, respectively. It has been specified that $\Pr(\mathcal{A}) = 0.93$ and $\Pr(\mathcal{B}) = 0.75$. Since the two events are independent, $\Pr(\mathcal{A} \cap \mathcal{B}) = \Pr(\mathcal{A})\Pr(\mathcal{B}) = 0.70$. From the general law of addition (1.32), $\Pr(\mathcal{A} \cup \mathcal{B}) = 0.93 + 0.75 - 0.70 = 0.98$.

The concept of independence can be extended to more than two events. For example, for three events to be independent, $\Pr(\mathcal{A}_i \cap \mathcal{A}_j) = \Pr(\mathcal{A}_i)\Pr(\mathcal{A}_j)$ and $\Pr(\mathcal{A}_i \cap \mathcal{A}_j \cap \mathcal{A}_k) = \Pr(\mathcal{A}_i)\Pr(\mathcal{A}_j)\Pr(\mathcal{A}_k)$ for $i, j, k = 1, \ldots, 3$.

Example 1.10 An inventor has built a system to automatically pick foraminifera from a sediment sample and sort them according to species. Suppose that the system produces a false pick with probability p and a correct pick with probability $q = 1 - p$. Further suppose that 10 picked forams selected at random are tested and that these tests are independent. What is the probability that two of the 10 forams were picked correctly?

The sample space contains all possible arrangements of 10 forams, whether correctly or incorrectly picked. For $j = 1, \ldots, 10$, let $\mathcal{I}_j$ be the event that the jth item is incorrectly picked and $\mathcal{C}_j$ that it is correctly picked. Because of independence, $\Pr(\mathcal{I}_1 \cap \mathcal{C}_1 \cap \mathcal{I}_2 \cap \mathcal{C}_2 \cap \cdots \cap \mathcal{I}_{10} \cap \mathcal{C}_{10}) = \Pr(\mathcal{I}_1) \cdots \Pr(\mathcal{C}_{10}) = p^2 q^8$. No matter in what order they occur, the probability that two picks are incorrect is $p^2 q^8$. Hence the probability that there will be two false picks in 10 samples is $p^2 q^8$ times the number of combinations producing them, or $\binom{10}{2} p^2 q^8$. For a 1% error rate, $p = 0.01$ and $q = 0.99$, and the probability of getting two false picks is 0.00415, or about 0.4%.

1.9 Bayes' Theorem

Let $\{\mathcal{A}_i\}$ be a set of k disjoint events that span the sample space $\mathcal{S}$. Let $\mathcal{B}$ be another event in the same sample space. The conditional probability of the jth disjoint event given $\mathcal{B}$ follows from its definition (1.42)

$$\Pr(\mathcal{A}_j|\mathcal{B}) = \frac{\Pr(\mathcal{A}_j \cap \mathcal{B})}{\Pr(\mathcal{B})} = \frac{\Pr(\mathcal{A}_j)\Pr(\mathcal{B}|\mathcal{A}_j)}{\sum_{i=1}^{k} \Pr(\mathcal{A}_j)\Pr(\mathcal{B}|\mathcal{A}_j)} \tag{1.44}$$

The numerator on the right is just the joint probability expressed in terms of the conditional probability. The denominator follows from the law of total probability (1.35). The result is called *Bayes' theorem*. It gives a simple rule for computing the conditional probabilities of each event $\mathcal{A}_j$ given $\mathcal{B}$ from the conditional probabilities of $\mathcal{B}$ given $\mathcal{A}_j$ and the unconditional probabilities for $\mathcal{A}_j$. It also leads to the concepts of prior and posterior probabilities that lie at the heart of Bayesian statistics.

Example 1.11 Suppose that a test exists for a serious disease. The known statistics for the test are that if one has the disease, the test will give a positive result 99.99% of the time, but if one does not have the disease, there is a 0.01% chance of a positive result. The chance of a given member of the population having the disease is about 1 in 1000 based on national incidence. An individual takes the test and learns that he got a positive test result. What is the probability that he has the disease?

Let $\mathcal{A}$ be the event that the individual has the disease and $\mathcal{B}$ be the event that the test result is positive. Using Bayes' theorem,

$$\begin{aligned}\Pr(\mathcal{A}|\mathcal{B}) &= \frac{\Pr(\mathcal{B}|\mathcal{A})\Pr(\mathcal{A})}{\Pr(\mathcal{B}|\mathcal{A})\Pr(\mathcal{A}) + \Pr(\mathcal{B}|\mathcal{A}^C)\Pr(\mathcal{A}^C)} \\ &= \frac{0.9999 \times 0.001}{0.9999 \times 0.001 + 0.0001 \times 0.999} = 0.909\end{aligned}$$

This is substantially less than the reliability of the test. If the incidence of the disease is lower, the probability of the test being right drops a lot. For instance, if the chance of having the disease is 1/10,000 or 1/100,000, the probability drops to 0.5 and 0.091, respectively. This example illustrates why it makes little sense to test a large segment of the population for a rare disease unless the accuracy of the test is very high.

2 Statistical Concepts

2.1 Overview

This chapter builds on the probability concepts of Chapter 1 to construct the theoretical foundation for computational data analysis. The key ideas that are introduced include the probability density function (pdf) for discrete, continuous, and mixed distributions; the cumulative distribution function (cdf), which is the sum or integral of the pdf; the quantile function, which is the inverse of the cdf; and the characteristic function, which serves as an alternate pathway for computing the pdf and cdf. A discussion of the bivariate distribution extends the prior univariate descriptions to two variables, with the full multivariate case deferred until Chapters 10 and 11 (although there will be some slight cheating in Chapter 3), and to independence of random variables (rvs). The formalism that enables transformation from one or more variables to another set (e.g., Cartesian to circular coordinates) is then described, and the distribution of the largest and smallest of a set of random variables is derived as an introduction to the order statistics that are covered in Chapter 4. A number of theoretical entities for location (e.g., the expected value), dispersion (e.g., the variance), shape (e.g., the skewness), direction (e.g., the mean direction), and the covariance between two rvs is described as a counterpart to the sample entities that are presented in Chapter 4. The concept of conditional probability from Chapter 1 is extended to the expected value and variance, leading to the laws of total expectation, variance, and covariance. Finally, the chapter closes by extending the concept of inequality to stochastic variables, leading to convergence relations for rvs that pervade the remainder of the book.

2.2 The Probability Density Function

2.2.1 Discrete Distributions

A distribution may be discrete (i.e., it exists only for specific values on the real line), continuous (i.e., it exists for any value on the real line between a given set of bounds that may include $\pm\infty$), or a mixture of continuous and discrete. An rv $\mathbf{X}$ is discrete or, equivalently, $\mathbf{X}$ has a discrete distribution if $\mathbf{X}$ can take only a countably finite number k of different values or at most an infinite sequence of distinct values (such as all of the integers). It may not take on all possible values in an interval and remain discrete.

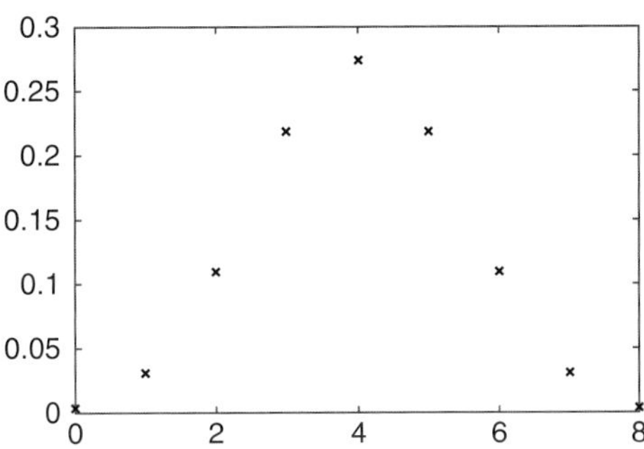

Figure 2.1 The binomial distribution for $p = 0.5$ and eight trials.

The *probability function* or *probability mass function* or *probability density function* or *pdf* of a discrete rv $\mathbf{X}$ is defined as the function $f(x)$ that assigns a real number x to each value of $\mathbf{X}$ with the following two properties:

$$f(x) = \Pr(\mathbf{X} = x) \qquad x = x_i, \ \ i = 1, \ldots, N \tag{2.1}$$

$$f(x) = 0 \qquad x \neq x_i, \ \ i = 1, \ldots, N \tag{2.2}$$

By the Kolmogrov axioms, $\sum_{i=1}^{N} f(x_i) = 1$, where it is understood that N can approach infinity in (2.1) and (2.2).

A discrete probability function is shown in Figure 2.1, where the sum of all the values is 1. A similar plot could be obtained using **disttool** in MATLAB, which allows one to plot the pdf of a wide range of discrete and continuous distributions and is a useful tool for the reader to become familiar with some standard distributions.

For $\mathbf{X} \in \mathcal{A}$, where $\mathcal{A}$ is a subset of the possible values of x, the corresponding probability is obtained by summing all values of $f(x)$ lying in $\mathcal{A}$. Mathematically, this is written

$$\Pr(\mathbf{X} \in \mathcal{A}) = \sum_{x_i \in \mathcal{A}} f(x_i) \tag{2.3}$$

As an example, suppose the rv $\mathbf{X}$ is equally likely to take on any of the integer values between 1 and N. The probability density function of $\mathbf{X}$ is

$$f(x) = \frac{1}{N} \qquad x = 1, \ldots, N$$

This represents the outcome of an experiment that results in one of the integers on the closed interval $[1, N]$ being chosen at random. The uniform distribution on integers only exists for a finite sequence of integers and cannot be extended to all the integers. In MATLAB, this is **unidpdf**(x, n).

An alternate way to represent a discrete pdf uses a weighted sequence of Dirac delta functions. The Dirac delta function $\delta(x - x')$ is a generalized function defined on the real line whose value is zero everywhere except when its argument is zero, where it is infinite. Rigorous derivation requires the use of measure theory, but a working definition follows from

$$\int_{-\infty}^{\infty} g(x)\,\delta(x - x')\,dx = g(x')$$

As a consequence, a discrete pdf can be represented using a comb of Dirac delta functions

$$f(x) = \sum_{i=1}^{N} a_i\,\delta(x - x_i) \tag{2.4}$$

where $\sum_{i=1}^{N} a_i = 1$. Computation of probabilities obtains by integration of (2.4). For example, let $x_1 \le a < b \le x_N$. Then the probability on the interval $[a, b]$ is

$$\int_a^b f(x)\;dx = \sum_{x_i \in [a,b]} a_i \tag{2.5}$$

The biggest advantage of the Dirac comb representation of a discrete pdf is that mathematical manipulation of it is identical to that for continuous distributions, so mixed distributions can be handled easily if the discrete part is represented by (2.4).

2.2.2 Continuous Distributions

A rv $\mathbf{X}$ has a continuous distribution if there exists a function $f(x) \ge 0$ defined on the real line such that, for every subset $\mathcal{A}$, the probability that $\mathbf{X}$ takes a value in $\mathcal{A}$ is the integral of f over the set $\mathcal{A}$. Mathematically, this is

$$\Pr(\mathbf{X} \in \mathcal{A}) = \int_{x \in \mathcal{A}} f(x)dx \tag{2.6}$$

where the Kolmogorov axioms require that $\int_{\mathcal{S}} f(x)dx = 1$. The bounds on the last integral cover all allowed values of x (or the sample space $\mathcal{S}$). The function f is the pdf of $\mathbf{X}$. It does not have dimensions of probability; rather, its integral over some part of the real line gives that quantity (see Figure 2.2). For example, the probability that $\mathbf{X}$ lies between a and b is

$$\Pr(a < x \le b) = \int_a^b f(x)\;dx \tag{2.7}$$

This probability is also given by $\Pr(a < x < b)$ or $\Pr(a \le x \le b)$ or $\Pr(a \le x < b)$, unlike for a discrete variable where inclusion of the endpoints of an interval is required to correctly estimate the probability.

It immediately follows that $\Pr(\mathbf{X} = c) = 0$ for any $c \in \mathcal{S}$, or else $\int_{\mathcal{S}} f(x)\,dx \to \infty$. This does not imply that $\mathbf{X} = c$ is an impossible abscissa value because if it were, then $\mathbf{X}$ could not take on any value. Rather, it is necessary to integrate $f(x)$ over a finite, but not infinitesimal, range to get the probability. As a consequence, the pdf is not unique; it can change at a finite or infinite number of points without changing the integral over some

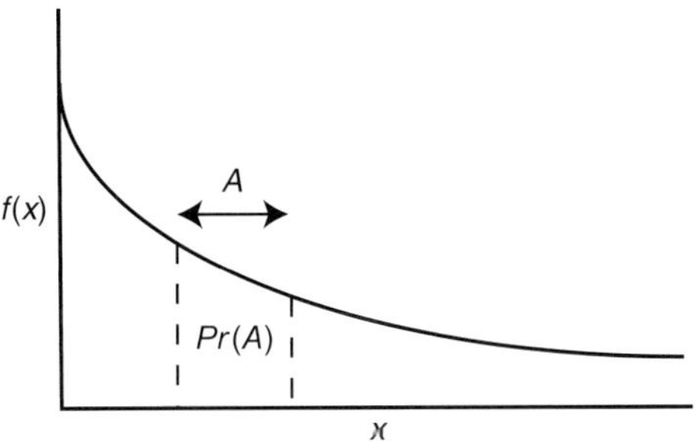

Figure 2.2 The probability density function for a continuous distribution, showing that the probability over a subset of the real axis is obtained from the area under the pdf.

range and hence altering the probability. This generally requires introducing discontinuities into an otherwise continuous pdf. The natural choice is the continuous version, although that is not a mathematical requirement.

Example 2.1 Let a and b be real numbers with $a < b$, and consider an experiment in which $\mathbf{X}$ is selected from the interval $(a, b]$ in such a way that the probability is proportional to the interval length $b - a$. The pdf is

$$f(x) = \frac{1}{b - a} \qquad a < x \le b$$

and zero otherwise. Then $\Pr(c < \mathbf{X} \le d) = (d - c)/(b - a)$ if $a \le c \le d \le b$, and $\Pr(c < \mathbf{X} \le d) = (d - a)/(b - a)$ if $c \le a \le d \le b$, and so on. $\Pr(\mathbf{X} \ge d) = (b - d)/(b - a)$ if $a \le d \le b$, 1 if $d \le a$, and 0 if $d \ge b$. In MATLAB, the uniform continuous distribution on the interval $(a, b]$ at the values in the vector x is given by **unifpdf** (x, a, b).

Example 2.2 The quantization noise from an analog-to-digital converter is uniformly distributed because while the voltage it measures is continuous, the converter has a minimum resolution of the least significant bit, and hence the uncertainty of the real voltage is uniformly distributed over −½ to ½ bit.

It is possible for a pdf to exist over an unbounded interval of the real line as long as integrals over parts of the real line give finite probability and the integral over the entire interval is 1. It is also possible for $f(x)$ to be unbounded at some point as long as its integral remains finite.

Example 2.3 Let $f(x) = 2x^{-1/3}/3$ on $[0, 1)$. This is unbounded at the origin, but the integral over its support is unity, and the integral over any part of $[0, 1)$ is bounded.

2.2.3 Mixed Distributions

Mixed distributions for variables that may be both continuous and discrete also exist. An example of a physical process with a mixed distribution is pressure fluctuations on the seafloor, which not only has a continuous power spectrum (that is analogous to the pdf with frequency or period as the abscissa) caused by processes such as stochastic wind forcing, but also has superimposed ocean tides that exist only at a countably finite set of frequencies and are zero elsewhere.

Continuous and discrete variables may be treated simultaneously by replacing the familiar Riemann integral $\int_a^b f(x)\,dx$ with the Riemann-Stieltjes integral $\int_a^b g(x)\,dH(x)$, where $g(x)$ and $H(x)$ are real-valued functions (or alternately, by going to the general definition of an integral under the Riemann-Lebesque theorem; see Doob 1993 or Carter & van Brunt 2000). Specifying the partition of the interval $[a, b]$ to be $a = x_0 < x_1 < \cdots < x_n = b$, the Riemann-Stieltjes integral is defined to be the limit as the interval approaches zero of

$$\int_a^b g(x)\,dH(x) \approx \sum_{i=0}^{n-1} g(y_i)[H(x_{i+1}) - H(x_i)] \tag{2.8}$$

where y_i is a value (e.g., the midpoint) on the subinterval $[x_i,\ x_{i+1}]$. If $H(x)$ is differentiable with derivative $h(x)$ [formally, if $H(x)$ has a pdf $h(x)$ under Lebesgue measure], then

$$\int_a^b g(x)\,dH(x) = \int_a^b g(x)\,h(x)\,dx \tag{2.9}$$

If $h(x) = 1$, meaning that $H(x) = x$, then the Riemann integral is recovered. If $H(x)$ contains a set of step function discontinuities, then $h(x)$ will include a sequence of delta functions, and the integral becomes the sum of a continuous part and a discrete part, yielding a mixture distribution. Riemann-Stieltjes notation is used in many textbooks on advanced statistics, such as the classics by Stuart & Ord (1994) and Stuart, Ord, & Arnold (1999).

Mixed distributions can occur in practical problems. Suppose that **X** and **Y** are the times when earthquakes occur on two subparallel faults, that p is the probability that earthquakes occur on both faults at the same time, and hence $1 - p$ is the probability that earthquakes do not occur simultaneously. If the earthquakes occur at the same time, then the pdf for x is $f_x(x)$, while if the faults fail at different times, their bivariate pdf (see Section 2.5) is $f(x, y)$. It immediately follows that $f(x, y)$ cannot be continuous because the probability that the paired rvs (**X**, **Y**) lie on the line $x = y$ is p and not 0, as would be required for a continuous distribution. The bivariate pdf becomes continuous in the limit $p \rightarrow 0$ and hence is a mixture distribution for nonzero p.

2.3 The Cumulative Distribution and Quantile Functions

The *distribution function* or *cumulative distribution function* or *cdf* is defined as

$$F(x) = \Pr(\mathbf{X} \le x) \tag{2.10}$$

This definition applies to continuous, discrete, or mixed distributions. The cdf has three key properties:

1. $F(x)$ is nondecreasing as x increases: if $x_1 < x_2$, then $F(x_2) \ge F(x_1)$;
2. $\lim_{x\to-\infty} F(x) = 0$ and $\lim_{x\to\infty} F(x) = 1$; and
3. $F(x)$ is always continuous from the right: $F(x) = F(x^+)$ at all points x, where $F(x^+) = \lim_{y\to x \forall y > x} F(y)$.

A cdf need not be continuous, and property 3 guarantees that probability is always defined even in the presence of discontinuities in $F(x)$. It follows that continuity of a cdf requires that $F(x) = F(x^-) = F(x^+)$, where $F(x^-) = \lim_{y\to x \forall y<x} F(y)$.

Probabilities may be determined directly from a known cdf. There are four types of intervals to consider

$$\Pr(\mathbf{X} > x) = 1 - \Pr(\mathbf{X} \le x) = 1 - F(x) \tag{2.11}$$

This is called the *complementary cumulative distribution function* or *tail distribution* or *survivor function*. In addition,

$$\Pr(x_1 < \mathbf{X} \le x_2) = F(x_2) - F(x_1) \tag{2.12}$$

$$\Pr(\mathbf{X} < x) = F(x^-) \tag{2.13}$$

$$\Pr(\mathbf{X} = x) = F(x^+) - F(x^-) \tag{2.14}$$

where (2.14) is zero for a continuous distribution.

For a discrete distribution, $F(x)$ will have a jump of magnitude $\Pr(\mathbf{X} = x_i) = f(x_i)$ at each point where $f(x)$ exists. A discrete rv can be represented equally well by a cdf or a pdf. Figure 2.3 shows the cdf for the same discrete distribution as in Figure 2.1.

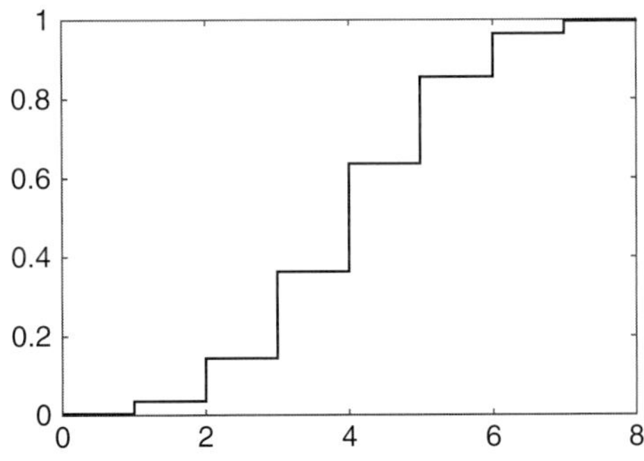

Figure 2.3 The cdf for the binomial distribution with $p = 0.5$ and eight trials.

Note the stairstep appearance, with discrete jumps in probability occurring at integer intervals.

For a discrete distribution, an alternate representation of the cdf uses *indicator functions*. An indicator function takes only the values 0 and 1 and is 1 when its argument $x \in \mathcal{A}$ and 0 when $x \notin \mathcal{A}$, where $\mathcal{A}$ is an event of interest. For example, $\mathcal{A}$ could consist of all the values on the real line where a discrete distribution is nonzero. Denote the indicator function by $\mathbf{1}_{\mathcal{A}}(x_i)$. The cdf is given by

$$F(x) = \sum_{x_i \in (-\infty, x]} a_i \mathbf{1}_{\mathcal{A}}(x_i) \tag{2.15}$$

where $\sum a_i = 1$.

For a continuous distribution, the cdf may be obtained by integration of the pdf

$$F(x) = \int_{-\infty}^{x} f(t)\, dt \tag{2.16}$$

$$f(x) = \partial_x F(x) \tag{2.17}$$

As for a discrete rv, a continuous rv can be represented equally well by a cdf or a pdf.

It is sometimes useful to turn a statistical problem around so that instead of asking for $\Pr(\mathbf{X} \le x)$, a probability level is chosen and used to determine the corresponding value for the distribution abscissa. This can be accomplished by searching through the values of $F(x)$ until the probability level of interest is found (and that is exactly what was done by eye in the days of statistical tables that are now gathering dust in university libraries). However, if $F(x)$ is continuous and uniquely invertible, $F^{-1}(p)$ can be defined such that

$$x = F^{-1}(p) \tag{2.18}$$

$F^{-1}(p)$ is called the *inverse cumulative distribution* or *quantile function*. Some properties of the quantile function include

1. $F^{-1}(p)$ is a nondecreasing function of p;
2. $F^{-1}[F(x)] \le x$; and
3. $F\left[F^{-1}(p)\right] \ge p$.

Like the cdf, the quantile function depends only on the underlying probability distribution. Any two rvs with the same distribution will have the same quantile function. With some adjustment, the concept of quantiles extends to discrete as well as continuous distributions.

For a given probability p, $F^{-1}(p)$ given by (2.18) is called the *p-quantile* or *100p percentile* of x. If x is the $100p$ percentile, then $100p$ percent of the distribution is at or below x. The 0.5 quantile or 50th percentile is called the *median*. The 0.25 quantile or 25th percentile is called the *lower quartile*. The 0.75 quantile or 75th percentile is called the *upper quartile*.

Example 2.4 Suppose $F(x) = e^x$ with support $(-\infty, 0]$. The quantile function is obtained by solving $p = F(x)$ for x, yielding $F^{-1}(p) = \log p$. The median is -0.6931. The lower and upper quartiles are -1.3863 and -0.2877, respectively.

The concepts in this section extend to rvs that represent directions on a circle (and can easily be extended to a sphere). Two standard references on directional distributions are Fisher (1995) and Mardia & Jupp (2000).

The cdf for rvs comprising random angles $\{\boldsymbol{\theta}_i\}$ is defined by

$$F(x) = \Pr(0 < \theta \le x) \qquad 0 \le x \le 2\pi \tag{2.19}$$

where

$$F(x + 2\pi) - F(x) = 0 \qquad -\infty < x < \infty \tag{2.20}$$

Equation (2.19) depends on a reference direction that simply adds a constant to $F(x)$. Equation (2.20) specifies that any set of angles with a length 2π has a probability of 1 because x is defined modulo 2π. The probability for an interval modulo 2π is given by (2.12), but the second property at the beginning of Section 2.3 must be replaced by $F(0) = 0$ and $F(2\pi) = 1$. Presuming that $F(x)$ is continuous, the pdf is defined as in Section 2.2.2, although the support becomes $(0, 2\pi)$, and the pdf is modulo 2π.

2.4 The Characteristic Function

An alternate way to specify a probability distribution of an rv $\mathbf{X}$ is through the complex-valued *characteristic function* or *cf* given by the inverse Fourier transform of the pdf

$$\phi(t) = \int_{-\infty}^{\infty} e^{itx} f(x)dx \tag{2.21}$$

or, conversely, if the cf is specified, the pdf is the dual given by

$$f(x) = \frac{1}{2\pi} \fint_{-\infty}^{\infty} \phi(t)\, e^{-itx} dt \tag{2.22}$$

where the integral is the Cauchy principal value. The cf converges both absolutely and uniformly in t. Consequently, it may be integrated or differentiated behind the integral sign. Further, the cf of a real-valued rv always exists because it is the integral of a bounded continuous function of finite measure. The cdf is given by

$$F(x) - F(0) = \frac{1}{2\pi} \fint_{0}^{\infty} \frac{1 - e^{-itx}}{it} \phi(t)\, dt \tag{2.23}$$

While in the past the cf was used as a tool to compute the moments of a distribution, in the modern world it is important because there are classes of probability distributions for which closed-form expressions for the pdf do not exist, but simple expressions for the cf are available. The most important of these are the stable distributions that will be introduced in Section 3.4.2.

Some key properties of the characteristic function include

1. $\phi(0) = 1$ follows from the requirement that the integral of the pdf over its support be 1;
2. $|\phi(t)| = \left|\int_{-\infty}^{\infty} e^{itx} f(x)\,dx\right| \leq \int_{-\infty}^{\infty} |e^{itx}| f(x)\,dx \leq \int_{-\infty}^{\infty} f(x)\,dx = 1$;
3. $|\phi(t)| \to 0$ as $|t| \to \infty$ implies that the pdf is continuous;
4. $|\phi(t)| \nrightarrow 0$ as $|t| \to \infty$ implies that the pdf is discrete;
5. $\phi(-t) = \phi(t)^*$ (Hermitian property, where the superscript * denotes the complex conjugate), so for a real rv that is symmetric about the origin, the cf is real-valued and even;
6. $\phi(x+y) = \phi(x)\phi(y)$ when x and y are independent; and
7. $\phi(t) = \int_0^1 e^{itF^{-1}(p)}dp$, where $F^{-1}(p)$ is the quantile function.

The real part of the cf is even, and the imaginary part is odd. As a consequence, (2.22) can be rewritten as

$$f(x) = \frac{1}{\pi}\int_0^\infty \mathrm{Re}[\phi(t)] \cos(xt)\,dt + \frac{1}{\pi}\int_0^\infty \mathrm{Im}[\phi(t)] \sin(xt)dt \tag{2.24}$$

which is often useful for numerical integration purposes.

Example 2.5 The Gaussian pdf is $e^{-(x-\mu)^2/(2\sigma^2)}/(\sqrt{2\pi}\sigma)$. Derive its cf.

$$\phi(t) = \frac{1}{\sqrt{2\pi}\sigma}\int_{-\infty}^{\infty} e^{itx} e^{-(x-\mu)^2/(2\sigma^2)} = e^{it\mu - t^2\sigma^2/2}$$

using Gradshteyn & Ryzhik (1980, 3.323-2). This becomes real when $\mu = 0$ so that the distribution is symmetric about the origin, and properties 1–3 and 5 hold.

For directional rvs, the requirement that the angle θ is defined modulo 2π means that the variable t in the cf must take on only integer values. Consequently, the directional cf is given by

$$\phi_k = \int_0^{2\pi} e^{ik\theta} f(\theta)d\theta \qquad k = 0, \pm 1, \pm 2, \ldots \tag{2.25}$$

for a continuous directional rv. It is obvious that $\phi_0 = 0$ and $\phi_{-k} = \phi_k^*$. The complex sequence given by (2.25) comprises the coefficients in the Fourier expansion for $F(\theta)$, and presuming that it is square integrable, the pdf is given by

$$f(\theta) = \frac{1}{2\pi} \sum_{k=-\infty}^{\infty} \phi_k e^{-ik\theta} \tag{2.26}$$

which is the directional analog to (2.22).

2.5 Bivariate Distributions

The bivariate distribution is the joint distribution of two variables and can be discrete, continuous, or mixed. The joint pdf is defined by

$$f(x,y) = \Pr(\mathbf{X} = x \cap \mathbf{Y} = y) \tag{2.27}$$

Where $f(x,y) \geq 0$ and $\iint f(x,y)\,dx\,dy = 1$ or $\sum_i \sum_j f\left(x_i, y_j\right) = 1$ (for continuous and discrete distributions, respectively), and the integration or summation takes place over the support of x and y.

The remaining probability concepts for bivariate variables are straightforward. For any subset $\mathcal{A}$ in the xy plane

$$\Pr[(\mathbf{X}, \mathbf{Y}) \in \mathcal{A}] = \iint_{x,y \in \mathcal{A}} f(x,y)\,dx\,dy \tag{2.28}$$

Probabilities may be computed by integrating the pdf over the relevant part of the real plane. For a discrete variable, probabilities can be arranged in table form, where each entry is the probability for an ordered pair of values $\left(x_i, y_j\right)$.

Example 2.6 An oil company wishes to take a statistical approach to finding new producing wells. For its existing wells, the company determines how many producing wells were drilled into formations of a given age. Let the rvs $\mathbf{X}$ and $\mathbf{Y}$ be formation type and geologic age, respectively. Table 2.1 gives the results.

Table 2.1 shows the joint probabilities; summing rows or columns gives the marginal probabilities. For example, summing the first row gives the probability for Permian formations of all facies types, while summing the first column gives the probability for

Table 2.1 Probability of Finding Oil as a Function of Formation Type and Age

Age	Sandstone	Limestone	Shale	Conglomerate	*Y*-Marginal
Permian	0.15	0.10	0.05	0	0.30
Triassic	0.10	0.08	0.02	0	0.20
Jurassic	0.20	0	0	0	0.20
Cretaceous	0.05	0.12	0.03	0	0.20
Tertiary	0	0	0.10	0	0.10
X-marginal	0.50	0.30	0.20	0	1

sandstone of any age. From the table it is clear that (1) conglomerates do not bear hydrocarbons, (2) sandstone offers the best probability for striking oil, (3) Tertiary drilling should be in shales, and (4) Permian formations are the best bet overall.

The bivariate cdf is defined by

$$F(x,y) = \Pr(\mathbf{X} \le x \cap \mathbf{Y} \le y) \tag{2.29}$$

and can be expanded for particular choices of intervals such as

$$\begin{aligned}\Pr(a \le \mathbf{X} \le b \cap c \le \mathbf{Y} \le d) &= \Pr(a \le \mathbf{X} \le b \cap \mathbf{Y} \le d) - \Pr(a \le \mathbf{X} \le b \cap \mathbf{Y} \le c) \\ &= [\Pr(\mathbf{X} \le b \cap \mathbf{Y} \le d) - \Pr(\mathbf{X} \le a \cap \mathbf{Y} \le d)] \\ &\quad -[\Pr(\mathbf{X} \le b \cap \mathbf{Y} \le c) - \Pr(\mathbf{X} \le a \cap \mathbf{Y} \le c)] \\ &= F(b,d) - F(a,d) - F(b,c) + F(a,c)\end{aligned} \tag{2.30}$$

The cdf of **X** alone (or the marginal distribution of **X**) can be obtained from the bivariate cdf by taking an appropriate limit

$$F_1(x) = \Pr(\mathbf{X} \le x) = \lim_{y\to\infty} \Pr(\mathbf{X} \le x \cap \mathbf{Y} \le y) = \lim_{y\to\infty} F(x,y) \tag{2.31}$$

Note that the marginal cdf for either x or y can be derived from the joint distribution, but the reverse is not true without additional information about the relationship between x and y.

The marginal pdf of **X** can be obtained from the bivariate pdf by integration

$$f_1(x) = \Pr(\mathbf{X} = x) = \int_{y\in S} f(x,y)dy \tag{2.32}$$

where the integral is over all possible values for y. The marginal pdf for either$x$ or y can be derived by integration of the joint pdf, but the reverse is not true without additional information about the covariance of x and y.

Example 2.7 Let $f(x,y) = e^{-(x+y)}$ with support $[0,\infty)$. The marginal distribution for x is $f(x) = e^{-x}\int_0^\infty e^{-y}dy = e^{-x}$.

The concept of a bivariate distribution extends to many variables as *multivariate distributions* and will be covered in Chapters 10 and 11.

2.6 Independent and Exchangeable Random Variables

Expanding on the concepts introduced in Section 1.8, if **X** and **Y** are independent rvs, then their probabilities factor

$$\Pr(\mathbf{X} \le x \cap \mathbf{Y} \le y) = \Pr(\mathbf{X} \le x)\Pr(\mathbf{Y} \le y) \tag{2.33}$$

The reverse is also true: if their probabilities factor, then a pair of variables is independent. More precisely, $\mathbf{X}$ and $\mathbf{Y}$ are independent if and only if $F(x,y) = F_1(x)F_2(y)$ or $f(x,y) = f_1(x)f_2(y)$. If the joint distribution is known and can be expressed as the product of the marginal distributions for two variables, then the variables are independent.

For a discrete variable where $\mathbf{X}$ can take on the values $\{x_i\}$, $i = 1, ..., r$, and $\mathbf{Y}$ can take on the values $\{y_j\}$, $j = 1, ..., s$, $\Pr(\mathbf{X} = x_i \cap \mathbf{Y} = y_j) = p_{ij}$. Let $\Pr(\mathbf{X} = x_i) = \sum_{j=1}^{s} p_{ij} = p_{i+}$ and $\Pr(\mathbf{Y} = y_j) = \sum_{i=1}^{r} p_{ij} = p_{+j}$. Then $\mathbf{X}$ and $\mathbf{Y}$ are independent if and only if $p_{ij} = p_{i+}p_{+j}$.

Example 2.8 Returning to Example 2.6 and Table 2.1, $p_{11} = 0.15$, $p_{1+} = 0.30$, and $p_{+1} = 0.50$, and hence sandstone and Permian are independent. However, $p_{12} = 0.10$, $p_{1+} = 0.30$, and $p_{+2} = 0.30$, and hence $p_{12} \ne p_{1+}p_{+2}$. Limestone and Permian are not independent.

The concept of independence extends to N rvs. If an N-variate set of rvs $\{\mathbf{X}_1, ..., \mathbf{X}_N\}$ is independent, and each has the same marginal distribution, then $\{\mathbf{X}_1, ..., \mathbf{X}_N\}$ are *independent and identically distributed* or *iid*. If F and f are, respectively, the cdf and pdf for the rvs, then $\mathbf{X}_1, ..., \mathbf{X}_N \sim F$ and $\mathbf{X}_1, ..., \mathbf{X}_N \sim f$, where the symbol ~ means "is distributed as" and not "is approximately." An equivalent statement is that $\{\mathbf{X}_1, ..., \mathbf{X}_N\}$ is a *random sample of size* N from F or f.

A closely related principle that underlies the nonparametric and resampling methods that will be covered in Chapters 7 and 8 is *exchangeability*. A set of rvs $\{\mathbf{X}_1, ..., \mathbf{X}_N\}$ (either finite or infinite in length) is exchangeable if any finite permutation of their indices (i.e., the permutation operates on a finite set of indices with the remainder fixed) results in no change to their joint probability distribution. A set of iid rvs is always exchangeable, as is a set of jointly Gaussian rvs with identical covariances. Exchangeability is a weaker assumption than independence; for example, if rvs are drawn from a finite population without replacement, they are clearly dependent but remain exchangeable.

A simple example of a set of data that are exchangeable but not independent follows from considering the empirical deviates from the grand mean for two data sets. Let $\{\mathbf{X}_i\}$ and $\{\mathbf{Y}_i\}$ comprise N_1 and N_2 rvs, respectively, and let $\mathbf{G} = \left(\sum_{i=1}^{N_1}\mathbf{X}_i + \sum_{i=1}^{N_2}\mathbf{Y}_i\right)/(N_1 + N_2)$ be their grand mean. The new rvs $\{\mathbf{X}_i' = \mathbf{X}_i - \overline{\mathbf{G}}\}$ and $\{\mathbf{Y}_i' = \mathbf{Y}_i - \overline{\mathbf{G}}\}$ are exchangeable but are dependent due to the grand mean term.

A simple transformation can often be applied to make a set of rvs exchangeable. Suppose that a set of rvs $\{\mathbf{X}_1, ..., \mathbf{X}_N\}$ has a population mean μ and a distribution $F(x - \mu)$ and an independent set of rvs $\{\mathbf{Y}_1, ..., \mathbf{Y}_M\}$ has a population mean λ and a distribution $F(y - \lambda)$. The transformed independent variables $\mathbf{X}_i' = \mathbf{X}_i - \mu$ and $\mathbf{Y}_i' = \mathbf{Y}_i - \lambda$ are exchangeable.

2.7 Conditional Probability Distributions

The conditional distribution is a generalization of the concept of conditional probability introduced in Section 1.7. Conditional distributions describe the probabilities for events determined by rvs conditional on the occurrence of other events described by other rvs. In other words, the probabilities for some rvs may change once other rvs are observed. Following on the definition of conditional probability in Section 1.7,

$$\Pr(\mathbf{X}|\mathbf{Y}=y)=\frac{\Pr(\mathbf{X}\cap\mathbf{Y}=y)}{\Pr(\mathbf{Y}=y)} \tag{2.34}$$

presuming that $\Pr(\mathbf{Y}=y)$ is nonzero. Equation (2.34) is called the *conditional probability density function* of $\mathbf{X}$ given that $\mathbf{Y}=y$. The rule to remember is that conditional distribution equals joint distribution divided by marginal distribution, just as it is for events.

There are two nonequivalent conditional pdfs for a specified joint distribution, given by

$$g_1(x|y)=\frac{f(x,y)}{f_2(y)} \tag{2.35}$$

$$g_2(y|x)=\frac{f(x,y)}{f_1(x)} \tag{2.36}$$

Equation (2.35) depends on all possible values of $\mathbf{X}$ for a given value of $\mathbf{Y}$, and equation (2.36) depends on all possible values of $\mathbf{Y}$ for a given value of $\mathbf{X}$. The sum or integral (depending on whether the distribution is discrete or continuous) over all possible values of x for g_1 and y for g_2 must be 1, which follows immediately from the definition.

If the joint distribution is known, the marginal and conditional distributions can be computed by summation or integration. Further, if the marginal and conditional distributions are known, the joint distribution can be recovered. It is often easier to obtain the conditional distribution, so this is sometimes a useful pathway to compute a joint distribution.

Since $f(x,y)=g_1(x\,|\,y)f_2(y)=g_2(y\,|\,x)f_1(x)$ from (2.35) and (2.36), a generalization of Bayes' theorem to distributions is given by

$$\begin{aligned} g_1(x\,|\,y)&=\frac{g_2(y\,|\,x)f_1(x)}{f_2(y)}\\ g_2(y\,|\,x)&=\frac{g_1(x\,|\,y)f_2(y)}{f_1(x)} \end{aligned} \tag{2.37}$$

From the definition of independence, $f(x,y)=f_1(x)f_2(y)$. It immediately follows that x and y are independent if and only if $g_1(x\,|\,y)=f_1(x)$ and $g_2(y\,|\,x)=f_2(y)$.

Example 2.9 A simple model of a crystalline rock is an aggregation of spherical crystals of varying radius, with $f(r)$ being the pdf for the radius r. Suppose that the rock is sliced to make a thin section, yielding a two-dimensional cross section of the three-dimensional rock in which the spherical crystals appear as circles of varying radius. Let the pdf of the circles be $g(\rho)$, where ρ is the radius of the circles (see Figure 2.4). What is the relationship between $f(r)$ and $g(\rho)$?

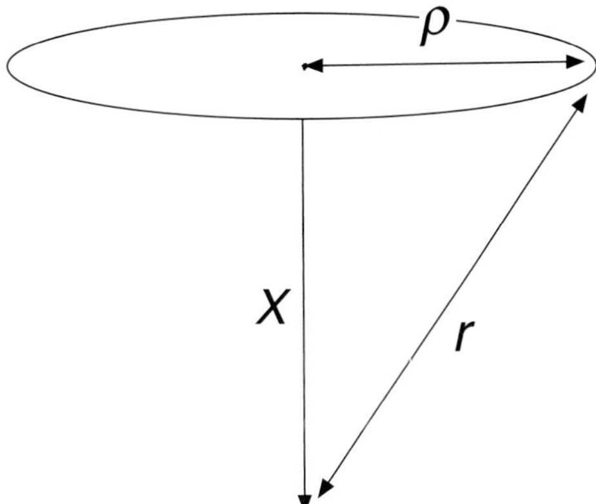

Figure 2.4 A planar slice of a sphere of radius r at a distance x from its center, producing a circle of radius ρ.

This problem can be solved by assuming that the orientation of the cross sections is random in space and finding the conditional distribution $h(\rho|r)$. For a given value of the rv $\mathbf{R} = r$, let $\mathbf{X}$ be the distance from the center of a sphere to the center of the corresponding circle (see Figure 2.4). Since the sampling is assumed to be random, it follows that $\mathbf{X}$ is uniformly distributed on $[0, r]$ and that $\mathbf{X} = \sqrt{r^2 - \mathbf{P}^2}$. The conditional cdf is

$$\begin{aligned} H(\rho\,|\,r) &= \Pr(\mathbf{P} \le \rho|\mathbf{R} = r) \\ &= \Pr\left(\sqrt{r^2 - \mathbf{X}^2} \le \rho\right) \\ &= \Pr\left(\mathbf{X} \ge \sqrt{r^2 - \rho^2}\right) \\ &= 1 - \frac{\sqrt{r^2 - \rho^2}}{r} \end{aligned}$$

The radius ρ cannot exceed r, ensuring that the cdf is bounded by 0 and 1. The conditional pdf follows by differentiation:

$$h(\rho\,|\,r) = \frac{\rho}{r\sqrt{r^2 - \rho^2}}$$

The *law of total probability* (Section 1.4) states that

$$g(\rho) = \int h(\rho\,|\,r)\, f(r)dr$$

where the integral is taken over all possible values of the radius of the spheres. Consequently,

$$g(\rho) = \int_0^\infty h(\rho\,|\,r)\, f(r)\, dr = \int_\rho^\infty \frac{\rho}{r\sqrt{r^2 - \rho^2}} f(r)\, dr$$

because r cannot be smaller than ρ. This is an Abel integral equation that occurs frequently in image analysis. In practice, one can estimate $g(\rho)$ from measurements on a series of randomly oriented thin sections and then estimate $f(r)$ by an approximate inversion of the integral equation.

Table 2.2 Conditional Probability of Finding Oil

Age	Sandstone	Limestone	Shale
Permian	0.50	0.33	0.17
Triassic	0.50	0.25	0.10
Jurassic	1.0	0	0
Cretaceous	0.25	0.60	0.15
Tertiary	0	0	1.0

Example 2.10 The concepts of conditional probability also extend to discrete distributions. For the oil model of Example 2.6, conditional probability tables can be constructed. Pr ($\mathbf{X} = x \mid \mathbf{Y} = y$) is the probability that a rock is of a particular type given a value for its age. The result is given in Table 2.2.

The rows add to 1. For this example, if it is Tertiary, it is shale; if it is Jurassic, it is sandstone.

2.8 Functions of a Random Variable

Suppose that an rv $\mathbf{X}$ has a pdf $f(x)$ and a cdf $F(x)$. Define a new rv $\mathbf{Y} = r(\mathbf{X})$, where r is a function. The cdf of $\mathbf{Y}$ is

$$G(y) = \Pr(\mathbf{Y} \le y) = \Pr[r(\mathbf{X}) \le y] = \int_{x \forall r(x) \le y} f(x)dx \tag{2.38}$$

The integration takes place over all values of x such that $r(x) \le y$.

Example 2.11 Let $\mathbf{X}$ be drawn from a uniform distribution with support $[-1, 1)$ so that $f(x) = 1/2$. Compute the pdf of $\mathbf{Y} = \mathbf{X}^2$.

The rv $\mathbf{Y}$ must lie on $[0, 1)$, so

$$G(y) = \Pr(\mathbf{Y} \le y) = \Pr\left(\mathbf{X}^2 \le y\right) = \Pr\left(-\sqrt{y} \le \mathbf{X} \le \sqrt{y}\right) = \int_{-\sqrt{y}}^{\sqrt{y}} f(x)dx$$

It follows immediately that $g(y) = 1/\left(2\sqrt{y}\right)$. The pdf of y is unbounded at $y = 0$.

It is not true in general that $\mathbf{Y} = r(\mathbf{X})$ will have a continuous distribution when $\mathbf{X}$ does, and in that instance, (2.38) is the way to obtain the distribution for $\mathbf{Y}$. However, when the distribution of $\mathbf{Y}$ is continuous, there is another approach that yields the pdf directly. Let $\mathbf{X}$ be a rv with pdf $f(x)$ for which $\Pr(a < \mathbf{X} \le b) = 1$. Let $\mathbf{Y} = r(\mathbf{X})$, where r is continuous and monotone (i.e., either strictly increasing or decreasing) on $(a, b]$. Suppose that $a < \mathbf{X} \le b$ if and only if $\alpha < \mathbf{Y} \le \beta$, and let $\mathbf{X} = s(\mathbf{Y})$ be the unique inverse mapping of $r(\mathbf{X})$. Then the pdf of $\mathbf{Y}$ is $g(y) = |\partial_y s(y)| f[s(y)]$ for $s(y)$ on $(a, b]$ (or y on $(\alpha, \beta]$) and zero otherwise.

Example 2.12 A standardized Gaussian variable has the pdf $f(x) = \left(1/\sqrt{2\pi}\right)e^{-x^2/2}$. Derive the pdf of $y = x^2$.

Since $-\infty < x \le \infty$, it follows that $0 < y \le \infty$. The inverse mapping is $s(y) = \sqrt{y}$. Then $g(y) = \sqrt{1/8\pi y e^{-y/2}}$, which is the chi square distribution with one degree of freedom. Note that no integral had to be computed to get this result.

These concepts easily extend to large numbers of rvs. For a set of rvs $\{\mathbf{X}_i\}$, $i = 1, ..., N$ with joint pdf $f(x_1, ..., x_N)$, define a new rv $\mathbf{Y} = r(\mathbf{X}_1, ..., \mathbf{X}_N)$. The cdf of $\mathbf{Y}$ can be computed from first principles. Let $\mathcal{A} \subset \mathcal{S}$ comprise the subset containing all values $\{x_1, ..., x_N\}$ such that $r(x_1, ..., x_N) \le y$. The cdf is

$$\begin{aligned} G(y) &= \Pr(\mathbf{Y} \le y) = \Pr[r(\mathbf{X}_1, ..., \mathbf{X}_N) \le y] \\ &= \int \cdots \int_{\mathcal{A}} f(x_1, ..., x_N)\ dx_1 \cdots dx_N \end{aligned} \tag{2.39}$$

If $G(y)$ is continuous, then the pdf can be found by differentiation.

Example 2.13 Suppose that the independent rvs $\{\mathbf{X}_i\}$, $i = 1, ..., N$, are a random sample from a distribution with pdf $f(x)$ and cdf $F(x)$. The largest value in the sample is $\mathbf{Y}_N = \max\{\mathbf{X}_i\}$, and the smallest value is $\mathbf{Y}_1 = \min\{\mathbf{X}_i\}$. Compute the pdf and cdf of $\mathbf{Y}_N$ and $\mathbf{Y}_1$.

For any allowed value of y,

$$\begin{aligned} G_N(\mathbf{Y}) &= \Pr(\mathbf{Y}_N \le y) = \Pr(\mathbf{X}_1 \le y \cap \cdots \cap \mathbf{X}_N \le y) = \Pr(\mathbf{X}_1 \le y) \cdots \Pr(\mathbf{X}_N \le y) \\ &= F(y) \cdots F(y) = [F(y)]^N \end{aligned}$$

The pdf follows by differentiation

$$g_N(y) = N[F(y)]^{N-1} f(y)$$

The distribution of the smallest value follows from the same reasoning

$$\begin{aligned} G_1(y) &= \Pr(\mathbf{Y}_1 \le y) = 1 - \Pr(\mathbf{Y}_1 > y) = 1 - \Pr(\mathbf{X}_1 > y \cap \cdots \cap \mathbf{X}_N > y) \\ &= 1 - \Pr(\mathbf{X}_1 > y) \cdots \Pr(\mathbf{X}_N > y) = 1 - [1 - F(y)] \cdots [1 - F(y)] = 1 - [1 - F(y)]^N \end{aligned}$$

The pdf is

$$g_1(y) = N[1 - F(y)]^{N-1} f(y)$$

The bivariate distribution of $\mathbf{Y}_1$ and $\mathbf{Y}_N$ can also be derived from first principles

$$\begin{aligned}\Pr(\mathbf{Y}_1 \le y_1 \cap \mathbf{Y}_N \le y_N) &= \Pr(1 - \mathbf{Y}_1 > y_1 \cap \mathbf{Y}_N \le y_N)\\ &= \Pr(\mathbf{Y}_N \le y_N) - \Pr(\mathbf{Y}_1 > y_1 \cap \mathbf{Y}_N \le y_N)\\ &= \Pr(\mathbf{Y}_N \le y_N) - \Pr(y_1 < \mathbf{X}_1 \le y_N, \ldots, y_1 < \mathbf{X}_N \le y_N)\\ &= G_N(y_N) - \prod_{i=1}^{N} \Pr(y_1 < \mathbf{X}_i \le y_N)\\ &= [F(y_N)]^N - [F(y_N) - F(y_1)]^N\end{aligned}$$

The bivariate joint pdf follows from differentiation as

$$g(y_1, y_N) = \partial^2_{y_1 y_N} G(y_1, y_N) = N(N-1)[F(y_N) - F(y_1)]^{N-2} f(y_1) f(y_N)$$

Since neither the pdf nor the cdf factors, it follows that the variables are dependent.

The minimum and maximum values are examples of order statistics that are obtained by ranking and ordering a set of rvs and are covered in more detail in Section 4.10. In fact, these are extreme order statistics and are the subject of a whole subfield of statistics – extreme value theory – that arises in the earth sciences in the study of floods or severe storms, among other areas. Note that once a set of data is ranked and ordered, the data are no longer independent or identically distributed even when the original data are iid.

2.9 Functions of Two or More Random Variables

Transformation of a bivariate (or multivariate, because the procedure is the same) pdf is a straightforward but tedious task. Suppose that $\mathbf{Y}_j = r_j(\mathbf{X}_i),\ i,j = 1, \ldots, p$, and that the mapping is uniquely invertible so that $\mathbf{X}_i = s_i(\mathbf{Y}_j)$. If the partial derivatives of the transformation functions s_i with respect to the variable y_j exist, then

$$g(y_1, \ldots, y_N) = |\det \mathbf{J}|\, f(s_1, \ldots, s_N) \tag{2.40}$$

where $\mathbf{J}$ is the Jacobian matrix whose i,j element is $\partial_{y_j} s_i$, and det is the determinant.

Example 2.14 Consider the distribution of two independent rvs $\mathbf{X}_1$ and $\mathbf{X}_2$. Derive the distribution of their product and quotient. What are these distributions when $\mathbf{X}_1$ and $\mathbf{X}_2$ are standard Gaussian?

Let the joint pdf of $\mathbf{X}_1$ and $\mathbf{X}_2$ be $f(x_1, x_2)$. Beginning with the quotient distribution, let $\mathbf{Y}_1 = \mathbf{X}_1/\mathbf{X}_2$ and $\mathbf{Y}_2 = \mathbf{X}_2$. The inverse transformations are $\mathbf{X}_1 = \mathbf{Y}_1\mathbf{Y}_2$ and $\mathbf{X}_2 = \mathbf{Y}_2$, and the Jacobian determinant is y_2, so the joint pdf of $\mathbf{Y}_1$ and $\mathbf{Y}_2$ is

$$g(y_1, y_2) = |y_2| f(y_1 y_2, y_2)$$

and the marginal pdf of $\mathbf{Y}_1$ is

$$g_1(y_1) = \int_{-\infty}^{\infty} |y_2| \, f(y_1 y_2, y_2) \, dy_2$$

If the variables are independent standardized Gaussian, then

$$g(y) = \frac{1}{2\pi} \int_{-\infty}^{\infty} |x| \, e^{-x^2/2} e^{-x^2 y^2/2} dx$$

This is symmetric about $x = 0$, so

$$g(y) = \frac{1}{\pi} \int_{0}^{\infty} x e^{-x^2\left[(y^2+1)/2\right]} dx = \frac{1}{\pi(1+y^2)}$$

which is the standardized Cauchy distribution (or equivalently, Student's t distribution with one degree of freedom). It falls to zero much more slowly with y than the Gaussian distribution because its tails are algebraic rather than exponential. The Cauchy distribution is often used as a simple analogue for a long-tailed normal distribution. It is also a member of the stable distribution family covered in Section 3.4.2.

For the product distribution, let $\mathbf{Y}_1 = \mathbf{X}_1 \mathbf{X}_2$ and $\mathbf{Y}_2 = \mathbf{X}_2$. The inverse transformations are $\mathbf{X}_1 = \mathbf{Y}_1/\mathbf{Y}_2$ and $\mathbf{X}_2 = \mathbf{Y}_2$, and the Jacobian determinant is $1/y_2$, so the joint pdf of $\mathbf{Y}_1$ and $\mathbf{Y}_2$ is

$$g(y_1, y_2) = \frac{1}{|y_2|} f(y_1/y_2, y_2)$$

and the marginal pdf of $\mathbf{Y}_1$ is

$$g_1(y_1) = \int_{-\infty}^{\infty} \frac{1}{|y_2|} f(y_1/y_2, y_2) \, dy_2$$

If the variables are independent standardized Gaussian, then

$$g(y) = \frac{1}{2\pi} \int_{-\infty}^{\infty} \frac{1}{|x|} \, e^{-x^2/2} e^{-(y/x)^2/2} dx = \frac{1}{\pi} K_0(|y|)$$

using Gradshteyn & Ryzhik (1980, 3.471-9), where $K_0(x)$ is a modified Bessel function of the second kind of order 0.

As a useful special case, consider linear transformations of the form $\mathbf{Y} = \mathbf{A} \cdot \mathbf{X}$, where $\mathbf{A}$ is an $N \times N$ matrix, and the dot denotes the inner product. Then $\mathbf{X} = \mathbf{A}^{-1} \cdot \mathbf{Y}$, presuming that $\mathbf{A}$ is invertible, and $\det \mathbf{J} = \det \mathbf{A}^{-1} = 1/\det \mathbf{A}$. The pdf of the transformed variables is

$$g(y_1, \ldots, y_N) = f(x_1, \ldots, x_N)/|\det \mathbf{A}| = f\left(\mathbf{A}^{-1} \cdot \mathbf{y}\right)/|\det \mathbf{A}| \tag{2.41}$$

Example 2.15 Let two rvs $\mathbf{X}_1$ and $\mathbf{X}_2$ have a joint pdf $f(x_1, x_2)$. Compute the pdf of their sum and difference.

Let $\mathbf{Y}_1 = \mathbf{X}_1 + \mathbf{X}_2$ and $\mathbf{Y}_2 = \mathbf{X}_2$, which inverts to $\mathbf{X}_1 = \mathbf{Y}_1 - \mathbf{X}_2$ and $\mathbf{X}_2 = \mathbf{Y}_2$. The transformation matrix

$$\mathbf{A} = \begin{bmatrix} 1 & 1 \\ 0 & 1 \end{bmatrix}$$

So

$$\mathbf{A}^{-1} = \begin{bmatrix} 1 & -1 \\ 0 & 1 \end{bmatrix}$$

and $\det \mathbf{A}^{-1} = 1$, yielding

$$g(y_1, y_2) = f(y_1 - y_2, y_2)$$

The marginal pdf of y_1 is

$$g_1(y_1) = \int_{-\infty}^{\infty} f(y_1 - y_2, y_2)\, dy_2$$

If $\mathbf{X}_1$ and $\mathbf{X}_2$ are independent, then $g_1(y_1) = \int_{-\infty}^{\infty} f_1(y_1 - z) f_2(z)\, dz$, which is just the convolution of the marginal pdfs.

Using the same procedure, it is easy to show that the pdf of the difference between two rvs is

$$g(y) = \int_{-\infty}^{\infty} f(y + z, z)\ dz = \int_{-\infty}^{\infty} f_1(y + z)\, f_2(z) dz$$

where the second integral holds for independent variables. The pdf of the difference between two rvs is the correlation of the marginal pdfs.

2.10 Measures of Location

Location parameters are a class of measurable factors that characterize the center of a probability distribution. For a given pdf f, θ is a location parameter if the form of the distribution is $f(x - \theta)$. There are many location parameters, but they divide into three classes: means, median, and mode.

Suppose that an rv has a discrete distribution with pdf $f(x)$. The *arithmetic mean* or *expectation* or *expected value* or *first moment* of the rv is

$$\mu = \mathcal{E}(\mathbf{X}) = \sum_x x f(x) \tag{2.42}$$

provided that the sum is absolutely convergent $\left(\sum_x |x| f(x) < \infty\right)$. For a continuous pdf,

$$\mu = \mathcal{E}(\mathbf{X}) = \int_{-\infty}^{\infty} x f(x)\, dx \tag{2.43}$$

provided that the integral is absolutely convergent $\left(\int_{-\infty}^{\infty} |x| f(x) dx < \infty\right)$. If $\mathbf{X}$ is bounded so that $\Pr(a < X \le b) = 1$, then $\mathcal{E}(\mathbf{X})$ always exists.

The expected value is in a loose sense the most probable value, or the center of gravity, or the mean of the distribution $f(x)$. Conceptually, it can be thought of as the arithmetic average of a large number of iid draws from the distribution. For a symmetric pdf, the expected value occurs at the point of symmetry x_o , where $f(x_o - \delta) = f(x_o + \delta)$ for all δ.

Example 2.16 The pdf of an rv that is the square root of the sum of the squares of two independent Gaussian rvs is Rayleigh. This is the distribution of the magnitude of complex rvs such as the Fourier transform at a given frequency. The pdf is $f(x) = x e^{-x^2/2}$ for $x \ge 0$. The expected value is

$$\mu = \int_0^{\infty} x^2 e^{-x^2/2} dx = \sqrt{\frac{\pi}{2}}$$

Example 2.17 The Cauchy distribution has the pdf $f(x) = 1/[\pi(1 + x^2)]$ with support $(-\infty, \infty)$. Because $f(x)$ is symmetric about $x = 0$, one might conclude that the expected value is zero. However, the Cauchy distribution is not absolutely integrable, and hence the expected value does not exist.

The expected value of a function follows directly from the relation $\mathbf{Y} = r(\mathbf{X})$, yielding

$$\mathcal{E}[r(x)] = \int_{-\infty}^{\infty} r(x) f(x) dx \tag{2.44}$$

and can be computed without finding the pdf of $r(\mathbf{X})$. The characteristic function of Section 2.4 is just the expected value of e^{itx}.

Example 2.18 Find $\mathcal{E}(\mathbf{X}^2)$ for a Rayleigh rv.

$\mathcal{E}(\mathbf{X}^2) = \int_0^{\infty} x^3 e^{-x^2/2} dx = 2$. This result could have been obtained by first finding the pdf for x^2 and then computing its expected value. That would definitely be the hard way to solve the problem.

The key properties of the expected value that can easily be derived from the definition are

1. If $\mathbf{Y} = a\mathbf{X} + b$, where a and b are constants, then $\mathcal{E}(\mathbf{Y}) = a\mathcal{E}(\mathbf{X}) + b$ (affine property).
2. If there exists a number a such that $\Pr(\mathbf{X} \geq a) = 1$, then $\mathcal{E}(\mathbf{X}) \geq a$. If there exists a number b such that $\Pr(\mathbf{X} \leq b) = 1$, then $\mathcal{E}(\mathbf{X}) \leq b$.
3. $\mathcal{E}(\mathbf{X}_1 + \cdots + \mathbf{X}_N) = \mathcal{E}(\mathbf{X}_1) + \cdots + \mathcal{E}(\mathbf{X}_N)$ if $\mathcal{E}(\mathbf{X}_i)$ exists. This linearity property holds regardless of any dependence between the rvs.
4. If a set of rvs $\{\mathbf{X}_i\}$ is independent, then $\mathcal{E}\left(\prod_{i=1}^{N}\mathbf{X}_i\right) = \prod_{i=1}^{N}\mathcal{E}(\mathbf{X}_i)$. This follows from the properties of the joint pdf of independent variables.

Example 2.19 A sample of rocks contains M rock types with the proportion of type 1 being p. Let N rocks be drawn at random without replacement. Find the expected value of the number of type 1 rocks drawn.

Let $\mathbf{X}_i$ be 1 if type 1 is drawn and 0 otherwise. Note that the set of rvs $\{\mathbf{X}_i\}$ is not independent because the sampling is without replacement. Then $\Pr(\mathbf{X}_i = 1) = p$, $\mathcal{E}(\mathbf{X}_i) = p$, and $\mathcal{E}(\mathbf{X}_1 + \cdots + \mathbf{X}_N) = Np$.

There are two additional types of mean location parameters that are less frequently used, and both require that the variables be nonnegative. The *harmonic mean* is the reciprocal of the expected value of $1/x$:

$$\frac{1}{\mu_H} = \int_0^\infty \frac{f(x)\,dx}{x} \tag{2.45}$$

The *geometric mean* is given by

$$\log \mu_G = \int_0^\infty \log(x) f(x)\,dx \tag{2.46}$$

Equation (2.46) has the important property that the geometric mean of a ratio is the ratio of the geometric means and hence is the only mean that can be used when one is dealing with normalized variables. Compositional data are a prominent example of normalized variables and will be examined in Chapter 11. Finally, the harmonic mean is always smaller than the geometric mean, which is, in turn, smaller than the expected value.

The median $\tilde{\mu}$ of the distribution of $\mathbf{X}$ is the point where $\Pr(\mathbf{X} \geq \tilde{\mu}) \geq 1/2$ and $\Pr(\mathbf{X} \leq \tilde{\mu}) \geq 1/2$. In terms of the pdf, this is

$$\int_{-\infty}^{\tilde{\mu}} f(x)\,dx = \int_{\tilde{\mu}}^{\infty} f(x)\,dx = \frac{1}{2} \tag{2.47}$$

In other words, the median is the point for which half the probability lies above and half lies below. The median is always unique for continuous distributions. However, for discrete distributions, the median may not be unique and may not coincide with an allowed value. In some respects, the median is a better representation of the average of a distribution because not all distributions have a mean (remember the Cauchy distribution), but all do have a median.

Example 2.20 The median of the Rayleigh distribution is found from $\int_0^{\tilde{\mu}} xe^{-x^2/2}dx = 1/2$, which yields $\tilde{\mu} = \sqrt{2\log 2}$.

A generalization of the median is obtained by defining the N quantiles of a distribution

$$\int_{-\infty}^{Q_j} f(x)\,dx = \frac{j - \frac{1}{2}}{N} \qquad j = 1, \ldots, N \tag{2.48}$$

The quantiles divide the area under the pdf into $N + 1$ probability intervals, with the first and last having probability increments of size $1/(2N)$, and the remainder having probability increments of $1/N$. If the quantiles are known, a lot of information about the distribution is available. The median is the middle quantile.

The mode μ^* is the maximum value on the pdf if it exists and hence in a very loose sense is the most probable value. The mode is easily obtained as the solution to $\partial_x f(x) = 0$ subject to $\partial_x^2 f(x) < 0$. There may be more than one mode for a distribution, in which case it is multimodal and the mode is not unique, and otherwise it is unimodal. For a symmetric unimodal distribution, the expected value, median, and mode coincide.

2.11 Measures of Dispersion

Dispersion is a measure of the spread or width or scale of a distribution. Dispersions divide into three classes: measures of the deviation from some central value (e.g., the variance), measures of the distance between specified representative values (e.g., the range), and measures of the deviation of the population among themselves (e.g., the mean difference).

The most widely used scale parameter is the *variance*, or mean of the square of the deviation from the mean, and is defined as

$$\sigma^2 = var(\mathbf{X}) = \mathcal{E}\left[(\mathbf{X} - \mu)^2\right] = \int_{-\infty}^{\infty} (x - \mu)^2 f(x)dx \tag{2.49}$$

Because the variance is the expected value of a nonnegative random variable, it follows that it must be nonnegative. The variance exists only when (2.49) is finite. If $\mathbf{X}$ is finitely bounded, then the variance always exists. If the variance is small, then the distribution is tight around μ, and if the variance is large, then the distribution is broad around μ. The *standard deviation* σ is the positive square root of the variance. The *precision* is the inverse of the variance and is sometimes taken as $1/(2\sigma^2)$.

Some key properties of the variance that are easily derived from the definition are

1. $var(\mathbf{X}) = 0$ if and only if $\Pr(\mathbf{X} = c) = 1$ (i.e., when the probability is concentrated at a single point);
2. $var(a\mathbf{X} + b) = a^2 var(\mathbf{X})$ (i.e., shifting the mean of a distribution by b does not change its spread and hence its variance);
3. $var(\mathbf{X}) = \mathcal{E}(\mathbf{X}^2) - \mathcal{E}(\mathbf{X})^2$, a formula that is often useful for computation; and
4. If $\{\mathbf{X}_i\}$ are independent, $var(\mathbf{X}_1 + \cdots + \mathbf{X}_N) = var(\mathbf{X}_1) + var(\mathbf{X}_N)$. However, when the rvs are dependent, how they covary must also be considered.

A less commonly used measure of scale is the *average absolute deviation from the median* given by

$$\tilde{\sigma} = \int_{-\infty}^{\infty} |x - \tilde{\mu}| f(x)\, dx \tag{2.50}$$

The average absolute deviation from the mean can also be applied. A more useful measure of scale is the *median absolute deviation from the median* or *MAD*, which is the solution σ_{MAD} to

$$F(\tilde{\mu} + \sigma_{\text{MAD}}) - F(\tilde{\mu} - \sigma_{\text{MAD}}) = \frac{1}{2} \tag{2.51}$$

The sample counterpart to (2.51) is especially useful when data contain a fraction of extreme values because it is relatively insensitive to such unusual data.

The *range* is the difference between the largest and smallest value in a population. However, this is not an especially useful measure of dispersion because it says nothing about the behavior of the population between its extremes, and obviously has no meaning if the distribution limits are infinite. A more useful measure of distance between representative points of the distribution is the *interquartile range*, which is the difference between the upper and lower quartiles and hence contains half the probability around the distribution center. It is given by

$$\sigma_{\text{IQ}} = F^{-1}(0.75) - F^{-1}(0.25) \tag{2.52}$$

The interquartile range exists when the variance is undefined and is twice the MAD for symmetric distributions.

The *mean difference* is defined by

$$\Delta = \int_{-\infty}^{\infty} \int_{-\infty}^{\infty} |x - y| f(x) f(y) dx\, dy \tag{2.53}$$

and is the mean deviation of the variable $x - y$ about zero when x and y are independently distributed as f. The mean difference is the average of the differences between all possible pairs of variables independent of sign, and hence depends on the dispersion of the variables among themselves rather than about some central value.

Coefficients of variation are combinations of scale and location parameters that are dimensionless and hence useful for comparing different populations measured in distinct units. The Pearson coefficient of variation is σ/μ and is the most widely used in practice. An alternative is the Gini coefficient of variation $\Delta/(2\mu)$. Both of these have the disadvantage that they depend on the expected value and hence are useful only when there is a natural choice of origin.

2.12 Measures of Shape

The mean and variance are summary properties about a distribution. However, higher-order moments may be required and serve as measures of distribution shape. Define the kth noncentral moment

$$\mu'_k = \mathcal{E}(\mathbf{X}^k) \tag{2.54}$$

and the kth central moment

$$\mu_k = \mathcal{E}\left[(\mathbf{X}-\mu)^k\right] \tag{2.55}$$

that both exist for order k provided that $\mathcal{E}\left(|\mathbf{X}|^k\right) < \infty$. If the kth moment is the first to be undefined, then all the j moments for $j < k$ exist, but the ones that are larger than k do not. The noncentral moments may be found using the cf

$$\partial_t^k \phi(t) = i^k \int_{-\infty}^{\infty} e^{itx} x^k f(x)\, dx \tag{2.56}$$

so

$$\mu'_k = (-i)^k \left[\partial_t^k \phi(t)\right]_{t=0} \tag{2.57}$$

A measure of the asymmetry of a distribution is the third central moment μ_3 normalized by σ^3 and is called the *skewness*. It vanishes (along with all higher-order odd moments) for a symmetric distribution. A positive skewness means that the right tail of the distribution is longer, whereas the opposite holds for a negative skewness.

A measure of the flatness of a distribution is given by the fourth central moment μ_4 normalized by σ^4 and is called the *kurtosis*. A high-kurtosis distribution has a sharp peak and longer tails, whereas a low-kurtosis distribution has a rounded peak and shorter tails.

The *excess kurtosis* is $\mu_4/\sigma^4 - 3$ but is frequently simply called the *kurtosis*, which is often a source of confusion. A distribution with zero excess kurtosis is *mesokurtic*, and the Gaussian is a classic example. A distribution with positive excess kurtosis is *leptokurtic*, whereas one with negative excess kurtosis is *platykurtic*.

The skewness and kurtosis are intuitive rather than precise. A better way of comparing the shapes of two probability distributions F and G is as follows:

1. G is more skewed to the right than F if $G^{-1}(F)$ is convex for all x; and
2. When F and G are both symmetric, G has greater kurtosis than F if $G^{-1}(F)$ is convex for all $x > \tilde{\mu}$, where a function on an interval is convex upward if the line segment between any two points on its graph lies above the function. This extends trivially to convex downward functions.

2.13 Measures of Direction

The trigonometric moments are given by

$$\begin{aligned} \alpha_k &= \mathcal{E}(\cos k\theta) = \int_0^{2\pi} \cos k\theta \, f(\theta)\, d\theta \\ \beta_k &= \mathcal{E}(\sin k\theta) = \int_0^{2\pi} \sin k\theta \, f(\theta)\, d\theta \end{aligned} \tag{2.58}$$

The sequence of trigonometric moments $\{\alpha_k, \beta_k\}$, for $k = 0, \pm 1, \pm 2, ...$, for the random variable θ completely specifies the cf, and because a probability distribution on a circle is uniquely specified by the cf, so are the cdf and (usually) the pdf. Consequently, in contrast to distributions on the real line, a distribution on a circle is completely defined by its moments.

Following on (2.25), the kth element in the cf may be written (for $k \ge 0$)

$$\phi_k = \rho_k e^{i\mu_k} \tag{2.59}$$

where ρ_k is the kth *resultant length*, and μ_k is the kth *direction*. The most important case is $k = 1$, and its trigonometric moments, resultant, and direction will be, respectively, α, β, ρ, and μ. These are called, in turn, the *mean cosine moment*, *mean sine moment*, *mean resultant length*, and *mean direction*. The mean direction is equivariant under rotation because the mean direction of $\theta' = \theta - \vartheta$ is given by $\mu' = \mu - \vartheta$. The mean resultant length is invariant under rotation and reflection. Further, the kth *trigonometric moment about the mean direction* is given by $\overline{\alpha}_k + i\overline{\beta}_k$, where

$$\begin{aligned} \overline{\alpha}_k &= \mathcal{E}[\cos k(\theta - \mu)] \\ \overline{\beta}_k &= \mathcal{E}[\sin k(\theta - \mu)] \end{aligned} \tag{2.60}$$

with $\overline{\alpha}_0 = 1$ and $\overline{\beta}_0 = \overline{\beta}_1 = 0$. Consequently, the mean resultant length is given by $\overline{\alpha}_1$. Finally, the median direction $\tilde{\mu}$ must satisfy

$$\int_{\tilde{\mu}-\pi}^{\tilde{\mu}} f(\theta)\, d\theta = \int_{\tilde{\mu}}^{\tilde{\mu}+\pi} f(\theta)\, d\theta = \frac{1}{2} \tag{2.61}$$

and is not unique unless the distribution is unimodal. For computational purposes, the median direction is better obtained by solving

$$\min_{\tilde{\mu}}\left[\pi - \int_0^{2\pi} |\pi - |\theta - \tilde{\mu}||\, f(\theta)d\theta\right] \tag{2.62}$$

The circular variance is given by

$$v = 1 - \rho \tag{2.63}$$

and lies on [0, 1]. The circular variance is zero if and only if the distribution is tightly concentrated around the mean direction μ. Conversely, the circular variance is one when there is no angular concentration around any given direction. The *circular standard deviation* is given by

$$\sigma = \sqrt{-2\log\rho} \tag{2.64}$$

and is not the square root of the circular variance (2.63). It has its origin in the definition of the Gaussian distribution wrapped onto the unit circle, as described in Mardia & Jupp (2000, chap. 3). The *circular dispersion* is an alternate measure of dispersion to the circular variance and standard deviation and is given by

$$\delta = \frac{1 - \rho_2}{2\rho^2} \tag{2.65}$$

and is important in statistical inference for the mean direction.

Shape parameters for directional distributions are analogous to the standard skewness and kurtosis. The directional skewness is given by

$$\varsigma = \frac{\bar{\beta}_2}{(1 - \rho)^{3/2}} \tag{2.66}$$

and the directional kurtosis is

$$\kappa = \frac{\bar{\alpha}_2 - \rho^4}{(1 - \rho)^2} \tag{2.67}$$

Finally, the *directional p-quantile* is the solution $\tilde{\mu}_p$ to

$$\int_{\tilde{\mu}-\pi}^{\tilde{\mu}_p} f(\theta)\, d\theta = p \tag{2.68}$$

where p is a probability. The *interquartile range for directional data* follows as the angular difference between $\tilde{\mu}_{0.25}$ and $\tilde{\mu}_{0.75}$.

2.14 Measures of Association

The mean and variance give information about the marginal distributions in the bivariate or multivariate cases but not about the joint distribution because they provide no information about how the different variables are related. Let $\mu_x = \mathcal{E}(\mathbf{X})$, $\mu_y = \mathcal{E}(\mathbf{Y})$, $\sigma_x^2 = var(\mathbf{X})$, and $\sigma_y^2 = var(\mathbf{Y})$. The *covariance* of **X** and **Y** is defined as

$$cov(\mathbf{X},\mathbf{Y}) = \mathcal{E}\left[(\mathbf{X}-\mu_x)\left(\mathbf{Y}-\mu_y\right)\right] \tag{2.69}$$

The covariance exists provided that the respective variances are finite, and may be positive or negative.

The *correlation* of **X** and **Y** is the normalized covariance given by

$$\rho(\mathbf{X},\mathbf{Y}) = \frac{cov(\mathbf{X},\mathbf{Y})}{\sigma_x\sigma_y} \tag{2.70}$$

and lies between -1 and 1. When ρ is positive or negative, the two rvs are said to be positively or negatively correlated.

Some key properties of the covariance that can be derived from the definition include

1. $cov(\mathbf{X},\mathbf{Y}) = \mathcal{E}(\mathbf{XY}) - \mathcal{E}(\mathbf{X})\mathcal{E}(\mathbf{Y})$.
2. If **X** and **Y** are independent with finite variance, $cov(\mathbf{X},\mathbf{Y}) = \rho(\mathbf{X},\mathbf{Y}) = 0$. However, $cov(\mathbf{X},\mathbf{Y}) = \rho(\mathbf{X},\mathbf{Y}) = 0$ does not prove independence, as can be demonstrated by example: let $\mathbf{X} = \{-1,0,1\}$ with equal probability. Then $\mathcal{E}(\mathbf{X}) = 0$. Let $\mathbf{Y} = \mathbf{X}^2$, which is clearly dependent on **X**. Then $\mathcal{E}(\mathbf{XY}) = \mathcal{E}\left(\mathbf{X}^3\right) = 0$; hence $cov(\mathbf{X},\mathbf{Y}) = 0$. The variables are uncorrelated but not independent.
3. $var(\mathbf{X}\pm\mathbf{Y}) = var(\mathbf{X}) + var(\mathbf{Y}) \pm 2\,cov(\mathbf{X},\mathbf{Y})$.
4. $var(a\mathbf{X}+b\mathbf{Y}+c) = a^2\,var(\mathbf{X}) + b^2\,var(\mathbf{Y}) + 2ab\;\;cov(\mathbf{X},\mathbf{Y})$.
5. $var\left(\sum_{i=1}^{N}\mathbf{X}_i\right) = \sum_{i=1}^{N} var(\mathbf{X}_i) + 2\sum_{j=1}^{N}\sum_{i=1}^{j-1} cov\left(\mathbf{X}_i,\mathbf{Y}_j\right)$.

These extend trivially to the correlation.

2.15 Conditional Expected Value and Variance

The conditional expected value and variance are just the expected value and variance computed using the conditional rather than the marginal distribution. The conditional expected value is

$$\mathcal{E}(\mathbf{X}|\mathbf{Y}=y) = \int_{-\infty}^{\infty} x f(x|y)\,dx \tag{2.71}$$

and the conditional variance is

$$var(\mathbf{X}|\mathbf{Y}=y)=\int_{-\infty}^{\infty}[x-\mu(y)]^2 f(x|y)dx \tag{2.72}$$

where $\mu(y)=\mathcal{E}(\mathbf{X}|\mathbf{Y}=y)$. The conditional expected value and variance are themselves rvs because they depend on a specific value for the rv **Y**. Therefore, they have a pdf, an expected value, and a variance of their own, as defined by

$$\mathcal{E}(\mathbf{X})=\mathcal{E}_Y[\mathcal{E}(\mathbf{X}|\mathbf{Y})] \tag{2.73}$$

$$var(\mathbf{X})=\mathcal{E}[var(\mathbf{X}|\mathbf{Y})]+var[\mathcal{E}(\mathbf{X}|\mathbf{Y})] \tag{2.74}$$

Equation (2.73) is the *law of total expectation*, whereas (2.74) is the *law of total variance*. A generalization of the latter is the *law of total covariance*

$$cov(\mathbf{X},\mathbf{Z})=\mathcal{E}[cov(\mathbf{X},\mathbf{Z}|\mathbf{Y})]+cov[\mathcal{E}(\mathbf{X}|\mathbf{Y}),\mathcal{E}(\mathbf{Z}|\mathbf{Y})] \tag{2.75}$$

which reduces to (2.74) when $\mathbf{X}=\mathbf{Z}$.

If **X** must be predicted without any information about **Y**, the best estimate for **X** is $\mathcal{E}(\mathbf{X})$ with variance $var(\mathbf{X})$. However, if a priori information about **Y** is available, then the expected value of the conditional expectation is a better estimate for **X** and will have a lower variance.

2.16 Probability Inequalities

Probability inequalities are useful for placing bounds on entities that are difficult to compute directly. There are many of these, and only a few useful ones will be presented.

Let an rv **X** be nonnegative (i.e., $\Pr(\mathbf{X}\geq 0)=1$) and have a finite expected value. For any number $t>0$, the *Markov inequality* states that

$$\Pr(\mathbf{X}\geq t)\leq\frac{\mathcal{E}(\mathbf{X})}{t}=\frac{\mu}{t} \tag{2.76}$$

The Markov inequality is applicable only for large values of t and in fact produces nothing useful if $t<\mu$.

The *Chebyshev inequality* applies to an rv **X** that can take any value but has a finite variance. For any number $t>0$, it holds that

$$\Pr(|\mathbf{X}-\mu|\geq t)\leq\frac{var(\mathbf{X})}{t^2}=\frac{\sigma^2}{t^2} \tag{2.77}$$

and is not meaningful unless $t>1$. The Chebyshev inequality is easily derived using the Markov inequality on the rv $\mathbf{X}^2$. Both of these inequalities hold for any distribution for the rvs.

Hoeffding (1963) introduced several inequalities that provide much sharper bounds than the Chebyshev inequality when rvs with zero expected value are bounded.

The most general of these applies when $0 \le \mathbf{X}_i \le 1$, such as obtains with Bernoulli rvs, and is given by

$$\Pr\left[\overline{\mathrm{X}}_N - \mathcal{E}\left(\overline{\mathrm{X}}_N\right) \ge t\right] \le e^{-2Nt^2} \tag{2.78}$$

where $\overline{\mathrm{X}}_N$ is the sample mean. Hoeffding generalized (2.78) to variables with arbitrary lower and upper bounds. Because $\Pr\left[\left|\overline{\mathrm{X}}_N - \mathcal{E}\left(\overline{\mathrm{X}}_N\right)\right| \ge t\right] = \Pr\left[\overline{\mathrm{X}}_N - \mathcal{E}\left(\overline{\mathrm{X}}_N\right) \ge t\right] + \Pr\left[-\overline{\mathrm{X}}_N - \mathcal{E}\left(\overline{\mathrm{X}}_N\right) \ge t\right]$, it follows that

$$\Pr\left[\left|\overline{\mathrm{X}}_N - \mathcal{E}\left(\overline{\mathrm{X}}_N\right)\right| \ge t\right] \le 2e^{-2Nt^2} \tag{2.79}$$

which is usually simply called *Hoeffding's inequality.*

Example 2.21 Let $t = 4\sigma$. A Chebyshev bound of 1/16 can be placed on the probability that an rv $\mathbf{X}$ differs from its expected value by more than four standard deviations regardless of the distribution for $\mathbf{X}$. However, this bound is relatively crude in some cases; for the Gaussian distribution, the actual value is about 10^{-4}.

A better bound than the Chebyshev inequality can be obtained using *Cantelli's inequality* when only one tail of the distribution is of interest. Let $\mathbf{X}$ be an rv with finite variance; compute the standardized rv $\mathbf{Z} = (\mathbf{X} - \mu)/\sigma$. Cantelli's inequality holds that

$$\Pr(\mathbf{Z} \ge t) \le \frac{1}{1 + t^2} \tag{2.80}$$

Cantelli's inequality can be used to prove that the mean and median are always within one standard deviation of each other. Setting $t = 1$ in (2.80) yields

$$\Pr(\mathbf{Z} \ge 1) = \Pr(\mathbf{X} - \mu \ge \sigma) \le \frac{1}{2}$$

Changing the sign of $\mathbf{X}$ and μ gives

$$\Pr(\mathbf{X} \le \mu - \sigma) \le \frac{1}{2}$$

and proves the result.

2.17 Convergence of Random Variables

Convergence theory for rvs is concerned with the characteristics of sequences of rvs and, in particular, their limiting behavior. In the absence of stochasticity, a sequence of real numbers $\{x_i\}$ converges to a value x if $|x_N - x| < \varepsilon$ as $N \to \infty$ for every number $\varepsilon > 0$. This becomes subtle when the variables are random because two continuous rvs are equal with probability zero [i.e., $\Pr\left(\mathbf{X}_i = \mathbf{X}_j\right) = 0$ for distinct i, j]. This leads to three concepts: convergence in distribution, convergence in probability, and convergence almost surely.

Let $\mathbf{X}_1, \mathbf{X}_2, ..., \mathbf{X}_N$ be a sequence of rvs, and let $\mathbf{Y}$ be another rv. Let $F_N(x)$ be the cdf of $\mathbf{X}_N$ and $G(y)$ be the cdf for $\mathbf{Y}$. Then the following definitions pertain:

1. $\mathbf{X}_N$ converges in distribution to $\mathbf{Y}$ (written symbolically as $\mathbf{X}_N \xrightarrow{d} \mathbf{Y}$) if $\lim_{N\to\infty} F_N(x) = G(y)$ at all points where $G(y)$ is continuous.
2. $\mathbf{X}_N$ converges in probability to $\mathbf{Y}$ (written symbolically as $\mathbf{X}_N \xrightarrow{p} \mathbf{Y}$) if $\Pr(|\mathbf{X}_N - \mathbf{Y}| > \varepsilon) \to 0$ as $N \to \infty$ for every number $\varepsilon > 0$.
3. $\mathbf{X}_N$ converges almost surely (written symbolically as $\mathbf{X}_N \xrightarrow{as} \mathbf{Y}$) if $\Pr\left(\lim_{N\to\infty} |\mathbf{X}_N - \mathbf{Y}| = 0\right) = 1$.

Caution: Convergence almost surely implies convergence in probability, which in turn implies convergence in distribution. However, the reverse relationships do not hold. There is one exception that is almost pathological: if $\mathbf{X}_N \xrightarrow{d} \mathbf{Y}$ and $\Pr(\mathbf{Y} = a) = 1$ for some number a, then $\mathbf{X}_N \xrightarrow{p} \mathbf{Y}$.

Some of these convergence properties are preserved under transformations. Let $\mathbf{X}_N, \mathbf{Y}, \mathbf{U}_N$, and $\mathbf{V}$ be rvs and g be a continuous function. Then the following relationships hold:

1. If $\mathbf{X}_N \xrightarrow{as} \mathbf{Y}$ and $\mathbf{U}_N \xrightarrow{as} \mathbf{V}$, then $\mathbf{X}_N + \mathbf{U}_N \xrightarrow{as} \mathbf{Y} + \mathbf{V}$.
2. If $\mathbf{X}_N \xrightarrow{as} \mathbf{Y}$ and $\mathbf{U}_N \xrightarrow{as} \mathbf{V}$, then $\mathbf{X}_N\mathbf{U}_N \xrightarrow{as} \mathbf{YV}$.
3. If $\mathbf{X}_N \xrightarrow{p} \mathbf{Y}$ and $\mathbf{U}_N \xrightarrow{p} \mathbf{V}$, then $\mathbf{X}_N + \mathbf{U}_N \xrightarrow{p} \mathbf{Y} + \mathbf{V}$.
4. If $\mathbf{X}_N \xrightarrow{p} \mathbf{Y}$ and $\mathbf{U}_N \xrightarrow{p} \mathbf{V}$, then $\mathbf{X}_N\mathbf{U}_N \xrightarrow{p} \mathbf{YV}$.
5. If $\mathbf{X}_N \xrightarrow{p} \mathbf{Y}$, then $g(\mathbf{X}_N) \xrightarrow{p} g(\mathbf{Y})$.
6. If $\mathbf{X}_N \xrightarrow{d} \mathbf{Y}$, then $g(\mathbf{X}_N) \xrightarrow{d} g(\mathbf{Y})$.
7. If $\mathbf{X}_N \xrightarrow{d} \mathbf{Y}$ and $\mathbf{U}_N \xrightarrow{d} a$, where a is a constant, then $\mathbf{X}_N + \mathbf{U}_N \xrightarrow{d} \mathbf{Y} + a$.
8. If $\mathbf{X}_N \xrightarrow{d} \mathbf{Y}$ and $\mathbf{U}_N \xrightarrow{d} a$, where a is a constant, then $\mathbf{X}_N\mathbf{U}_N \xrightarrow{d} a\mathbf{Y}$.

The last two of these constitute Slutsky's theorem which holds for convergence in probability as well as convergence in distribution. Note that if $\mathbf{X}_N \xrightarrow{d} \mathbf{Y}$ and $\mathbf{U}_N \xrightarrow{d} \mathbf{V}$, then $\mathbf{X}_N + \mathbf{U}_N$ does not necessarily converge in distribution to $\mathbf{Y} + \mathbf{V}$.

3 Statistical Distributions

3.1 Overview

This chapter builds on the foundations in Chapters 1 and 2 and provides an overview of the main statistical distributions that occur in many science fields, with emphasis on those that are important in the earth sciences, such as the lognormal and generalized extreme value distributions. The treatment of distributions is divided between discrete and continuous types, both of which are important for the analysis of earth sciences data. MATLAB support for each of the distributions is described and used to illustrate many of the characteristics and applications of the distributions. Much of the material in this chapter is provided for reference purposes in subsequent chapters or for general use.

The classic reference on univariate discrete distributions is Johnson, Kotz, & Kemp (1993), and the classic references on univariate continuous distributions are Johnson, Kotz, & Balakrishnan (1994, 1995). Johnson, Kotz, & Balakrishnan (1997) and Kotz, Balakrishnan, & Johnson (2000) extend the discrete and continuous treatments to the multivariate case.

3.2 MATLAB Support for Distributions

MATLAB has provided support for a large set of distributions since early releases of the statistics toolbox. These were implemented using a mnemonic for the distribution name followed by "pdf," "cdf," "inv," or "rnd" for, respectively, the probability density function, the cumulative distribution function, the quantile function, and random draws. For example, the Gaussian distribution is obtained by prepending "norm" to one of these postfixes and supplying parameters such as the argument range and location/scale parameters, with default values for the latter if omitted.

In later releases of MATLAB, support was added for object-oriented programming, and an additional method to access probability distributions was implemented based on object-oriented principles. An increasingly large fraction of the supported distributions can only be accessed using this approach, so it behooves the user to understand how it functions. This chapter will focus on the use of distribution objects for this reason.

MATLAB enables the creation of a distribution object consisting of its parameters and a model description through the **makedist** function. For example, to create a normal distribution object, use

```
pd = makedist('Normal');
```

which defaults to a mean of 0 and a standard deviation of 1 because these parameters were not supplied. To create a normal distribution object with a specified mean and standard deviation, provide keyword-value pairs for them

```
pd = makedist('Normal', 'mu', 1, 'sigma', 2);
```

Once a probability distribution object has been created, numerous derived quantities can be obtained using the distribution handle *pd*. To obtain the pdf, cdf, quantile function, or random draws from the distribution specified by *pd*, the methods **pdf**, **cdf**, **icdf**, and **random** are used. For example, to define the pdf of a Gaussian distribution with mean 1 and standard deviation 2,

```
y = pdf(pd, x);
```

returns the pdf values at the arguments contained in x. The cdf of the same distribution is given by

```
y = cdf(pd, x);
```

and the complementary cdf (see Section 2.2.2) follows from

```
y = cdf(pd, x, 'upper');
```

The quantile function and random draws operate in an analogous manner. MATLAB also implements a number of distribution parameters using additional methods. For example, the mean is obtained from

```
mean(pd)
```

The standard deviation, variance, median, and interquartile range are obtained by replacing mean with **std**, **var**, **median**, and **iqr**, respectively.

MATLAB also provides a GUI called **disttool** that presents the pdf or cdf of a large number of probability distributions and facilitates exploration of their behavior as the distribution parameters are changed.

3.3 Discrete Distributions

3.3.1 Bernoulli Distribution

Consider an experiment having only two possible outcomes: live or dead, success or failure, hit or miss, and so on, for which the variable 0 or 1 can be assigned as a proxy. A random variable (rv) with this characteristic is called an *indicator* or *Bernoulli* variable.

Let $\Pr(\mathbf{X} = 1) = p$ and $\Pr(\mathbf{X} = 0) = 1 - p$. The pdf of a Bernoulli variable is

$$f(x; p) = p^x (1 - p)^{1-x} \tag{3.1}$$

for $x = 0$ or 1 (and 0 otherwise). The parameter p is both the probability that a sample with outcome 1 will be drawn and the population proportion with outcome 1. The expected value and variance are $\mathcal{E}(\mathbf{X}) = p$ and $var(\mathbf{X}) = p(1 - p)$. If a set of rvs $\{\mathbf{X}_i\}$ is iid and each has a Bernoulli distribution with parameter p, then the sequence is called a *set of Bernoulli trials*.

Example 3.1 If a coin is tossed, and 1 is assigned for heads and 0 for not heads (or tails), then the outcome of N coin tosses is a set of Bernoulli trials with parameter p = 0.5 for a fair coin.

3.3.2 Binomial Distribution

The *binomial distribution* arises whenever the underlying independent events in a process have two possible outcomes whose probabilities remain constant. It applies when a sample of fixed size N is drawn from an infinite population where each population element is independent and has the same probability for some attribute. It also applies when a sample of fixed size is drawn from a finite population where each element has the same probability for some attribute and the sampling is done randomly with replacement. Consequently, the binomial distribution applies to random sampling of an rv that is in some loose sense not rare.

Suppose that the outcome of an experiment is a Bernoulli variable and that the experiment is repeated independently N times. Let the rv $\mathbf{X}$ be the number of successes in the N Bernoulli trials. The probability of a given sequence of N trials containing exactly x successes and $N - x$ failures is $p^x (1 - p)^{N-x}$ because of independence. Further, there are $c_{N,x}$ possible combinations of the N trials. Consequently, $\mathbf{X}$ has the probability density function given by

$$\mathrm{bin}(x; N, p) = \binom{N}{x} p^x (1 - p)^{N-x} \quad x = 0, \ldots, N \tag{3.2}$$

and zero elsewhere. Equation (3.2) is called the *binomial distribution* with parameters N and p and has finite support on the integers between 0 and N. The Bernoulli distribution is a special case of the binomial distribution with $N = 1$.

The expected value and variance of the binomial distribution are $\mathcal{E}(\mathbf{X}) = Np$ and $var(X) = Np(1 - p)$. The skewness and kurtosis are $(1 - 2p)/\sqrt{Np(1 - p)}$ and $3 + [1 - 6p(1 - p)]/[Np(1 - p)]$, respectively. The skewness is positive (negative) if $p < (>) 1/2$, and the distribution is symmetric (skewness of zero) if and only if $p = 1/2$. The mode is $\lfloor (N + 1)p \rfloor$, where the floor function $\lfloor x \rfloor$ is the greatest integer less than or equal to x, and hence the binomial distribution is unimodal. When $p < 1/(N + 1)$, the mode occurs at the origin. The binomial median may not be unique, and there is no single formula for it, although there are numerous special cases.

The binomial cdf is

$$\mathrm{Bin}(x;N,p) = \sum_{i=0}^{\lfloor x \rfloor} \binom{N}{i} p^i (1-p)^{N-i} \tag{3.3}$$

which is also called the *regularized incomplete beta function*, denoted by $I_{1-p}(N-x,\ x+1)$. The pdf and cdf are accessed as described in Section 3.2. However, the quantile function given by

```
pd = makedist('Binomial', 'N', n, 'p', p);
y = icdf(pd, pp)
```

returns the least integer y such that the binomial cdf evaluated at pp equals or exceeds y because of the discrete form of the distribution, as is also true of other discrete distributions described in this section. Figure 3.1 shows the binomial distribution for several pairs of parameters.

Suppose that the binomial variable $\mathbf{X}$ is standardized by subtracting the expected value and dividing by the standard deviation

$$\mathbf{X}' = \frac{\mathbf{X} - Np}{\sqrt{Np(1-p)}} \tag{3.4}$$

This standardization is sometimes called *studentization* for reasons that will become apparent in Chapter 4. Then the *De Moivre-Laplace theorem* states that $\mathbf{X}' \xrightarrow{p} \mathrm{N}(0,1)$, where $\mathrm{N}(0,1)$ is the standardized Gaussian distribution (see Section 3.4.1), so

$$\lim_{N\to\infty} \Pr(\alpha < \mathbf{X}' < \beta) = \frac{1}{\sqrt{2\pi}} \int_{\alpha}^{\beta} e^{-u^2/2} du \tag{3.5}$$

Equation (3.5) is an early form of the classic central limit theorem that is described in Section 4.7.2. This was important for computation of binomial probabilities in the past because the factorials in the pdf become tedious to compute for large N. The Gaussian approximation becomes quite accurate for $N > 50$.

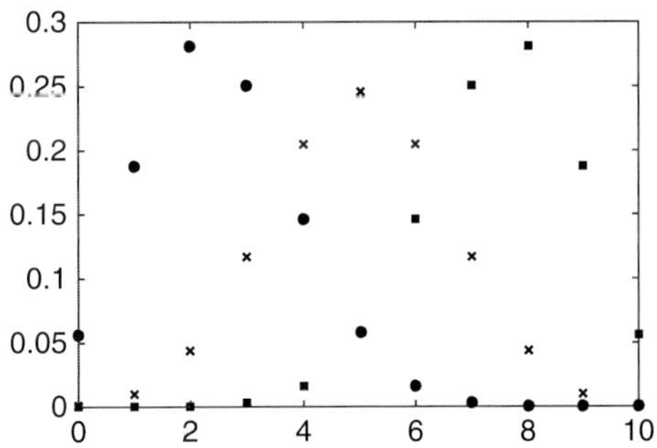

Figure 3.1 The binomial pdf for $N = 10$ and $p = 0.25$ (circles), 0.5 (x), and 0.75 (squares).

A key property of the binomial distribution is that of additivity. If a set of N Bernoulli rvs $\{\mathbf{X}_i\}$ has parameter p, then the rv $\mathbf{Y} = \mathbf{X}_1 + \cdots + \mathbf{X}_N$ is binomial with parameters N and p. Further, if the $\{\mathbf{X}_i\}$ are k independent binomial rvs with parameters N_i and p, then their sum is also binomial with parameters $N = N_1 + \cdots + N_k$ and p.

A few examples of binomial distributions in the earth sciences include

- Abundant mineral occurrence in samples of fixed size (0 = not present, 1 = present);
- Abundant fossil occurrence in samples of fixed size (0 = not present, 1 = present);
- Incidence of producing versus dry holes in the petroleum industry (0 = dry, 1 = producing); and
- Occurrence of cross-beds in sandstone (0 = not present, 1 = present).

Note the keyword "abundant" in the preceding; if the occurrence is rare, then the distribution is Poisson.

Example 3.2 Suppose that there is a 10% chance that a drilled well will produce oil. Twenty-five holes are randomly drilled. What is the probability that at most, at least and exactly three of the wells will produce oil?

Let the rv **X** represent well production, with 1 indicating a producing hole and 0 indicating a dry one. **X** is a binomial random variable with parameters (25, 0.1). First, create a distribution object with these parameters:

```
pd = makedist('Binomial', 'N', 25, 'p', 0.1);
```

The MATLAB function **pdf** can be used to compute $\Pr(X \le 3)$, which is the probability that at most three wells will produce oil

```
sum(pdf(pd, 0:3))
ans =
   0.7636
```

The same result can be obtained using **cdf**(*pd*, 3). The probability that at least three wells will be productive is $\Pr(X \ge 3)$

```
sum(pdf(pd, 3:25))
ans =
   0.4629
```

The same result can be obtained using **cdf**(*pd*, 25) – **cdf**(*pd*, 2) or 1 – **cdf**(*pd*, 2) because **cdf**(*pd*, 25) $\equiv$ 1. The probability that exactly three wells will produce is $\Pr(X = 3)$

```
pdf(pd, 3)
ans =
   0.2265
```

In contrast, the probability that exactly 25 wells will be productive is vanishingly small

```
pdf(pd, 25)
ans =
   1.0000e-25
```

while the probability of a complete strikeout is surprisingly large

```
pdf(pd, 0)
ans =
   0.0718
```

3.3.3 Negative Binomial Distribution

The *negative binomial distribution* is a generalization of the binomial distribution to the case where, instead of counting successes in a fixed number of trials, the trials are observed until a fixed number of successes is obtained. Let p be the probability of success in a given trial, and let the fixed integer k denote the number of successes observed after x trials. Let the rv $\mathbf{X}$ denote the number of trials. Because of independence, a given sequence of trials has probability $p^k(1-p)^{x-k}$. The last trial succeeded, and the remaining $k-1$ successes that already occurred are assigned to the remaining $x-1$ trials as $c_{x-1,k-1}$, yielding

$$\text{nbin}(x;p,k) = \binom{x-1}{k-1} p^k (1-p)^{x-k} \quad x = 0, 1, \ldots \tag{3.6}$$

As a cautionary note, the negative binomial distribution is sometimes defined so that the first parameter r is the exact number of failures (rather than the number of trials) that occur before the kth success is achieved. This is the approach used in MATLAB. Equation (3.6) can be easily transformed into this version, yielding

$$\text{nbin}(r;p,k) = \binom{k+r-1}{r} p^k (1-p)^r \quad r = 0, 1, \ldots \tag{3.7}$$

Note that the support of (3.6) and (3.7) is infinite, in contrast to the binomial distribution. Figure 3.2 illustrates the negative binomial distribution for several parameter values.

The expected value and variance for the negative binomial distribution are $(1-p)r/p$ and $(1-p)r/p^2$, respectively, and hence the negative binomial distribution is

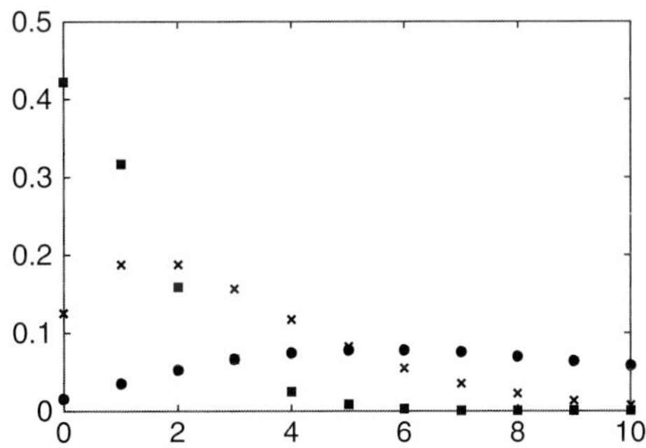

Figure 3.2 The negative binomial pdf for $k = 3$ successes with $p = 0.25$ (circles), 0.5 (x), and 0.75 (square). Note the reverse behavior as compared to the binomial pdf in Figure 3.1.

overdispersed (i.e., the standard deviation is larger than the mean). The skewness and kurtosis are $(2-p)/\sqrt{r(1-p)}$ and $3+[p^2+6(1-p)]/[r(1-p)]$, respectively.

The negative binomial cdf is

$$\mathrm{Nbin}(r;p,k)=I_p(k,r+1) \tag{3.8}$$

The keyword to create a negative binomial distribution object in MATLAB is "NegativeBinomial."

The geometric distribution is a special case of the negative binomial distribution with $r=1$. It possesses a memory-less property $\Pr(\mathbf{X}=x+s|\mathbf{X}\geq x)=\Pr(\mathbf{X}=s)$. As for the binomial distribution, the negative binomial and geometric distribution quantile functions return the least integer x such that the cdf evaluated at x equals or exceeds y because of the discrete form of the distribution. There is no separate keyword to specify a geometric distribution object, but it can be accessed directly [i.e., the geometric pdf is obtained from **geopdf**(x, p)]. A distribution object can otherwise be created for it by setting $k=1$.

Example 3.3 For Example 3.2, what is the probability that exactly 10 failures are observed before three successes are achieved? What is the probability that no more than 20 failures are observed before three successes are achieved?

```
pd = makedist('NegativeBinomial', 'R', 3, 'p', 0.1);
pdf(pd, 10)
ans =
   0.0230
cdf(pd, 20)
ans =
   0.4080
```

3.3.4 Multinomial Distribution

The *multinomial distribution* is the multivariate generalization of the binomial and is introduced here rather than in Chapter 10 because it plays an important role in goodness-of-fit testing, as described in Chapter 7. Suppose that a population contains K different classes of items ($K\geq 2$), with the proportion of items of the ith class given by $p_i>0$ for $i=1,\ldots,K$ such that $\sum_{i=1}^{K}p_i=1$. Further suppose that N items are selected at random with replacement, and let the rv $\mathbf{X}_i$ denote the number of selected items that are of the ith class. Then the random vector $\{\mathbf{X}_1,\ldots,\mathbf{X}_K\}$ has the multinomial distribution with parameters N and $\{p_1,\ldots,p_K\}$.

Define the multinomial coefficient

$$\binom{N}{x_1,x_2,\ldots,x_K}=\frac{N!}{x_1!x_2!\ldots x_K!} \tag{3.9}$$

If $x_1, ..., x_K$ are nonnegative integers such that $\sum_{i=1}^{K} x_i = N$, then the multinomial pdf is given by

$$\text{mnom}(\mathbf{x}; N, \mathbf{p}) = \binom{N}{x_1, ..., x_K} p_1^{x_1} \cdots p_K^{x_K} \tag{3.10}$$

Equation (3.10) reduces to the binomial pdf (3.2) when $K = 2$. The marginal distribution for any one class may be found by summing over the remaining classes but would be a formidable task. However, note that x_i can be interpreted as the number of successes in N trials, each of which has probability of success p_i and probability of failure $1 - p_i$, and hence x_i is a binomial rv with pdf

$$\text{bin}(x_i; N, p_i) = \binom{N}{x_i} p_i^{x_i} (1 - p_i)^{N - x_i} \tag{3.11}$$

The multinomial distribution has the expected value $\mathcal{E}(\mathbf{X}_i) = Np_i$, the variance $var(\mathbf{X}_i) = Np_i(1 - p_i)$ (both of which follow from the binomial marginal distribution) and the covariance $cov(\mathbf{X}_i, \mathbf{X}_j) = -Np_ip_j$. The negative value for the covariance of a multinomial distribution makes sense because there are only N outcomes to be shared among K classes, and hence, if one of the classes becomes large, most of the remaining classes will have to decrease. The covariance matrix is singular with rank $K - 1$ because of the sum constraint on the class probabilities.

The multinomial distribution is important in goodness-of-fit testing. It is the joint distribution of the heights of the bins in a histogram. It is also the joint distribution for counted entities in geology, such as point counts or fossil counts, as long as the counts are absolute rather than expressed as a proportion.

Example 3.4 Based on the historical record, an oil company knows that drilled holes produce oil and gas with probabilities 0.05 and 0.1, and dry holes are encountered with probability 0.85. The oil company drills 50 new holes. What is the probability that it will drill exactly two oil- and five gas-bearing holes? What is the probability of obtaining exactly two oil-bearing and no more than five gas-bearing holes?

This problem will be solved using the function **mnpdf** rather than the distribution object for the multinomial distribution. The probability vector is

```
p = [0.05 0.1 0.85];
```

The pdf for a multinomial distribution with 50 trials will be constructed as follows:

```
n = 50;
x1 = 0:n;
x2 = 0:n;
[y1, y2] = meshgrid(x1, x2);
y3 = n + 1 - y1 - y2;
y = mnpdf([y1(:) y2(:) y3(:)],repmat(p, (n + 1)^2, 1));
```

```
y = reshape(y, n + 1, n + 1);
sum(sum(y))
ans = 1.0000
```

Consequently, the result is a pdf. The probability of obtaining exactly two oil- and five gas-bearing holes is

```
y(3, 6)
ans =
   0.0060
```

The probability of obtaining exactly two oil- and no more than five gas-bearing holes is

```
sum(y(2, 1:6))
ans =
   0.0248
```

3.3.5 Hypergeometric Distribution

Suppose that there is a collection of objects containing a of type 1 and b of type 2. Further suppose that N objects are selected at random from the sample without replacement. Let the rv $\mathbf{X}$ be the number of type 1 objects obtained in the draw. It can neither be larger than N nor a, so $x \le \min(a, N)$. The number of type 2 objects is $N - x$ and cannot exceed b; hence x must be at least $N - b$. Since x is also nonnegative, $x \ge \max(0, N - b)$. Taken together, the value of x must be an integer lying on the interval $[\max(0, N - b), \min(N, a)]$. The hypergeometric pdf for exactly x type 1 objects is given by the ratio of combinations

$$\text{hyge}(x; a, b, N) = \frac{\binom{a}{x}\binom{b}{N-x}}{\binom{a+b}{N}} \tag{3.12}$$

Equation (3.12) can be recast in terms of the probability p of a type 1 object being drawn and the total number of objects M by replacing a and b with Mp and $M(1 - p)$, respectively. The resulting pdf is

$$\text{hyge}(x; M, N, p) = \frac{\binom{Mp}{x}\binom{M(1-p)}{N-x}}{\binom{M}{N}} \tag{3.13}$$

and is the more commonly encountered form. The hypergeometric distribution has already been encountered in Example 1.6.

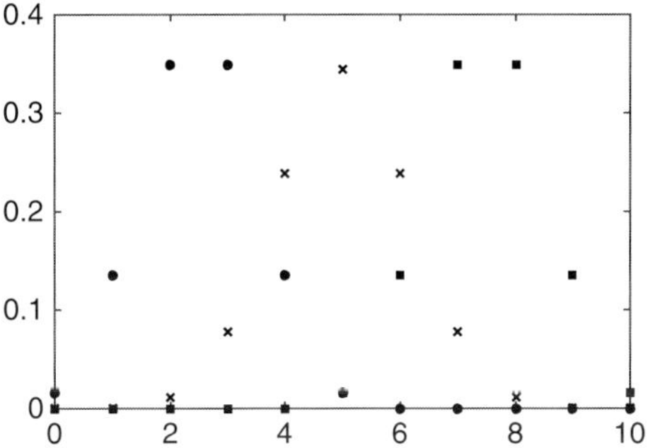

Figure 3.3 The hypergeometric pdf for $N = 20$, $M = 15$, and $p = 0.33$ (circles), $M = 20$ and $p = 0.5$ (x), and $M = 25$ and $p = 0.75$ (square).

The expected value and variance for the hypergemetric distribution are $\mathcal{E}(\mathbf{X}) = aN/(a+b) = Np$ and $var(\mathbf{X}) = Nab(a+b-N)/[(a+b)^2(a+b-1)] = Np(1-p)(M-N)/(M-1)$. The variance of a hypergeometric distribution is smaller than that of a binomial distribution by the factor $(M-N)/(M-1)$ as a consequence of sampling without replacement. As $M \to \infty$, the factor becomes 1, and their variances become identical. Figure 3.3 illustrates the hypergeometric pdf. The hypergeometric cdf cannot be expressed in terms of elementary functions and in fact is proportional to a generalized hypergeometric function.

The hypergeometric distribution is appropriate for studying sampling from finite populations without replacement, where the binomial is the distribution for studying sampling from finite populations with replacement, or else infinite populations. The hypergeometric distribution reduces to the binomial distribution for large M because the distinction between sampling with and without replacement becomes moot. Consequently, it also reduces to the Gaussian for large M by the DeMoivre-Laplace theorem (3.5).

Example 3.5 Suppose that the 100 staff members of the geophysics department are sharply divided in their taste for alcoholic beverages, with 45 identifying themselves as beer drinkers, 40 as wine drinkers, and 15 as teetotalers. The director is appointing the members of a committee to choose the drinks for the St. Patrick's Day party by selecting eight members at random and in a batch (so without replacement). Find the joint distribution of the number of beer drinkers, wine drinkers, and teetotalers on the committee, the marginal distributions for each group, and the probability that a majority will be beer drinkers.

There are i beer drinkers, j wine drinkers, and k teetotalers on the committee, where $i+j+k=8$. They are composed of i beer drinkers from the group of 45, j wine drinkers from the group of 40, and k teetotalers from the group of 15. By combinatorics, there are $\binom{100}{8}$ possible and equally likely committees. The joint distribution is $\Pr(i$ beer drinkers $\cap\, j$ wine drinkers $\cap\, k$ teetotalers$)$

$$f(i,j,k) = \frac{\binom{45}{i}\binom{40}{j}\binom{15}{k}}{\binom{100}{8}}$$

where i,j,k lie on $[0, 8]$ and $i+j+k=8$. The marginal distribution for beer drinkers is

$$f_1(i) = \frac{\binom{45}{i}\binom{55}{8-i}}{\binom{100}{8}}$$

and for wine drinkers is

$$f_2(j) = \frac{\binom{40}{j}\binom{60}{8-j}}{\binom{100}{8}}$$

and for teetotalers is

$$f_3(k) = \frac{\binom{15}{k}\binom{85}{8-k}}{\binom{100}{8}}$$

The marginal distributions $f_1 - f_3$ are all hypergeometric distributions, while the joint distribution f is a multivariate generalization of the hypergeometric. Note that the joint distribution is not the product of the marginals, so the numbers of each type of drinker are not independent. This makes sense because (for example) knowing the number of beer drinkers changes the distribution of the number of wine drinkers.

The probability that there will be a majority of beer drinkers is the probability that there are at least five beer lovers on the committee. This is easily obtained as the sum from 5 through 8 of $f_1(i)$ or using MATLAB (since **makedist** does not support the hypergeometric distribution)

```
sum(hygepdf(5:8, 100, 45, 8))
ans =
  0.2518
```

or this could be obtained from 1 – **hygecdf**(4, 100, 45, 8).

The probability of a beer drinker majority occurring by chance is only about 25% despite the fact that beer drinkers comprise 45% of the staff. The probability of getting a majority of teetotalers is only 0.0017. The imbibers are pretty safe.

3.3.6 Poisson Distribution

The Poisson distribution applies to rare independent events from infinite samples, where the binomial is the distribution for common independent events from large samples. It can be derived as a limiting form of the binomial distribution. Take the binomial distribution (3.2), and let $\lambda = Np$ so that

$$\begin{aligned} \mathrm{bin}(x; N, \lambda) &= \binom{N}{x}\left(\frac{\lambda}{N}\right)^x\left(1-\frac{\lambda}{N}\right)^{N-x} \\ &= \left(\frac{\lambda^x}{x!}\right)\left(1-\frac{\lambda}{N}\right)^N \frac{N(N-1)\cdots(N-x+1)}{N^x(1-\lambda/N)^{-x}} \end{aligned} \tag{3.14}$$

Let $N \to \infty$ and $p \to 0$ (i.e., consider the limit of a large sample and a rare event) such that $Np = \lambda$ is a constant. Note that no condition is being placed on the size of λ. Evaluating the last two terms in (3.14) gives

$$\lim_{N\to\infty} \frac{N(N-1)\cdots(N-x+1)}{N^x(1-\lambda/N)^{-x}} = 1 \tag{3.15}$$

$$\lim_{N\to\infty}\left(1-\frac{\lambda}{N}\right)^N = e^{-\lambda} \tag{3.16}$$

$$\therefore \quad \lim_{\substack{N\to\infty \\ p\to 0}} \mathrm{bin}(x; N, \lambda) = \mathrm{pois}(x; \lambda) = \frac{e^{-\lambda}\lambda^x}{x!} \quad x = 0, 1, \ldots \tag{3.17}$$

and 0 otherwise. This result was first obtained by De Moivre (1711) and was subsequently rederived by Poisson (1837). The first comprehensive study of the Poisson distribution is due to von Bortkiewicz (1898). The support of the Poisson distribution is infinite, in contrast to the finite support of the binomial distribution. As for the binomial distribution, there is an addition law: if $\{\mathbf{X}_i\}$ are independent and each is Poisson with mean λ_i, their sum is also Poisson with a mean given by $\sum_i \lambda_i$.

The expected value and variance for the Poisson distribution are $\mathcal{E}(\mathbf{X}) = \lambda$ and $var(\mathbf{X}) = \lambda$ and are identical. These are, respectively, Np and $Np(1-p)$ for the binomial distribution. The first immediately follows from the definition of λ, and the second follows when p is very small so that p^2 may be neglected compared to p. The skewness is $1/\sqrt{\lambda}$, and the kurtosis is $3 + 1/\lambda$. The mode is $\lfloor\lambda\rfloor$ in general and reduces to two equal maxima at $\lambda - 1$ and λ if λ is an integer. The median lies between $\lfloor\lambda - 1\rfloor$ and $\lfloor\lambda + 1\rfloor$ but does not have an exact representation. Equation (3.17) increases monotonically with λ for fixed x when $\lambda \le x$ and decreases monotonically thereafter. Figure 3.4 illustrates the Poisson pdf.

The Poisson cdf is

$$\mathrm{Pois}(x; \lambda) = \frac{\Gamma(x+1,\ \lambda)}{\Gamma(x)} \tag{3.18}$$

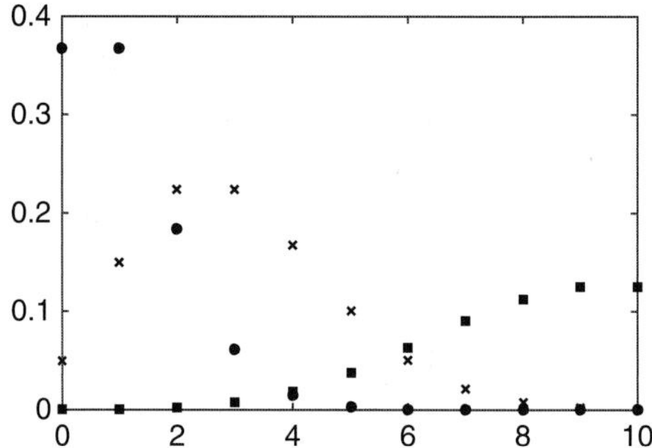

Figure 3.4 The Poisson pdf for $\lambda = 1$ (circles), 3 (x), and 5 (squares) out to $x = 10$.

where

$$\Gamma(x, \alpha) = \int_x^\infty t^{\alpha-1} e^{-t} dt \quad x > 0 \tag{3.19}$$

is the *complementary incomplete gamma function* [the *incomplete gamma function* $\gamma(x, \lambda)$ is given by (3.19) with integration limits of 0 and x]. Equation (3.19) reduces to $\Gamma(\alpha)$ when $x = 0$. Equation (3.18) is the *complementary incomplete gamma function ratio*.

Poisson distributions are common in the earth sciences:

- Radioactive decay where **X** is the number of (α, β, γ) particles emitted per unit time is Poisson; and
- The occurrence of rare mineral grains in point counting, where **X** is the number of observed grains.

They are also common in everyday life:

- The number of telephone calls coming into an exchange during a unit of time is Poisson if the exchange services customers who act independently;
- The number of accidents (such as falls in the shower) for a large population in a given period of time is Poisson because such accidents are presumably rare and independent (provided that there was only one person in the shower). Insurance companies frequently use a Poisson model; and
- The number of vehicles that pass a marker on a road in a given unit of time in light traffic (so that they operate independently) is Poisson.

The Poisson distribution is the counting distribution for Poisson processes that occur frequently in nature. The topic of Poisson processes is an area of active research. However, any physical process $X(t)$ that satisfies five conditions will be Poisson (Johnson, Kotz, & Kemp 1993):

1. $X(0) = 0$;
2. For $t > 0$, $0 < \Pr[X(t) > 0] < 1$;

3. The number of occurrences of a phenomenon in any pair of disjoint intervals must be independent (e.g., the fact that a lot of radioactive decays are observed in one time interval has no effect on the number observed in the next time interval, presuming that one is not dealing with a nuclear reactor or weapon);
4. The probability of an event during a particular time interval is proportional to the length of that interval; and
5. The probability of two or more events in a particular interval must be smaller than that for just one or, formally,

$$\lim_{\delta \to 0} \frac{\Pr[X(t+\delta) - X(t) \geq 2]}{\Pr[X(t+\delta) - X(t) = 1]} = 0 \tag{3.20}$$

If all five conditions are met, the result is a Poisson process, and the number of events in a time interval Δt will follow the Poisson distribution with mean $\lambda \Delta t$.

Example 3.6 Radioactive decay of ^{40}K in seawater gives an average of five γ particles per second. What is the probability of observing 10 decays in 2 seconds? 20 decays in 1 second?

The process is Poisson with $\lambda = 5$ in any 1-second interval. In a 2-second interval, it will be Poisson with $\lambda = 10$. Using MATLAB,

```
pd = makedist('Poisson', 'lambda', 10);
pdf(pd, 10)
ans =
  0.1251
pd = makedist('Poisson', 'lambda', 5);
pdf(pd, 20)
ans =
  2.6412e-07
```

Example 3.7 In a classic study (von Bortkiewicz 1898; reproduced as example 4 in Andrews & Herzberg 1985), the number of fatalities resulting from being kicked by a horse was recorded for 10 corps of Prussian cavalry over a period of 20 years in the late 1800s, giving 200 corps-years of data. These data and the probabilities from a Poisson model with $\lambda = 0.61$ are listed in Table 3.1. The first column gives the number of deaths per year. The second column lists the number of times that number of deaths was observed in 200 corps-years. The third column is the relative frequency obtained by dividing the second column by 200. The fourth column is the Poisson probability with $\lambda = 0.61$. The fit is quite pleasing. One could use this result to predict the probability of a given number of deaths from horse kicking per year if one ran an insurance company, for example.

Table 3.1 Fatalities from Horse Kicking in the Prussian Cavalry

No. of deaths/year	Observed	Relative frequency	Poisson probability
0	109	0.545	0.543
1	65	0.325	0.331
2	22	0.110	0.101
3	3	0.015	0.021
4	4	0.005	0.003

3.4 Continuous Distributions

3.4.1 Normal or Gaussian Distribution

The Gaussian distribution is the most important distribution in statistics for a number of reasons:

1. The Gaussian distribution may be derived from a number of simple and widely applicable models.
2. The Gaussian distribution is the central model behind the mathematical theory of errors.
3. The classic central limit theorem (Section 4.7.2) states that if a sufficiently large sample is drawn from any distribution with a finite variance, the sample mean will be Gaussian. While this is often true, it breaks down sometimes in earth science problems because the actual data variance is infinite, and in any case the amount of data required to reach the asymptotic limit is ill defined.
4. The mathematical form of the Gaussian distribution leads to simple expressions and simple forms. This is more of an excuse than a reason; mathematical convenience is no replacement for realistic descriptions, and with computers, simulations can replace analytic calculations.

As an example of point 1, Herschel (1850) considered the two-dimensional probability distribution for errors in measuring the position of a star. Let x and y be the zonal and meridional errors, respectively, with marginal distributions $f(x)$ and $g(y)$. Herschel postulated two conditions that flow from the underlying assumption of directional homogeneity of the errors

1. The errors in x and y are independent, so the joint distribution $h(x, y) = f(x)g(y)$.
2. If this expression is written in polar coordinates (r, θ), then the transformed joint distribution $h(r, \theta)$ must be independent of θ.

The second condition leads to the relation

$$f(x)g(y) = h\left(\sqrt{x^2 + y^2}\right) \tag{3.21}$$

Selecting $y = 0$, this simplifies to $f(x)g(0) = h(|x|)$, and selecting $x = 0$, it simplifies to $f(0)g(y) = h(|y|)$. Dividing both equations by $f(0)g(0)$ and noting that a change of sign for either x or y does not change the relationships shows that $f = g = h$. Taking logarithms gives

$$\log\left[f(x)/f(0)\right] + \log\left[f(y)/f(0)\right] = \log\left[f\left(\sqrt{x^2+y^2}\right)/f(0)\right] \tag{3.22}$$

Taking y to be zero so that the second term in (3.22) vanishes, the only possible solution is

$$\log\left[f(x)/f(0)\right] = ax^2 \tag{3.23}$$

where a is a constant, and therefore

$$f(x) = f(0)e^{ax^2} \tag{3.24}$$

Equation (3.24) can represent a finite probability only if $a < 0$, and $f(0)$ is determined by making the total probability equal to 1. Letting $a = -1/2\sigma^2$ and $f(0) = 1/\sqrt{2\pi}\sigma$, Equation (3.24) becomes

$$f(x) = \frac{1}{\sqrt{2\pi}\sigma} e^{-x^2/2\sigma^2} \tag{3.25}$$

which is the Gaussian pdf. Thus the only marginal pdf satisfying Herschel's postulates is the Gaussian.

Let $\mathbf{X}$ be a continuous rv on the real line with mean μ and variance σ^2, where $-\infty \le \mu \le \infty$ and $\sigma^2 > 0$. Then the normal distribution in standard form is

$$N\left(\mu, \sigma^2\right) = \frac{1}{\sqrt{2\pi}\sigma} e^{-(x-\mu)^2/2\sigma^2} \tag{3.26}$$

The expected value is $\mathcal{E}(\mathbf{X}) = \mu$, and since $\mathcal{E}\left(\mathbf{X}^2\right) = \mu^2 + \sigma^2$, $var(\mathbf{X}) = \mathcal{E}\left(\mathbf{X}^2\right) - \left[\mathcal{E}(\mathbf{X})\right]^2 = \sigma^2$. Figure 3.5 shows the Gaussian distribution for several parameter pairs.

The odd central moments higher than second order (e.g., the skewness and higher-order odd moments) are all zero, and hence the Gaussian distribution is symmetric around the mean. The even central moments are nonzero, and in particular, the kurtosis is 3. The

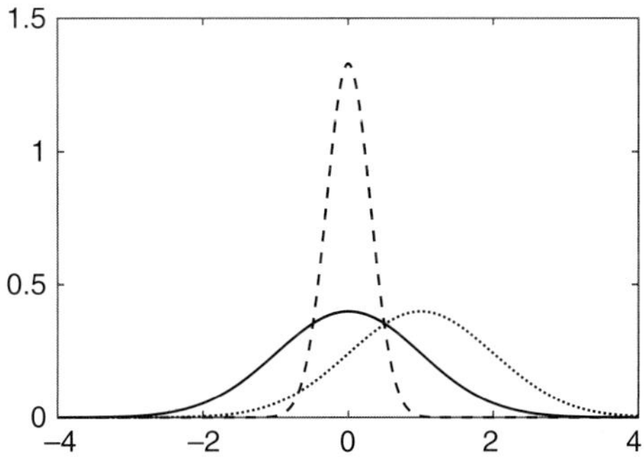

Figure 3.5 The Gaussian distributions $N(0,1)$ in solid line, $N(0, 0.3)$ in dashed line, and $N(1, 1)$ in dotted line.

symmetry point for the Gaussian is the mean μ. The mean μ, median $\tilde{\mu}$, and mode μ^* are identical. Inflection points where the second derivative is zero occur at $\mu \pm \sigma$. The Gaussian is unimodal and infinitely differentiable. Four other key characteristics of the Gaussian distribution are

1. If **X** and **Y** are two independent Gaussian rvs with means μ_x and μ_y and variances σ_x^2 and σ_y^2, then the linear combination $a\mathbf{X} + b\mathbf{Y}$, where a and b are constants, will also be Gaussian with mean $a\mu_x + b\mu_y$ and variance $a^2\sigma_x^2 + b^2\sigma_y^2$.
2. If **X** and **Y** are independent and their sum $\mathbf{X} + \mathbf{Y}$ is Gaussian distributed, then **X** and **Y** are each Gaussian (Cramér's theorem).
3. If **X** and **Y** are independent and $\mathbf{X} + \mathbf{Y}$ and $\mathbf{X} - \mathbf{Y}$ are also independent, then **X** and **Y** must be Gaussian (Bernstein's theorem).
4. If **X** and **Y** are jointly Gaussian and uncorrelated, then they are independent. The Gaussian is the only distribution for which this property holds. For non-Gaussian rvs, lack of correlation does not imply independence.

If an rv $\mathbf{X} \sim \mathrm{N}(\mu, \sigma^2)$, then $\mathbf{Y} = a\mathbf{X} + b \sim N(a\mu + b, a^2\sigma^2)$. If $a = 1/\sigma$ and $b = -\mu/\sigma$, **Y** is the standardized rv that is distributed as $\mathrm{N}(0, 1)$. This is why only $\mathrm{N}(0, 1)$ is tabulated (insofar as tabulation makes any sense in the modern world) because any other Gaussian distribution can be easily derived from it. The standardized normal distribution is

$$\mathrm{N}(0,1) = \varphi(x) = \frac{1}{\sqrt{2\pi}} e^{-x^2/2} \tag{3.27}$$

This is the default using MATLAB when the parameters are not specified.

The standardized Gaussian cdf is

$$\Phi(x) = \int_{-\infty}^{x} \varphi(t)\,dt = \left[1 + \mathrm{erf}(x/\sqrt{2})\right]/2 \tag{3.28}$$

where erf is the error function [**erf**(x) in MATLAB] that cannot be expressed in closed form. This is called the *probability integral* and is the default in MATLAB when the parameters are not specified. The Gaussian cf was derived in Section 2.4. The quantile function is

$$\Phi^{-1}(p) = \sqrt{2}\,\mathrm{erf}^{-1}(2p - 1) \tag{3.29}$$

where $\mathrm{erf}^{-1}(x)$ is the inverse error function. By definition, $\Phi(x) = \Pr(\mathbf{X} \le x)$, and since $\Pr(\mathbf{X} \ge -x) = 1 - \Phi(-x)$, $\Phi(x) + \Phi(-x) = 1$, so tables only list probabilities for $x \ge 0$.

Example 3.8 Suppose that an estimate is wanted for $\Pr(0 < x \le 10)$ for a $\mathrm{N}(1,\ 3)$ rv. Standardize the variable as $\mathbf{Z} = (\mathbf{X} - 1)/\sqrt{3}$, and consider the equivalent expression

$$\begin{aligned}\Pr(-0.5774 < \mathbf{Z} \le 5.1962) &= \Phi(5.1962) - \Phi(-0.5774)\\ &= \Phi(5.1962) + \Phi(0.5774) - 1\end{aligned}$$

Therefore, $\Pr(0 < \mathrm{X} \le 10) = 0.7181$. Of course, with MATLAB, this transformation becomes unnecessary, and the answer can be obtained with two lines of code.

Table 3.2 Gaussian Probabilities as a Function of Standard Deviations from the Mean

k	p_k	p_{cheb}
1	0.6826	0
2	0.9544	0.5
3	0.9974	0.8889
4	0.99997	0.9375
5	0.9999994	0.9600

```
pd = makedist('Normal', 'mu', 1, 'sigma', sqrt(3));
cdf(pd, 10) - cdf(pd, 0)
ans =
  0.7181
```

Let $p_k = \Pr(|\mathbf{X} - \mu| \le k\sigma) = \Pr(\mathbf{Z} \le k)$. Table 3.2 shows the fraction of the probability within k standard deviations of the mean compared to the result from the Chebyshev inequality, which is of little value for Gaussian variates.

The mean absolute deviation (MAD) for the Gaussian is given by

$$\tilde{\sigma} = \frac{1}{\sqrt{2\pi}} \int_{-\infty}^{\infty} |x| e^{-x^2/2} dx = \sqrt{\frac{2}{\pi}} \int_{0}^{\infty} x e^{-x^2/2} dx = \sqrt{\frac{2}{\pi}} \tag{3.30}$$

The MAD is the solution of $\Phi(\sigma_{\text{MAD}}) - \Phi(-\sigma_{\text{MAD}}) = 1/2$ or $\Phi(\sigma_{\text{MAD}}) = 3/4$, yielding $\sigma_{\text{MAD}} \cong 0.6595$. The interquartile distance is twice the MAD for the Gaussian because the distribution is symmetric. Half of its probability lies within 0.6595 of the mean.

3.4.2 Stable Distributions

A random variable is stable if a linear combination of two independent realizations of it has the same distribution. Let a and b be positive constants, let $\mathbf{X}$ be stable, and let $\mathbf{X}_1$ and $\mathbf{X}_2$ be independent copies of $\mathbf{X}$. The condition for stability is

$$a\mathbf{X}_1 + b\mathbf{X}_2 \stackrel{d}{=} c\mathbf{X} + d \tag{3.31}$$

Where $c > 0$ and $\stackrel{d}{=}$ denotes "equal in distribution," meaning that the rvs on either side of it have the same distribution. The term *stable* is used because the shape is unchanged under a transformation such as (3.31). This characteristic clearly applies to the Gaussian distribution with $c^2 = a^2 + b^2$ and $d = (a + b - c)\mu$ by the addition rule. However, there is an important extended family of stable distributions for which (3.31) also holds.

Stable distributions were first described by Lévy (1925) and may be defined through their cf

$$\begin{aligned} \phi_{st}(t) &= e^{-\gamma^\alpha|t|^\alpha\left[1+i\beta\tan\left(\frac{\pi\alpha}{2}\right)\mathrm{sgn}(t)\left(\gamma^{1-\alpha}|t|^{1-\alpha}-1\right)\right]+i\delta t} & \alpha \neq 1 \\ &= e^{-\gamma|t|\left[1+i\beta\frac{2}{\pi}\mathrm{sgn}(t)\log(\gamma|t|)\right]+i\delta t} & \alpha = 1 \end{aligned} \quad (3.32)$$

Stable distributions are parameterized by the tail thickness $\alpha \in (0, 2]$, skewness (not to be confused with the third normalized moment) $\beta \in [-1,\ 1]$, scale $\gamma \in (0, \infty)$, and location $\delta \in (-\infty, \infty)$; the latter two parameters are analogous to the standard deviation and mean. The pdf $\mathrm{st}(x; \alpha, \beta, \gamma, \delta)$ obtained as the Fourier transform of (3.32) cannot be expressed in closed form using elementary functions except for three special cases:

1. When $\alpha = 2$, the distribution is $\mathrm{N}(\delta, 2\gamma^2)$, and β has no meaning;
2. When $\alpha = 1,\ \beta = 0$, the distribution is Cauchy (see Section 2.9); and
3. When $\alpha = 1/2,\ \beta = 1$, the distribution is Lévy.

There are several alternate expressions to (3.32) for the stable cf that frequently cause confusion, but the present version (the 0-parameterization of Nolan [1998]) has the advantage of being continuous in the parameters. It is also a location-scale parameterization, in the sense that if $\mathbf{X} \sim \mathrm{st}(x; \alpha, \beta, \gamma, \delta)$, then $(\mathbf{X} - \delta)/\gamma \sim \mathrm{st}(x; \alpha, \beta, 1, 0)$.

Stable distributions are unimodal, have an infinitely differentiable pdf, and their support is doubly infinite except when $\beta = \pm 1$ and $\alpha < 1$, where they are bounded on (totally skewed to) the left (right) or right (left), respectively. The left (right) bound for $\beta = -1(1)$ is $\delta + \gamma \tan(\pi\alpha/2)$ $[\delta - \gamma \tan(\pi\alpha/2)]$. Stable distributions possess a reflection property: $\mathrm{st}(x; \alpha, \beta, \gamma, \delta) = -\mathrm{st}(x; \alpha, -\beta, \gamma, \delta)$. Further information may be found in Feller (1971), Samorodnitsky & Taqqu (1994), and Uchaikin & Zolotarev (1999). However, these books are written at an advanced level and are not readily accessible to the applied statistician. There is a serious need for a modern textbook on stable distributions.

Stable distributions possess finite variance only when $\alpha = 2$. For $1 < \alpha < 2$, stable distributions have finite mean but infinite variance, whereas for $0 < \alpha \leq 1$, both the mean and the variance are undefined. The tails of stable distributions are algebraic except for the Gaussian end member, with the tail thickness decreasing with α. For example, the Cauchy distribution has $1/|x|^2$ tails, whereas the Lévy distribution has a $1/x^{1.5}$ right tail. More generally, $\mathrm{st}(x; \alpha, \beta, \gamma, \delta) \to |x|^{-(\alpha+1)}$ as $x \to \pm\infty$ for $0 < \alpha < 2$ as long as $\beta \neq \pm 1$. As a consequence, stable data with infinite variance exhibit more extreme values and much more variability than Gaussian ones. Figures 3.6 and 3.7 show symmetric $(\beta = 0)$ and skewed standardized $(\gamma = 1,\ \delta = 0)$ examples of stable distribution pdfs. Very long tails are quite evident.

Aside from the empirical observation that stable distributions fit many types of data in fields ranging from physics to finance, as described by Uchaikin & Zolotarev (1999), they also have significance due to the generalized central limit theorem described in Section 4.7.3. Further, the existence of stably distributed real-world data is intimately intertwined with governing physics that contains fractional derivatives; Oldham & Spanier (1974) provide a description of fractional calculus. Spatial and temporal fractional derivatives reflect the existence of long-range ordering in the respective domain. Meerschaert (2012) summarizes the arguments. For example, in geophysics, some of the nonlinear and nonequilibrium processes that occur in the ionosphere and magnetosphere where short-period geomagnetic variations originate do display evidence of fractional derivative and multifractal behavior along with self-organized criticality, and as a consequence, magnetotelluric

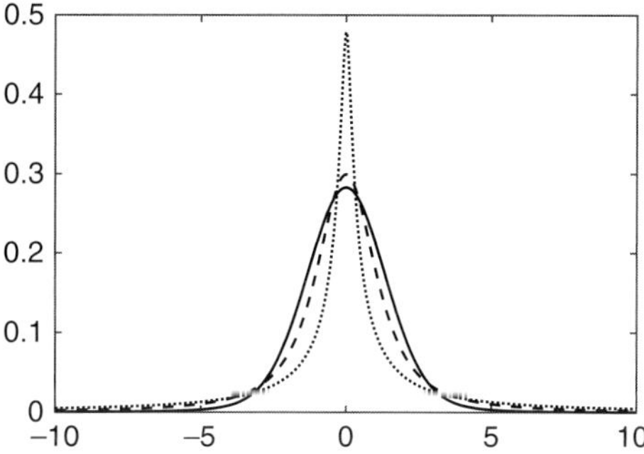

Figure 3.6 Symmetric ($\beta = 0$), standardized ($\gamma = 1$, $\delta = 0$) stable distributions with tail thickness parameters α of 1.8 (solid line), 1.2 (dashed line), and 0.6 (dotted line).

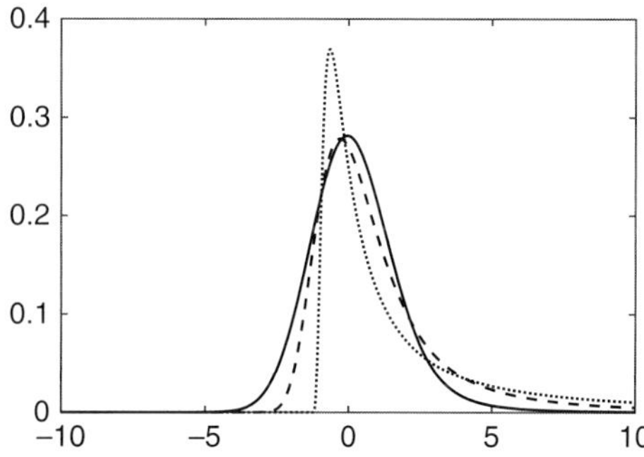

Figure 3.7 Skewed ($\beta = 1$), standardized ($\gamma = 1$, $\delta = 0$) stable distributions with tail thickness parameters α of 1.8 (solid line), 1.2 (dashed line), and 0.6 (dotted line).

data are pervasively stably distributed. Chave (2014) described a method to analyze such data that takes advantage of their stable nature. The importance of the stable distributions will certainly grow as their presence in nature becomes increasingly recognized.

The paucity of closed-form expressions for the density/distribution functions and their derivatives means that numerical methods must be used both to compute those entities and to estimate the stable parameters from data. Nolan (1997) describes algorithms to compute univariate stable density functions. Maximum likelihood estimation (covered in Chapter 5) of the stable parameters was introduced by DuMouchel (1975) and refined by Nolan (2001).

MATLAB did not provide support for stable distributions other than the Gaussian and Cauchy (which is equivalent to the Student's t distribution with one degree of freedom) until R2016a, when stable distribution objects were implemented. This appears to be based on third-party MATLAB software from www.robustanalysis.com.

3.4.3 Rayleigh Distribution

The Rayleigh distribution describes the magnitude of a vector quantity such as current speed in the ocean computed from its two horizontal vector components, assuming that the latter are uncorrelated and Gaussian distributed with a common variance and zero mean. Alternately, the Rayleigh distribution describes the magnitude of proper (meaning that they are uncorrelated with the complex conjugate; see Section 10.2.4) complex Gaussian data

The Rayleigh pdf is given by

$$\text{rayl}(x;\gamma) = \frac{x}{\gamma^2} e^{-x^2/2\gamma^2} \qquad x \geq 0 \tag{3.33}$$

where the scale parameter $\gamma > 0$. The pdf is easily derived from the distribution of two independent zero-mean Gaussian variates with a common variance by transforming to polar coordinates and integrating over all possible values of the angle. Figure 3.8 shows the Rayleigh pdf for a range of parameters.

The Rayleigh cdf is

$$\text{Rayl}(x;\gamma) = 1 - e^{-x^2/2\gamma^2} \tag{3.34}$$

The expected value is $\gamma\sqrt{\pi/2}$, and the variance is $\gamma^2(4-\pi)/2$. The skewness is $2\sqrt{\pi}(\pi-3)/(4-\pi)^{3/2} \approx 0.63$, so the right tail is longer than the left one, and the kurtosis is $(32-3\pi^2)/(4-\pi)^2 \approx 3.25$; hence the distribution is leptokurtic. The geometric and harmonic means are $\sqrt{2}\gamma e^{-e/2}$ (where $e \approx 0.5772$ in the exponent is the Euler-Mascheroni constant) and $\sqrt{2/\pi}\gamma$, respectively. The median $\gamma\sqrt{\log 4}$ is larger than the mean, and the mode γ lies below both. The mean, variance, and median can be verified using MATLAB for a given value of $\gamma = 2$ as follows:

```
pd = makedist('Rayleigh', 'b', 2);
[mean(pd) sqrt(pi/2)*2;
var(pd) 2*(4 - pi)
median(pd) 2*sqrt(log(4))]
ans =
   2.5066    2.5066
   1.7168    1.7168
   2.3548    2.3548
```

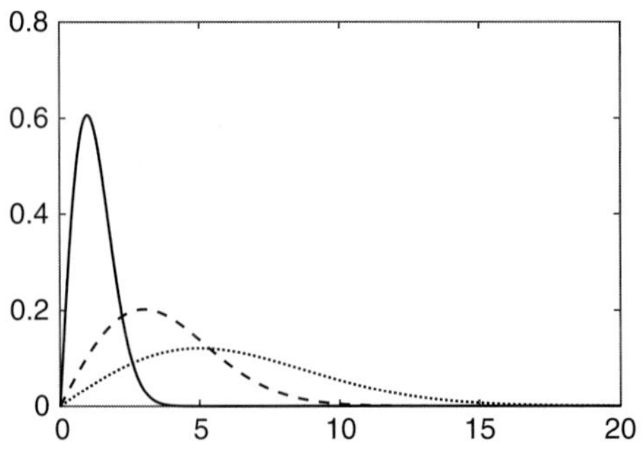

Figure 3.8 The Rayleigh pdf with scale parameters 1 (solid line), 3 (dashed line), and 5 (dotted line).

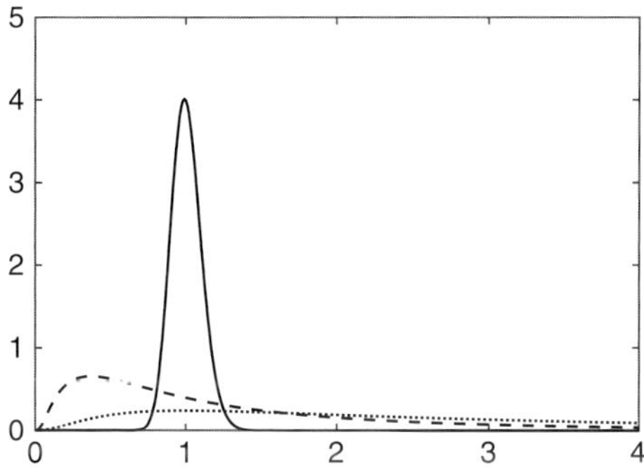

Figure 3.9 The lognormal pdf logn(0, 0.1) in solid line, logn(0, 1) in dashed line, and logn(1, 1) in dotted line.

3.4.4 Lognormal Distribution

The lognormal distribution is the distribution of a variable whose logarithm is Gaussian. It is the distribution of an rv that is the product of a large number of iid variables, just as the Gaussian distribution is that for an rv that is the sum of a large number of iid variables. It plays an important role in fields ranging from the earth sciences to finance.

Let $\mathbf{X} \sim \mathrm{N}(\mu, \sigma^2)$, and define a new rv $\mathbf{Y} = e^{\mathbf{X}}$. Then the distribution of $\mathbf{Y}$ is the lognormal distribution $\mathrm{logn}(y; \mu, \sigma^2) = \mathrm{N}(\log y; \mu, \sigma^2)$. The lognormal pdf has the form

$$\mathrm{logn}(y; \mu, \sigma) = \frac{1}{\sqrt{2\pi}\sigma y} e^{-(\log y - \mu)^2/2\sigma^2} \qquad y > 0 \tag{3.35}$$

This follows through a simple change of variables, as in Section 2.8. The lognormal cdf is

$$\mathrm{Logn}(y; \mu, \sigma) = \frac{1}{2}\mathrm{erfc}\left(-\frac{\log y - \mu}{\sqrt{2}\sigma}\right) = \Phi\left(\frac{\log y - \mu}{\sigma}\right) \tag{3.36}$$

where $\mathrm{erfc}(x)$ is the complementary error function [**erfc**(x) in MATLAB].

The expected value is $e^{\mu+\sigma^2/2}$, and the variance is $\left(e^{\sigma^2} - 1\right)e^{2\mu+\sigma^2}$. The skewness and kurtosis are $\sqrt{e^{\sigma^2} - 1}\,\left(e^{\sigma^2} + 2\right)$ and $e^{4\sigma^2} + 2e^{3\sigma^2} + 3e^{2\sigma^2} - 3$, respectively. The skewness is always positive, and consequently, the right tail is longer than the left one; hence the distribution is leptokurtic. The lognormal distribution falls toward zero quickly and toward infinity slowly (Figure 3.9). The geometric and harmonic means are e^{μ} and $e^{\mu-\sigma^2/2}$, respectively. The median and mode are e^{μ} (hence the same as the geometric mean) and $e^{\mu-\sigma^2}$, respectively, and the mean exceeds the median, which, in turn, exceeds the mode.

The mean, variance, and median formulas can be verified using MATLAB.

```
pd = makedist('LogNormal','mu',1,'sigma',2);
[mean(pd) exp(1+2^2/2);
var(pd) (exp(2^2)-1)*exp(2+2^2);
```

```
median(pd) exp(1)]
ans =
  1.0e+04 *
   0.0020    0.0020
   2.1623    2.1623
   0.0003    0.0003
```

The lognormal distribution is of major importance in the earth sciences. Examples include

1. The φ-fraction in sedimentology that is defined from the lognormal distribution. Let $\mathbf{X}$ be grain diameter in millimeters, and let $\varphi = -\log_2 x$. This choice is based on the empirical observation that $\mathbf{X}$ is lognormal;
2. Sedimentary bed thickness; and
3. Permeability and pore size.

Many geophysical processes are nonlinear (e.g., magnetospheric processes or their magnetic effects) and hence are the result of many multiplicative steps. The statistical distribution of such processes will tend to lognormal rather than Gaussian as a result.

Example 3.9 Suppose that a particle of initial size y_0 is subject to repeated impacts and that after each impact, a proportion $\mathbf{X}_i$ of the particle remains, where $\mathbf{X}_i$ is an rv. If the $\{\mathbf{X}_i\}$ are modeled as iid, after the first impact, the particle size is $\mathbf{Y}_1 = \mathbf{X}_1 y_0$; after the second impact, $\mathbf{Y}_2 = \mathbf{X}_1\mathbf{X}_2 y_0$; and after the Nth impact,$\mathbf{Y}_N = \mathbf{X}_1\mathbf{X}_2\cdots\mathbf{X}_N y_0$. Then $\log\mathbf{Y}_N = \log y_0 + \sum_{i=1}^{N}\log\mathbf{X}_i$. The distribution of particle size will be approximately lognormal.

3.4.5 Gamma Distribution

The gamma distribution is actually a family of distributions rather than a single one, of which the exponential and chi square distributions are the most important members. It is frequently used to model waiting times in reliability and life testing. The pdf has the general form

$$\mathrm{gam}(x;\alpha,\beta) = \frac{\beta^\alpha}{\Gamma(\alpha)} x^{\alpha-1} e^{-\beta x} \qquad x \ge 0 \tag{3.37}$$

where α is the shape parameter and β is the rate parameter (sometimes stated as $1/\beta$, when it is called the *scale parameter*), both of which must be positive. In standard form, $\beta = 1$. The gamma distribution is sometimes expressed as a three-parameter distribution with x replaced by $x - \gamma$, where γ is a location parameter, and $x > \gamma$.

The gamma cdf is

$$\mathrm{Gam}(x;\alpha,\beta) = \frac{\gamma(\beta x,\alpha)}{\Gamma(\alpha)} \tag{3.38}$$

where $\gamma(x, a)$ is the incomplete gamma function given by

$$\gamma(x, a) = \int_0^x t^{a-1} e^{-t} dt \tag{3.39}$$

Note the distinct versions of the complementary incomplete and incomplete gamma function in the cdf for the Poisson distribution (3.18) and (3.38).

The expected value and variance are α/β and α/β^2, respectively. The skewness and kurtosis are $2/\sqrt{\alpha}$ and $3 + 6/\alpha$, so the right tail of the distribution is always longer than the left one; hence the distribution is leptokurtic. The geometric and harmonic means are $e^{\psi(\alpha)}/\beta$ [where $\psi(x)$ is the digamma function] and $\beta/(\alpha - 1)$. The mode is $(\alpha - 1)/\beta$ if $\alpha \geq 1$, which is smaller than the mean, whereas the distribution tends to infinity as $x \to 0$ for $\alpha < 0$, so the mode occurs at the origin. There is no simple expression for the median. The standard gamma distribution tends to $\mathrm{N}(0, 1)$ as $\alpha \to \infty$.

The mean and variance formulas can be verified using MATLAB [but note that the parameter β in (3.37) is inverted in MATLAB, or is a scale parameter]:

```
a = 1;
b = 2;
pd = makedist('Gamma', 'a', a, 'b', b);
[mean(pd) a*b;
var(pd) a*b^2]
ans =
    2    2
    4    4
```

A key property of the gamma distribution is a reproductive one. If a set of N rvs $\{\mathbf{X}_i\}$ is independently gamma distributed with parameters (α_i, β), then $\sum_{i=1}^N \mathbf{X}_i$ is also gamma distributed with parameters $\left(\sum_{i-1}^N \alpha_i, \beta\right)$. Note the similarity to the binomial distribution. A second property is the scaling property. If $\mathbf{X} \sim \mathrm{gam}(x; \alpha, \beta)$, then $a\mathbf{X} \sim \mathrm{gam}(x; a\alpha, \beta)$.

There are a number of important special cases of the gamma distribution:

1. The exponential distribution obtained when $\alpha = 1$ is used in lifetime, reliability, and failure analysis and is important enough that it deserves its own subsection.
2. The chi square distribution family obtained when $\alpha = \nu/2$, $\beta = 1/2$ is the distribution of the sum of squared central (i.e., zero-mean) normal variates used widely in statistical inference. It will be covered in Section 4.9.1.
3. The Erlang distribution obtained when α is an integer is the distribution of the waiting time of the αth occurrence of a Poisson process with parameter $\lambda = \beta$. This is a generalization of the role of the exponential distribution in Poisson processes that will not be pursued further.

3.4.6 Exponential Distribution

The exponential distribution is used in geophysics to model the time between earthquakes or geomagnetic reversals and in engineering to model component or system lifetimes. It has the pdf (Figure 3.10)

$$\text{expe}(x;\beta) = \beta e^{-\beta x} \qquad x \geq 0 \tag{3.40}$$

The exponential cdf is

$$\text{Expe}(x;\beta) = 1 - e^{-\beta x} \tag{3.41}$$

The mean, variance, skewness, and kurtosis all follow from the definitions for the gamma distribution, except that the harmonic mean is not defined. The median is $\log 2/\beta$, whereas the mode is at the origin.

Caution: The MATLAB definition of the exponential distribution uses $1/\beta$ in place of β in (3.40) and (3.41).

The survival or reliability function is the probability that a component or system will survive beyond a given time and is just the complementary cdf $\Pr(\mathbf{X} > x)$. For the exponential distribution, the survival function is

$$\text{R}(x;\,\beta) = 1 - \text{expe}(x;\,\beta) = e^{-\beta x} \tag{3.42}$$

The hazard or failure rate function is the rate at which failures occur and has units of failures per unit time. It is given by $-\partial_x \text{R}(x;\beta)/\text{R}(x;\beta)$, which for the exponential distribution is

$$\text{h}(x;\,\beta) = \frac{\text{expe}(x;\,\beta)}{\text{R}(x;\,\beta)} = \beta \tag{3.43}$$

and hence the hazard is independent of x. The exponential is the only distribution used to model failure that has this property, and this is due to the following important behavior:

$$\Pr(\mathbf{X} > x + d | \mathbf{X} > d) = \frac{\Pr(\mathbf{X} > x + d \cap \mathbf{X} > d)}{\Pr(\mathbf{X} > d)} = \frac{\Pr(\mathbf{X} > x + d)}{\Pr(\mathbf{X} > d)} = \text{R}(d;\beta) \tag{3.44}$$

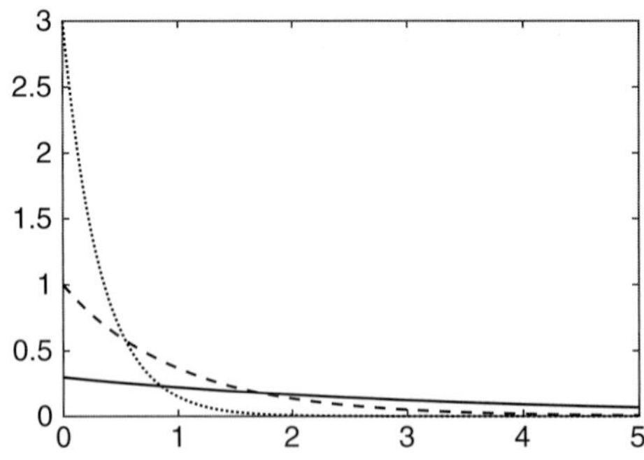

Figure 3.10 The exponential pdfs $\text{expe}(x;\,0.3)$ in solid line, $\text{expe}(x;\,1)$ in dashed line, and $\text{expe}(x;\,3)$ in dotted line.

The conditional probability is independent of x, so the exponential distribution has a memory-less property. If $\mathbf{X}$ were the time between a pair of events, this says that if the event has not occurred in t units of time, then the probability that the event will occur over the next d units of time is $\mathrm{R}(d;\beta)$, which is the same as the probability that it will occur in a time interval d from 0. Hence it does not matter (for example) that a large earthquake occurred at time t because the system has no memory of that event. It is not necessary to consider past occurrences of an event to predict future ones. The exponential distribution is the only continuous distribution with this memory-less property. However, the discrete geometric distribution is also memory-less.

Example 3.10 Suppose that geomagnetic reversals occur independently and that the lifetime of normal polarity intervals is $\mathrm{expe}(x;\beta)$. Let the rv $\mathbf{X}_i$ be the length of the ith normal interval, and assume that there are N normal intervals. What is the distribution of the length of time until the first normal to reverse transition occurs?

Let $\mathbf{Y}_1 = \min(\mathbf{X}_i)$ be an rv representing the time to the first normal to reverse transition. Then

$$\begin{aligned}\Pr(\mathbf{Y}_1 > t) &= \Pr(\mathbf{X}_1 > t \cap \cdots \cap \mathbf{X}_N > t)\\ &= \Pr(\mathbf{X}_1 > t)\cdots\Pr(\mathbf{X}_N > t)\\ &= e^{-N\beta t}\end{aligned}$$

The distribution of $\mathbf{Y}_1$ is $\mathrm{expe}(x;N\beta)$. It is easy to show that $\mathbf{Y}_2$ is distributed as $\mathrm{expe}(x;(N-1)\beta)$ and so on. The pdf $\mathrm{expe}(x;\ N\beta)$ falls off more rapidly than $\mathrm{expe}(x;(N-j)\beta)$ for $j = 1, \ldots, N-1$, and hence it is more probable that the first normal to reverse transition occurs earlier than the last one does.

Example 3.11 The rate of radioactive decay of a certain isotope is Poisson with parameter λ. Suppose that the variable of interest is how long it will take until a radioactive decay is observed. Let $\mathbf{Y}_1$ be the time until the first radioactive decay, and let $\mathbf{X}$ be the number of particles observed in time t. It follows that $\mathbf{Y}_1 \le t$ if and only if $\mathbf{X} \ge 1$ (i.e., the first radioactive decay is observed by time t if and only if at least one radioactive decay has occurred). The rv $\mathbf{X}$ is Poisson with mean λt, so, for $t > 0$, $\Pr(\mathbf{Y}_1 \le t) = \Pr(\mathbf{X} \ge 1) = 1 - \Pr(\mathbf{X} = 0) = 1 - e^{-\lambda t}$, where the last term is just the Poisson pdf when $x = 0$. This is the cdf for $\mathbf{Y}_1$, where it is understood that t is positive. The pdf follows by differentiation and is the exponential distribution.

The exponential distribution describes the elapsed time before the occurrence of an event that is Poisson (i.e., a rare event). This is the reason that it serves as a reasonable starting model for the intervals between earthquakes and geomagnetic reversals. In this sense, it is the continuous counterpart to the geometric distribution that describes the number of Bernoulli trials required until a discrete process changes state.

Example 3.12 The occurrence of floods in a given region is a Poisson process of μ floods per year, whereas their amplitude $\mathbf{X}$ ($\mathbf{X} > 0$) is a random variable with cdf $H(x)$. Find the cdf for the magnitude of the largest flood over an interval of duration t using an exponential distribution for $H(x)$.

The distribution of flood occurrence is $f(n;\, \mu t) = e^{-\mu t}(\mu t)^n/n!$ for nonnegative integer n. By the law of total probability, the distribution for the magnitude of the largest flood is

$$G(x) = \sum_n f(n;\mu t)\, H(x, n)$$

where the sum is over all possible values of n. Recall (Example 2.13) that the cdf of the largest element in a set of n rvs is $F^n(x)$. Substituting, the distribution for the magnitude of the largest flood can be written

$$G(x) = e^{-\mu t}\sum_{n=1}^{\infty}\frac{[\mu t H(x)]^n}{n!} = \exp\{-\mu t[1 - H(x)]\}$$

Let $H(x) = 1 - e^{-\beta(x-\lambda)}$ for the exponential cdf with a scale and location parameter. The flood occurrence cdf becomes

$$G(x) = \exp\left(-\mu t e^{-\beta(x-\lambda)}\right)$$

This is Gumbel's extreme value distribution and will be further described in Section 3.4.9. Note that this example mixes discrete and continuous distributions, as is appropriate for the problem under consideration.

3.4.7 Weibull Distribution

The Weibull distribution is used to model failures that are more complicated than can be described by the exponential distribution, such as when the conditions for strict randomness break down. For example, one might expect the time interval between earthquakes to be small immediately after a major earthquake and increase with time thereafter. In this case, the time interval between aftershocks might be better modeled by a Weibull distribution with shape parameter $\varsigma < 1$ than by an exponential distribution.

The Weibull pdf is

$$\text{weib}(x;\xi,\gamma,\varsigma) = \frac{\varsigma}{\gamma}\left(\frac{x-\xi}{\gamma}\right)^{\varsigma-1} e^{-[(x-\xi)/\gamma]^{\varsigma}} \tag{3.45}$$

for $x \geq \xi$, with location parameter ξ, scale parameter $\gamma > 0$, and shape parameter $\varsigma > 0$. It reduces to the exponential distribution when $\varsigma = 1$, $\xi = 0$, and to the Rayleigh distribution when $\varsigma = 2, \xi = 0$, and consequently can be viewed as an interpolator that spans these two distributions. The Weibull pdf is frequently seen without the location parameter ξ.

The Weibull cdf is

$$\mathrm{Weib}(x; \xi, \gamma, \varsigma) = 1 - e^{-[(x-\xi)/\gamma]^{\varsigma}} \tag{3.46}$$

so the survival function is

$$\mathrm{R}(x; \xi, \gamma, \varsigma) = e^{-[(x-\xi)/\gamma]^{\varsigma}} \tag{3.47}$$

and the hazard function is

$$\mathrm{h}(x; \xi, \gamma, \varsigma) = \frac{\varsigma}{\gamma}\left(\frac{x-\xi}{\gamma}\right)^{\varsigma-1} \tag{3.48}$$

The hazard function is independent of x when $\varsigma = 1$ (the exponential distribution) but is a decreasing function of x when $\varsigma < 1$ and an increasing function of x when $\varsigma > 1$. This leads to a simple interpretation for the shape parameter ς in a reliability context: a value less than 1 means that the failure rate decreases over time due to the removal of early problems from infant mortality, and a value greater than 1 means that an aging process is in effect so that failures increase with time. This defines the *bathtub curve* in reliability theory, where failures decrease rapidly at the start, are relatively constant through the design life, and then rise when the end of design life is reached.

When $\xi = 0$, the expected value and variance are $\gamma\Gamma(1/\varsigma)/\varsigma$ and $\gamma^2\{2\Gamma(2/\varsigma) - \varsigma[\Gamma(1+1/\varsigma)]^2\}/\varsigma$, respectively. The expressions for the skewness and kurtosis are complicated. The geometric and harmonic means are $\gamma e^{-e/\varsigma}$ and (when $\varsigma > 1$) $\gamma^{-\varsigma/(\varsigma-1)}(\varsigma-1)^2/\{\varsigma\ \Gamma[1/(\varsigma-1)]\}$. The median and mode are $\gamma(\log 2)^{1/\varsigma}$ and $\gamma[(\varsigma-1)/\varsigma]^{1/\varsigma}$ for $\varsigma > 1$. The mode is at the origin when $\varsigma \le 1$. Figure 3.11 shows the Weibull distribution for a scale parameter of 1 and a variety of shape parameters.

The mean, variance, and median formulas can be verified using MATLAB.

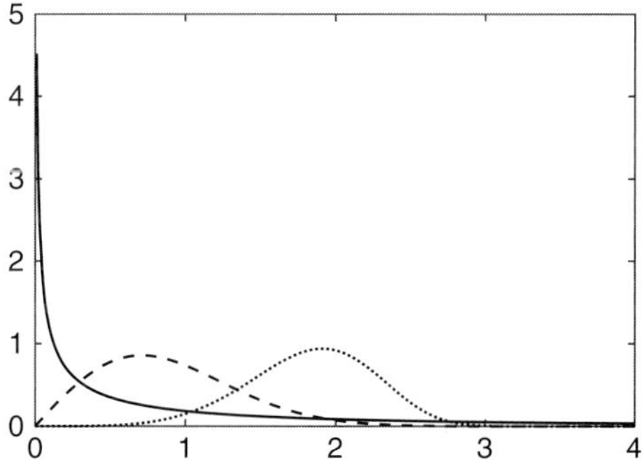

Figure 3.11 The Weibull pdf with $\gamma = 1, \varsigma = 0.5$ (solid line), $\gamma = 1, \varsigma = 2$ (dashed line), and $\gamma = 2, \varsigma = 5$ (dotted line).

```
s = 1;
c = 0.5;
pd = makedist('Weibull', 'a', s, 'b', c);
[mean(pd) s*gamma (1+1/c);
var(pd) s^2*(gamma(1+2/c) - gamma(1+1/c)^2);
median(pd) s*log(2)^(1/c)]
ans =
   2.0000    2.0000
  20.0000   20.0000
   0.4805    0.4805
c = 1.5;
pd = makedist('Weibull', 'a', s, 'b', c);
[mean(pd) s*gamma(1 + 1/c);
var(pd) s^2*(gamma(1 + 2/c) - gamma(1 + 1/c)^2);
median(pd) s*log(2)^(1/c)]
ans =
   0.9027    0.9027
   0.3757    0.3757
   0.7832    0.7832
```

3.4.8 Beta Distribution

The beta distribution has support [0, 1] and plays a major role in the distribution of the order statistics, as will be shown in Section 4.10. It is also used to describe probabilities for events that are constrained to happen within a specified time window and in likelihood ratio testing. The beta pdf in standard form is

$$\mathrm{beta}(x;\beta_1,\beta_2) = \frac{\Gamma(\beta_1+\beta_2)}{\Gamma(\beta_1)\Gamma(\beta_2)}x^{\beta_1-1}(1-x)^{\beta_2-1} = \frac{x^{\beta_1-1}(1-x)^{\beta_2-1}}{\mathrm{B}(\beta_1,\beta_2)} \qquad 0 \le x \le 1 \tag{3.49}$$

where $\beta_1 > 0$ and $\beta_2 > 0$ are shape parameters, and $\mathrm{B}(a, b)$ is the beta function. MATLAB implements the beta function as **beta**(*b*1, *b*2). Figure 3.12 shows the beta pdf for a few values of the shape parameters. The beta cdf is the regularized incomplete beta function ratio introduced in Section 3.3.2:

$$\mathrm{Beta}(x;\beta_1,\beta_2) = I_x(\beta_1,\beta_2) \tag{3.50}$$

The expected value and variance are $\beta_1/(\beta_1+\beta_2)$ and $\beta_1\beta_2/[(\beta_1+\beta_2)^2(\beta_1+\beta_2+1)]$. The expressions for the skewness and kurtosis are complicated. The geometric and harmonic means are $e^{\psi(\beta_1)-\psi(\beta_1+\beta_2)}$ and (when $\beta_1 > 1$) $(\beta_1-1)/(\beta_1+\beta_2-1)$. The mode is $(\beta_1-1)/(\beta_1+\beta_2-2)$ when β_1 and β_2 are larger than 1, and occurs at the distribution endpoints otherwise. When it is unique, the mode is always smaller than the median, which is, in turn, smaller than the mean. There is no general closed form expression for the

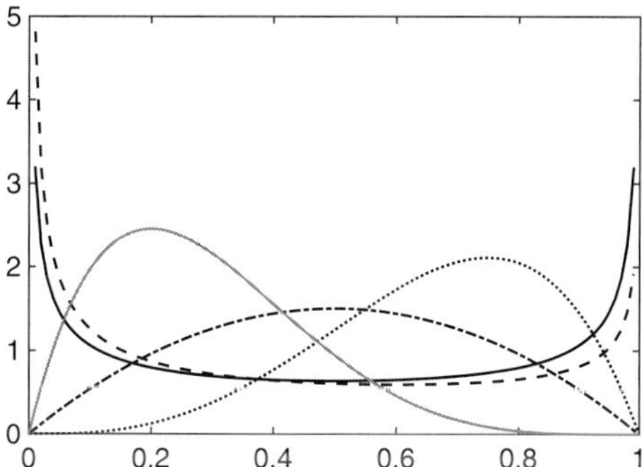

Figure 3.12 The beta pdf with parameters (0.5, 0.5) in solid line, (0.4, 0.6) in dashed line, (2, 2) in dotted line, (4, 2) in dash-dot line, and (2, 5) in gray line. The arcsine distribution corresponds to (0.5, 0.5), and the parabolic distribution corresponds to (2, 2).

median, but there are numerous special cases, and it can be obtained numerically by solving $I_x(\beta_1, \beta_2) = 1/2$.

The mean, variance, and median can be verified in MATLAB after defining a function handle for the numerical solution for the median

```
b1 = 0.5;
b2 = 0.5;
pd = makedist('Beta', 'a', b1, 'b', b2);
fun = @(x) betainc (x, b1, b2) - 0.5;
[mean(pd) b1/(b1 + b2);
var(pd) b1*b2/((b1 + b2)^2*(b1 + b2 + 1));
median(pd) fzero(fun,0.5)]
ans =
   0.5000    0.5000
   0.1250    0.1250
   0.5000    0.5000
```

Repeating using different values for the shape parameters:

```
b1 = 1.5;
b2 = 0.5;
pd = makedist('Beta', 'a', b1, 'b', b2);
fun = @(x) betainc(x, b1, b2) - 0.5;
[mean(pd) b1/(b1 + b2);
var(pd) b1*b2/((b1 + b2)^2*(b1 + b2 + 1));
median(pd) fzero(fun, 0.5)]
ans =
   0.7500    0.7500
   0.0625    0.0625
   0.8368    0.8368
```

Some symmetry properties of the beta distribution include

1. $\text{beta}(x;\beta_1,\beta_2) = \text{beta}(1-x;\beta_2,\beta_1)$ and $\text{Beta}(x;\beta_1,\beta_2) = \text{Beta}(1-x;\beta_2,\beta_1)$ (reflection symmetry); and
2. Let $\mathbf{X} \sim \text{beta}(x;\beta_1,\beta_2)$ and $\mathbf{Y} \sim \text{beta}(x;\beta_2,\beta_1)$. Then $\mathcal{E}(\mathbf{X}) = 1 - \mathcal{E}(\mathbf{Y})$ and $var(\mathbf{X}) = var(\mathbf{Y})$.

In addition, the beta distribution can take on a wide range of shapes depending on its parameters. When the shape parameters are identical, then

1. The pdf is symmetric about ½;
2. The expected value and median both equal ½; and
3. The skewness is zero.
4. When the shape parameters are both smaller than 1, the pdf is U-shaped and bimodal, with the modes at the endpoints (see Figure 3.12).
5. When the shape parameters equal 1, the result is the uniform distribution on (0, 1).
6. When the shape parameters are both larger than 1, the pdf is symmetric and unimodal, with the mode at ½.

When the shape parameters are not identical, a nonexhaustive list of properties include

1. When β_1 and β_2 are both smaller than 1, the pdf is U-shaped and bimodal, with the modes at the endpoints, and has positive skew when $\beta_1 < \beta_2$ and negative skew otherwise.
2. When β_1 and β_2 are both larger than 1, the pdf is unimodal and has positive skew when $\beta_1 < \beta_2$ and negative skew otherwise.
3. When $\beta_1 < 1$ and $\beta_2 \geq 1$, the pdf is J-shaped with a right tail, has a positive skew, and is strictly decreasing. The mode is zero.
4. When $\beta_1 \geq 1$ and $\beta_2 < 1$, the pdf is J-shaped with a left tail, has negative skew, and is strictly increasing. The mode is 1.

3.4.9 Generalized Extreme Value Distribution

Extreme value distributions are limit distributions for the maximum or minimum of a set of rvs, and are very important in the earth sciences, especially in hydrology, where they are used to model flood or river outflow data, and in meteorology, where they are used to model rainfall or extremes of wind speed. It can be proved that there are only three types of extreme value distributions: the Gumbel distribution, which was derived in Example 3.12; the Fréchet distribution; and a transformed form of the Weibull distribution that was covered in Section 3.4.7. These are sometimes referred to as the *Type I, II, and III extreme value distributions*, respectively. These three distributions can be combined into a single form, the generalized extreme value (gev) or Fisher-Tippett distribution.

The gev pdf is given by

$$\begin{aligned}\mathrm{gev}(x;\xi,\gamma,\varsigma) &= \frac{1}{\gamma}\left[1+\varsigma\left(\frac{x-\xi}{\gamma}\right)\right]^{-(1+1/\varsigma)} e^{-\left[1+\varsigma\left(\frac{x-\xi}{\gamma}\right)\right]^{-1/\varsigma}} & \varsigma \neq 0 \\ &= \frac{1}{\gamma} e^{-(x-\xi)/\gamma} e^{-e^{-(x-\xi)/\gamma}} & \varsigma = 0\end{aligned} \quad (3.51)$$

where ξ is the location parameter, $\gamma > 0$ is the scale parameter, and ς is the shape parameter. The support of (3.51) is $x > \xi - \gamma/\varsigma$ when $\varsigma > 0$ and $x < \xi - \gamma/\varsigma$ when $\varsigma < 0$, and the pdf is zero outside those ranges. Consequently, the pdf support has a lower limit for positive shape parameters and an upper bound for negative shape parameters. When $\varsigma = 0$, the support is $(-\infty, \infty)$. When $\varsigma < 0$, the gev pdf is a transformed Weibull. When $\varsigma = 0$, it is a transformed Gumbel, and when $\varsigma > 0$, it is Fréchet. The first two are not the same as the standard Weibull and Gumbel distributions. Rather than using this terminology, it is safer to work only with the generalized extreme value distribution. It is more useful to note that for $\varsigma > 0$, the right tail is algebraic, the distribution has infinite variance for $\varsigma \geq 1/2$ and infinite mean for $\varsigma \geq 1$. For $\varsigma = 0$, the tails are exponential, whereas for $\varsigma < 0$, the upper tail is truncated. Figure 3.13 illustrates this behavior.

The cdf is

$$\begin{aligned}\mathrm{Gev}(x;\xi,\gamma,\varsigma) &= e^{-\left[1+\varsigma\left(\frac{x-\xi}{\gamma}\right)\right]^{-1/\varsigma}} & \varsigma \neq 0 \\ &= e^{-e^{-(x-\xi)/\gamma}} & \varsigma = 0\end{aligned} \quad (3.52)$$

with the same support as the pdf.

The preceding forms are appropriate for maximal extreme values. To obtain their versions for minimal ones, replace x with $-x$ in (3.52), and subtract the result from 1. The pdf follows by differentiation and will be distinct from (3.51).

The expected value is $\xi + \gamma[\Gamma(1-\varsigma) - 1]/\varsigma$ when $\varsigma < 1$ and $\varsigma \neq 0$, $\xi + e\gamma$ (where $e \approx 0.5772$ is the Euler-Mascheroni constant) when $\varsigma = 0$, and ∞ otherwise. The variance is $\gamma^2\left\{\Gamma(1-2\varsigma) - [\Gamma(1-\varsigma)]^2\right\}/\varsigma^2$ when $\varsigma < 1/2$ and $\varsigma \neq 0$, $\gamma^2\pi^2/6$ when $\varsigma = 0$, and ∞ otherwise. The skewness and kurtosis are complicated expressions that exist only for $\varsigma < 1/3$ and $\varsigma < 1/4$, respectively, when $\varsigma \neq 0$. For $\varsigma = 0$, the skewness is $4\sqrt{54}z(3)/$

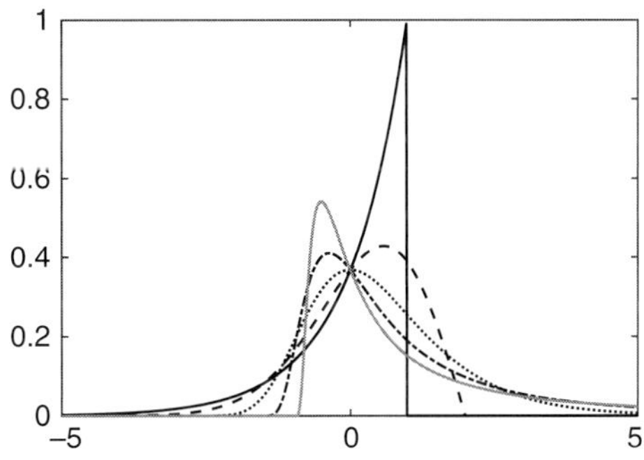

Figure 3.13 The generalized extreme value distribution with $\xi = 0$ and $\gamma = 1$ for shape parameters of -1.0 (solid line), -0.5 (dashed line), 0.0 (dotted line), 0.5 (dash-dot line), and 1.0 (gray line).

$\pi^3 \approx 1.14$ [where $z(x)$ is the Riemann zeta function; $z(3) \approx 1.2021$], and the kurtosis is 27/5. The geometric and harmonic means cannot be expressed in closed form. The median is $\xi + \gamma[(\log 2)^{-\varsigma} - 1]/\varsigma$ when $\varsigma \neq 0$ and $\xi - \gamma \log(\log 2)$ when $\varsigma = 0$. The mode is $\xi + \gamma[(\log 2)^{-\varsigma} - 1]/\varsigma$ when $\varsigma \neq 0$ and ξ when $\varsigma = 0$.

The mean, variance, and median can be verified using MATLAB

```
xi = 0;
gam = 2;
sig = 0.4;
pd = makedist('GeneralizedExtremeValue', 'k', sig, 'sigma',
gam, 'mu', xi);
[mean(pd) xi + gam* (gamma(1 - sig) - 1)/sig;
var(pd) gam^2* (gamma(1 - 2* sig) - gamma(1 - sig)^2)/sig^2;
median(pd) xi + gam* (log(2)^(-sig) - 1)/sig]
ans =
     2.4460     2.4460
    59.3288    59.3288
     0.7895     0.7895
sig = 0;
euler = -psi(1); %Euler's constant using the digamma function
pd = makedist('GeneralizedExtremeValue', 'k', sig, 'sigma',
gam, 'mu', xi);
[mean(pd) xi + euler* gam;
var(pd) gam^2* pi^2/6;
median(pd) xi - gam* log(log(2))]
ans =
   1.1544     1.1544
   6.5797     6.5797
   0.7330     0.7330
```

3.4.10 Bivariate Gaussian Distribution

Let $\mathbf{Z}_1$ and $\mathbf{Z}_2$ be independent N(0, 1) rvs. Their joint pdf is the product of two marginal Gaussian distributions

$$f(z_1, z_2) = \frac{1}{2\pi} e^{-\left(z_1^2 + z_2^2\right)/2} \tag{3.53}$$

Consider two new rvs $\mathbf{X}_1$ and $\mathbf{X}_2$, where $\mathbf{X}_1 = \sigma_1 \mathbf{Z}_1 + \mu_1$ and $\mathbf{X}_2 = \sigma_2 \left(\rho \mathbf{Z}_1 + \sqrt{1-\rho^2}\mathbf{Z}_2\right) + \mu_2$. Their joint pdf may be found by linear transformation as in Section 2.9. The relationship given by

$$\begin{pmatrix} \mathbf{X}_1 \\ \mathbf{X}_2 \end{pmatrix} = \begin{pmatrix} \sigma_1 & 0 \\ \sigma_2 \rho & \sigma_2 \sqrt{1-\rho^2} \end{pmatrix} \begin{pmatrix} \mathbf{Z}_1 \\ \mathbf{Z}_2 \end{pmatrix} + \begin{pmatrix} \mu_1 \\ \mu_2 \end{pmatrix} \tag{3.54}$$

has the inverse

$$\begin{pmatrix} \mathbf{Z}_1 \\ \mathbf{Z}_2 \end{pmatrix} = \frac{1}{\sigma_1\sigma_2\sqrt{1-\rho^2}} \begin{pmatrix} \sigma_2\sqrt{1-\rho^2} & 0 \\ -\sigma_2\rho & \sigma_1 \end{pmatrix} \begin{pmatrix} \mathbf{X}_1 - \mu_1 \\ \mathbf{X}_2 - \mu_1 \end{pmatrix} \tag{3.55}$$

The determinant of the matrix in (3.54) is $\sigma_1\sigma_2\sqrt{1-\rho^2}$, so the joint pdf of $\mathbf{X}_1$ and $\mathbf{X}_2$ is

$$\mathrm{N}_2(x_1, x_2; \mu_1, \mu_2, \sigma_1, \sigma_2, \rho) = \frac{1}{2\pi\sigma_1\sigma_2\sqrt{1-\rho^2}} \times e^{\frac{-1}{2(1-\rho^2)}\left[\left(\frac{x_1-\mu_1}{\sigma_1}\right)^2 - 2\rho\left(\frac{x_1-\mu_1}{\sigma_1}\right)\left(\frac{x_2-\mu_2}{\sigma_2}\right) + \left(\frac{x_2-\mu_2}{\sigma_2}\right)^2\right]} \tag{3.56}$$

which is the bivariate Gaussian distribution with $\mathcal{E}(\mathbf{X}_i) = \mu_i$, $var(\mathbf{X}_i) = \sigma_i^2$, and population correlation coefficient ρ. It generalizes to N terms to give the multivariate Gaussian, as described in Chapter 10. Since $\mathbf{Z}_1$ and $\mathbf{Z}_2$ are independently $\mathrm{N}(0,1)$ and $\mathbf{X}_1$ and $\mathbf{X}_2$ are linear combinations of $\mathbf{Z}_1$ and $\mathbf{Z}_2$, the marginal distributions for $\mathbf{X}_1$ and $\mathbf{X}_2$ are $\mathrm{N}(\mu_i, \sigma_i^2)$.

If $\mathbf{X}_1$ and $\mathbf{X}_2$ are uncorrelated, $\rho = 0$. The joint pdf of $\mathbf{X}_1$ and $\mathbf{X}_2$ then becomes the product of the marginal ones, and hence $\mathbf{X}_1$ and $\mathbf{X}_2$ are independent. The Gaussian is the only distribution for which this holds. Lack of correlation does not prove independence unless the rvs are Gaussian.

MATLAB does not explicitly implement the bivariate normal distribution but does provide the multivariate normal distribution as functions rather than distribution objects. The pdf and cdf are given by **mvnpdf**(*x*, *mu*, *sigma*) and **mvncdf**(*x*, *mu*, *sigma*), where *mu* is a *p*-vector and *sigma* is the $p \times p$ covariance matrix. The bivariate Gaussian distribution obtains when $p = 2$. MATLAB does not provide the quantile function for the multivariate normal distribution. Random draws are given by **mvnrnd**(*mu*,*sigma*).

3.4.11 Directional Distributions

There are a wide variety of directional distributions on the circle and sphere, as thoroughly described in Mardia & Jupp (2000, chap. 3). The most fundamental distribution on a unit circle (i.e., a circle with a radius of unity) is the *uniform distribution* given by

$$f(\theta) = \frac{1}{2\pi} \tag{3.57}$$

such that probability is proportional to arc length, and there is no concentration of probability about any given direction.

The simplest transformation of the uniform distribution on a circle (3.57) is its perturbation by a cosine function to give the *cardioid distribution*

$$\mathrm{card}(\theta; \rho, \mu) = \frac{1}{2\pi}[1 + 2\rho\cos(\theta - \mu)] \qquad |\rho| < \frac{1}{2} \tag{3.58}$$

whose name ensues because $r = \mathrm{card}(\theta; \rho, \mu)$ defines a cardioid in polar coordinates. The mean resultant length of (3.58) is ρ, the mean direction is μ, and it reduces to (3.57) when $\rho = 0$.

The most useful distribution on the unit circle is the von Mises (1918) distribution, which is easily derivable from the bivariate Gaussian distribution (3.56). Suppose that $\mathbf{X}_1$ and $\mathbf{X}_2$ are independent Gaussian variables with population means μ_1 and μ_2 and common population variance σ^2. Their joint pdf is given by (3.56) with correlation coefficient $\rho = 0$ and $\sigma_1 = \sigma_2 = \sigma$. The joint pdf can be transformed to polar coordinates with $x_1 = r\cos\theta$ and $x_2 = r\sin\theta$ using the method of Section 2.9. Defining $\beta = 1/\sigma$, $\lambda = \sqrt{\mu_1^2 + \mu_2^2}$, and $\mu = \tan^{-1}(\mu_2/\mu_1)$, the joint pdf in polar coordinates is

$$g(r, \theta; \beta, \lambda, \mu) = \frac{\beta^2 r}{2\pi} e^{-\beta^2(r^2+\lambda^2)/2} e^{\beta^2 \lambda r \cos(\theta-\mu)} \tag{3.59}$$

The marginal distribution for r is obtained by integrating (3.59) over all possible azimuths, which, in turn, requires the integration of an exponential with a cosine argument. That term can be expanded as an infinite series of modified Bessel functions of the first kind using the generating function

$$e^{t\cos\vartheta} = I_0(t) + 2\sum_{k=1}^{\infty} I_k(t)\cos(k\vartheta) \tag{3.60}$$

Performing the integration gives

$$g_1(r; \beta, \lambda) = \beta^2 r e^{-\beta^2(r^2+\lambda^2)/2} I_0(\beta^2 \lambda r) \tag{3.61}$$

Define the concentration parameter $\kappa = \beta^2\lambda^2$, and let the dimensionless radius be $r' = r/\lambda$. The distribution for θ conditional on a particular value of the dimensionless radius $r' = r^*$ is obtained by dividing (3.59) by (3.61):

$$g(\theta; \kappa, \mu | r' = r^*) = \frac{e^{\kappa r^* \cos(\theta-\mu)}}{2\pi I_0(\kappa r^*)} \tag{3.62}$$

Setting $r^* = 1$ yields the distribution for direction on the unit circle, or the *von Mises distribution*:

$$\text{mises}(\theta; \kappa, \mu) = \frac{e^{\kappa\cos(\theta-\mu)}}{2\pi I_0(\kappa)} \tag{3.63}$$

As $\kappa \to 0$, $I_0(\kappa)$ and the exponential term go to 1, so the distribution is just the uniform distribution (3.57). For small κ, the exponential term in (3.63) may be replaced with the first two terms in its series expansion to give $\text{mises}(\theta;\ \kappa, \mu) \cong \text{card}(\theta;\ \kappa/2, \mu)$, and the von Mises distribution becomes cardioidal. As $\kappa \to \infty$, the distribution approaches $\sqrt{\kappa/2\pi} e^{\kappa[\cos(\theta-\mu)-1]}$. For any value of the argument of the cosine term save 0, the exponential is decreasing with κ faster than $\sqrt{\kappa}$ increases, and the result is zero. When $\theta - \mu = 0$, the distribution is infinite and hence becomes the point distribution $\delta(\theta - \mu)$ in the large κ limit. The von Mises distribution is unimodal and symmetric about the mean direction μ. The cdf cannot be expressed in closed form and must be computed numerically. Figure 3.14 illustrates the von Mises distribution and suggests the end member behaviors.

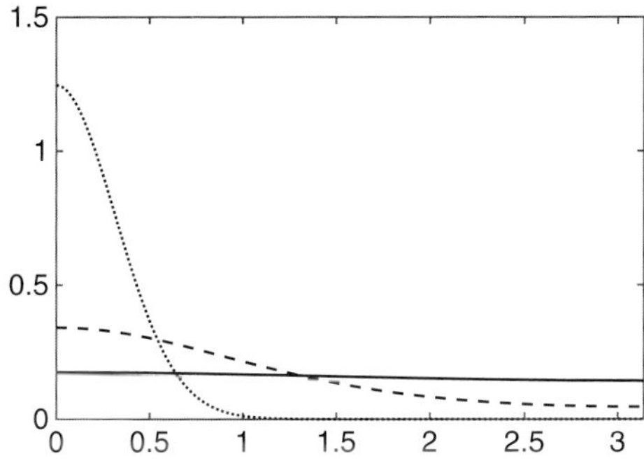

Figure 3.14 The von Mises pdf for θ in the range $[0, \pi]$ for $\kappa = 0.1$ (solid line), 1 (dashed line), and 10 (dotted line).

The kth sine moment about the mean direction μ is zero by symmetry, while the kth cosine moment about the mean is given by

$$\overline{\alpha}_k = \frac{1}{2\pi\ I_0(\kappa)} \int_0^{2\pi} \cos k(\theta - \mu)\, e^{\kappa \cos(\theta-\mu)} d\theta = \frac{I_k(\kappa)}{I_0(\kappa)} \tag{3.64}$$

because the modified Bessel function of the first kind is defined by

$$I_k(\kappa) = \frac{1}{2\pi} \int_0^{2\pi} \cos k\vartheta\ e^{\kappa \cos \vartheta} d\vartheta \tag{3.65}$$

Consequently, the mean resultant length ρ is given by (3.64) with $k = 1$.

MATLAB does not provide support for the von Mises distribution. The von Mises distribution is quite distinct from the marginal distribution for angle obtained by integrating (3.59) over all possible values of the radius. Since for positive a,

$$\int_0^{\infty} re^{-ar^2} e^{br} dr = \frac{2\sqrt{a} + \sqrt{\pi} b e^{b^2/(4a)} \{1 + \mathrm{erf}[b/(2\sqrt{a})]\}}{4a^{3/2}} \tag{3.66}$$

by Gradshteyn & Ryzhik (1980, 3.462-5), the marginal distribution for the angle is given by

$$g_2(\theta; \kappa, \mu) = \frac{1}{2\pi} \left(\begin{array}{l} e^{-\kappa/2} + \sqrt{\frac{\pi\kappa}{2}} \cos(\theta - \mu) e^{-\kappa \sin^2(\theta-\mu)/2} \times \\ \left\{ 1 + \mathrm{erf}\left[\sqrt{\frac{\kappa}{2}} \cos(\theta - \mu) \right] \right\} \end{array} \right) \tag{3.67}$$

and reduces to the uniform distribution on a circle (3.57) as $\kappa \to 0$. Equation (3.67) is the distribution of the angle regardless of the value of the radius (which is the fundamental definition of a marginal distribution), whereas (3.63) is the distribution of the angle conditional on the radius being equal to 1. It is not possible to derive (3.63) from (3.67) because neither explicitly includes the joint distribution (3.59). However, they do have a relationship in the large concentration parameter, small angular difference limit. For large κ, the asymptotic form of $I_0(\kappa) \sim e^\kappa/\sqrt{2\pi\kappa}$, so (3.63) reduces to

$$\text{mises}(\theta;\kappa,\mu) \cong \sqrt{\frac{\kappa}{2\pi}} e^{\kappa[\cos(\theta-\mu)-1]} \tag{3.68}$$

For large x, $\text{erf}(x) \sim 1 - e^{-x^2}/\sqrt{\pi}x$, and (3.67) reduces to

$$g_2(\theta;\kappa,\mu) \cong \sqrt{\frac{\kappa}{2\pi}} \cos(\theta-\mu) e^{-\kappa\sin^2(\theta-\mu)/2} \tag{3.69}$$

and bears little resemblance to (3.68). However, for small values of $\theta-\mu$, $\sin(\theta-\mu) \approx \theta-\mu$ and $\cos(\theta-\mu) \approx 1-(\theta-\mu)^2/2$, and (3.68) becomes the angular Gaussian distribution

$$\text{mises}(\theta;\kappa,\mu) \cong \sqrt{\frac{\kappa}{2\pi}} e^{-\kappa(\theta-\mu)^2/2} \tag{3.70}$$

whereas with the further condition that $(\theta-\mu)^2/2 \ll 1$, (3.69) becomes identical.

If the angle on a circle must be characterized without a priori information about the magnitude, the expected value and variance using the angular marginal distribution (3.67) will provide the best estimate. However, if it is known (or desired) that the magnitude be such that the parameter space is confined to the unit circle, then the expected value using the von Mises distribution will yield a better estimate and will have a lower variance,

The generalization of the von Mises distribution to a spherical geometry is important in paleomagnetism and plate tectonics. As a firstprinciples derivation, consider a three Cartesian component geomagnetic vector $\mathbf{x}$, and model it using a Gaussian pdf with a mean three-vector $\boldsymbol{\mu}$ and a common scalar variance σ^2

$$f(\mathbf{x};\boldsymbol{\mu},\sigma^2) = \frac{e^{-(\mathbf{x}-\boldsymbol{\mu})^T\cdot(\mathbf{x}-\boldsymbol{\mu})/(2\sigma^2)}}{(2\pi)^{3/2}\sigma^3} \tag{3.71}$$

where • denotes the inner product. This must be transformed from Cartesian coordinates (x,y,z), where by convention x points north, y points east, and z points down, to (F,θ,ϕ), where $F=\sqrt{x^2+y^2+z^2}$ is the scalar magnitude, $\theta=\tan^{-1}\left(z/\sqrt{x^2+y^2}\right)$ is the inclination, and $\phi=\tan^{-1}(x/y)$ is the declination. The inverse transformation is $x=F\cos\theta\cos\phi$, $y=F\cos\theta\sin\phi$, and $z=F\sin\theta$. The Jacobian determinant is $-F^2\cos\theta$. Consequently, the transformed pdf is

$$\begin{aligned} f\left(F,\theta,\phi;\mu_F,\mu_\theta,\mu_\phi,\sigma\right) &= \frac{F^2\cos\theta}{(2\pi)^{3/2}\sigma^3} e^{-\left(F\cos\theta\cos\phi-\mu_F\cos\mu_\theta\cos\mu_\phi\right)^2/(2\sigma^2)} \\ &\quad\times e^{-\left(F\cos\theta\sin\phi-\mu_F\cos\mu_\theta\sin\mu_\phi\right)^2/(2\sigma^2)} e^{-(F\sin\theta-\mu_F\sin\mu_\theta)^2/(2\sigma^2)} \\ &= \frac{F^2\cos\theta}{(2\pi)^{3/2}\sigma^3} e^{-\left(F^2+\mu_F^2\right)/(2\sigma^2)} e^{F\mu_F\cos\xi/\sigma^2} \end{aligned} \tag{3.72}$$

where $\cos\xi = \cos\theta\cos\mu_\theta\cos\left(\phi - \mu_\phi\right) + \sin\theta\sin\mu_\theta$ is the cosine of the off-axis angle between a particular unit magnetic vector and the mean unit vector, and $\left(\mu_F, \mu_\theta, \mu_\phi\right)$ are the mean intensity, inclination, and declination, respectively.

Love & Constable (2003) give the following for the marginal distribution for intensity obtained by integrating (3.72) over both the angular terms:

$$f_F(F;\mu_F,\sigma) = \sqrt{\frac{2}{\pi}\frac{F}{\mu_F\sigma}}e^{-\left(F^2+\mu_F^2\right)/\left(2\sigma^2\right)}\,\sinh\left(\frac{\mu_F F}{\sigma^2}\right) \tag{3.73}$$

Consequently, the conditional distribution for inclination and declination given a particular value for the intensity F^* is obtained by dividing the joint distribution (3.72) by the intensity marginal distribution (3.73). Defining $k = \mu_F^2/\sigma^2$ to be the concentration parameter, the result is

$$f\left(\theta,\phi;\kappa,\mu_\theta,\mu_\phi|F = F^*\right) = \frac{\kappa F^*/\mu_F\cos\theta}{4\pi\sinh\left(\kappa F^*/\mu_F\right)}e^{\kappa F^*\cos\xi/\mu_F} \tag{3.74}$$

Setting the dimensionless intensity $F^*/\mu_F = 1$ gives the distribution of inclination and declination on the unit sphere

$$\text{fisher}\left(\theta,\phi;\kappa,\mu_\theta,\mu_\phi\right) = \frac{\kappa\cos\theta}{4\pi\sinh\kappa}e^{\kappa\cos\xi} \tag{3.75}$$

Equation (3.75) is called the *Fisher distribution* or the *von Mises-Fisher distribution* and was given by Fisher (1953), although it was actually derived much earlier in a statistical mechanics context by Langevin (1905). Equation (3.75) is exact, commensurate with the model assumptions implicit to (3.71). Chave (2015) carried out a comparison of (3.75) with the inclination-declination marginal distribution, showing that they are similar only in the limit of large concentration parameter and very small off-axis angle, as was the case for the von Mises distribution on a circle.

4 Characterization of Data

4.1 Overview

This chapter marks the transition from the theory to the practice of data analysis. It first introduces the concept of an estimator, or formula for computing a statistic, and then elaborates on the sample counterparts to the population entities for location, dispersion, shape, direction, and association that were covered in Chapter 2. Using this information, the key limit theorems for random variables (rvs) are described, of which the most important are the two laws of large numbers and the classic central limit theorem (CLT), which pertains to rvs whose distributions have a finite variance. Extensions of the CLT to directional data and to an rv whose distribution has infinite variance are then described. A wide variety of visualization tools to characterize rvs are introduced, for which the key ones are the percent-percent and quantile-quantile plots and the kernel density estimator. Their implementations using MATLAB are presented by example. The three arguably most important distributions in statistics, the Student's t, chi square, and F distributions, which are the sampling distributions for the sample mean and variance, are then defined. The concept of an order statistic obtained by sorting a set of rvs into ascending order and their distributions is described, leading to the sample median and interquartile range. Finally, the joint distribution of the sample mean and sample variance that is one of the most important relationships in statistics is derived and explained.

4.2 Estimators of Location

A *parameter* is an attribute of a statistical distribution, such as the population mean or variance in the Gaussian distribution of Section 3.4.1. Under the frequentist philosophy of statistics, a parameter is a fixed entity without statistical properties. Suppose there are N iid samples $\{\mathbf{X}_i\}$ of an rv with population mean μ and variance σ^2. Form the *estimator*

$$\overline{\mathrm{X}}_N = \frac{1}{N}\sum_{i=1}^{N}\mathbf{X}_i \tag{4.1}$$

An *estimator* is a *statistic* (or function that depends on a given set of data but not on any other unknown parameters) that specifies how to compute a parameter for a specific set of measurements of an rv $\{\mathbf{X}_i\}$. The probability distribution of a statistic is called the *sampling distribution* and is discussed in Section 4.9. An estimator changes when an element $\mathbf{X}_i$ is added or removed and hence is itself an rv. An *estimate* is a particular realization of an estimator, such as $\overline{X}_{10}$. Equation (4.1) is the *sample mean*. The sample mean is an example of a *point estimator* that yields a specific value for the parameter (although point estimators can also be vector valued). Alternately, an estimator can be an *interval estimator* that yields a range of values for the parameter of interest or a *ratio estimator* that describes the ratio of two sets of rvs. These concepts will be explored further in Chapter 5.

The expected value of the sample mean is

$$\mathcal{E}\left(\overline{X}_N\right) = \mu \tag{4.2}$$

using (4.1). The expected value of the sample mean is the population mean and hence it is an *unbiased estimator*.

Presuming that the data are independent, the variance of the sample mean is

$$var\left(\overline{X}_N\right) = \frac{1}{N^2} var\left(\sum_{i-1}^{N} \mathbf{X}_i\right) = \frac{\sigma^2}{N} \tag{4.3}$$

The distribution of the sample mean becomes increasingly concentrated about the population mean as N increases and hence in some sense becomes a better estimate for it as the number of data rises. This statement can be made more precise using the Chebyshev inequality (2.77). Because the mean and variance of the sample mean are μ and σ^2/N, it follows that

$$\Pr\left(\left|\overline{X}_N - \mu\right| \geq t\right) \leq \frac{\sigma^2}{Nt^2} \tag{4.4}$$

This can be used to ask questions such as how many samples are required to get the sample mean close to the true mean at some specified probability level? For example, if $t = 0.1\,\sigma$, then the right side of (4.4) is $100/N$. Consequently, 100 data values are required to even make (4.4) meaningful, and to achieve a value of 0.05 requires 2000 data values.

Equation (4.3) gives the variance of the sample mean if the rvs used to compute it are independent. If the rvs are correlated, then the variance of the sample mean is given by the sum of their covariances

$$\begin{aligned} var\left(\overline{X}_N\right) &= \frac{1}{N}\sum_{i=1}^{N}\sum_{j=1}^{N} cov\left(\mathbf{X}_i, \mathbf{X}_j\right) \\ &= \frac{1}{N}\sum_{i=1}^{N} var(\mathbf{X}_i) + \frac{2}{N}\sum_{i=1}^{i-1}\sum_{j=1}^{N} cov\left(\mathbf{X}_i, \mathbf{X}_j\right) \end{aligned} \tag{4.5}$$

If the rvs have a common or scalar variance σ^2 and a common correlation ρ, then (4.5) reduces to

$$var\left(\overline{X}_N\right) = \frac{\sigma^2}{N} + \frac{N-1}{N}\rho\sigma^2 \tag{4.6}$$

and hence the variance of the sample mean increases (decreases) as the correlation becomes more positive (negative). Moreover,

$$\lim_{N \to \infty} \overline{X}_N = \rho \sigma^2 \tag{4.7}$$

so if the variables are standardized, the sample mean divided by the variance is approximately the correlation of the rvs in the large sample limit.

A variant on the sample mean is the *weighted sample mean*. Let $\{w_i\}$ be a set of weights. Then the weighted sample mean is given by

$$\overline{X}'_N = \frac{\sum_{i=1}^{N} w_i x_i}{\sum_{i=1}^{N} w_i} \tag{4.8}$$

If the weights are already normalized, then the denominator in (4.8) is unity, and the weighted sample mean is given by its numerator. A typical choice for the weights is the inverse of a measure of the reliability for a given rv.

In MATLAB, the sample mean of a vector of data is given by **mean**(x). If x is a matrix, then the mean of each column is returned. If the row means are required, then **mean**(x, 1) will compute them, and the mean along any dimension of an N-dimensional array follows from **mean**(x, *ndim*). The weighted mean is easily computed by **mean**($w.*x$), where w is a matrix of weights that is already normalized, or else **mean**($w.*x$)/**sum**(w) if they are not.

MATLAB also implements the trimmed mean y = **trimmean**(x, *percent*) that removes the largest and smallest values according to the parameter *percent*. If there are N values in x, then the $k = N*(percent/100)/2$ highest and lowest values in x are trimmed before the arithmetic mean is computed. This is sometimes useful to reduce the effect of outlying data lying at the top or bottom of the data distribution. A third parameter *flag* controls how trimming is accomplished when k is not an integer: "round" (default) rounds k to the nearest integer, and "floor" rounds k down to the next smallest integer. A fourth parameter *ndim* operates as for **mean**.

MATLAB also provides the function **nanmean**(x) that computes the arithmetic mean while skipping over NaN (not a number) values. It operates exactly as **mean** does.

The harmonic mean estimator is given by

$$\overline{H}_N = \frac{N}{\sum_{i=1}^{N} \frac{1}{x_i}} = \frac{N \prod_{i=1}^{N} x_i}{\sum_{i=1}^{N} \frac{\prod_{j=1}^{N} x_j}{x_i}} \tag{4.9}$$

A weighted version of the harmonic mean is straightforward. The harmonic mean provides the best average when the random variables are rates. For example, if a ship travels a fixed distance at a speed a and then returns at a speed b, then the harmonic mean of a and b is the average speed and will return the actual transit time when divided into the total distance traveled. Further, the harmonic mean is dominated by the smallest values in a set of rvs and cannot be made arbitrarily large by the addition of large values to the set. As a consequence, it is robust to large values but strongly influenced by small ones. MATLAB implements the harmonic mean as **harmmean**(x). Its attributes are identical to those for mean.

The geometric mean estimator is a measure of central tendency based on the product of a set of rvs, where the sample mean uses their sum, and is given by

$$\overline{\mathrm{G}}_N = \left(\prod_{i=1}^{N} x_i\right)^{1/N} = e^{\frac{1}{N}\sum_{i=1}^{N}\log x_i} \tag{4.10}$$

The geometric mean pertains only to positive numbers due to the logarithm in the second form of (4.10). If $\{\mathbf{X}_i\}$ and $\{\mathbf{Y}_i\}$ are two sets of rvs, then the geometric mean of the ratio $\mathbf{X}/\mathbf{Y}$ is the ratio of the geometric means of $\mathbf{X}$ and $\mathbf{Y}$. This makes the geometric mean the appropriate estimator when averaging normalized entities, such as compositions covered in Chapter 11. Further, recall from Section 2.10 that the harmonic mean is always smaller than the geometric mean, which is, in turn, smaller than the arithmetic mean, provided that a set of rvs has entries with different values. MATLAB implements the geometric mean as **geomean**(x). Its behaviors are identical to **mean**.

If a set of N random variables $\{\mathbf{X}_i\}$ is arranged in ascending order so that

$$\mathbf{X}_{(1)} \leq \mathbf{X}_{(2)} \leq \cdots \leq \mathbf{X}_{(N)} \tag{4.11}$$

Then $\mathbf{X}_{(i)}$ is the ith *order statistic*. Even if the unordered $\{\mathbf{X}_i\}$ are iid, the $\{\mathbf{X}_{(i)}\}$ are necessarily dependent because they have been placed in rank order. The order statistics for a set of data are easily obtained using the MATLAB **sort**(x) function.

The sample median is the middle-order statistic given by

$$\tilde{\mathrm{X}}_N = \mathbf{X}_{(\lfloor N/2 \rfloor + 1)} \tag{4.12}$$

The median (4.12) may not be uniquely defined if N is even. A simple workaround in this instance is to define the median as the average of the two adjoining middle-order statistics. MATLAB implements the sample median as **median**(x), with additional options as for mean.

A more general location estimator than the sample median is the pth *sample quantile* given by interpolating the order statistics:

$$q = \hat{F}_N^{-1}(p) = (1-\lambda)x_{(i)} + \lambda x_{(i+1)} \tag{4.13}$$

where $i = \lfloor Np \rfloor$ and $\lambda = Np - \lfloor Np \rfloor$, and $\hat{F}_N^{-1}(x)$ is the inverse of the empirical cdf that is discussed further in Section 4.8.2. The sample median is obtained when $p = 0.5$, and the sample minimum (maximum) is obtained when $p = 0(1)$. The lower and upper sample quartiles are given when p is 0.25 or 0.75.

The *sample mode* is defined as the most frequent or common value in a sample and is most useful for discrete variables and moderate to large data sets. MATLAB implements **mode**(x) that operates on vectors or matrices as for **mean**. If there are multiple values occurring with equal frequency, **mode** returns the smallest of them, and consequently, the function is not suitable for use with multimodal data. It is also unlikely to be accurate with continuous data, where the occurrence of rvs with the same value is rare.

Example 4.1 The file cavendish.dat contains 29 measurements of the density of the Earth as obtained by Henry Cavendish in 1798 using a torsion balance. These data are given in Stigler (1977, table 8), except that the value 4.88 has been replaced with 5.88. The data are presented as a multiple of the density of water and so are dimensionless. Compare the harmonic, geometric, and arithmetic means with the median. What do they suggest? Use the trimmed mean to further investigate the issue.

```
data = importdata('cavendish.dat');
histogram(data, 10)
```

A histogram of the data in Figure 4.1 suggests that there are outlying values at the bottom and top of the distribution, and the remainder of the data appear to be skewed to the left, although the number of data is too small to be definitive.

```
harmmean(data)
ans =
    5.4805
geomean(data)
ans =
   5.4845
mean(x)
ans =
   5.4886
median(data)
ans =
   5.4700
```

The three types of means are in the appropriate order and all cluster around 5.48, which appears to be near the center of the histogram. By contrast, the median is smaller than any of the means. Applying a trimmed mean with a 5% cutoff and the default mode of round will remove one data value at the top and bottom of the distribution.

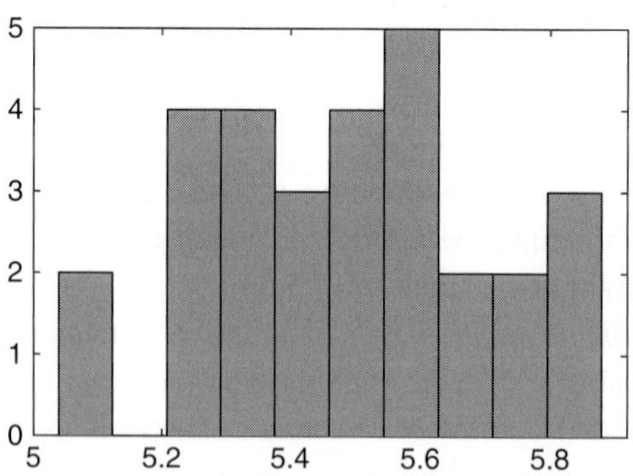

Figure 4.1 Histogram of the Cavendish density of Earth data.

```
trimmean(data, 5)
ans =
  5.4896
```

The result does not shift the mean very much, but the histogram in Figure 4.1 shows that there are multiple data at the extremes of the distribution. Increasing the *percent* parameter in **trimmean** to 20 removes three data each at the top and bottom of the distribution and gives a value of 5.4848.

4.3 Estimators of Dispersion

The estimator for the *sample variance* is given by the sum of squared differences between an rv and its sample mean

$$\hat{s}_N^2 = \frac{1}{N}\sum_{i=1}^{N}\left(\mathbf{x}_i - \bar{\mathbf{x}}_N\right)^2 = \frac{1}{N}\sum_{i=1}^{N}\mathbf{x}_i^2 - \left(\bar{\mathbf{x}}_N\right)^2 \tag{4.14}$$

Since $\mathcal{E}\left(\hat{s}_N^2\right) = (N-1)\sigma^2/N$, the sample variance is biased, but asymptotically unbiased. For this reason, (4.14) is sometimes called the *biased sample variance*. An unbiased estimator for finite N can be obtained by dividing it by $N-1$ instead of N. This is the *unbiased sample variance*

$$\hat{s}'^2_N = \frac{1}{N-1}\sum_{i=1}^{N}\left(\mathbf{x}_i - \bar{\mathbf{x}}_N\right)^2 \tag{4.15}$$

Conceptually, the denominator in (4.15) is smaller by 1 because the sample mean has been computed from the same set of rvs, and in the process imposing a linear relationship among them.

The standard deviation is the square root of (4.14) or (4.15). However, because the square root is not a linear function, the value computed from (4.15) is not unbiased. For Gaussian data, a correction factor is given by

$$\mathcal{E}(\hat{s}_N) = \sqrt{\frac{2}{N-1}}\frac{\Gamma(N/2)}{\Gamma[(N-1)/2]}\sigma \tag{4.16}$$

For $N = 3$, 10, 30, and 100, the correction factors are 0.886, 0.973, 0.991, and 0.998, respectively, and hence the bias is not a serious issue except in the presence of a very small number of data. Note also that the bias in the estimator $\hat{s}_N$ is upward. Alternate correction factors exist for some non-Gaussian distributions but will not be described further.

MATLAB implements the sample variance as **var**(x) and by default is the unbiased estimator (4.15). Adding a second attribute **var**(x, 1) returns the sample estimator (4.14), and **var**(x, w) returns the variance using weights in w that should be positive and sum to unity. When x is a matrix, **var** returns the variance of the columns, but this can be changed by calling it with a third parameter *ndim* that specifies the dimension over which the variance is to be computed. The function **nanvar** ignores NaN values and otherwise operates as for **var**. The function **std** returns the standard deviation given by the square root of the variance and operates as for **var** but does not implement the bias correction factor (4.16). The function **nanstd** ignores NaN values and otherwise operates as **std** does.

The sample median absolute deviation (MAD) is given by

$$\hat{s}_{\mathrm{MAD}} = \left| \mathbf{X} - \mathbf{X}_{(\lfloor N/2 \rfloor + 1)} \right|_{(\lfloor N/2 \rfloor + 1)} \tag{4.17}$$

and can be obtained by two sorts of the rvs. Because it operates on the medians, it is insensitive to extreme data in the tails of the sampling distribution and hence is said to be *robust*.

MATLAB implements **mad**(x). However, the default is to return the mean deviation from the mean given by **mean**(**abs**(x - **mean**(x))), which is not useful as a measure of dispersion unless the data are free of extreme values, in which case the variance or standard deviation operates well and is more familiar (and more efficient). To compute the MAD using medians, it must be called as **mad**(x, 1). A third parameter may be passed to specify the dimension along which the statistic is to be computed as for **var**. The function **mad** treats NaN values as missing data. There is a significant lack of consistency in how MATLAB handles NaNs across different functions, and hence it is important to carefully check the documentation.

An estimator for the range is given by

$$\hat{s}_r = \mathbf{X}_{(N)} - \mathbf{X}_{(1)} \tag{4.18}$$

MATLAB implements the sample range as **range**(x) with the same behaviors as for **mean**. A sample estimator for the interquartile range is

$$\hat{s}_{IQ} = \mathbf{X}_{(3N/4)} - \mathbf{X}_{(N/4)} \tag{4.19}$$

using the order statistics (4.11) and is implemented in MATLAB as **iqr**(x).

MATLAB also provides the function **zscore**(x, *flag*), which returns the standardized form of the rvs in x by subtracting the mean and dividing by the standard deviation. The parameter *flag* is 0 or 1 as the standard deviation is based on the unbiased (default) or biased version of the variance. Matrices are treated as for **var**.

Example 4.2 Returning to the Cavendish density of the Earth data, compare the standard deviation, MAD, and interquartile range for them. Remove the two smallest and largest data values and repeat. What does the result suggest?

```
std(data)
ans =
   0.2154
n = length(data);
sqrt(2/(n - 1))*gamma(n/2)/gamma((n - 1)/2)*std(data, 1)
ans =
   0.2098
```

```
mad(data, 1)
ans =
   0.1500
iqr(data)
ans =
   0.2925
```

The MAD is slightly larger than half the interquartile range, suggesting a distribution that is approximately symmetric. The bias correction to the standard deviation is very small.

```
data = sort(data);
data = data(3:n - 2);
std(data)
ans =
   0.1685
mad(data,1)
ans =
   0.1300
iqr(data)
ans =
   0.2725
```

The standard deviation has shifted downward by about 20% due to the removal of four data points. By contrast, the MAD and interquartile range are changed only slightly, but the interquartile range is now larger than twice the MAD, suggesting that asymmetry has increased.

4.4 Estimators of Shape

Recall from Section 2.12 that skewness is a measure of the asymmetry of a distribution. The sample counterpart measures the distributional asymmetry of a set of rvs and is given by

$$\hat{s}_3 = \frac{\frac{1}{N}\sum_{i=1}^{N}\left(\mathbf{X}_i - \overline{\mathbf{X}}_N\right)^3}{\left[\frac{1}{N}\sum_{i=1}^{N}\left(\mathbf{X}_i - \overline{\mathbf{X}}_N\right)^2\right]^{3/2}} \tag{4.20}$$

Equation (4.20) is a biased estimator. A bias correction is often applied, yielding the so-called unbiased estimator of skewness

$$\hat{s}_3' = \frac{\sqrt{N(N-1)}}{N-2}\hat{s}_3 \tag{4.21}$$

For $N = 3$, 10, 30, and 100, the bias corrections are 4.08, 1.18, 1.05, and 1.02, and hence the bias correction pertains only to small numbers of data but requires at least three elements in the data set. Further, the bias in (4.20) is downward.

Recalling that the kurtosis is a measure of the flatness of a distribution, the sample estimator for the kurtosis of a set of rvs is

$$\hat{s}_4 = \frac{\frac{1}{N}\sum_{i=1}^{N}\left(\mathbf{X}_i - \overline{\mathbf{X}}_N\right)^4}{\left[\frac{1}{N}\sum_{i=1}^{N}\left(\mathbf{X}_i - \overline{\mathbf{X}}_N\right)^2\right]^2} \tag{4.22}$$

As for the skewness, there is a bias-corrected form of the kurtosis given by

$$\hat{s}_4' = \frac{N-1}{(N-2)(N-3)}\left[(N+1)\hat{s}_4 - 3(N-1)\right] + 3 \tag{4.23}$$

The bias correction in (4.23) is affine, in contrast to the skewness, where it is multiplicative, and requires at least four elements in the data set. For a value in (4.22) of 3 (the kurtosis of a Gaussian distribution) and $N = 10$, 30, and 100, the bias-corrected values (4.23) are 4.13, 3.23, and 3.06, respectively, and so the bias in (4.22) is downward.

MATLAB provides functions **skewness** and **kurtosis** to compute these statistics. The function **skewness**(*x*, *flag*) returns (4.20) when *flag* = 1 (default when *flag* is absent) and (4.21) when *flag* = 0. The function **kurtosis** behaves the same way. Both functions handle matrices as for **mean**, and both treat NaNs as missing data.

MATLAB also includes the function **moment**(*x*, *order*), which returns the *order*th central sample moment of the rvs in *x*. The normalization is by *N*, so the function gives the sample variance for *order* = 2. Note that **moment** gives only the numerator of (4.20) or (4.22) and so does not generalize the skewness or kurtosis to higher orders.

A serious issue using the skewness, kurtosis, and moments higher than 4 with real data (and especially earth sciences data) is robustness. As the order of a moment increases, the influence of extreme data rises dramatically. For this reason, it is statistical best practice to examine a given data set using some of the tools in Section 4.8 to characterize their distribution and look for outlying data before blindly using the variance, skewness, or kurtosis as summary statistics.

Example 4.3 Again returning to the Cavendish density of Earth data from Example 4.1, compute and compare the skewness and kurtosis with and without the two smallest and largest data with and without applying the bias correction.

```
data = importdata('cavendish.dat');
skewness(data)
ans =
  0.1067
skewness(data, 0)
```

```
ans =
   0.1126
kurtosis(data)
ans =
   2.3799
kurtosis(data, 0)
ans =
   2.4974
```

Note the ~10% effect that the bias correction has on the skewness. The skewness is positive, implying that the right tail of the data distribution is longer than the left one, which is consistent with Figure 4.1. The excess kurtosis (see Section 2.12) for the bias-corrected sample kurtosis is about −0.6, so the data distribution is *platykurtic*, meaning that it has a broad peak and short tails, again consistent with Figure 4.1.

```
data = sort(data);
data = data(3:n -2);
skewness(data)
ans =
   0.4140
skewness(data, 0)
ans =
   0.4409
kurtosis(data)
ans =
   2.2672
kurtosis(data, 0)
ans =
   2.3809
```

The effect of removing four data is substantial. The skewness shifts upward by about a factor of 4, remaining positive and implying a long right tail, consistent with Figure 4.1. The kurtosis changes downward slightly.

4.5 Estimators of Direction

When a set of rvs represents direction on a circle or sphere, the classical estimators for mean or variance break down because of discontinuities (i.e., 359° is not 358° distant from 1°) and because circular or spherical distributions can be bimodal or multimodal (e.g., consider paleomagnetic data from a rock formation that record both normal and reversed polarities). The natural way to deal with directional data is via trigonometric moments as in Section 2.13. Suppose that N rvs $\{\mathbf{X}_i\}$ are unit vectors with corresponding angles $\{\theta_i\}$. Then the mean direction is the resultant of adding all of the unit vectors together. The Cartesian coordinates of a given $\mathbf{X}_i$ are $(\cos\theta_i, \sin\theta_i)$, and their *vector resultants* are

$$\hat{C} = \sum_{i=1}^{N} \cos\theta_i \tag{4.24}$$

and similarly for $\hat{S}$. Note that the ordering of the rvs in the resultants is immaterial (i.e., exchangeability holds). The *sample mean direction* for θ is

$$\overline{\theta} = \tan^{-1}\left(\frac{\hat{S}}{\hat{C}}\right) \tag{4.25}$$

The *sample median direction* is obtained by finding the value of $\tilde{\theta}$ that minimizes the function

$$\Phi\left(\tilde{\theta}\right) = \pi - \frac{1}{N}\sum_{i=1}^{N} \left|\pi - \left|\theta_i - \tilde{\theta}\right|\right| \tag{4.26}$$

Since (4.26) is not a smooth function of its argument, this can present numerical challenges.

Define $\hat{R}^2 = \hat{C}^2 + \hat{S}^2$. The estimator $\hat{R}$ is the *resultant length* and lies in the range $[0, N]$. The *mean resultant length* is $\overline{R} = \hat{R}/N$ and lies on $[0, 1]$. When $\overline{R} = 1$, all the rvs are coincident, but $\overline{R} = 0$ does not always imply that the rvs are uniformly scattered around the unit circle, and consequently, it is not useful as a measure of dispersion.

The *sample circular variance* is defined as

$$\hat{V} = 1 - \overline{R} \tag{4.27}$$

and has similar limitations to the mean resultant length. However, the *circular standard deviation* is not the square root of (4.27) but rather is given by

$$\hat{v} = \sqrt{-2\log\left(1 - \hat{V}\right)} \tag{4.28}$$

The kth *noncentral trigonometric moment* of the rvs is given by

$$\hat{\mu}'_k = \hat{C}'_k + i\hat{S}'_k = \hat{R}'_k e^{i\hat{\theta}'_k} \tag{4.29}$$

where

$$\hat{C}'_k = \frac{1}{N}\sum_{i=1}^{N} \cos\left(k\theta_i\right) \tag{4.30}$$

and similarly for $\hat{S}_k$. Consequently,

$$\hat{R}'_k = \sqrt{\hat{C}'^2_k + \hat{S}'^2_k} \tag{4.31}$$

and

$$\hat{\theta}'_k = \tan^{-1}\left(\frac{\hat{S}'_k}{\hat{C}'_k}\right) \tag{4.32}$$

The kth *central trigonometric moment* of the rvs is calculated relative to the sample mean direction (4.25)

$$\hat{\mu}_k = \hat{\mathrm{C}}_k + i\hat{\mathrm{S}}_k = \hat{\mathrm{R}}_k e^{i\hat{\theta}_k} \tag{4.33}$$

where

$$\hat{\mathrm{C}}_k = \frac{1}{N}\sum_{i=1}^{N} \cos\left[k\left(\theta_i - \bar{\theta}\right)\right] \tag{4.34}$$

and similarly for the remaining estimators in (4.33). Using trigonometric identities, it can be shown that the first and second central trigonometric moments are given by $\hat{\mu}_1 = \hat{\mathrm{R}}_1 = \overline{\mathrm{R}}$ and $\hat{\mu}_2 = \hat{\mathrm{R}}_2 = \sum_{i=1}^{N} \cos\left[2\left(\theta_i - \overline{\theta}\right)\right]/N$. The *sample circular dispersion* is defined as

$$\hat{\delta} = \left(1 - \hat{\mathrm{R}}_2\right)/\left(2\overline{\mathrm{R}}^2\right) \tag{4.35}$$

and plays a role in hypothesis testing about directional data.

The sample skewness for circular data is given by

$$\hat{s}_3 = \frac{\overline{\mathrm{R}}_2 \sin\left(\overline{\theta}_2 - 2\overline{\theta}\right)}{\left(1 - \overline{\mathrm{R}}\right)^{3/2}} \tag{4.36}$$

and the sample kurtosis is

$$\hat{s}_4 = \frac{\overline{\mathrm{R}}_2 \cos\left(\overline{\theta}_2 - 2\overline{\theta}\right) - \overline{\mathrm{R}}^4}{\left(1 - \overline{\mathrm{R}}\right)^2} \tag{4.37}$$

For the von Mises distribution (3.63), the best estimator (actually, the maximum likelihood estimator, or mle, defined in Section 5.4) for the mean direction $\hat{\mu}$ is (4.25), while the mle for the concentration parameter is obtained by solving

$$\frac{I_1(\hat{\kappa})}{I_0(\hat{\kappa})} = \overline{\mathrm{R}} \tag{4.38}$$

for $\hat{\kappa}$.

All of these estimators generalize to a spherical geometry, although the algebra gets more complicated. For a three-dimensional (3D) rv, there are three Cartesian unit vectors represented as $(\mathbf{X}_i, \mathbf{Y}_i, \mathbf{Z}_i)$ that will be transformed to polar coordinates $(\cos\theta\cos\phi, \cos\theta\sin\phi, \sin\theta)$, where θ is the latitude and ϕ is the longitude. The mean direction on a sphere is obtained by summing the Cartesian unit vectors, computing the resultant length, and obtaining the direction cosines $\hat{s}_x = \sum_{i=1}^{N} x_i/\hat{R}$, and so on. The mean polar coordinates then follow as $\overline{\theta} = \cos^{-1}\hat{s}_z$ and $\overline{\phi} = \tan^{-1}\left(\hat{s}_y/\hat{s}_x\right)$. MATLAB does not provide support for directional data.

Example 4.4 The file paleocurrent.dat (taken from Fisher [1995, app. B.6]) contains 30 values of the cross-bed azimuths of paleocurrents measured in the Bedford Anticline

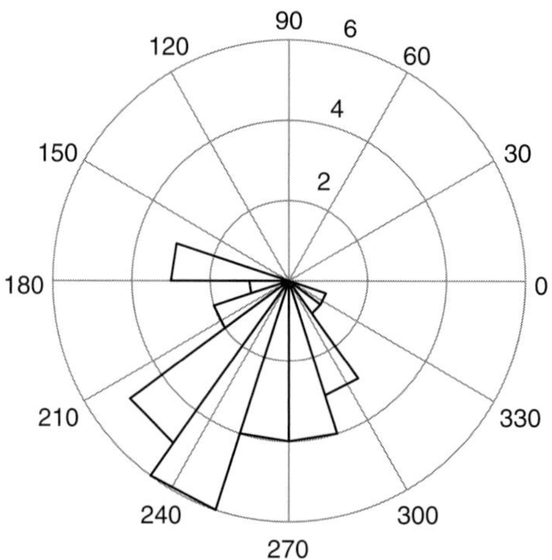

Figure 4.2 Rose diagram of the paleocurrent data.

in New South Wales, Australia. Evaluate the data set and compute the vector resultants, sample mean direction, mean resultant length, circular variance, sample skewness, and sample kurtosis. What do these tell you about the data? Fit a von Mises distribution to the data and qualitatively evaluate the fit.

The MATLAB graphic function **rose**(x) produces a circular histogram of the data shown in Figure 4.2. The data are clustered around 250°, and there may be outliers below 180°.

```
data=importdata('paleocurrent.dat');
rose(pi*data/180)
chat = sum(cosd(data))
chat =
  -8.9432
shat = sum(sind(data))
shat =
  -21.7155
thetabar = atand(shat/chat)
thetabar =
   67.166
```

The values for *shat* and *chat* indicate that the mean direction is in the third quadrant, so 180° needs to be added to this value.

```
thetabar = thetabar + 180
thetabar =
   247.6166
```

This value is consistent with Figure 4.2.

```
rhat=sqrt(chat^2 + shat^2);
rbar = rhat/length(data)
rbar =
   0.7828
vhat = 1 - rbar
vhat =
   0.2172
c2hat = sum(cosd(2*(data - thetabar)))
c2hat =
   10.9547
s2hat = sum(sind(2*(data - thetabar)))
s2hat =
   0.7323
theta2bar = atand(s2hat/c2hat)
theta2bar =
   3.8244
r2bar=sqrt(c2hat^2+s2hat^2)/length(data)
r2bar =
   0.3660
skew = r2bar*sind(theta2bar - 2*thetabar)/(1 - rbar)^1.5
skew =
   -2.7122
kurt = (r2bar*cosd(theta2bar - 2*thetabar) - rbar^4)/(1 - rbar)^2
kurt =
   -13.0959
```

The skewness and kurtosis suggest a distribution that is heavier toward small values of angle and that is platykurtic. A von Mises distribution fit to the data has a mean direction of *thetabar*, and the concentration parameter is easily determined using the nonlinear root finder function **fzero**.

```
fun=@(x) besseli(1, x)./besseli(0, x) - rbar;
kappabar = fzero(fun, 1)
kappabar =
   2.6757
histogram(data, 10, 'Normalization', 'pdf')
hold on
theta = 160:.1:340;
plot(theta, exp(kappabar*cosd(theta - thetabar))/ ...
    (360*besseli(0, kappabar)))
ylabel('Probability Density')
xlabel('Angle')
```

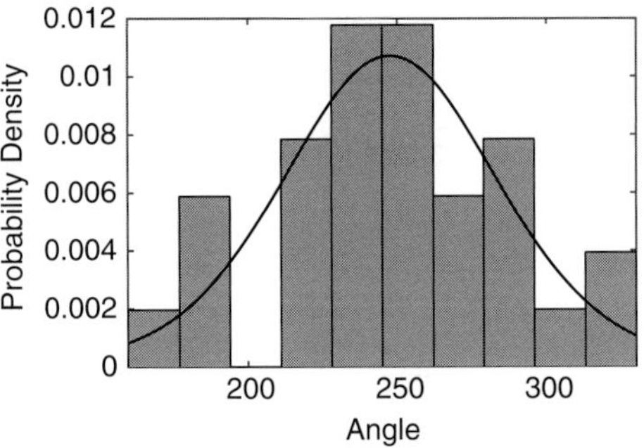

Figure 4.3 Empirical pdf of the paleocurrent data compared to a von Mises distribution with parameters $\bar{\kappa} = 2.6757$, $\bar{\mu} = 247.6166$.

Figure 4.3 compares the fitted von Mises distribution to a histogram of the data with pdf normaliztion. The fit is not particularly good, probably because the data set is fairly small.

4.6 Estimators of Association

The *unbiased sample covariance matrix* computed from a matrix $\overleftrightarrow{\mathbf{x}}$ whose N rows are observations and p columns are variables is

$$cov\left(\overleftrightarrow{\mathbf{x}}\right) = \left(\overleftrightarrow{\mathbf{x}} - \mathbf{j}_N \cdot \overleftrightarrow{\mathbf{x}} / N\right)^H \cdot \left(\overleftrightarrow{\mathbf{x}} - \mathbf{j}_N \cdot \overleftrightarrow{\mathbf{x}} / N\right) / (N - 1) \tag{4.39}$$

where $\mathbf{j}_N$ is a $1 \times N$ vector of ones, and the superscript H denotes the Hermitian (i.e., complex conjugate) transpose so that (4.39) encompasses complex data. The Hermitian transpose becomes the ordinary transpose when the rvs are real. The $p \times p$ covariance matrix is Hermitian for complex and symmetric for real rvs. The quantity $\mathbf{j}_N \cdot \overleftrightarrow{\mathbf{x}} / N$ is a $1 \times p$ vector of the sample means of the columns of $\overleftrightarrow{\mathbf{x}}$. The *biased sample covariance matrix* obtains by replacing the numerator in (4.39) with N.

The function **cov**(x) returns the variance if x is a vector and the covariance matrix if x is a matrix oriented as in (4.39). For matrices, **diag**(**cov**(x)) returns a vector of p variances. When called with two vectors or matrices of the same size, **cov**(x, y) returns their covariance. The default is the unbiased version, but **cov**(x, 1) returns the biased covariance or second-moment matrix of the rvs about their mean. The function **nancov** ignores NaN values and otherwise operates like **cov**. The function **corrcov**(x) converts the covariance matrix in x into the correlation matrix.

The function **corrcoef**(x, y) returns the correlation coefficient with the same behaviors as for **cov**. It has additional options for hypothesis testing that will be described in Chapters 6 and 7.

4.7 Limit Theorems

4.7.1 The Laws of Large Numbers

The *laws of large numbers* are one of the major achievements of probability theory, stating that the mean of a large sample of rvs is close to the expected value of their distribution. This is easy to demonstrate using MATLAB.

```
x = unidrnd(10, 1000, 1);
y = [];
for i = 1:1000
    y = [y mean(x(1:i))];
end
plot(y)
hold on
x = 5.5*ones(size(y));
plot(x)
```

The MATLAB script draws 1000 numbers from the integers between 1 and 10 and then computes the mean for 1 through 1000 variables. The result is shown in Figure 4.4. The population mean is 5.5, and the sums converge to that value fairly rapidly (at about 180 rvs) and then fluctuate around the population value.

There are two versions of the law of large numbers. The *weak law of large numbers* states that if $\{\mathbf{X}_i\}$ are iid with a common population mean μ, then $\overline{\mathrm{X}}_N \xrightarrow{p} \mu$. In other words, the distribution of $\overline{\mathrm{X}}_N$ becomes increasingly concentrated around μ as N increases. The weak law of large numbers is easily proved using Chebyshev's inequality.

The *strong law of large numbers* states that if $\{\mathbf{X}_i\}$ are iid with a common population mean μ, then $\overline{\mathrm{X}}_N \xrightarrow{as} \mu$. The proof of the strong law is more complicated than that for the weak law. If the variables are independent but not identically distributed, then it can be shown that $\overline{\mathrm{X}}_N - \mathcal{E}(\overline{\mathrm{X}}_N) \xrightarrow{as} 0$ under the condition that each rv has finite variance. This is called *Kolmogorov's strong law of large numbers*.

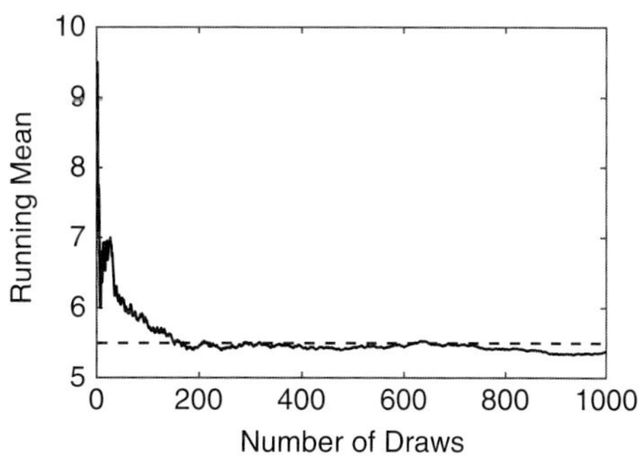

Figure 4.4 Running mean of 1000 random draws from the integers between 1 and 10. The dashed line is the population value of 5.5.

4.7.2 Classic Central Limit Theorems

The *classic central limit theorem* states that a large iid sample drawn from any distribution with a population mean μ and finite variance σ^2 will yield a sample mean $\overline{X}_N$ that is approximately normal with mean μ and variance σ^2/N. More precisely, the Lindeberg-Lévy form of the CLT is given by

$$\frac{\overline{X}_N - \mu}{\sqrt{var\left(\overline{X}_N\right)}} = \sqrt{N}\frac{\overline{X}_N - \mu}{\sigma} \xrightarrow{d} N(0, 1) \tag{4.40}$$

This means that probability statements about the sample mean for rvs from any distribution can be made using the Gaussian distribution provided that the sample is large enough. The simplest proof of the classic CLT uses characteristic functions. A remaining issue is the definition of large, which is difficult to quantify.

Example 4.5 An oceanographer makes 25 independent measurements of the salinity of seawater. She knows that the salinity data are identically distributed but does not know what distribution they are drawn from and wants to know (1) a lower bound for the probability that the average salinity will differ from the true mean by less than $\sigma/5$ and (2) a CLT estimate of the probability.

From the Chebyshev inequality (2.77), $\Pr\left(\left|\overline{X}_N - \mu\right| \geq t\right) \leq \sigma^2/\left(Nt^2\right)$. Let $t = \sigma/5$. Then $\Pr\left(\left|\overline{X}_N - \mu\right| \geq \sigma/5\right) \leq \left(\sigma^2/25\right)/(\sigma/5)^2 = 1$. Therefore, $\Pr\left(\left|\overline{X}_N - \mu\right| < \sigma/5\right) = 0$, which is not a particularly useful bound.

From the central limit theorem (4.40),

$$\begin{aligned}\Pr\left(\left|\overline{X}_N - \mu\right| \leq \sigma/5\right) &= \Pr\left(\sqrt{25}\left|\overline{X}_N - \mu\right|/\sigma \leq 1\right)\\ &= \Phi(1) - \Phi(-1) = 2\Phi(1) - 1 = 0.6826\end{aligned}$$

Example 4.6 Obtain 1000 realizations of 1000 random draws from the uniform distribution on the interval $[-1, 1]$. Plot the empirical pdf for 1, 10, 100, and 1000 point averages. Does the CLT apply? Repeat for 100 realizations of 100 random draws from the Cauchy distribution in standard form. Does the CLT apply? Why or why not?

$\mathcal{E}(\mathbf{X}) = 0$ from the support of the uniform distribution. A MATLAB script and output that evaluates the first part is

```
x = unifrnd(-1, 1, 1000, 1000);
subplot(2, 2, 1); histogram(x(1, :))
subplot(2, 2, 2); histogram(mean(x(1:10, :)))
subplot(2, 2, 3); histogram(mean(x(1:100, :)))
subplot(2, 2, 4); histogram(mean(x))
```

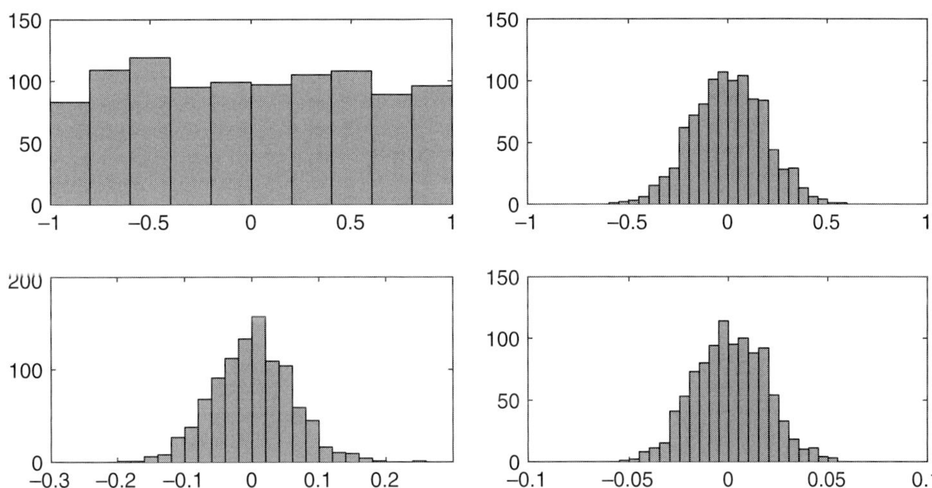

Figure 4.5 Empirical pdfs for 1000 realizations of 1000 random draws from the uniform distribution on $[-1, 1]$ averaged 1 time (upper left), 10 times (upper right), 100 times (lower left), and 1000 times (lower right).

The result is shown in Figure 4.5 and looks like the uniform distribution in the absence of averaging but does look more Gaussian as the number of variables being averaged rises. In addition, the variance decreases, so the distribution is tighter around the mean.

The Cauchy distribution is equivalent to the t distribution with one degree of freedom. The MATLAB script is

```
x = trnd(1, 100, 100);
subplot(1, 3, 1); histogram(x(1, :));
subplot(1, 3, 2); histogram(mean(x(1:10, :)));
subplot(1, 3, 3); histogram(mean(x));
```

The result is shown in Figure 4.6. The peak is not increasingly concentrated as the number of data rises, and in fact, the reverse is true, because more extreme values appear when the averaging rises. The problem is that μ and σ^2 are undefined for the Cauchy distribution, so the classic CLT does not apply.

A more general form of the classic CLT is the *Lyapunov CLT*. Let $\{\mathbf{X}_i\}$ be N independent but not necessarily identically distributed rvs, each of which has mean μ_i and variance σ_i^2. Under certain conditions on the central moments higher than 2, this version of the CLT is

$$\frac{\sum_{i=1}^{N}(X_i - \mu_i)}{\sum_{i=1}^{N}\sigma_i^2} \xrightarrow{d} \mathrm{N}(0,1) \tag{4.41}$$

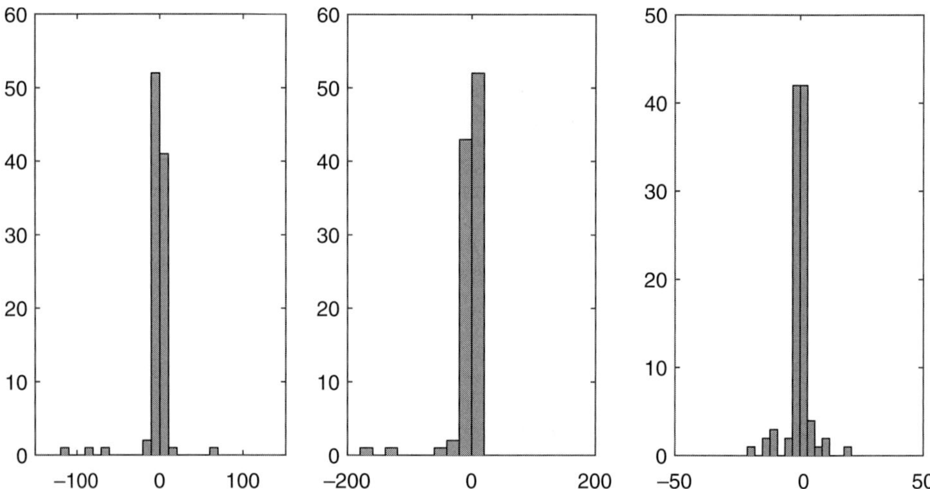

Figure 4.6 Empirical pdfs for 100 realizations of 100 random draws from the standardized Cauchy distribution averaged (from left to right) 1, 10, and 100 times.

The Lyapunov CLT states that the sum of independent but not necessarily identically distributed rvs is approximately Gaussian with mean $\sum_{i=1}^{N} \mu_i$ and variance $\sum_{i=1}^{N} \sigma_i^2$. When the rvs are identically distributed, then this reduces to (4.40). In practice, if the third central moment exists, then the Lyapunov CLT holds.

The CLTs provide an explanation for the empirical observation that the distribution of many physical rvs is Gaussian. However, the CLT should be used with caution as a justification for treating a problem with Gaussian statistics. In particular, in the earth sciences, it is very common to have data that are mostly Gaussian but that have a small fraction of extreme data (in the sense that they would not be expected under a Gaussian model) that often are drawn from distributions with infinite variance, such as the generalized extreme value distribution or the stable distribution family. In that instance, the classic CLT does not apply because the data variance is not finite.

Note that these CLTs give an asymptotic distribution for the sample mean. Convergence is most rapid when the data values are near the Gaussian peak and is much slower in the tails. However, if the third central moment of the rvs exists, then the *Berry-Esseen theorem* shows that convergence occurs at least as fast as $1/\sqrt{N}$. This is not particularly rapid; note that the sample variance converges to the population value as $1/N$.

4.7.3 Other Central Limit Theorems

There are specialized CLTs for the order statistics and directional data, and in addition, there is a generalized CLT that applies to rvs with infinite variance.

Consider the rth order statistic $\mathrm{X}_{(r)}$ from a sample of size N obtained from a continuous cdf $F(x)$. Let $x_p = F^{-1}(p)$ be the p-quantile, where $\lim_{N\to\infty} r/N = p$. Under these assumptions, the CLT for the order statistics is

$$\sqrt{\frac{N}{p(1-p)}} f(x_p)\left(\mathrm{X}_{(r)} - x_p\right) \xrightarrow{d} \mathrm{N}(0,1) \tag{4.42}$$

where $f(x)$ is the pdf corresponding to $F(x)$. Equation (4.42) was proved by Mosteller (1946).

The CLT for directional rvs is a specialization of the CLT for bivariate rvs. Define $\overline{\mathrm{C}}_1 = \hat{\mathrm{C}}/N$ where $\hat{\mathrm{C}}$ is given by (4.24) and similarly for $\overline{\mathrm{S}}_1$. Let C_1 and S_1 denote the population values for $\overline{\mathrm{C}}_1$ and $\overline{\mathrm{S}}_1$. Then, the CLT for circular data is

$$\left(\overline{\mathrm{C}}_1, \overline{\mathrm{S}}_1\right) \xrightarrow{d} \mathrm{N}_2\left[(\mathrm{C}_1, \mathrm{S}_1), \overleftrightarrow{\mathbf{\Sigma}}/N\right] \tag{4.43}$$

Where N_2 is the bivariate Gaussian distribution that was described in Section 3.4.10, and $\overleftrightarrow{\mathbf{\Sigma}}$ is the covariance matrix whose elements can be expressed in terms of the population noncentral trigonometric moments

$$\overleftrightarrow{\mathbf{\Sigma}} = \begin{Bmatrix} \left[1 + \mathrm{C}_2' - 2\left(\mathrm{C}_1'\right)^2\right]/2 & \left(\mathrm{S}_2' - 2\mathrm{C}_1'\mathrm{S}_1'\right)/2 \\ \left(\mathrm{S}_2' - 2\mathrm{C}_1'\mathrm{S}_1'\right)/2 & \left[1 - \mathrm{C}_2' - 2\left(\mathrm{S}_1'\right)^2\right]/2 \end{Bmatrix} \tag{4.44}$$

Let $\{\mathbf{X}_i\}$ be a set of iid variables drawn from $F(x)$ that possibly possesse infinite mean and/or infinite variance. The cdf $F(x)$ is in the *domain of attraction* of a stable distribution with $0 < \alpha < 2$ if there exist constants $a_N > 0$ and $b_N \in \mathcal{R}$ such that (Feller 1971, chap. 6)

$$\frac{\sum_{i=1}^{N} \mathbf{X}_i - b_N}{a_N} \xrightarrow{d} st(\alpha, \beta, 1, 0) \tag{4.45}$$

In other words, a distribution has a domain of attraction if and only if it is stable. Further, the norming constant a_N is proportional to $N^{1/\alpha}$. The *generalized central limit theorem* holds that the sum of rvs drawn from symmetric distributions with power law tails proportional to $-1-\alpha$, where $0 < \alpha < 2$, will converge to a symmetric stable distribution with tail thickness parameter α. When $\alpha > 2$, the sum of rvs converges to a Gaussian distribution.

4.7.4 The Delta Method

The *delta method* is a way to provide an approximation to the distribution of a function of a variable that has an asymptotic Gaussian distribution and hence builds on the CLT. Suppose that the classic CLT (4.40) holds for a set of rvs $\{\mathbf{X}_i\}$ with sample mean $\overline{\mathrm{X}}_N$. Let $\eta(x)$ be a function. The delta method holds that

$$\sqrt{N}\frac{\eta\left(\overline{\mathrm{X}}_N\right) - \eta(\mu)}{\sigma\eta'(\mu)} \xrightarrow{d} \mathrm{N}(0,1) \tag{4.46}$$

The proof follows by expanding $\eta\left(\overline{\mathrm{X}}_N\right)$ in a first-order Taylor series to get

$$\eta\left(\overline{\mathrm{X}}_N\right) \approx \eta(\mu) + \eta'(\mu)\left(\overline{\mathrm{X}}_N - \mu\right) \tag{4.47}$$

The terms can be rearranged to yield

$$\overline{X}_N - \mu \approx \frac{\eta(\overline{X}_N) - \eta(\mu)}{\eta'(\mu)} \tag{4.48}$$

Substituting into (4.40) gives (4.46).

Example 4.7 Suppose that $\{\mathbf{X}_i\}$ are a random sample from a Poisson distribution. Remembering that the mean and variance of the Poisson distribution are identical, the classic CLT states that $\sqrt{N}(\overline{X}_N - \lambda)/\sqrt{\lambda}$ is approximately standard Gaussian for large N. Consider the function $\eta(x) = x^2$. Applying the delta method, $(\overline{X}_N)^2$ is approximately normal with mean λ^2 and variance $\sqrt{N/\lambda^2}/2$.

4.8 Exploratory Data Analysis Tools

4.8.1 The Probability Integral Transform

An rv $\mathbf{X}$ has a continuous distribution for which the cdf is $F(x)$. Define a new rv $\mathbf{U}$ through the transformation

$$u = F(x) \tag{4.49}$$

Then $\mathbf{U}$ has the uniform distribution on [0, 1]. Equation (4.49) is the *probability integral transform*. Its inverse is

$$x = F^{-1}(u) \tag{4.50}$$

because both the cdf and the quantile function are monotone. The proof of the probability integral transformation is straightforward

$$\begin{aligned} \Pr\left[F^{-1}(u) \le x\right] &= \Pr[u \le F(x)] \\ &= F(x) \end{aligned} \tag{4.51}$$

where the last step holds because $\Pr(u \le y) = y$ for a uniformly distributed rv. The probability integral transform is important for computing random variables from arbitrary distributions, in goodness-of-fit testing, and in plotting data against a target distribution.

A simple way to generate random variables from any distribution is *inverse transform sampling*, in which a random number is generated from the uniform distribution on [0, 1] using one of several algorithms that will not be described and applying the inverse

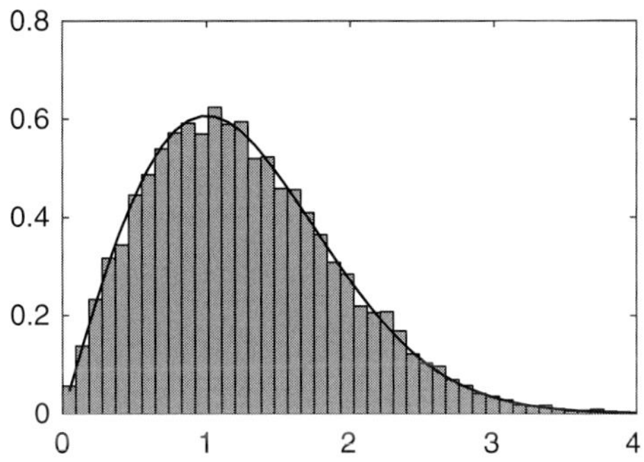

Figure 4.7 Probability histogram of 10,000 random draws from a standardized Rayleigh distribution using the inverse probability integral transform compared with the Rayleigh pdf (solid line).

probability integral transform (4.50) to get random numbers from $F(x)$. As a demonstration that this works, it is easy to generate a set of uniform rvs in MATLAB and apply the inverse transform to get Rayleigh rvs, with the result shown in Figure 4.7.

```
x = unifrnd(0, 1, 10000, 1);
y = raylinv(x, 1);
h = histogram(y, 50, 'Normalization', 'pdf');
hold on
xx = h.BinEdges + h.BinWidth/2;
plot(xx, raylpdf(xx, 1))
```

A more advanced method to generate random variables from any distribution is the *acceptance-rejection method*, which is especially applicable when the target pdf $f(x)$ is difficult to compute. Let $g(x)$ be an auxiliary or instrumental pdf such that $f(x) < \alpha g(x)$, where $\alpha > 1$ is an upper bound on the ratio $f(x)/g(x)$. Samples are taken from $\alpha g(x)$ and accepted or rejected based on a probability test. For example, an implementation due to Von Neumann (1951) uses samples $\mathbf{X}$ from $g(x)$ and $\mathbf{U}$ from the uniform distribution, computation of $f(x)/[\alpha g(x)]$, and then testing $u < f(x)/[\alpha g(x)]$. If the test holds, then x is accepted as a random draw from $f(x)$ and is otherwise rejected. The limitations of acceptance-rejection sampling are that a large number of rejections ensue unless α is chosen accurately or if $f(x)$ is sharply peaked, and inefficiency as the dimensionality of the distribution rises.

4.8.2 The Histogram and Empirical CDF

Let $\{\mathbf{X}_i\}$ be a data sample of size N. Define a set of r *bins* b_i that partition the *data range* $[\min(x_i), \max(x_i)]$ according to some rule $[b_1^L, b_1^U], \ldots, [b_r^L, b_r^U]$. For example, the rule could partition the data range into equal-sized intervals. The midpoints of the bin partitions are called the *bin centers* $(b_1^L + b_1^U)/2, \ldots, (b_r^L + b_r^U)/2$, and their common boundaries are called *bin edges* $b_1^U = b_2^L, \ldots, b_{r-1}^U = b_r^L$. A *frequency histogram* is a plot of the counts of data in each defined bin against a characteristic of the bins such as the bin centers or, more commonly, a bar plot of the data counts against the bin ranges. A *relative frequency histogram* comprises the data counts normalized by the number of data against the bin characteristics. Other normalization algorithms can also be applied.

MATLAB implements the frequency histogram as the function **histogram**(x), and it has been used repeatedly since Chapter 1. It produces a bar plot with the bins uniformly distributed across the data range by default. The number of bins can be controlled using a scalar second-argument *nbins*, and the bin edges can be managed by providing an alternative vector second-argument *edges*. In addition, a series of name-value pairs can be used to more finely control the binning algorithm. A useful one is "Normalization" followed by "count" to produce a frequency histogram (default), "probability" to produce a relative frequency histogram, and "pdf," where the height of each bar is the number of observations divided by N times the width of each bin. There are additional options to produce other types of plots. There are numerous other name-value pairs that provide considerable flexibility to the user.

Let $\{\mathbf{X}_i\}$ be an iid sample of size N from a cdf $F(x)$, and rank them to obtain the order statistics (4.11). A nonparametric sample or *empirical cdf* $\hat{F}_N(x)$ can be constructed by defining it to be the proportion of the observed values that are less than or equal to x. The empirical cdf is easily constructed in terms of the order statistics as

$$\hat{F}_N(x) = \begin{matrix} 0 & x < x_{(1)} \\ \dfrac{k}{N} & x_{(k)} \le x < x_{(k+1)} \\ 1 & x_{(N)} > x \end{matrix} \tag{4.52}$$

An alternative but equivalent definition of the empirical cdf obtains from

$$\hat{F}_N(x) = \frac{\sum_{i=1}^{N} \mathbf{1}(x_i \le x)}{N} \tag{4.53}$$

In either case, the result is a monotonically increasing set of step functions with step size $1/N$ assuming that the data are unique and larger steps if they are not.

The empirical cdf is useful for exploratory data analysis because it approximates the true cdf increasingly well as the number of data increases and hence provides critical input about the data distribution needed for statistical inference. It can also be compared to standard cdfs such as those described in Chapter 3 to assess whether the data are consistent with a given distribution.

MATLAB provides the function $[f, x]$ = **ecdf**(y) that produces the empirical cdf in f at the values in x from the data in y. If it is called without the left-hand side, **ecdf** produces a plot of the empirical cdf. MATLAB also implements the function **ecdfhist**(f, x, m) that produces a probability histogram (hence an empirical pdf) from the output of **ecdf** using m bins (which defaults to 10 if not provided). Alternately, **histogram** will create an empirical cdf if the name-value pair "Normalization," "cdf"is used.

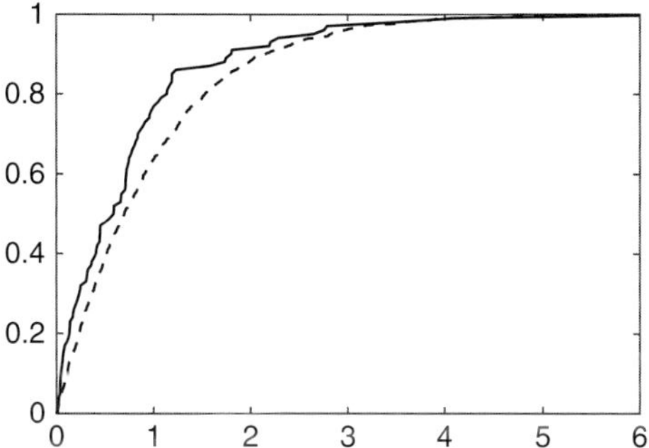

Figure 4.8 The empirical cdf computed from 100 (solid line) and 1000 (dashed line) random draws from the standardized exponential distribution.

Example 4.8 Obtain 100 and 1000 random draws from the standard exponential distribution, and plot the empirical cdf and pdf for each.

```
x100 = exprnd(1, 100, 1);
x1000 = exprnd(1, 1000, 1);
[f1, x1] = ecdf(x100);
[f2, x2] = ecdf(x1000);
plot(x1, f1, x2, f2)
ecdfhist(f1, x1, 20)
ecdfhist(f2, x2, 20)
```

Figures 4.8 and 4.9 show the results. The empirical cdfs and pdfs are much smoother with the larger number of samples, which ought not to be surprising. Figure 4.10 shows the empirical pdf computed using histogram and is identical to Figure 4.9.

Because it depends on a random sample, the empirical cdf is itself an rv and hence has a probability distribution. It is readily apparent that $k = N\hat{F}_N(x) \sim \text{bin}(k; N, F(x))$, so

$$\Pr\left[\hat{F}_N(x)\right] = \frac{k}{N} = \frac{1}{N}\binom{N}{k}[F(x)]^k[1 - F(x)]^{N-k} \tag{4.54}$$

for $k = 0, \ldots, N$, and hence the pdf of the empirical cdf is different for each value of k. Equation (4.54) is $\text{bin}[k; N, F(x)]/N$. The expected value and variance follow directly from the characteristics of the binomial distribution described in Section 3.3.2, yielding $\mathcal{E}\left[\hat{F}_N(x)\right] = F(x)$ and $var\left[\hat{F}_N(x)\right] = F(x)[1 - F(x)]/N$. Consequently, the empirical cdf is an unbiased estimator for the population cdf.

Using Chebyshev's inequality (2.77), it follows that

$$\Pr\left(\left|\hat{F}_N(x) - F(x)\right| \geq t\right) \leq \frac{F(x)[1 - F(x)]}{Nt^2} \tag{4.55}$$

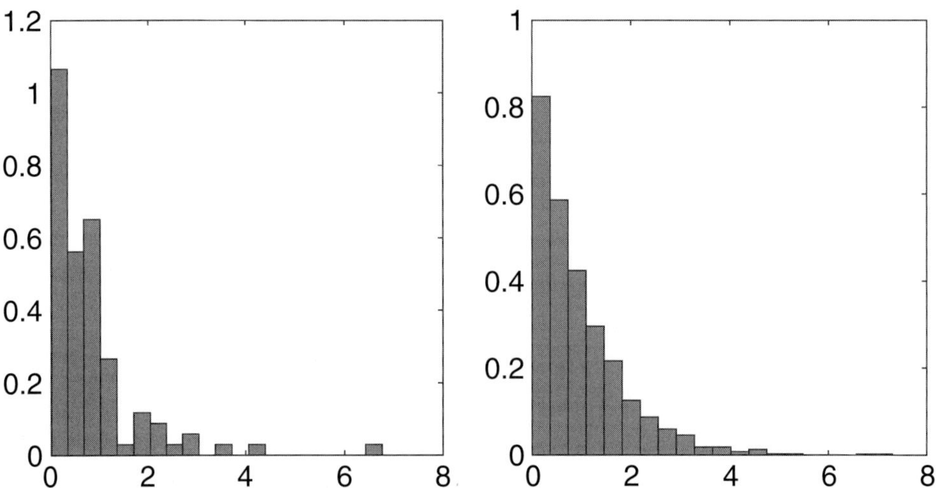

Figure 4.9 Empirical pdfs computed using **ecdfhist** for 100 (left) and 1000 (right) random draws from the standardized exponential distribution.

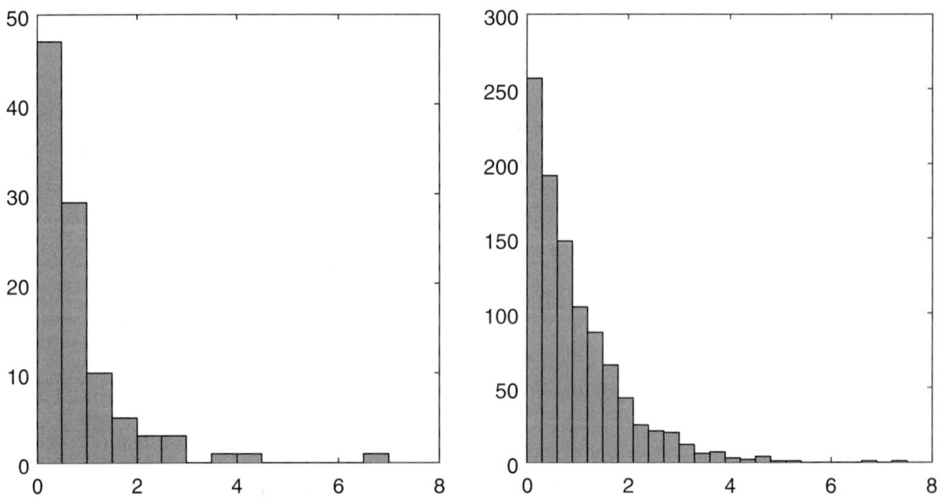

Figure 4.10 Empirical pdfs computed using **histogram** for 100 (left) and 1000 (right) random draws from the standardized exponential distribution.

so $\hat{F}_N(x) \xrightarrow{p} F(x)$. In fact, the stronger statement $\hat{F}_N(x) \xrightarrow{as} F(x)$ holds because of the strong law of large numbers. An even stronger statement follows from the Glivenko-Cantelli theorem, $\sup_x |F_N(x) - F(x)| \xrightarrow{as} 0$, where $\sup_x$ is the *supremum* or the least element that is greater than or equal to all the elements (or, equivalently, the least upper bound). This shows that convergence of the empirical cdf to the true one occurs uniformly. Consequently, the empirical cdf serves as a well-behaved estimator for the population cdf.

A much tighter bound than (4.55) can be placed using Hoeffding's inequality (2.79) because the empirical cdf is a binomial variable:

$$\Pr[|F_N(x) - F(x)| \geq t] \leq 2e^{-2Nt^2} \quad (4.56)$$

From the classic CLT, it follows that the asymptotic distribution of the empirical cdf is Gaussian because

$$\sqrt{N}\left[\hat{F}_N(x) - F(x)\right] \xrightarrow{d} N\{0, F(x)[1 - F(x)]\} \quad (4.57)$$

However, given the existence of the exact distribution for the empirical cdf and the ease with which binomial entities can be computed, it makes limited sense to use the asymptotic form (4.57).

4.8.3 Kernel Density Estimators

Kernel density estimators differ from regular or probability histograms in that they produce a smoothed rather than a binned representation of a distribution. Let $\{\mathbf{X}_i\}$ be a set of iid rvs drawn from an unknown pdf $f(x)$. The kernel density estimator that represents the shape of $f(x)$ is

$$\hat{f}_\delta(x) = \frac{1}{N\delta}\sum_{i=1}^{N} K\left(\frac{x - x_i}{\delta}\right) \quad (4.58)$$

where $K(x)$ is a kernel function that is symmetric and integrates to unity over its support, and $\delta > 0$ is the smoothing bandwidth. The smoothing bandwidth is a free parameter, and it is intuitive to make it as small as possible, but there is a tradeoff between bias and variance that must also be considered. A wide range of kernel functions is in common use, but the standard Gaussian works well in most cases. Alternatives include the Epanechnikov $K(x) = 3(1 - x^2)/4$ on $[-1, 1]$, which is optimally minimum variance, and the half cosine $K(x) = \pi \cos(\pi x/2)/4$ on $[-1, 1]$, among others.

The choice of bandwidth has a significant effect on the result, as will be shown later. For a Gaussian kernel and Gaussian data, the optimal value of the bandwidth is

$$\delta = \left(\frac{4\hat{s}_N^5}{3N}\right)^{1/5} \quad (4.59)$$

More complex analyses in terms of mean integrated squared error can be performed, but in the end they depend on the unknown $f(x)$ and hence are at best approximations.

MATLAB implements the kernel density estimator as **ksdensity**(x). It produces a plot if no arguments are returned and otherwise returns the ordinate and abscissa as $[f, xi]$ = **ksdensity**(x). By default, the function evaluates the kernels at 100 equally spaced points that cover the range of x and uses a Gaussian kernel function. However, there are a wide range of options:

1. Adding a third parameter *bw* to the left-hand side returns the bandwidth.
2. Adding the argument "support" followed with "unbounded" (default) or "positive" or $[a, b]$ specifies the support of the pdf.

3. Adding the argument "kernel" followed by "normal" (default), "box," "triangle," and "epanechnikov" chooses the kernel function. Alternately, a function handle to a user-defined kernel function can be provided.
4. Adding the argument "function" followed by "pdf" (default), "cdf," "icdf," "survivor," or "cumhazard" chooses the type of distribution function to fit.
5. Adding the argument "npoints" followed by a number specifies the number of equally spaced points where the kernel function will be centered. The default is 100.

Example 4.9 Obtain 10,000 random draws, with half from N(0, 1) and half from N(3, 1), compute the kernel density estimator for the data, and then vary the bandwidth and compare with the actual distribution.

```
x = normrnd(0, 1, 1, 5000);
x = [ x normrnd(3, 1, 1, 5000)];
[f, xi, bw]  = ksdensity(x);
plot(xi,f)
bw
bw =
   0.3716
hold on
[f, xi]  = ksdensity(x, 'bandwidth', 0.1);
plot(xi, f, 'k--')
[f, xi]  = ksdensity(x, 'bandwidth', 1);
plot(xi, f, 'k:')
plot(xi, 0.5* (normpdf(xi, 0, 1) + normpdf(xi, 3,1)), 'k-.')
```

In Figure 4.11, the default bandwidth produces a reasonable approximation to the actual pdf and resolves the bimodality well. When the bandwidth is too low, the pdf

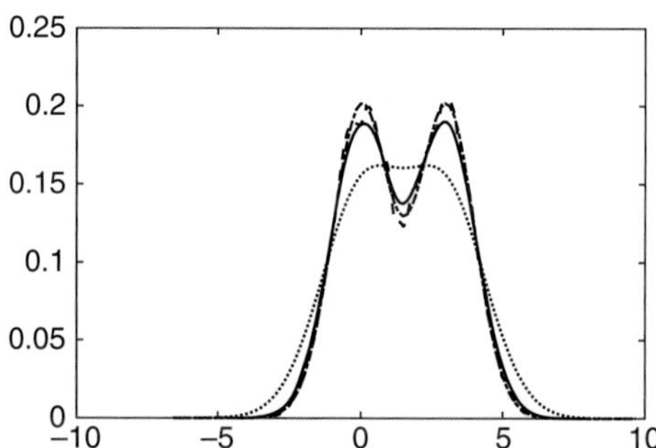

Figure 4.11 Kernel density pdf estimates computed for 5000 N(0, 1) + 5000 N(3,1) random draws using a Gaussian kernel function with default bandwidth of 0.3716 (solid line), 0.1 (dashed line), and 1.0 (dotted line) compared with the actual pdf evaluated at the kernel density estimator abscissa values (dash-dot line).

appears erratic, whereas if it is too high, it fails to show the bimodality. The actual pdf appears as a set of line segments because there are only 100 points along the abscissa.

Example 4.10 Obtain 10,000 random draws from the arcsine distribution, and compute the pdf with support that is unbounded and confined to [−1, 1]. The arcsine distribution is the beta distribution with parameters ½, ½. Repeat for the cdf.

```
x = betarnd .5, .5, 10000, 1);
[f, xi] = ksdensity(x);
plot(xi, f)
hold on
[f, xi] = ksdensity(x, 'support', [0 1]);
plot(xi, f, 'k--')
plot(xi, betapdf(xi, .5, .5), 'k:')
```

In the absence of any information about the support of the distribution, tapered tails appear at the bottom and top of the distribution in Figure 4.12, and it violates the properties of the beta distribution. When the support is limited to [0, 1], the distribution looks almost like the analytic form. However, the left and right endpoints of the arcsine distribution are infinite, and there is a mismatch between the kernel density and the exact pdf at these points that is inevitable for a numerical result. Consequently, using the kernel density estimator for any sort of quantitative inference could be dangerous, and the kernel density estimator certainly will not integrate to unity. Further, forcing it to integrate to unity will introduce a large mismatch into the center of the distribution.

The kernel density estimator for the cdf in Figure 4.13 exceeds the lower and upper support bounds in the absence of constraints but agrees very well with the analytic values once they are in place.

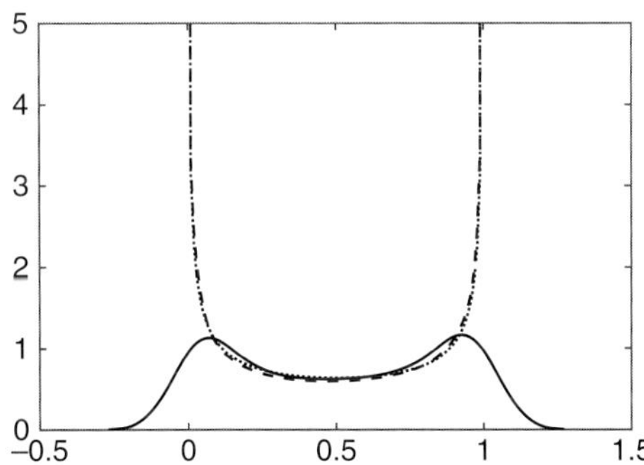

Figure 4.12 Random draws from the arcsine distribution fit by a kernel density estimator for the pdf using a Gaussian kernel function with unbounded support (solid line) and support bounded to [0,1] (dashed line) compared with the actual pdf (dotted line).

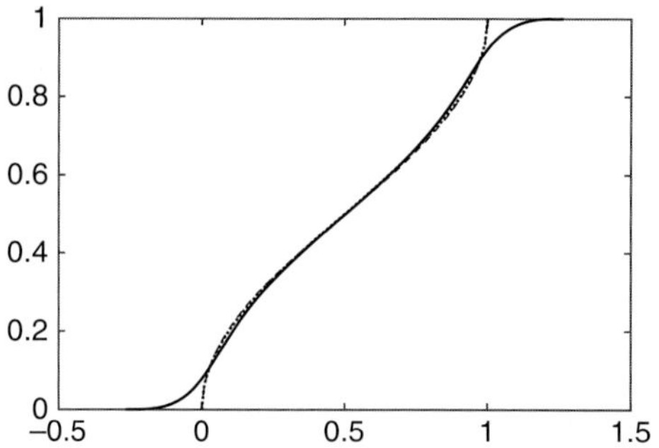

Figure 4.13 Random draws from the arcsine distribution fit by a kernel density estimator for the cdf using a Gaussian kernel function with unbounded support (solid line) and support bounded to [0,1] (dashed line) compared with the actual cdf (dotted line).

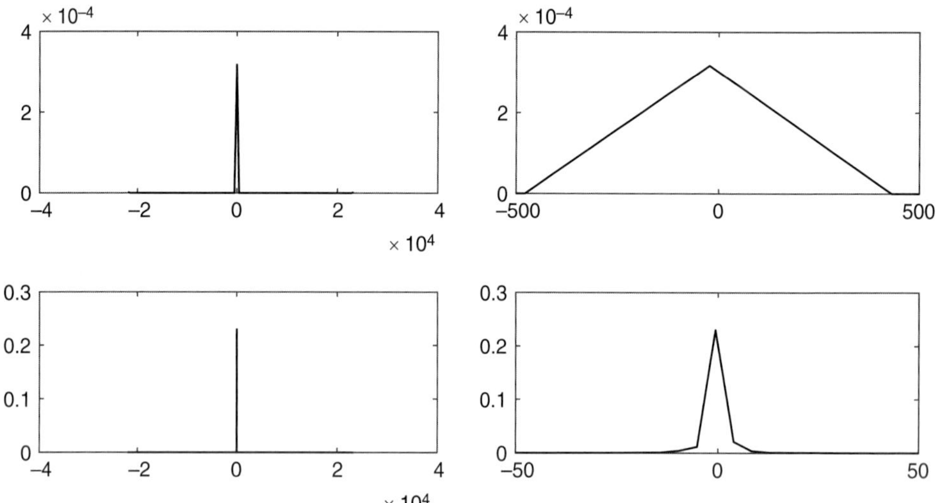

Figure 4.14 Kernel density estimators for 10,000 random draws from the Cauchy distribution using the default 100 points estimated evenly across the support (top left), the center of that kernel density estimate (top right), and the kernel density estimator using 10,000 points estimated evenly across the support (bottom left) and the center of that estimate (bottom right).

Example 4.11 Obtain 10,000 random draws from the Cauchy distribution, and compute a kernel density estimator for them. The Cauchy distribution is equivalent to the t distribution with one degree of freedom.

Figure 4.14 shows the result. One would anticipate difficulty because the Cauchy distribution has algebraic tails, and hence a small number of kernel density estimates evenly distributed across the support is not adequate for its characterization. The top two panels of the figure confirm this assertion. Increasing the number of kernel density estimates to 10,000 produces a much cleaner result with the penalty of more computation.

4.8.4 The Percent-Percent and Quantile-Quantile Plots

The *percent-percent* or *p-p plot* is a very useful qualitative tool for assessing the distribution of a data sample. It can be made quantitative by adding error bounds and assessing the goodness-of-fit, as will be described in Chapter 7. The p-p plot is based on the probability integral transform of Section 4.8.1, consisting of a plot of the uniform quantiles against the order statistics of a data set transformed using a target cdf. If the target cdf is the correct one for the data, then the plot should approximate a straight line. Deviations from a straight line are suggestive of the nature of the distribution and hence can be used to refine the target distribution. The p-p plot is most sensitive at the mode of the distribution and hence is appropriate for evaluating heavy-tailed distributions but is not very useful for detecting outliers.

Let $\{\mathbf{X}_{(i)}\}$ be the N-order statistics computed for a set of rvs, and let the quantiles of the uniform distribution be

$$u_i = (i - 1/2)/N \qquad i = 1, \ldots, N \tag{4.60}$$

If $F^*(x)$ is a target cdf of the location-scale type, then the p-p plot consists of (4.60) plotted against $F^*\left[\left(x_{(i)} - \hat{\mu}\right)/\hat{\sigma}\right]$, where $\hat{\mu}$ and $\hat{\sigma}$ are location and scale estimates.

A useful variant on the p-p plot was introduced by Michael (1983). The standard p-p plot has its highest variance near the distribution mode and its lowest variance at the tails. An arcsine transformation applied to (4.60) gives

$$r_i = 2\sin^{-1}\left(\sqrt{u_i}\right)/\pi \tag{4.61}$$

whose probability distribution is

$$r_i \sim \pi \sin(r)/2 \tag{4.62}$$

for $0 \le r \le 1$. The order statistics of data distributed as (4.62) have the same asymptotic variance. Consequently, use of a sine transformation realizes a stabilized p-p plot that has constant variance at all points. The variance-stabilized p-p plot consists of (4.61) plotted against

$$s_i = \frac{2}{\pi}\sin^{-1}\sqrt{F^*\left[\left(x_{(i)} - \hat{\mu}\right)/\hat{\sigma}\right]} \tag{4.63}$$

The variance-stabilized p-p plot is superior to the ordinary p-p plot and is recommended for standard use. MATLAB does not provide support for either the standard or variance-stabilized p-p plot.

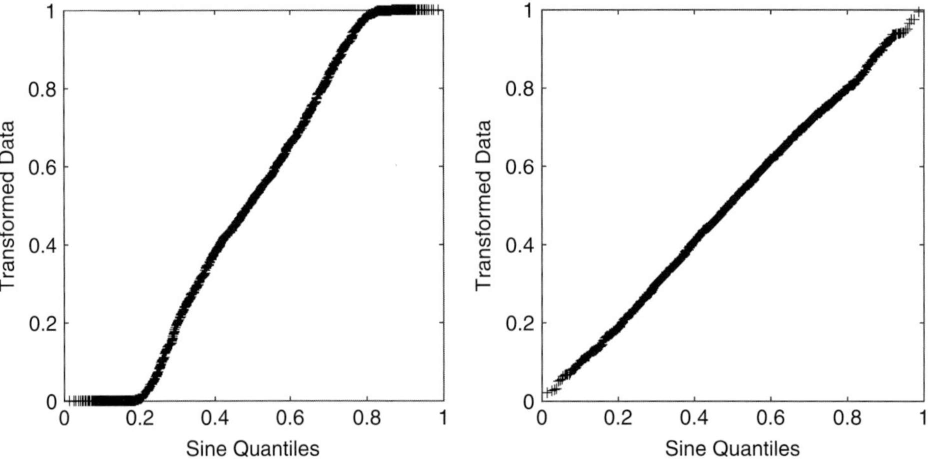

Figure 4.15 Variance-stabilized p-p plots for 1000 random draws from the standard Cauchy distribution using the standard Gaussian (left) and the standard Cauchy (right) as the target.

Example 4.12 Compute 1000 random draws from the standardized Cauchy distribution, and produce stabilized p-p plots of them against the standardized normal and Cauchy distributions. What differences do you see?

```
x = trnd(1, 1, 1000);
u = ((1:1000) - 0.5)/1000;
r = 2/pi*asin(sqrt(u));
s = 2/pi*asin(sqrt(normcdf(sort(x))));
subplot(1, 2, 1); plot(r, s, 'k+ ')
axis square
s = 2/pi*asin(sqrt(tcdf(sort(x), 1)));
subplot(1, 2, 2); plot(r, s, 'k+ ')
axis square
```

Figure 4.15 shows the result. The stabilized p-p plot against the normal cdf is long tailed, as evidenced by a sharp downturn at the distribution tails, and there are more subtle differences near the distribution mode. The plot using the Cauchy as the target is a straight line, as expected.

Example 4.13 Compute 1000 random draws from the standard Rayleigh distribution. Produce variance-stabilized p-p plots using the lognormal and generalized extreme value distributions with mle parameters obtained from **lognfit** and **gevfit** and the standard Rayleigh cdf. What are the differences, and why?

Figure 4.16 shows the variance-stabilized p-p plots. The lognormal distribution has a deficit of probability at its lower end and an excess of probability around the mode compared with the Rayleigh. It then has a deficit of probability above the mode but

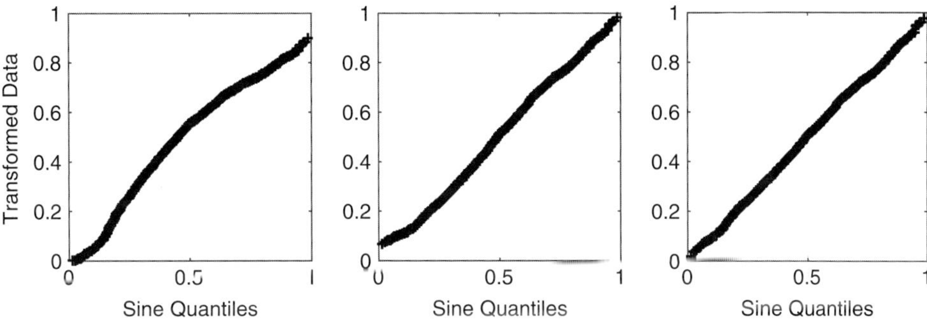

Figure 4.16 Variance-stabilized p-p plots for 1000 random draws from the standard Rayleigh distribution plotted against a target standard lognormal distribution (left), a target standard generalized extreme value distribution (middle), and a standard Rayleigh target distribution (right).

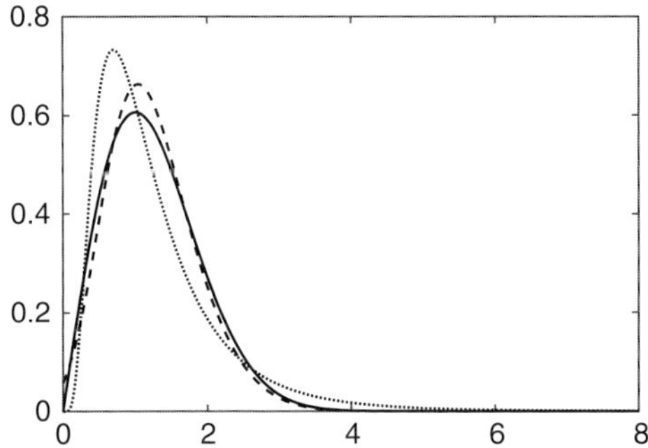

Figure 4.17 Probability density functions for the three target distributions of Figure 4.16. The standard Rayleigh pdf is shown as a solid line, the lognormal with mle parameters is shown as a dashed line, and the generalized extreme value distribution with mle parameters is shown as a dotted line.

ultimately has a fatter tail. Consequently, the initial slope is less than 1 and then increases toward the mode, resulting in increasing slope, but the slope decreases beyond that point and then turns upward. The generalized extreme value distribution is very similar to the Rayleigh but has a slight deficit of probability at the bottom of the distribution followed by an excess around the mode. This results in subtle upward curvature at the bottom of the distribution and downward curvature at the middle. The Rayleigh plot is, of course, a straight line. Figure 4.17 shows the corresponding pdfs.

The *quantile-quantile* or *q-q plot* is a very useful qualitative tool for assessing the distribution of a data sample. It can be made quantitative by adding error bounds and assessing the goodness-of-fit, as will be described in Chapter 7. It consists of a plot of the N quantiles of a target distribution against the order statistics computed from the sample, where the latter may be scaled for easier interpretation. The target distribution quantiles are given by the target quantile function applied to the uniform quantiles (4.60), and the order statistics follow from a sort and optional scaling of the data. The q-q plot emphasizes the distribution tails and hence is a complement to the p-p plot that is very useful for studying light-tailed data and detecting outliers, but it is less useful for systematically heavy-tailed data. If a q-q plot is a reasonable approximation to a straight line, then the data are drawn from the distribution corresponding to the quantile function. Curvature or other departures from linearity are clues to defining the correct data distribution. Finally, a small fraction of unusual data will typically plot as departures from linearity at the extremes of the q-q plot, and hence it is a useful tool to detect data that are not consistent with the bulk of the sample.

MATLAB provides the function **qqplot**(x) that produces a plot of the standard normal quantiles against the order statistics of the data in x. It also allows two samples to be compared using **qqplot**(x, y). For other distributions, a q-q plot is easily obtained from the script

```
n = length(x)
u = ((1:n) - 0.5)/n;
q = expinv(p);
plot(q, sort(x), '+ ')
```

where any quantile function can be substituted for the exponential one. The probability as defined in the first line ascribes $1/(2N)$ to the first and last intervals and $1/N$ to the remaining ones so that **sum**(u) = 1.

Example 4.14 Compute 1000 random draws from the standardized Cauchy distribution, and produce q-q plots of them against the standardized normal and Cauchy distributions. What differences do you see?

```
u = ((1:10000) - 0.5)/10000;
x = trnd(1, 1, 10000);
subplot(1, 2, 1); plot(norminv(u), sort(x), 'k+')
axis square
subplot(1, 2, 2); plot(tinv(u, 1), sort(x), 'k+')
axis square
```

Figure 4.18 shows the results. The Cauchy distribution has algebraic tails, whereas the Gaussian has exponential ones; hence the q-q plot against normal quantiles shows sharp upward curvature at the tail extremes. Note that the q-q plot is dominated by a small number of extreme values in this case and that the sense of curvature caused by long tails has the opposite sense for q-q than for p-p plots (i.e., compare Figures 4.15 and 4.18). The Cauchy q-q plot is close to a straight line, but note that the bulk of the data appears over a small part of the center of the plot. This is a limitation of the q-q plot for heavy-tailed data.

Example 4.15 Compute 1000 random draws from the standard Rayleigh distribution. Produce q-q plots using the lognormal and generalized extreme value distributions with mle parameters and the standard Rayleigh cdf. What are the differences, and why?

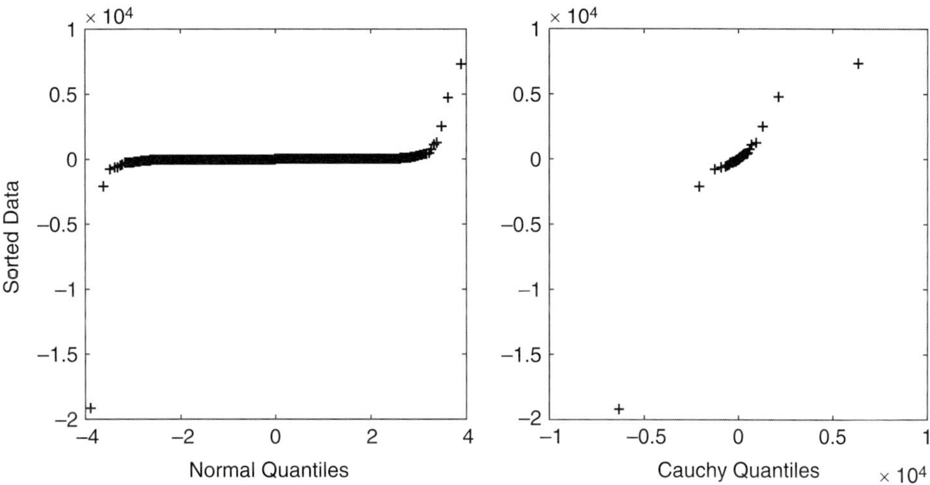

Figure 4.18 Quantile-quantile plots for 10,000 random draws from the standard Cauchy distribution using the standard Gaussian cdf as the target (left) and the standard Cauchy cdf as the target (right).

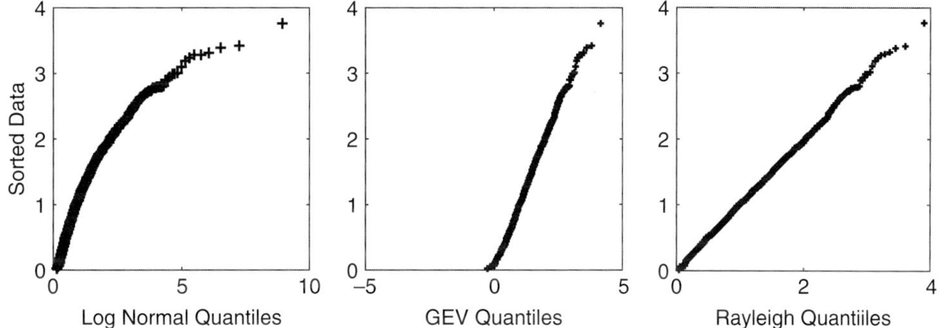

Figure 4.19 Quantile-quantile plots for 1000 random draws from the standard Rayleigh distribution plotted against a target lognormal distribution with mle parameters (left), a target generalized extreme value distribution with mle parameters (middle), and a standard Rayleigh target distribution (right).

Figure 4.19 shows the result. The lognormal distribution initially has a surplus of probability followed by a deficit of probability that is manifest as an initial steep positive slope that then decreases. The generalized extreme value distribution exists for negative argument, although this is masked in the p-p plot of Figure 4.15. An initial probability deficit is seen as upward curvature at the bottom of the distribution. The Rayleigh plot is a straight line.

In many instances, the outcome of a statistical procedure requires the identification and elimination of anomalous data. This process is called *censoring*, and accurate statistical inference requires that the distribution of the data must be truncated to reflect it. Let $f_X(x)$ be the pdf of a random variable $\mathbf{X}$ before censoring. After truncation, the pdf of the censored random variable $\mathbf{X}'$ becomes

$$f_{x'}(x') = \frac{f_x(x)}{F_x(d) - F_x(c)} \tag{4.64}$$

where $c \leq x' \leq d$, and $F_X(x)$ is the cdf for $\mathbf{X}$. Let the original number of rvs in $\mathbf{X}$ be N, the number of data in $\mathbf{X}'$ after censoring be m, and the number of data censored from the lower and upper ends of the distribution be m_1 and m_2, respectively, so that $m = N - m_1 - m_2$. Then suitable choices for c and d are the m_1th and $N - m_2$th quantiles of the original distribution $f_X(x)$. The m quantiles of the truncated distribution can be computed from that of the original distribution using

$$F_x^{-1}\left(Q_j\right) = [F_x(d) - F_x(c)]\frac{j - 1/2}{m} + F_x(c) \tag{4.65}$$

Equation (4.65) should be used to produce p-p or q-q plots if data have been censored.

4.8.5 Simulation

In many cases, the distribution of combinations of variables like products or quotients are of theoretical interest, but in most cases, obtaining an analytic expression for the distribution is intractable. In such a case, a qualitative or quantitative understanding can be achieved by simulation.

Example 4.16 Suppose that the distribution of the ratio of a standardized Rayleigh to a standardized Gaussian rv is of interest. By drawing an equally sized set of Rayleigh and Gaussian rvs and then computing their point-by-point quotient, a set of rvs is generated from the distribution of interest and can then be characterized using a kernel density estimator or other assessment tool.

```
x = raylrnd(1, 10000, 1);
y = normrnd(0, 1, 10000, 1);
z = x./y;
ksdensity(z, 'npoints', 10000)
```

Figure 4.20 shows the result truncated at abscissa values of ± 75 (the actual simulation extends from -2500 to 17,500). The resulting pdf is very heavy tailed and sharply peaked at the origin, and even with 10,000 kernel centers, the middle of the distribution is not adequately characterized.

This example illustrates an important principle: in many cases the ratio of two rvs will result in a distribution with algebraic tails and infinite variance. This occurs because the ratio of two distributions is the same as the product of a first distribution with the inverted form of a second distribution. Lehmann & Schaffer (1988) reviewed the characteristics of

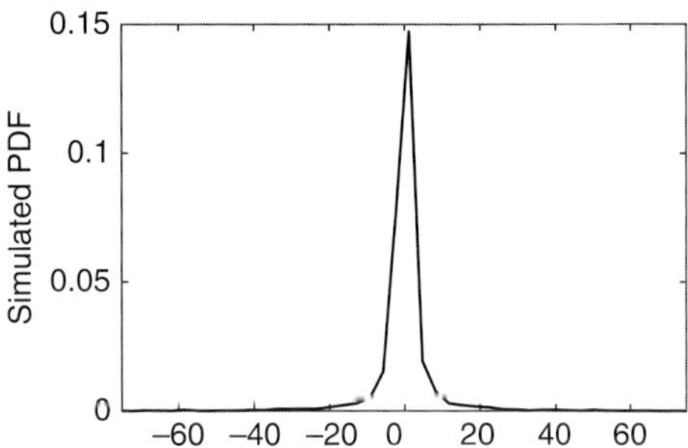

Figure 4.20 Simulation of the ratio of standardized Rayleigh and Gaussian variables presented as a kernel density estimate with 10,000 kernel centers. The abscissa has been truncated at ±75.

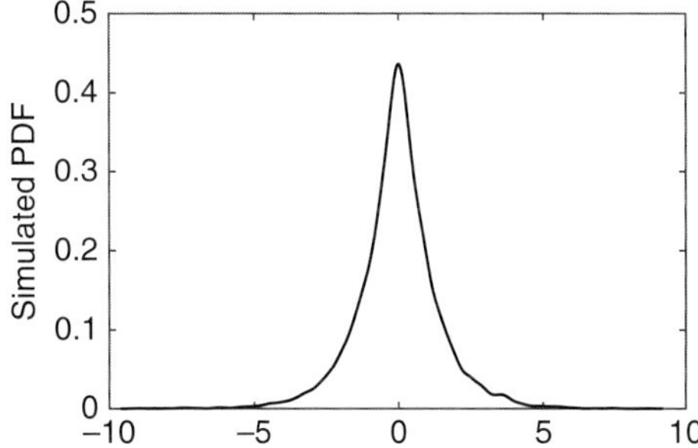

Figure 4.21 Simulation of the product of standardized Rayleigh and Gaussian variables presented as a kernel density estimate with 100 kernel centers.

inverted distributions, the most relevant of which is a proof that the inverted probability density function $g(y)$ of the random variable $\mathbf{Y} = 1/\mathbf{X}$ given the original distribution $f(x)$ has the property

$$\lim_{y\to\infty} \frac{g(y)}{1/(1+y^2)} = f(0^+) \tag{4.66}$$

Consequently, when the left side of (4.66) tends to a finite constant, the inverted distribution has a right Cauchy tail, and it is lighter or heavier than Cauchy when $f(0^+)$ is zero or infinity. An analogous relationship holds for the left tail. This establishes Cauchy tails (hence infinite mean and variance) for the inverted distributions of many common random variables such as Gaussian, Student's t, and exponential ones. However, when $f(0^+) = 0$, the infinite variance property does not always hold; suitable examples with finite variance include many inverted gamma and beta distributions.

Repeating the simulation for the product rather than the ratio of the two distributions gives a well-behaved distribution with exponential tails, as in Figure 4.21, and uses the default value of 100 for the number of kernel centers.

4.9 Sampling Distributions

A *sampling distribution* is the distribution of a statistic based on a set of observations of an rv and can, in principle, be derived from the joint distribution of the rvs. It is the distribution of the statistic for all possible samples of a specified size from a given population and hence depends on the distribution of the population, the sample size, and the statistic of interest. Sampling distributions are important because they allow statistical inferences to be drawn from them rather than from the joint distribution of the rvs, which usually simplifies calculations considerably.

Example 4.17 If N rvs $\{\mathbf{X}_i\}$ are drawn from $\mathrm{N}(\mu, \sigma^2)$, then $\overline{\mathrm{X}}_N$ is distributed as $\mathrm{N}(\mu, \sigma^2/N)$. The sampling distribution for the sample mean is $\mathrm{N}(\mu, \sigma^2/N)$.

Example 4.18 If N rvs $\{\mathbf{X}_i\}$ have the cdf $F(x)$, then the sampling distribution for $\max(\mathbf{X}_i)$ is $NF(x)^{N-1}f(x)$.

Example 4.19 If N_1 rvs $\{\mathbf{X}_{1i}\}$ are drawn from $\mathrm{N}(\mu_1, \sigma_1^2)$ and N_2 rvs $\{\mathbf{X}_{2i}\}$ are drawn from $\mathrm{N}(\mu_2, \sigma_2^2)$, then the sampling distribution for $\overline{\mathrm{X}}_{N,1} - \overline{\mathrm{X}}_{N,2}$ is $\mathrm{N}(\mu_1 - \mu_2, \sigma_1^2/N_1 + \sigma_2^2/N_2)$.

Example 4.20 If N rvs $\{\mathbf{X}_i\}$ are each the square root of the sum of the squares of two uncorrelated Gaussian variables with zero mean and a common variance, then their sampling distribution is Rayleigh.

There are four important sampling distributions that are widely used in applied statistics: the chi square, Student's t, F, and correlation coefficient distributions, along with their noncentral variants.

4.9.1 Chi Square Distributions

The *chi square distribution* (sometimes chi squared or χ^2) is the distribution of the sum of squares of random samples from the standard normal distribution and was first derived by Pearson (1900). When there is a need to distinguish it from the noncentral chi square distribution, it is called the *central chi square distribution*. It plays a major role in likelihood ratio hypothesis testing, goodness-of-fit testing, and in the construction of

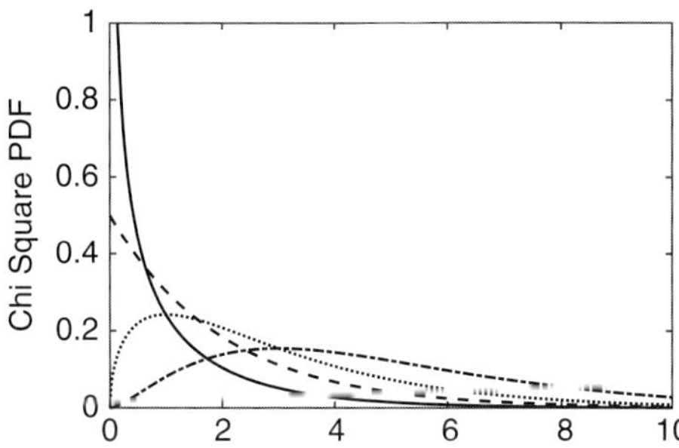

Figure 4.22 The chi square pdf with $v = 1$ (solid line), 2 (dashed line), 3 (dotted line), and 5 (dash-dot line).

confidence intervals. As noted in Section 3.4.5, the chi square distribution is the gamma distribution with $\alpha = v/2, \beta = 1/2$, where v is usually an integer in applications, although this is not a theoretical requirement. Consequently, it shares all the properties of the gamma distribution.

If the means of the normal rvs used to compute the chi square statistic are not zero, the corresponding distribution becomes *noncentral chi square*, as derived by Fisher (1928). The noncentral chi square distribution plays an important role in hypothesis testing, notably in defining the alternate hypothesis to estimate the power of a test when the null distribution is chi square. These two distributions will be considered together in this section.

Let $\{\mathbf{X}_i\}$ be N independent rvs from $\mathrm{N}(0, 1)$. Then the rv $\mathbf{Y}_N = \sum_{i=1}^N \mathbf{X}_i^2$ is distributed as chi square with N degrees-of-freedom and is denoted as $\mathbf{Y}_N \sim \chi_N^2$. The chi square pdf is

$$\mathrm{chi2}(x; v) = \frac{1}{2^{v/2}\Gamma(v/2)} x^{(v/2)-1} e^{-x/2} \qquad x \geq 0 \tag{4.67}$$

where v is called the *degrees-of-freedom*. The gamma function $\Gamma(x)$ can be expressed in closed form when v is an integer because $\Gamma(k) = (k-1)!$ and $\Gamma(1/2) = \sqrt{\pi}$, and hence (4.67) can be simplified in that instance. Figure 4.22 shows representative chi square pdfs.

Let $\mathbf{X}_i \sim \mathrm{N}\left(\mu_i, \sigma_i^2\right)$, and define $\mathbf{Y}_v' = \sum_{i=1}^v (\mathbf{X}_i/\sigma_i)^2$. The distribution of $\mathbf{Y}_v'$ is

$$\mathrm{nchi2}(x; v, \lambda) = \frac{1}{2} e^{-(x+\lambda)/2} \left(\frac{x}{\lambda}\right)^{(v-2)/4} I_{v/2-1}\left(\sqrt{\lambda x}\right) \tag{4.68}$$

where the noncentrality parameter is

$$\lambda = \sum_{i=1}^{v} \left(\frac{\mu_i}{\sigma_i}\right)^2 \tag{4.69}$$

and $I_v(x)$ is a modified Bessel function of the first kind. Figure 4.23 shows the noncentral chi square distribution for three degrees-of-freedom and a variety of noncentrality parameters.

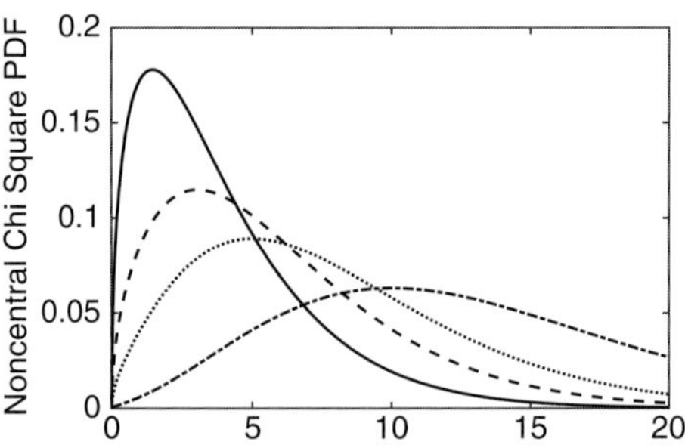

Figure 4.23 The noncentral chi square distribution with three degrees-of-freedom and noncentrality parameters of 1 (solid line), 3 (dashed line), 5 (dotted line), and 10 (dash-dot line).

The central chi square cdf is

$$\mathrm{Chi2}(x; \nu) = \frac{\gamma(x/2, \nu/2)}{\Gamma(\nu/2)} \tag{4.70}$$

which is the incomplete gamma function ratio. The noncentral chi square cdf is a weighted average of central chi square distributions, resulting in a fairly complicated expression that is omitted.

The expected value of the central chi square distribution is ν, and the variance is 2ν. The skewness and kurtosis are $\sqrt{8/\nu}$ and $3 + 12/\nu$. Hence the distribution is leptokurtic, and the heaviness of the right tail decreases with rising ν. The geometric mean is $2e^{\psi(\nu/2)}$ [where $\psi(x)$ is the digamma function], and the harmonic mean is $2\Gamma(\nu/2)/\Gamma(\nu/2 - 1)$ for $\nu > 2$ and is otherwise undefined. The mode is at the origin for $\nu \le 2$, and the distribution is J-shaped, whereas the distribution is unimodal with the mode at $\nu - 2$ thereafter.

The expected value of the noncentral chi square distribution is $\nu + \lambda$, and the variance is $2(\nu + 2\lambda)$. The skewness and kurtosis are $2\sqrt{2}(\nu + 3\lambda)/(\nu + 2\lambda)^{3/2}$ and $3 + 12(\nu + 4\lambda)/(\nu + 2\lambda)^2$, respectively. Each of these entities reduces to the central value when $\lambda = 0$. The mode of the noncentral chi square distribution occurs at the value of x that satisfies $\mathrm{nchi2}(x; \nu, \lambda) = \mathrm{nchi2}(x; \nu - 2, \lambda)$.

By the CLT, a standardized central chi square variate satisfies

$$\frac{\chi_N^2 - \nu_N}{\sqrt{2\nu_N}} \xrightarrow{d} \mathrm{N}(0, 1) \tag{4.71}$$

with $\nu > 50$ being sufficiently large for the Gaussian approximation to be accurate. The limiting distribution of a standardized noncentral chi square variable is $\mathrm{N}(0, 1)$ for either fixed λ as $\nu \to \infty$ or fixed ν as $\lambda \to \infty$.

By the additivity property of the gamma distribution family, if N rvs $\{\mathbf{X}_i\}$ are independent, and each is chi square distributed with ν_i degrees-of-freedom, then $\sum_{i=1}^{N} \mathbf{X}_i$ is also chi square with total degrees-of-freedom $\nu_N = \sum_{i=1}^{N} \nu_i$. This makes it straightforward to calculate statistics that combine many chi square entities, such as when assessing the fit of a model to data where the misfit of a given datum is measured as the squared difference between it and the model value. The noncentral chi square distribution shares the additivity property with the central version: if N rvs $\{\mathbf{X}_i\}$ are independent and each is noncentral chi square distributed with noncentrality parameter λ_i and v_i degrees-of-freedom, then $\sum_{i=1}^{N}\mathbf{X}_i$ is also noncentral chi square with total noncentrality parameter $\lambda_N = \sum_{i=1}^{N}\lambda_i$ and degrees-of-freedom $v_N = \sum_{i=1}^{N} v_i$.

If p linear constraints are imposed on the random sample $\{\mathbf{X}_i\}$, then the degrees-of-freedom are reduced to $N-p$. This applies to both the central and noncentral chi square distributions. An example of a linear constraint is estimation of the mean from the data and could be something more complicated.

The chi square distribution can be accessed in MATLAB using the gamma distribution object. MATLAB also supports the chi square distribution pdf, cdf, quantile function, and random draws through **chi2pdf**(x, v), **chi2cdf**(x, v), **chi2inv**(p, v), and **chi2rnd**(v), respectively. It also implements the noncentral chi square distribution as **ncx2pdf**(x, v, *lambda*), **ncx2cdf**(x, v, *lambda*), **ncx2inv**(p, v, *lambda*), and **ncx2rnd**(v, *lambda*).

4.9.2 Student's *t* Distributions

The t distribution is the work of William S. Gossett, who was employed by the Guinness Brewing Company in Dublin, Ireland. He wrote under the pseudonym Student because Guinness would not allow him to publish using the company name since the company did not want competitors to know that it was using statistical techniques. For a historical perspective, see Hanley, Julian, & Moodie (2008). Gossett (1908) also published the first statistical paper to use simulation to verify a theoretical result and did so at least 40 years before computers became available.

The t distribution applies in sampling and statistical inference when neither the mean nor the variance of the rvs is known a priori and hence must be estimated from the data. Prior to Gossett's work, the normal distribution was used interchangeably but incorrectly when σ^2 was known or unknown. The difference between the Gaussian and t distributions is especially important for small samples, and they are essentially indistinguishable when the degrees-of-freedom exceed 30.

Define two independent rvs $\mathbf{Z} \sim \mathrm{N}(0,1)$ and $\mathbf{Y} \sim \chi_v^2$. Define a new rv $\mathbf{X} = \mathbf{Z}/\sqrt{\mathbf{Y}/v} = \sqrt{v}\mathbf{Z}/\sqrt{\mathbf{Y}}$. The distribution of $\mathbf{X}$ may be derived through a change of variables, as described in Section 2.9. The distribution of $\mathbf{X}$ is Student's t distribution and has the pdf

$$\mathrm{tee}(x; v) = \frac{\Gamma[(v+1)/2]}{\sqrt{v\pi}\,\Gamma(v/2)}\left(1 + x^2/v\right)^{-(v+1)/2} \tag{4.72}$$

where v is the degrees-of-freedom, and the support is $(-\infty, \infty)$. As for the chi square distribution, the t distribution exists for real as well as integer v. When $v = 1$, it is a Cauchy distribution. Figure 4.24 shows the t pdf over a range of values for v.

As for the chi square distribution, there is a noncentral form of the t distribution. Define two independent rvs $\mathbf{Z} \sim \mathrm{N}(0,1)$ and $\mathbf{Y} \sim \chi_v^2$, and let $\mathbf{X} = \sqrt{v}(\mathbf{Z} + \lambda)/\sqrt{\mathbf{Y}}$, where λ is the noncentrality parameter. The rv $\mathbf{X}$ is distributed as the noncentral t distribution. The

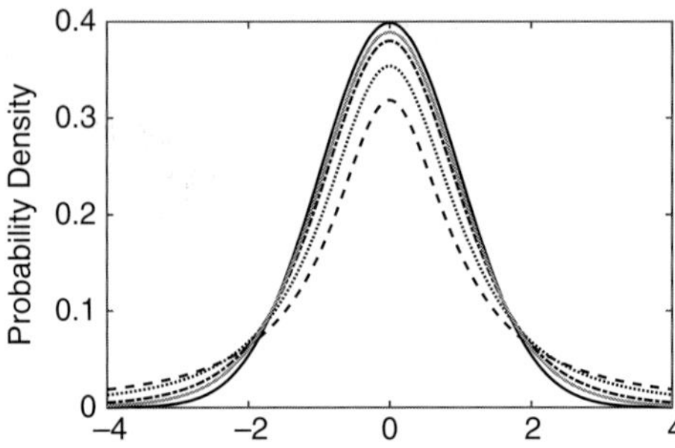

Figure 4.24 The standard normal distribution (solid line) compared with the t distribution with 1 (dashed line), 2 (dotted line), 5 (dash-dot line), and 10 (gray line) degrees-of-freedom.

noncentral version of the t distribution is quite important in hypothesis testing, especially when the alternate hypothesis or power is of interest for a null hypothesis that is t distributed. Its pdf is given by

$$\begin{aligned} \text{ntee}(t;\nu,\lambda) &= \frac{1}{2^{(\nu-1)/2}\sqrt{\pi\nu}\,\Gamma(\nu/2)} e^{-\nu\lambda^2/\left(\nu+t^2\right)}\left(\frac{\nu}{\nu+t^2}\right)^{(\nu-1)/2} \\ &\quad\times \int_0^\infty x^\nu e^{-\frac{1}{2}\left(x-\lambda t/\sqrt{\nu+t^2}\right)^2} dx \end{aligned} \tag{4.73}$$

The integral term in (4.73) was originally derived by R. L. Fisher as the νth repeated integral of the Gaussian pdf and can be expressed in closed form as

$$\begin{aligned} \int_0^\infty x^\nu e^{-(x-\mu)^2/2} dx &= 2^{\nu/2}\mu\,\Gamma[(\nu/2)+1]\,{}_1\text{F}_1\left(\frac{1-\nu}{2},\frac{3}{2};-\frac{\mu^2}{2}\right) \\ &\quad + 2^{(\nu-1)/2}\Gamma[(\nu+1)/2]\,{}_1\text{F}_1\left(-\frac{\nu}{2},\frac{1}{2};-\frac{\mu^2}{2}\right)\Bigg\} \end{aligned} \tag{4.74}$$

where ${}_1\text{F}_1(a,b;x)$ is Kummer's hypergeometric function. Equation (4.74) can also be expressed in terms of parabolic cylinder functions, but that is no more enlightening. Solutions to Equation (4.73) must be obtained numerically.

Johnson, Kotz, & Balikrishnan (1995) give a number of alternative expressions for the noncentral t pdf, none of which are substantially simpler than (4.73). The noncentral t distribution is unimodal but asymmetric when the noncentrality parameter is nonzero, with the right tail heavier when $\mu > 0$, and vice versa. Figure 4.25 shows the noncentral t distribution for five degrees-of-freedom over a range of noncentrality parameters. When $\lambda < 0$, the pdf is a mirror image around the origin of the pdf for the same value of $\lambda > 0$.

The central t cdf is

$$\text{Tee}(x;\nu) = I_x(\nu/2,\nu/2) \tag{4.75}$$

Table 4.1 Comparison of Gaussian and Student *t* Distribution Quantiles

k	Pr (|Z| < *k*)	Pr (|t| < *k*, 2)	Pr (|t| < *k*, 5)	Pr (|t| < *k*, 10)	Pr (|t| < *k*, 20)
1	0.6827	0.5774	0.6368	0.6591	0.6707
2	0.9545	0.8165	0.8981	0.9266	0.9407
3	0.9974	0.9045	0.9699	0.9867	0.9929
4	0.99994	0.9428	0.9897	0.9975	0.9993

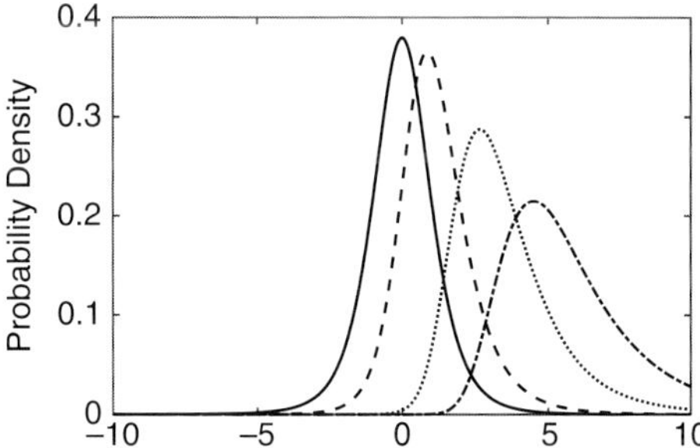

Figure 4.25 The noncentral *t* distribution with five degrees-of-freedom and noncentrality parameters of 1 (dashed line), 3 (dotted line), and 5 (dash-dot line) compared with the central *t* distribution with five degrees-of-freedom (solid line).

where $I_x(a, b)$ is the regularized incomplete beta function ratio and is functionally the same as the cdf for the beta distribution. Table 4.1 shows the difference between standard Gaussian and *t* distribution probabilities for small values of ν and illustrates the large differences that ensue for small degrees-of-freedom.

Expressions for the noncentral *t* cdf are complicated and are omitted.

The expected value of the central *t* distribution is zero for $\nu > 1$ and otherwise does not exist. The variance is $\nu/(\nu - 2)$ when $\nu > 2$ and does not exist for $\nu \le 2$. The skewness is zero for $\nu > 3$ and otherwise does not exist. The kurtosis is $3 + 6/(\nu - 4)$ for $\nu > 4$ and otherwise does not exist. In general, the first $\nu - 1$ moments of the central *t* distribution exist but not moments of higher order. The mode and median are both zero.

In contrast to the central *t* distribution, the odd moments of the noncentral *t* distribution are nonzero, as is apparent from Figure 4.25. The expected value is $\mu = (\nu/2)^{1/2}\lambda\Gamma[(\nu - 1)/2]/\Gamma(\nu/2)$ for $\nu > 1$ and otherwise does not exist. The variance is $\nu(1 + \lambda^2)/(\nu - 2) - \mu^2$ for $\nu > 2$ and otherwise does not exist. Both of these expressions reduce to the central value when $\lambda = 0$.

MATLAB does not implement Student's *t* distribution through distribution objects. MATLAB supports Student's *t* pdf, cdf, quantile function, and random draws through **tpdf**(*x*, *v*), **tcdf**(*x*, *v*), **tinv**(*p*, *v*), and **trnd**(*v*), respectively. It also implements the noncentral *t* as **nctpdf**(*x*, *v*, *lambda*), **nctcdf**(*x*, *v*, *lambda*), **nctinv**(*p*, *v*, *lambda*), and **nctrnd**(*p*, *v*, *lambda*).

4.9.3 The *F* Distributions

The central F distribution arises in many hypothesis testing problems where the variances of two independent random samples from normal distributions are compared. It also frequently occurs as the null distribution in likelihood ratio tests. Its noncentral counterpart is used in assessing the power of a hypothesis test where the null distribution is F.

Consider two independent random variables $\mathbf{Y}$ and $\mathbf{Z}$ such that $\mathbf{Y} \sim \chi_\mu^2$ and $\mathbf{Z} \sim \chi_\nu^2$. Form the new rv $\mathbf{X} = (\mathbf{Y}/\mu)/(\mathbf{Z}/\nu)$, where μ and ν are the degrees-of-freedom for the numerator and denominator, respectively. The distribution of $\mathbf{X}$ is the F distribution and can be obtained as a specific case of the quotient distribution derived in Section 2.9 when the numerator and denominator are chi square distributions. The resulting pdf is

$$\mathrm{eff}(x;\mu,\nu) = \frac{\mu^{\mu/2}\nu^{\nu/2}}{\mathrm{B}(\mu/2,\nu/2)} x^{(\mu/2)-1}(\nu+\mu x)^{-(\mu+\nu)/2} \qquad x \ge 0 \tag{4.76}$$

where $\mathrm{B}(\alpha,\beta)$ is the beta function.

Figure 4.26 illustrates the F pdf for a few choices of parameters. Note the tendency toward Gaussianity as both the numerator and denominator degrees-of-freedom get large. The pdf always tends to zero for large x and does the same as x approaches zero provided that $\mu > 2$. It is always positively skewed. The order of the paired parameters (μ,ν) is important. The F distribution for (μ,ν) is not the same as for (ν,μ) unless $\mu = \nu$. Further, if an rv $\mathbf{X} \sim F_{\mu,\nu}$, then $1/\mathbf{X} \sim F_{\nu,\mu}$, and if an rv $\mathbf{X} \sim t_\nu$, then $\mathbf{X}^2 \sim F_{1,\nu}$.

As for the chi square and t distributions, there is a noncentral form of the F distribution. Actually, there are two of them for the cases where the numerator or both the numerator and denominator are noncentral chi square, and these are called the *singly* and *doubly noncentral F distributions*, respectively. The noncentral F distributions are important in analysis of variance and hypothesis testing involving linear models. Considering the singly noncentral case, let $\mathbf{Y} \sim \mathrm{nchi2}(x;\mu,\lambda)$ with $\lambda > 0$ and $\mathbf{Z} \sim \chi_\nu^2$ be independent rvs, and take $\mathbf{X} \sim (\mathbf{Y}/\mu)/(\mathbf{Z}/\nu)$. The rv $\mathbf{X}$ is distributed as the singly noncentral F distribution whose pdf is given by

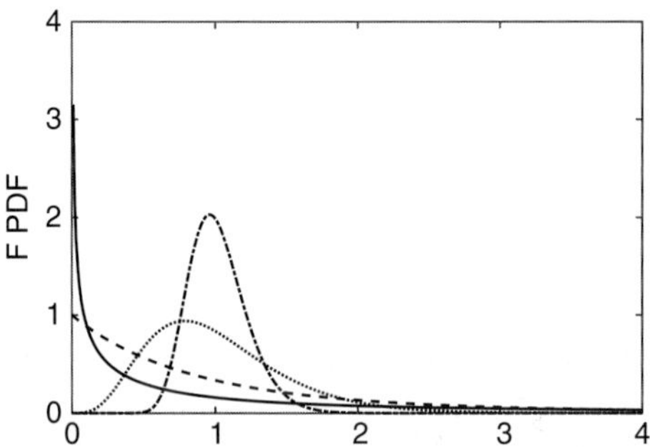

Figure 4.26 The F pdf for paired degrees-of-freedom (μ,ν) of 1, 1 (solid line), 2, 10 (dashed line), 10, 100 (dotted line), and 100, 100 (dash-dot line).

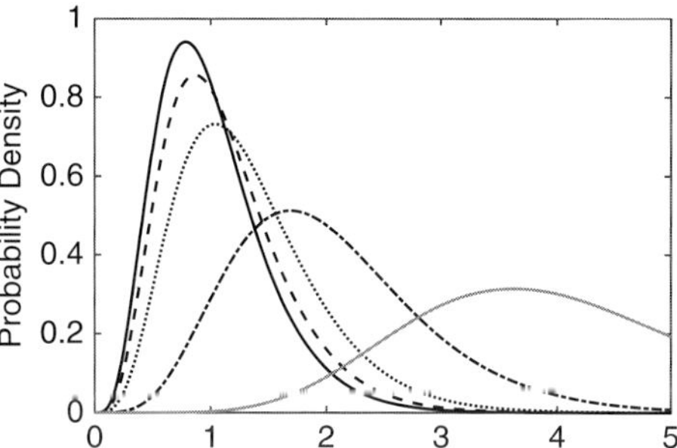

Figure 4.27 The central F distribution with (μ, v) set to 10,100 (solid line) compared with the noncentral F distribution with the same degrees-of-freedom parameters and a noncentrality parameter of 1 (dashed line), 3 (dotted line), 10 (dash-dot line), and 30 (gray line).

$$\begin{aligned}\text{nceff}(x;\mu,v,\lambda) &= \frac{\mu^{\mu/2}v^{v/2}x^{(\mu/2)-1}}{B(\mu/2,v/2)(v+\mu x)^{(\mu+v)/2}}e^{-\lambda/2}\frac{\Gamma(\mu/2)}{\Gamma[(\mu+v)/2]} \\ &\times \sum_{k=0}^{\infty}\frac{(\lambda/2)^k\mu^k\,\Gamma[(\mu+v)/2+k]}{k!\Gamma[(\mu/2)+k]}\left(\frac{x}{v+\mu x}\right)^k \quad x\geq 0\end{aligned} \tag{4.77}$$

The leading term in (4.77) is the central F distribution (4.76). The remaining terms in (4.77) reduce to unity when $\lambda = 0$. Figure 4.27 shows some representative examples of the noncentral F pdf. The pdf for the doubly noncentral F distribution is even more complicated than (4.77), involving a double rather than a single sum, and is less frequently used in practice.

The expected value for the central F distribution is $v/(v-2)$ for $v>2$ and otherwise does not exist. The variance is $2v^2(\mu+v-2)/[\mu(v-2)^2(v-4)]$ for $v>4$ and is otherwise undefined. The mode is given by $v(\mu-2)/[\mu(v+2)]$ for $\mu>2$. The mode is at the origin for $\mu=2$, whereas if $\mu<2$, the pdf is infinite at the origin, and hence the mode is undefined.

The expected value of the noncentral F distribution is $v(\mu+\lambda)/[\mu(v-2)]$ for $v>2$ and does not exist otherwise. The variance is $2v^2[(\mu+\lambda)^2+(\mu+2\lambda)(v-2)]/[\mu^2(v-2)^2(v-4)]$ for $v>4$ and does not exist otherwise.

The F cdf is the regularized incomplete beta function ratio

$$\text{Eff}(x;\mu,v) = I_{\mu x/(\mu x+v)}\left(\frac{\mu}{2},\frac{v}{2}\right) \tag{4.78}$$

and is functionally the same as the cdf for the beta and t distributions. Consequently, if an rv $\mathbf{X}\sim\text{Beta}(\mu/2,v/2)$, then $v\mathbf{X}/[\mu(1-\mathbf{X})]\sim\text{Eff}(\mu,v)$, and if $\mathbf{X}\sim\text{Eff}(\mu,v)$, then $(\mu\mathbf{X}/v)/(\mu\mathbf{X}/v+1)\sim\text{Beta}(\mu/2,v/2)$.

MATLAB supports the F pdf, cdf, quantile function, and random draws as **fpdf**(x, $v1$, $v2$), **fcdf**(x, $v1$, $v2$), **finv**(p, $v1$, $v2$), and **frnd**($v1$, $v2$). The singly noncentral F distribution is implemented as **ncfpdf**(x, $v1$, $v2$, *lambda*), **ncfcdf**(x, $v1$, $v2$, *lambda*), **ncfinv**(p, $v1$, $v2$, *lambda*), and **ncfrnd**($v1$, $v2$, *lambda*). MATLAB does not implement the doubly noncentral F distribution.

4.9.4 The Correlation Coefficient

The distribution for the bivariate correlation coefficient (often called the *Pearson correlation coefficient* after Karl Pearson) obtained by normalizing the off-diagonal element of (4.39) when $p = 2$ by the square root of the product of its diagonal elements was first derived by Fisher (1915), although it received considerable subsequent attention due to the complexity of the statistics for even this simple estimator. See Johnson, Kotz, & Balikrishnan (1995, chap. 32) for details.

Let a given pair of rvs be $\{\mathbf{X}_i, \mathbf{Y}_i\}$ under the condition of mutual independence for the N distinct values of the index. The correlation coefficient is given by

$$\hat{r} = \frac{\sum_{i=1}^{N} \left(\mathbf{X}_i - \overline{\mathbf{X}}_N\right)\left(\mathbf{Y}_i - \overline{\mathbf{Y}}_N\right)}{\sqrt{\sum_{i=1}^{N} \left(\mathbf{X}_i - \overline{\mathbf{X}}_N\right)^2 \sum_{i=1}^{N} \left(\mathbf{Y}_i - \overline{\mathbf{Y}}_N\right)^2}} \tag{4.79}$$

Assume that the joint distribution of $\{\mathbf{X}_i, \mathbf{Y}_i\}$ is the bivariate normal distribution (3.56), so the marginal distributions of the two variables are Gaussian. The pdf of the correlation coefficient can be expressed in at least six forms, as shown in Johnson, Kotz & Balikrishnan (1995, chap. 32). The most useful version is due to Hotelling (1953)

$$\begin{aligned} \mathrm{corr}(r; \rho, N) &= \frac{(N-2)\Gamma(N-1)\left(1-\rho^2\right)^{(N-1)/2}\left(1-r^2\right)^{(N-4)/2}}{\sqrt{2\pi}\,\Gamma(v-1/2)(1-\rho r)^{N-3/2}} \\ &\times {}_2\mathrm{F}_1[1/2, 1/2; N-1/2; (1+\rho r)/2] \qquad -1 \le r \le 1 \end{aligned} \tag{4.80}$$

where ρ is the population correlation coefficient, and N is the number of data. The Gauss hypergeometric function admits the series representation

$${}_2\mathrm{F}_1(a, b, c; x) = \sum_{k=0}^{\infty} \frac{(a)_k (b)_k}{(c)_k\, k!} x^k \tag{4.81}$$

where the Pochhammer symbol is

$$(x)_k = \frac{\Gamma(x+k)}{\Gamma(x)} \tag{4.82}$$

It is relatively easy to implement a function that computes the series expression (4.81), although over part of parameter space the series is slow to converge. A numerical trick to speed up the convergence of the series uses a continued fraction implementation of the Padé approximants (Hänggi, Roesel, & Trautmann 1978).

```
function [Sum] = Hypergeometric2f1(a, b, c, x)
%Computes Gauss hypergeometric function using Padé approximants to
%accelerate convergence
```

```
rerr = 1e-8; %relative error
aerr = 1e-35; %absolute error
last = 0;
s = [];
if a < 0 && a - fix(a) == 0
    kup = -a;
elseif b < 0 && b - fix(b) == 0
    kup = -b;
else
    kup = 1000; %upper limit is arbitrary and should not be reached
end
if x ~= 1
    for k = 0:kup
        pocha = Pochhammer(a, k);
        pochb = Pochhammer(b, k);
        pochc = Pochhammer(c, k);
        s = [ s pocha* pochb* x^k/(pochc* gamma(k + 1))];
        Sum = Padesum(s);
        if abs(Sum - last) <= rerr* abs(Sum) + aerr
            return
        end
        last = Sum;
    end
else
    Sum = gamma(c)*gamma(c - a - b)/(gamma(c - a)*gamma(c - b));
end
end
function [Result] = Pochhammer(x, k)
% Computes Pochhammer's symbol
if k == 0
    Result = 1;
    return
else
    i = 0:k - 1;
    s = x + i;
    Result = prod(s);
    return
end
end
function [Cf] = Padesum(s)
%Computes sum from 1 to n of s(i) using Padé approximant
implemented with
%continued fraction expansion; see Z. Naturforschung, 33a,
402-417, 1978.
%
```

```
%Input variable
%s--series of values to be summed, may be complex
%Output variable
%Cf--sum of the series
%
n = length(s);
D = [];
x = zeros(1, n);
d = [];
t = zeros(1, n);
D(1) = s(1);
d(1) = D(1);
if n == 1
    Cf = d(1);
    return
end
D(2) = s(2);
d(2) = -D(2)/D(1);
if n == 2
    Cf = d(1)/(1 + d(2));
    return
end
for i = 3:n
    L = 2*fix((i - 1)/2);
%update x vector
    x(L:-2:4) = x(L - 1:-2:3) + d(i - 1)*x(L - 2:-2:2);
    x(2) = x(1) + d(i - 1);
% interchange odd and even parts
    t(1:2:L -1) = x(1:2:L - 1);
    x(1:2:L - 1) = x(2:2:L);
    x(2:2:L) = t(1:2:L - 1);
% compute cf coefficient
    D(i) = s(i) + s(i - 1:-1:i - L/2)*(x(1:2:L - 1)).';
    d(i) = -D(i)/D(i - 1);
end
%evaluate continued fraction
Cf = 1;
for k = n: -1:2
    Cf = 1 + d(k)/Cf;
end
Cf = d(1)/Cf;
end
```

A complete closed form expression for the cdf does not exist, although approximations for small degrees-of-freedom are extant, and numerical integration of (4.80) is

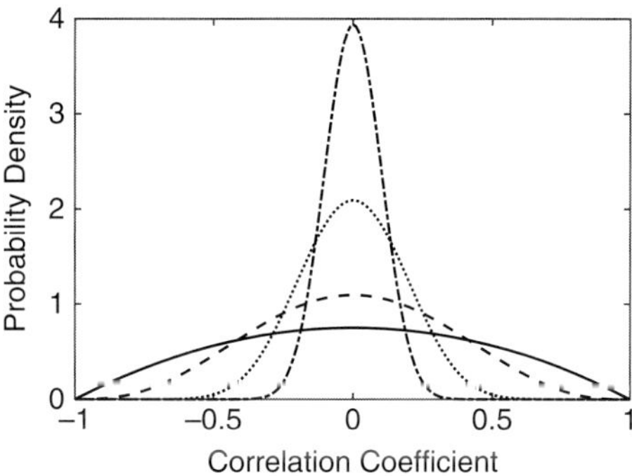

Figure 4.28 The pdf of the correlation coefficient for a zero population value given by (4.83) when the degrees-of-freedom are 6 (solid line), 10 (dashed line), 30 (dotted line), and 100 (dash-dot line).

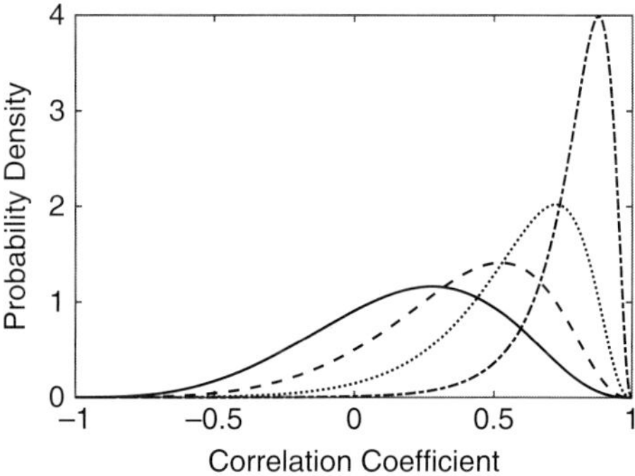

Figure 4.29 The pdf for the correlation coefficient with 10 degrees-of-freedom and a population correlation coefficient of 0.2 (solid line), 0.4 (dashed line), 0.6 (dotted line), and 0.8 (dash-dot line).

straightforward. The expected value and variance can be expressed in terms of Gauss hypergeometric functions, although in practice they are easier to obtain by numerical integration. A reasonable approximation for the expected value is $\rho - \rho(1-\rho^2)/[2(\nu+6)]$ and for the variance is $(1-\rho^2)^2/(\nu+6)$.

When the population correlation coefficient is zero so that the variables are uncorrelated, the bivariate normal distribution can be transformed into polar coordinates and integrated over the azimuth to yield

$$\mathrm{corr}(r;0,N) = \frac{\Gamma[(N-1)/2](1-r^2)^{(N-4)/2}}{\sqrt{\pi}\,\Gamma[(N/2)-1]} \tag{4.83}$$

Equation (4.83) is symmetric about $r = 0$ and gives a measure of the variability or spread of the sample correlation coefficient when there is no actual correlation. Figure 4.28 presents (4.83) for several values of the degrees-of-freedom, showing the large uncertainty in the correlation coefficient when that quantity is small.

Figure 4.29 shows (4.80) evaluated for fixed degrees-of-freedom of 10 and the population correlation coefficient set to 0.2, 0.4, 0.6, and 0.8. The pdfs for negative values of the population correlation coefficient are mirror images about the origin of these curves. The pdfs for positive (negative) values of the population correlation coefficient are increasingly skewed to the left (right) as the magnitude increases. When the population correlation coefficient is ± 1, the pdf is a Dirac delta function at ± 1.

```
r = -1:.01:1;
N = 10;
rho = 0.2;
Pdf = [];
for i = 1:length(r)
    Pdf = [Pdf (N - 2)*gamma(N - 1)*(1 - rho^2)^((N - 1)/2)* ...
            (1 - r(i)^2)^((N - 4)/2)* ...
            Hypergeometric2f1(1/2, 1/2, N - 1/2, (1 +rho* r
            (i))/2)...
            /(sqrt(2*pi)*gamma(N - 1/2)*(1 - rho* r(i))^ ...
            (N - 3/2))];
end
plot(r,Pdf)
```

The central moments can all be obtained by numerically by integrating the pdf with suitable weighting. First, define a function handle for the expected value integrand.

```
fun = @(r, rho, N) r*(N - 2)*gamma(N - 1)*(1 - rho^2)^ ...
        ((N - 1)/2)*(1 - r^2)^((N - 4)/2)* ...
        Hypergeometric2f1(1/2, 1/2, N - 1/2, ...
        (1 + rho* r)/2)/(sqrt(2*pi)*gamma(N - 1/2)* ...
        (1 - rho* r)^(N - 3/2));
```

Because the function *Padecf* does not accept vector-valued arguments, it is necessary to call the MATLAB numerical integration routine with the “ArrayValued second”, “true” name-value pair.

```
integral(@(r) fun(r,.2,10), -1, 1, 'ArrayValued', true)
ans =
  0.1896
```

Repeating with the population correlation coefficient set to 0.4, 0.6, and 0.8 gives the expected value for the four curves in Figure 4.29 of 0.19, 0.38, 0.58, and 0.78, all of which are smaller than the modes of the pdfs. The variance can also be obtained numerically in an analogous manner.

```
fun1 = @(r, rho, N, xm) (r - xm)^2*(N - 2)*gamma(N - 1)* ...
            (1 - rho^2)^((N - 1)/2)*(1 - r^2)^ ...
            ((N - 4)/2)* ...
            Hypergeometric2f1(1/2,1/2,N-1/2,...
            (1 + rho* r)/2)/(sqrt(2*pi)* ...
            gamma(N - 1/2)*(1 - rho* r)^(N - 3/2));
```

```
xm = integral(@(r) fun(r, .2, 10), -1, 1, 'ArrayValued', true);
Var = integral(@(r) fun1(r, .2, 10, xm), -1, 1, 'ArrayValued',
true)
ans =
  0.1045
```

Repeating for the remaining values of *rho*, the variances of the four curves in Figure 4.29 are 0.105, 0.085, 0.055, and 0.021, respectively. This is consistent with the gradual sharpening of the peaks as the population correlation coefficient rises

4.10 Distributions for Order Statistics

The order statistics were introduced in (4.11) and are obtained by sorting a set of rvs into ascending order. As was discussed in Section 2.8, once a set of independent rvs is converted to order statistics through sorting, dependence is introduced. The classic reference on order statistics is due to H. A. David; the most recent version is David & Nagaraja (2003).

4.10.1 Distribution of a Single Order Statistic

Let $\{\mathbf{X}_i\}$ be iid with cdf $F(x)$. Let $F_r(x)$ denote the cdf of the rth order statistic $\mathbf{X}_{(r)}$. In Section 2.8 it was established that $F_N(x) = [F(x)]^N$ and $F_1(x) = 1 - [1 - F(x)]^N$ and are special instances of the distribution of a single order statistic. For the general case, it follows that

$$\begin{aligned} F_r(x) &= \Pr\left(\mathbf{X}_{(r)} \le x\right) = \Pr(\text{at least } r \text{ of the } \mathbf{X}_i \le x) \\ &= \sum_{i=r}^{N} \mathrm{bin}(i; N, F(x)) \\ &= \sum_{i=r}^{N} \binom{N}{i} F^i(x)[1 - F(x)]^{N-i} \\ &= \mathrm{Beta}(F(x), r, N - r + 1) \end{aligned} \tag{4.84}$$

where the last term is the beta cdf (3.50) or, equivalently, the regularized incomplete beta function ratio. Equation (4.84) holds for any discrete or continuous parent distribution. When the parent distribution is continuous, the pdf follows by differentiation

$$f_r(x) = \frac{1}{B(r, N - r + 1)} F^{r-1}(x)[1 - F(x)]^{N-r} f(x) \tag{4.85}$$

Since (4.84) and (4.85) depend on $F(x)$ elevated to integer powers, analytic solutions may not be available except in the simplest cases, and recourse to numerical integration will be required.

Example 4.21 Find the distribution of the order statistics when the parent population is uniform on [0, 1].

The cdf for the uniform distribution is $F(x) = x$ on [0, 1] and zero elsewhere. The pdf for the rth order statistic can immediately be written using (4.85)

$$f_r(x) = \frac{1}{\mathrm{B}(r, N-r+1)} x^{r-1}(1-x)^{N-r} \qquad 0 \le x \le 1$$

which is the beta distribution with parameters $(r, N-r+1)$. The pdfs for $N = 10$ and $r = 1$, 3, 7, and 10 appear in Figure 4.30. Note that the spread depends on the order statistic of interest.

The expected value of the rth order statistic is

$$\mathcal{E}\left(\mathbf{X}_{(r)}\right) = \int x\, f_r(x)dx = \frac{1}{B(r, N-r+1)} \int_{-\infty}^{\infty} x F^{r-1}(x)[1-F(x)]^{N-r} f(x)dx \qquad (4.86)$$

The higher order moments can be computed in a like manner. Closed-form solutions are usually difficult to derive.

Example 4.22 Find the expected value of the rth order statistic for the uniform distribution on [0, 1].

$$\hat{\mu}_{(r)} = \frac{1}{B(r, N-r+1)} \int_0^1 x^r(1-x)^{N-r}dx = \frac{B(r+1, N-r+1)}{B(r, N-r+1)} = \frac{r}{N+1}$$

For $N = 10$ and $r = 1$, 3, 7, and 10, the expected values are 0.0909, 0.2727, 0.6364, and 0.9091. Compare these values to Figure 4.30.

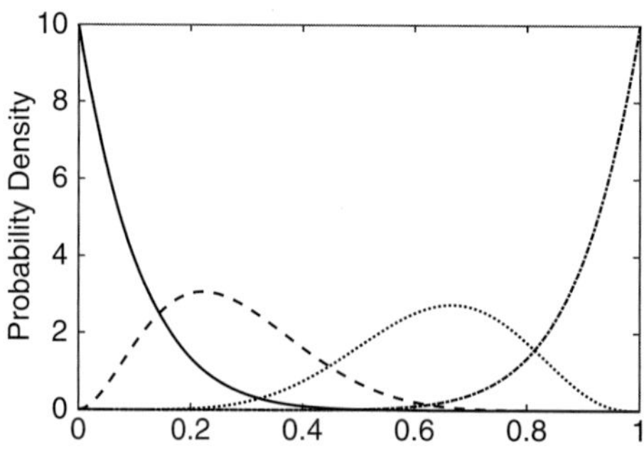

Figure 4.30 The pdf for the first (solid line), third (dashed line), seventh (dotted line), and tenth (dash-dot line) order statistics for a population of 10 whose parent distribution is the uniform distribution on [0, 1].

4.10.2 Distribution of the Sample Median

From (4.85), the pdf of the median is given by

$$f_{(\lfloor N/2 \rfloor+1)}(x) = \frac{1}{B(\lfloor N/2 \rfloor + 1, N - \lfloor N/2 \rfloor)} F^{\lfloor N/2 \rfloor}(x)[1 - F(x)]^{N-\lfloor N/2 \rfloor-1} f(x) \quad (4.87)$$

Figure 4.31 shows the median pdf for $N = 3$, 11, and 101 for a standardized Gaussian parent population. Note the increasing concentration as N rises, suggesting a decreasing variance with increasing sample size, as for the sample mean.

The expected value of the median follows from the definition. This cannot be easily obtained in closed form but can be evaluated numerically. The following MATLAB script implements the expected value.

```
fun = @(x, N) x.*normcdf(x).^floor(N/2).*(1 - normcdf(x)) ...
        .^(N - floor(N/2) - 1).*normpdf(x)/ ...
        beta(floor(N/2) + 1, N - floor(N/2));
integral(@(x) fun(x, N), -Inf, Inf)
```

The result is floating point zero. This can be verified at larger or smaller values of N and suggests that the sample median is unbiased.

The variance of the median is just the second moment of the distribution (4.87) because the first moment vanishes. Using the same numerical approach with a leading x.^2 in the definition of *fun*, computation gives 0.4487 for $N = 3$, 0.1372 for $N = 11$, 0.0155 for $N = 101$, and 0.0016 for $N = 1001$. Presuming that the variance goes like $\alpha\sigma^2/N$, where α is a constant and σ^2/N is the variance of the sample mean, yields estimates for α of 1.3461, 1.5092, 1.5655, and 1.5701, respectively. The exact asymptotic value is $\pi / 2 \approx 1.5708$. Note that the variance of the sample median is about 60% larger than that of the sample mean for a normal parent, which means that one needs substantially more data to get the same variance for the sample median as for the sample mean.

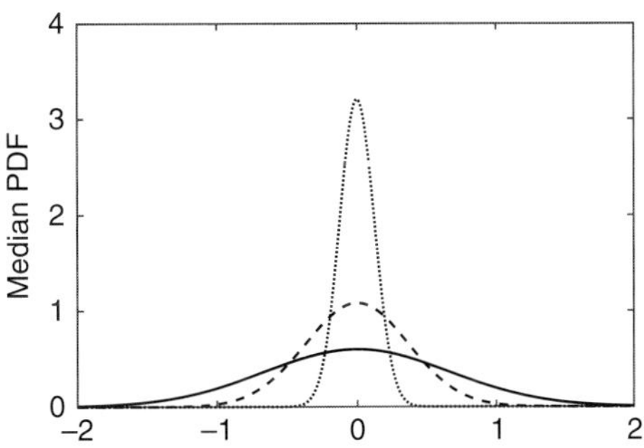

Figure 4.31 The sampling distribution for the median for a standard normal parent population of sizes 3 (solid line), 11 (dashed line), and 101 (dotted line).

4.10.3 Joint Distribution of a Pair of Order Statistics

The joint pdf of $\mathbf{X}_{(r)}$ and $\mathbf{X}_{(s)}$, where $1 \le r < s \le N$, can be derived using similar reasoning. Assuming that $x < y$, the joint cdf of the rth and sth order statistics is given by

$$\begin{aligned} F_{rs}(x,y) &= \Pr(\text{at least } r\ \mathbf{X}_i \le x \cap \text{at least } s\ \mathbf{X}_i \le y) \\ &= \sum_{j=s}^{N}\sum_{i=r}^{j} \Pr(\text{exactly } i\, X_i \le x \cap \text{exactly } j\, X_i \le y) \\ &= \sum_{j=s}^{N}\sum_{i=r}^{j} \frac{N!}{i!(j-i)!(N-j)!} F^i(x)[F(y)-F(x)]^{j-i}[1-F(y)]^{N-j} \end{aligned} \quad (4.88)$$

The pdf for a continuous distribution follows by differentiation

$$\begin{aligned} f_{rs}(x,y) &= \frac{N!}{(r-1)!(s-r-1)!(N-s)!} F^{r-1}(x) f(x) [F(y)-F(x)]^{s-r-1} \\ &\quad \times [1-F(y)]^{N-s} f(y) \end{aligned} \quad (4.89)$$

To find the pdf of the difference of two order statistics $\mathbf{W} = \mathbf{X}_{(s)} - \mathbf{X}_{(r)}$, the variables in (4.89) must be changed from (x,y) to (x,w), where $w = y - x$, and then integrated over the support of x. The Jacobian for the transformation is unity. Define Φ_{rs} to be the leading constant in (4.89). The pdf of the difference of two order statistics is given by

$$f(w) = \Phi_{rs} \int_{-\infty}^{\infty} F^{r-1}(x) f(x) [F(x+w) - F(x)]^{s-r-1} f(x+w)[1 - F(x+w)]^{N-s} dx \quad (4.90)$$

4.10.4 Distribution of the Interquartile Range

The pdf of the interquartile range follows from (4.90) by setting $s = 3N/4$ and $r = N/4$

$$f_{IQ}(x) = \Phi_{3N/4,N/4} \int_{-\infty}^{\infty} F^{N/4-1}(x) f(x) [F(x+w) - F(x)]^{N/2-1} f(x+w)[1 - F(x+w)]^{N/4} dx \quad (4.91)$$

Analytic integration of (4.91) is intractable except for the simplest distributions, such as the uniform distribution, and hence numerical approaches are required. Rather than attempting to compute the leading constant term, it is easier to compute the integral and then normalize the result such that its integral is unity. Let $F(x)$ be the standardized Gaussian distribution whose interquartile range is approximately 1.3161. The following script will compute the pdf for the interquartile range for any choice of N. Figure 4.32 shows the

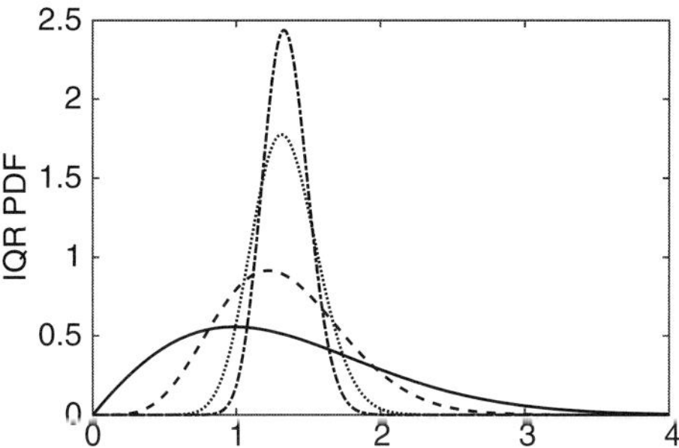

Figure 4.32 The sampling distribution for the interquartile range of a standard Gaussian rv for a sample size of 4 (solid line), 12 (dashed line), 48 (dotted line), and 100 (dash-dot line).

distribution for $N = 4$, 12, 48, and 100. Note the increasingly tight distribution about the Gaussian value for large N.

```
fun = @(x, w, N) normcdf(x).^(N/4 - 1).*normpdf(x).*...
(normcdf(x + w) -normcdf(x)).^(N/2 - 1).* ...
normpdf(x + w).*(1 - normcdf(x + w)).^(N/4);
f = [];
s = integral2(@(x,w) fun(x, w, N), -Inf, Inf, 0, Inf);
for w = 0:.01:5
    f = [f integral(@(x) fun(x, w, N), -Inf, Inf)/s];
end
```

The expected value can be computed using the following script:

```
fun = @(x, w, N) normcdf(x).^(N/4 - 1).*normpdf(x).* ...
        (normcdf(x + w) - normcdf(x)).^ ...
        (N/2 - 1).*normpdf(x + w).* ...
        (1 - normcdf(x + w)).^(N/4);
fun1 = @(x, w, N) w.* normcdf(x).^(N/4 - 1).*normpdf(x).* ...
        (normcdf(x + w) - normcdf(x)).^(N/2 - 1).* ...
        normpdf(x + w).*(1 - normcdf(x + w)).^(N/4);
e = integral2(@(x, w) fun1(x, w, N), -Inf, Inf, 0, Inf, ...
    'AbsTol', 1e-12, 'RelTol', 1e-8);
e = e/integral2(@(x, w) fun(x, w, N), -Inf, Inf, 0, Inf)
```

The result is 1.3264, 1.3297, 1.3406, and 1.3316 and is not affected by changing the tolerances. These values are all slightly larger than the population value of 1.3161, but it is unclear whether this represents bias or numerical round off. The cause will be revealed in Section 8.2.2.

Repeating this for the variance using the population expected value gives 0.5363, 0.1934, 0.0510, and 0.0169, respectively.

4.11 Joint Distribution of the Sample Mean and Sample Variance

Let N independent rvs $\{\mathbf{X}_i\}$ be distributed as $\mathrm{N}(\mu, \sigma^2)$ so that

$$\begin{aligned}\sum_{i=1}^{N}(\mathbf{X}_i-\mu)^2 &= \sum_{i=1}^{N}\left(\mathbf{X}_i-\overline{\mathbf{X}}_N+\overline{\mathbf{X}}_N-\mu\right)^2 \\ &= \sum_{i=1}^{N}\left(\mathbf{X}_i-\overline{\mathbf{X}}_N\right)^2+\sum_{i=1}^{N}\left(\overline{\mathbf{X}}_N-\mu\right)^2+2\sum_{i=1}^{N}\left(\mathbf{X}_i-\overline{\mathbf{X}}_N\right)\left(\overline{\mathbf{X}}_N-\mu\right)\end{aligned} \tag{4.92}$$

For fixed N, the second term is the constant $\left(\overline{\mathbf{X}}_N-\mu\right)^2$ summed N times, and the third term is the constant $2\left(\overline{\mathbf{X}}_N-\mu\right)$ times

$$\sum_{i=1}^{N}\left(\mathbf{X}_i-\overline{\mathbf{X}}_N\right)=N\overline{\mathbf{X}}_N-N\overline{\mathbf{X}}_N=0 \tag{4.93}$$

As a result, (4.92) divided by the population variance becomes

$$\sum_{i=1}^{N}\left(\frac{\mathbf{X}_i-\mu}{\sigma}\right)^2=\sum_{i=1}^{N}\left(\frac{\mathbf{X}_i-\overline{\mathbf{X}}_N}{\sigma}\right)^2+N\left(\frac{\overline{\mathbf{X}}_N-\mu}{\sigma}\right)^2 \tag{4.94}$$

Denote the first term on the right side of (4.94) as Q_1 and the second as Q_2.

According to Cochran's theorem (Cochran 1934), the sum of the squares of standardized Gaussian variables $\{\mathbf{U}_i\}$ can be written

$$\sum_{i=1}^{N}\mathbf{U}_i^2=\sum_{i=1}^{k}Q_i \tag{4.95}$$

where each Q_i is a sum of squares of linear combinations of the $\{\mathbf{U}_i\}$. The concept of *ranks* will be discussed in Chapter 7, but for the moment it suffices to define $r_i=\operatorname{rank}\left(x_{(i)}\right)=i$. Denote the ranks of the Q_i in (4.95) by r_i, and require that the sum of all of them be N. Cochran's theorem states that the $\{Q_i\}$ are independent and that each one has a chi square distribution with r_i degrees-of-freedom.

Returning to (4.94), the rank of Q_1 is $N-1$ because there are N rvs with one linear constraint among them to determine the sample mean. The rank of Q_2 is 1 because it is the square of just one linear combination of standard Gaussian rvs. Consequently, the conditions of Cochran's theorem are met, and Q_1 and Q_2 are independent. Since $\sigma^2 Q_1/N=\hat{s}_N^2$, where $\hat{s}_N^2$ is the sample variance (4.14) and $\overline{\mathbf{X}}_N=\sqrt{\sigma^2 Q_2/N}+\mu$, it follows that the sample mean and the sample variance are independent. The Gaussian distribution is the only distribution for which this is true. This result was first proved using geometric arguments by Fisher (1915) and constituted a core concern in the seminal paper by Gossett (1908). David (2009) provides a historical perspective.

Cochran's theorem also gives

$$N\left(\overline{\mathbf{X}}_N-\mu\right)^2\sim\sigma^2\chi_1^2 \tag{4.96}$$

and

$$\sum_{i=1}^{N} \left(\mathbf{X}_i - \overline{\mathrm{X}}_N\right)^2 \sim \sigma^2 \chi^2_{N-1} \tag{4.97}$$

Consequently, the ratio of (4.96) and (4.97), with each divided by its respective degrees-of-freedom, is given by

$$\frac{N\left(\overline{\mathrm{X}}_N - \mu\right)^2}{\left(\mathbf{X}_i - \overline{\mathrm{X}}_N\right)^2/(N-1)} \sim F_{1,N-1} = \left(t_{N-1}\right)^2 \tag{4.98}$$

Taking the square root of both sides of (4.98) gives

$$\frac{\sqrt{N}\left(\overline{\mathrm{X}}_N - \mu\right)}{\hat{s}'_N} = \frac{\sqrt{N-1}\left(\overline{\mathrm{X}}_N - \mu\right)}{\hat{s}_N} \sim t_{N-1} \tag{4.99}$$

Note that the difference between the sample mean and population mean is scaled by $\sqrt{N}$ when the unbiased estimate of the standard deviation is used but is scaled by $\sqrt{N-1}$ when the sample standard deviation is used. In either case, the ratio is distributed according to Student's t distribution with $N-1$ degrees-of-freedom. From (4.96), it follows that

$$\frac{\sum_{i=1}^{N} \left(\mathbf{X}_i - \overline{\mathrm{X}}_N\right)^2}{\sigma^2} = \frac{N\hat{s}^2_N}{\sigma^2} \sim \chi^2_{N-1} \tag{4.100}$$

There is an interesting analogue for uniformly distributed directional data. If a sample of directional data is independent and uniformly distributed, then the joint distribution of $\theta_i + \alpha$ is the same as the joint distribution of θ_i. Consequently, the sample mean direction (4.25) is equivariant under rotation (the sample mean of $\theta_i + \alpha$ is the same as the sample mean of θ_i), and the sample mean resultant length $\overline{\mathrm{R}}$ is invariant. As a result, the joint distribution of $\overline{\theta} + \alpha$ and $\overline{\mathrm{R}}$ is the same as the joint distribution of $\overline{\theta}$ and $\overline{\mathrm{R}}$, and the conditional distribution of $\overline{\theta}$ given $\overline{\mathrm{R}}$ is invariant under rotation. The uniform distribution is the only distribution that is invariant under rotation; hence $\overline{\theta}$ is uniformly distributed, and $\overline{\theta}$ and $\overline{\mathrm{R}}$ are independent.

5 Point, Interval, and Ratio Estimators

5.1 Overview

There are numerous types of estimators available for statistical inference, of which the method of moments, the maximum likelihood estimator (mle), Bayesian estimators, and the method of least squares are the most widely used. Each of these classes of estimators has its own characteristics, and different ones may be the "best" choice for distinct statistical inference scenarios. Choosing an estimator first requires a set of objective criteria to define the term *best*, and consistency, unbiasedness, efficiency, robustness, and sufficiency are introduced for this purpose. All these criteria save robustness have their origin in Fisher (1922). In practice, estimators that simultaneously meet the five described criteria do not exist, and hence the data analyst is left to make subjective decisions about which ones deserve the most emphasis. In the process of describing efficiency, the Crámer-Rao lower bound on the variance will be derived and is widely used in practice to characterize the dispersion of estimators. The method of moments and the maximum likelihood estimator (mle) are discussed in detail, leaving the method of least squares to Chapter 9. Finally, confidence and tolerance intervals are described, and the issue of estimating the ratios of random variables (rvs) is covered.

5.2 Optimal Estimators

In the sequel, estimators or statistics will be denoted using hatted Greek letters, such as $\hat{\lambda}_k$. An estimate will be shown with a particular value for the index and no hat, such as λ_N. The only exceptions to these practices are the sample mean and sample median that were introduced in Chapter 4 and are among the most commonly used statistics.

5.2.1 Consistency

Optimal estimators should become more precise as the number of data increases. Let $\hat{\lambda}_k$ denote a sequence of estimators for the parameter λ indexed by k for an increasingly large number of values from a set of rvs $\{\mathbf{X}_i\}$. A *consistent estimator* has the property that $\hat{\lambda}_k \xrightarrow{P} \lambda$ or, equivalently, for arbitrarily small but positive ε

$$\lim_{k\to\infty} \Pr\left(\left|\hat{\lambda}_k - \lambda\right| > \varepsilon\right) \to 0 \tag{5.1}$$

An estimator that is not consistent is said to be *inconsistent.*

Example 5.1 The sample mean estimator $\overline{X}_k$ has variance σ^2/k and $\mathcal{E}\left(\overline{X}_k\right) = \mu$. As k increases, the pdf for $\overline{X}_k$ is more tightly clustered around μ. This makes it a consistent estimator.

More formally, for Gaussian rvs,

$$\Pr\left(\left|\overline{X}_N - \mu\right| \geq \varepsilon\right) = \Pr\left(\frac{\sqrt{N}\left|\overline{X}_N - \mu\right|}{\sigma} \geq \frac{\sqrt{N}\varepsilon}{\sigma}\right) = 2\left[1 - \Phi\left(\frac{\sqrt{N}\varepsilon}{\sigma}\right)\right]$$

Since any cdf approaches 1 for large values of the argument, the last term approaches zero as $N \to \infty$.

Consistency can be demonstrated using any theorem that establishes convergence in probability, such as the Markov inequality (2.76) for nonnegative rvs or the Chebyshev inequality (2.77) when the measure of the difference between the sample and population values is the absolute value. Alternatively, if the estimator contains sums of rvs (as is often the case), the law of large numbers can be invoked.

The *continuous mapping theorem* holds that continuous functions preserve limits even when their arguments are sequences of rvs. Let $g(x)$ be a function that is continuous at λ. Then $g\left(\hat{\lambda}_k\right)$ is a consistent estimator for $g(\lambda)$.

The sample mean, sample median, and sample variance are all examples of consistent estimators. However, consistency is an asymptotic property, and demonstrating that an estimator is consistent says nothing about its behavior for finite k. Further, if there is one consistent estimator, then an infinite number of consistent estimators can be constructed from it because $(k - \alpha)\hat{\lambda}_k/(k - \beta)$ is also consistent for any finite constants α and β.

The definition of consistency given in this section is sometimes called *weak consistency.* If convergence in probability is replaced by convergence almost surely, then the estimator possesses *strong consistency.*

5.2.2 Unbiased Estimators

A consistent estimator will have a value in the vicinity of the true value for large numbers of data. Define the bias to be

$$B\left(\hat{\lambda}_k, \lambda\right) = \mathcal{E}\left(\hat{\lambda}_k\right) - \lambda \tag{5.2}$$

An unbiased estimator has $B\left(\hat{\lambda}_k, \lambda\right) = 0$ for all possible values of λ. An asymptotically unbiased estimator has $B\left(\hat{\lambda}_k, \lambda\right) \to 0$ as $k \to \infty$. While consistent estimators are not necessarily unbiased, a consistent estimator with a finite mean value for large k must be asymptotically unbiased.

Example 5.2 The sample mean $\overline{X}_k$ is an unbiased estimator for the population mean μ for any distribution for which it exists.

It is possible for an estimator to be unbiased but not consistent. For an iid random sample $\{\mathbf{X}_i\}$, define an estimator for the mean to be any particular fixed value of the random variables $\hat{\lambda}_k = x_k$. Since the sampling distribution of $\hat{\lambda}_k$ is the same as the distribution of the rvs, and this holds for any size $N \geq k$ for the sample because only the kth sample is used in the estimator, it follows that $\mathcal{E}(\hat{\lambda}_k) = \mathcal{E}(\mathbf{X})$, and the estimator is unbiased. However, it is not consistent because it does not converge to anything. It is also possible for an estimator to be biased but consistent. An example is the sample variance $\hat{s}_N^2$.

There may be no unbiased estimators, or there may be more than one unbiased and consistent estimator. For example, the sample mean is a consistent and unbiased estimator for the population mean of a Gaussian sample. The asymptotic variance of the sample median $\tilde{X}_k$ was shown to be $\pi\sigma^2/(2k)$ in Section 4.10.2, so the sample median is consistent. The sample median is also an unbiased estimator of the population mean, as can be seen from the symmetry of its distribution. Thus there are at least two consistent and unbiased estimators for the population mean of a Gaussian sample. Further optimality criteria are required to define *best estimator*.

5.2.3 Efficiency and the Cramér-Rao Lower Bound

Qualitatively, it would seem that the estimator whose sampling distribution is most tightly concentrated around the true parameter would be a "best estimator." This can be quantified using the mean squared error (mse) criterion:

$$mse\left[\hat{\lambda}_k\right] = \mathcal{E}\left[\left(\hat{\lambda}_k - \lambda\right)^2\right] = var\left[\hat{\lambda}_k\right] + \left[\mathcal{E}\left(\hat{\lambda}_k\right) - \lambda\right]^2 \tag{5.3}$$

The second term on the right is the square of the bias (5.2), and it vanishes if the estimator is unbiased, in which case the mse equals the variance. Consequently, different unbiased estimators can be compared through their variances.

Example 5.3 The sample median and sample mean are both unbiased estimators for the population mean. However, the variance of the sample median is $\pi/2$ times the variance of the sample mean and hence has a higher mean square error.

It has been shown previously that there are at least two estimators for the variance, the sample variance where the sum of squared differences between a set of rvs and their sample mean $\hat{s}^2$ is divided by N and the unbiased variance divided by $N-1$. A third estimator can be defined that is normalized by $N+1$, and it can be shown for Gaussian data that

$$mse\left[\hat{s}^2/(N+1)\right] < mse\left(\hat{s}_N^2\right) < mse\left(\hat{s}'^2_N\right) \tag{5.4}$$

Consequently, even for something as simple as the variance, there is no estimator that is simultaneously unbiased and minimum mse.

When the joint pdf of the observations $f(x_i;\lambda)$ is regarded as a function of λ for a given set of samples $\{x_i\}$, it is called a *likelihood function* $\mathcal{L}(\lambda;x_i)$. The *score function* is the derivative of the log likelihood with respect to the parameter:

$$\partial_\lambda \log \mathcal{L}(\lambda;x) = \frac{\partial_\lambda \mathcal{L}(\lambda;x)}{\mathcal{L}(\lambda;x)} \tag{5.5}$$

and has zero expected value, as can be easily shown from the definition. The variance of the score is

$$\mathcal{I}(\lambda) = var[\partial_\lambda \log \mathcal{L}(\lambda;x)] = \mathcal{E}\left\{[\partial_\lambda \log \mathcal{L}(\lambda;x)]^2\right\} \tag{5.6}$$

and is the *Fisher information* for a single rv. Presuming that the log likelihood function is twice differentiable, the Fisher information is also given by

$$\mathcal{I}(\lambda) = -\mathcal{E}\left[\partial_\lambda^2 \log \mathcal{L}(\lambda;x)\right] \tag{5.7}$$

The Fisher information measures the amount of information that an rv carries about an unknown parameter. An rv with high Fisher information corresponds to one whose squared score function is frequently large. From (5.7), the Fisher information is also a measure of the curvature of the log likelihood near its maximum value. If the log likelihood is sharply peaked around its maximum, then the Fisher information is high, whereas if it is very broad, the Fisher information is low.

Let $\hat{\mathrm{P}}$ be an estimator for some function of the parameter $\rho(\lambda)$. It can be shown that

$$var\left(\hat{\mathrm{P}}\right) \ge \frac{[\rho'(\lambda)]^2}{\mathcal{I}_N(\lambda)} \tag{5.8}$$

where $\mathcal{I}_N(\lambda) = N\mathcal{I}(\lambda)$ is the sample Fisher information. Equation (5.8) is the *Cramér-Rao inequality* that was introduced independently by Cramér (1945) and Rao (1945) and gives a lower bound for the variance of any estimator.

When the estimator $\hat{\mathrm{P}}$ is biased, let $\rho(\lambda) = B\left(\hat{\lambda},\lambda\right) + \lambda$, where the first term on the right is the bias (5.2), so

$$var\left(\hat{\mathrm{P}}\right) \ge \frac{\left[1 + B'\left(\hat{\lambda},\lambda\right)\right]^2}{\mathcal{I}_N(\lambda)} \tag{5.9}$$

When $\rho(\lambda) = \lambda$, the numerator of (5.8) is 1, and the Cramér-Rao bound becomes the inverse of the sample Fisher information. The variance of an unbiased estimator cannot be smaller than this value.

When the parameter λ is a p-vector, the Fisher information becomes a $p \times p$ matrix called the *Fisher information matrix*. Extension to the matrix form is straightforward. The elements of the Fisher information matrix are

$$\mathcal{I}_{ij} = \mathcal{E}\left[\partial_{\lambda_i} \log \mathcal{L}(\boldsymbol{\lambda}; x) \partial_{\lambda_j} \log \mathcal{L}(\boldsymbol{\lambda}; x)\right] \tag{5.10}$$

and the Cramér-Rao bound is

$$cov(\hat{\mathbf{P}}) \geq \overleftrightarrow{\mathbf{J}} \cdot \overleftrightarrow{\mathcal{I}}^{-1} \cdot \overleftrightarrow{\mathbf{J}}^T \tag{5.11}$$

where $\overleftrightarrow{\mathbf{J}} = \partial_\lambda \rho$ is the Jacobian matrix whose i, j element is $\partial_{\lambda_j} \rho_i$, and the inequality $\overleftrightarrow{\mathbf{X}} \geq \overleftrightarrow{\mathbf{Y}}$ means that $\overleftrightarrow{\mathbf{X}}$ is more positive definite than $\overleftrightarrow{\mathbf{Y}}$. The matrix form of the Fisher information occurs frequently. For example, if one were estimating the three parameters of the generalized extreme value distribution, the Fisher information matrix is 3×3, and in the absence of independence of the parameters, the off-diagonal elements define their covariance.

An unbiased estimator that achieves the minimum-variance Cramér-Rao lower bound is said to be *efficient*. An efficient estimator is also the *minimum variance unbiased* (mvu) estimator and has the lowest possible mse. If the asymptotic variance of an unbiased estimator is equal to the lower bound, it is *asymptotically efficient*. This may not hold for a finite sample, and other estimators may also be asymptotically efficient. A measure of efficiency for the univariate case is

$$e\left(\hat{\lambda}\right) = \frac{1}{\mathcal{I}_N(\lambda) var\left(\hat{\lambda}\right)} \tag{5.12}$$

with obvious matrix extensions when the estimator is multivariate. Equation (5.12) must be less than or equal to 1, with equality achieved when the estimator is mvu.

Example 5.4 Let $\{\mathbf{X}_i\}$ be N random samples from the exponential distribution with parameter β as given by (3.40). Determine the variance for the estimator $\hat{\lambda} = (N-1)/\left(\sum_{i=1}^{n} x_i\right)$ using the Cramér-Rao bound.

The log likelihood is $\log \mathcal{L}(\beta|x) = \log \beta - \beta x$. The Fisher information for a single datum is given by (5.7) as $1/\beta^2$. The Fisher information for the entire sample is N/β^2. By the reproductive property of the gamma distribution (Section 3.4.5), $\sum_{i=1}^{N} x_i$ is distributed as $\mathrm{gam}(x; N, \beta)$. The mean of $1/\sum_{i=1}^{N} x_i$ is $\beta/(N-1)$, so the estimator $\hat{\lambda}$ is unbiased. The variance of $1/\sum_{i=1}^{N} x_i$ is $\beta^2/\left[(N-1)^2(N-2)\right]$. The Cramér-Rao variance bound is the inverse sample Fisher information, or β^2/N, so the variance of $\hat{\lambda}$ is larger than the theoretical lower limit. The estimator efficiency (5.12) is $(N-2)/N$, which is smaller than one for three or more samples.

A measure of the *relative efficiency* of a pair of unbiased estimators is the ratio of their variances that must lie between 0 and 1.

Example 5.5 The sample mean is the minimum variance unbiased estimator for the population mean of a Gaussian distribution and has variance σ^2/N. The relative efficiency of the sample median with variance $\pi\sigma^2/2N$ is $2/\pi \approx 0.637$.

For a mvu estimator to exist, the log likelihood function must be factorable as

$$\log \mathcal{L}(\lambda; x_i) = S(\lambda)\eta(\lambda, x_i) \tag{5.13}$$

where $S(\lambda)$ is independent of the observations. If the log likelihood function cannot be written in this form, then an estimator that reaches the Cramér-Rao lower bound does not exist.

Example 5.6 The likelihood function for the Cauchy distribution is

$$\mathcal{L}(\lambda; x_i) = \prod_{i=1}^{N} \frac{1}{\pi\left[1 + (x_i - \lambda)^2\right]}$$

The log likelihood function is

$$\log \mathcal{L}(\lambda; x_i) = -N \log \pi - \sum_{i=1}^{N} \log\left[1 + (x_i - \lambda)^2\right]$$

This does not have the form of (5.13), and hence a mvu estimator for the location parameter does not exist.

While efficient estimators are desirable, they do not always exist. In some circumstances, small sample variance may be a more important criterion for "best" than bias, and hence estimators with bias may be acceptable. This tradeoff can be quantified by examining the mean square error (5.3).

5.2.4 Robustness

The most commonly used statistics (e.g., the sample mean and sample variance) are not robust to the presence of a small fraction of outlying data. To see this, use the **normrnd** function in MATLAB to create 10 random numbers drawn from the $N(1, 1)$ distribution. Their sample mean is 1.0729, their median is 0.9818, and their sample variance is 0.9785. Now replace the last value with a single random number from $N(1, 100)$. A new sample mean of 8.3497, median of 1.0866, and variance of 507.7205 are obtained. The mean has increased by a factor of ~8, the variance has increased by a factor of more than 500, but the median is almost unchanged. This relative insensitivity of the median to outlying data is termed *robustness*.

Define the *p-norm* by

$$\|y\|_p = \left(\sum_{i=1}^{N} |y_i|^p\right)^{1/p} \qquad p \geq 1 \tag{5.14}$$

The sample median and mean are obtained, respectively, by minimizing (5.14) after replacing y_i with $x_i - \tilde{X}_N$ for $p = 1$ and $x_i - \overline{X}_N$ for $p = 2$ or by minimizing the L_1 and L_2 norms. The preceding example suggests that an L_1 estimator like the median will prove to be of better utility than an L_2 estimator like the sample mean or variance because of robustness. However, L_1 estimators are typically less efficient than L_2 estimators (e.g., recall that the variance of the sample median is $\pi/2$ times the variance of the sample mean). In addition, computation of L_1 estimators is much more resource intensive than for L_2 estimators.

Robustness is typically more of an issue for estimates of dispersion such as the variance than for estimates of location such as the sample mean and becomes even more important for higher-order moments. It is possible to produce robust estimates of scale – in the preceding example, the median absolute deviation (MAD) is 0.4414 and 0.4490, respectively, without and with the outlying data point. The MAD is relatively immune to up to 50% outlying data and is an especially robust estimator for scale. However, like the median, it has lower efficiency than its L_2 counterparts.

A key concept in robustness is that of a *finite sample breakdown point*, defined as the smallest fraction of anomalous data that can render the estimator useless. The smallest possible breakdown point is $1/N$ (i.e., a single anomalous point can destroy the estimate). The sample mean and the sample variance both have breakdown points of $1/N$.

5.2.5 Sufficient Statistics

Sufficient statistics play a key role in statistical inference because, if they exist, a minimum variance estimator must be a function of the sufficient statistics. Further, if it exists, the efficient estimator must be a function of the sufficient statistics, and any function of the sufficient statistics will be asymptotically efficient.

Suppose that two scientists A and B want to estimate the salinity of seawater from N measurements of its electrical conductivity. Scientist A has access to all the raw data. Scientist B has access only to a summary statistic $\hat{\lambda}_N$. Suppose that, for all practical purposes, scientist B can estimate salinity as well as scientist A despite not having all the data. Then $\hat{\lambda}_N$ is a *sufficient statistic* for salinity. Sufficient statistics provide all the information necessary to estimate a parameter and provide a summary description of the measurements.

Example 5.7 The sample mean is sufficient for the population mean of a Gaussian distribution because no further information about the population mean can be derived from the sample itself. However, the sample median is not sufficient for the population mean. For example, if the values above the sample median are only slightly larger than it but those below the sample median extend to very small values, then having this information would provide additional information about the population mean.

A more formal definition holds that a statistic $\hat{\lambda}_k$ is sufficient for λ if the conditional distribution of $\{\mathbf{X}_i\}$ given that $\hat{\lambda}_k$ takes on a particular value λ^* does not depend on λ, or

$$\Pr\left(\mathbf{X};\lambda|\hat{\lambda}_k = \lambda^*\right) = \Pr\left(\mathbf{X}|\hat{\lambda}_k = \lambda^*\right) \tag{5.15}$$

A necessary and sufficient condition for $\hat{\lambda}_k$ to be a sufficient statistic is that the joint pdf (or likelihood function) be factorable as follows

$$f_N(x_1,\ldots,x_N;\lambda) = U(x_1,\ldots,x_N)V\left(\hat{\lambda}_k;\lambda\right) \tag{5.16}$$

where U and V are nonnegative functions. Equation (5.16) is called the *Fisher-Neyman factorization theorem*. Conversely, if the joint distribution cannot be factored as in (5.16), then a sufficient statistic does not exist.

Example 5.8 Let $\{\mathbf{X}_i\}$ be a random sample from a Poisson distribution with unknown mean λ. Show that $\sum_{i=1}^{N} x_i$ is a sufficient statistic for λ.

The Poisson pdf is $\text{pois}(x;\lambda) = \lambda^x e^{-\lambda}/x!$ for $x = 0, 1\ldots$. The joint pdf is

$$\text{pois}_N(x_1,\ldots,x_N;\lambda) = \prod_{i=1}^{N} \frac{e^{-\lambda}\lambda^{x_i}}{x_i!} = \left(\prod_{i=1}^{N} \frac{1}{x_i!}\right)\left(e^{-N\lambda}\lambda^{\sum_{i=1}^{N} x_i}\right)$$

The first term in parentheses on the right is U, and the second is V, so the joint pdf factors as in (5.16), and hence $\sum_{i=1}^{N} x_i$ is a sufficient statistic for λ. However, $\sum_{i=1}^{N} x_i$ does not achieve full efficiency because the sample mean has a lower variance.

Example 5.9 Let $\{\mathbf{X}_i\}$ be a random sample from the uniform distribution on $(0, \lambda]$ with unknown parameter λ. Find a sufficient statistic for λ.

The uniform pdf is $\text{unif}(x;\lambda) = 1/\lambda$ on $(0, \lambda]$. Consider the order statistics $x_{(i)}$. None of the order statistics, including the largest one $x_{(N)}$, can exceed λ. Therefore, the joint pdf is

$$\text{unif}(x_1,\ldots,x_N;\lambda) = \frac{1}{\lambda^N}$$

if $x_{(N)} = \lambda$ and zero otherwise. This may be succinctly written as

$$\text{unif}(x_1,\ldots,x_N;\lambda) = \frac{1}{\lambda^N}\mathbf{1}\left(\lambda - x_{(N)}\right)$$

where $\mathbf{1}(x)$ is the indicator function defined in Section 2.3. Since the joint pdf can be factored into a function of λ alone and a function of $x_{(N)}$, the largest order statistic is a sufficient statistic for λ.

Example 5.10 Show that the data $x_1, \ldots, x_k$ are a sufficient statistic.

Let $\hat{\lambda}_k = x_1, \ldots, x_k$, and set $U = 1$ in (5.16). Then $f(x_i;\lambda) = V\left(\hat{\lambda}_k;\lambda\right)$, and hence the data are a sufficient statistic.

At a minimum, the data are always a sufficient statistic but are hardly a parsimonious representation. The order statistics are a more useful sufficient statistic that always exists, and sometimes order statistics are the only sufficient statistic other than the data that is possible, such as for the Cauchy distribution. In fact, outside the exponential family of distributions, it is rare for a sufficient statistic to exist whose dimensionality is smaller than the sample size.

A *minimal sufficient statistic* is a sufficient statistic that is a function of all other possible sufficient statistics so that it is the simplest one possible. Minimal sufficient statistics are not unique because any monotone function of one is also minimal sufficient. However, they are important because a minimal sufficient statistic achieves the most parsimonious representation of a data sample of any possible sufficient statistic.

Suppose that a set of rvs has distribution $f(x;\lambda)$ and that there is an estimator $\hat{\lambda}$ for the distribution parameter. The ratio $f(x;\lambda)/f(x';\lambda)$ is constant for any pair of data values x and x' if and only if $\hat{\lambda}(\mathbf{X}_i)=\hat{\lambda}(\mathbf{X}_i')$. Then $\hat{\lambda}(\mathbf{X}_i)$ is a minimal sufficient statistic for λ. This is the *Lehmann-Scheffé theorem for minimal sufficient statistics*.

A statistic $\hat{\lambda}$ for λ is a *complete sufficient statistic* if $\mathcal{E}\left[g(\hat{\lambda})\right]=0$ for any λ means $\Pr\left[g(\hat{\lambda})=0\right]=1$ where $g(x)$ is a function that does not depend on any unknown parameters. Any complete sufficient statistic must also be minimal sufficient. Any one-to-one function of a sufficient, minimal sufficient, or complete sufficient statistic is sufficient, minimal sufficient, or complete sufficient, respectively.

Suppose that $x_1,\ldots,x_N$ are iid data drawn from a k-parameter exponential family distribution whose joint distribution is parameterized as

$$f_N(x_i|\boldsymbol{\lambda})=e^{\sum_{i=1}^{k}\left[A_j(\boldsymbol{\lambda})B_j(\mathbf{x}_i)+C_j(\mathbf{x}_i)+D_j(\boldsymbol{\lambda})\right]} \tag{5.17}$$

Univariate distributions in the exponential family, including the normal, binomial, Poisson, and gamma distributions, meet this condition. Then the vector $\left[\sum_{i=1}^{N}B_1(x_i),\ldots,\sum_{i=1}^{N}B_k(x_i)\right]$ is

1. A joint sufficient statistic for $\boldsymbol{\lambda}$; and
2. A minimal joint sufficient statistic for $\boldsymbol{\lambda}$ provided that there is not a linear constraint on its k elements.

This is a more explicit result than obtains from the Fisher-Neyman factorization theorem (5.16).

Example 5.11 Find a sufficient statistic for the gamma distribution with α unknown and β known.

In this case, $k=1$ in (5.17). The gamma pdf may be written as

$$\text{gam}(x;\alpha,\beta)=\frac{\beta^\alpha}{\Gamma(\alpha)}x^{\alpha-1}e^{-\beta x}=e^{\alpha(\log\beta+\log x)}e^{-(\log x+\beta x)}e^{-\log\Gamma(\alpha)}$$

Let $A=\alpha$, $B=\log\beta+\log x$, $C=-(\log x+\beta x)$, and $D=-\log\Gamma(\alpha)$. The pdf factors according to (5.17), and hence a sufficient statistic exists. It is given by $N\log\beta+\sum_{i=1}^{N}\log x_i$. The statistic is minimal and complete sufficient.

Example 5.12 Find a joint sufficient statistic for the Gaussian distribution $N(\mu, \sigma^2)$.

The pdf factors as

$$N(x; \mu, \sigma^2) = \frac{1}{\sqrt{2\pi}\sigma} e^{-\mu^2/(2\sigma^2)} e^{-x^2/(2\sigma^2)} e^{\mu x/\sigma^2}$$

Let $A_1 = \mu/\sigma^2$, $A_2 = -1/(2\sigma^2)$, $B_1 = x$, $B_2 = x^2$, $C = 0$, and $D = -\mu^2/(2\sigma^2) - \log(\sqrt{2\pi}\sigma)$. This has the form of (5.17); hence the joint sufficient statistic is the vector $\left(\sum_{i=1}^N x, \sum_{i=1}^N x_i^2\right)$ and is both minimal and complete joint sufficient. The one-to-one functions of the joint sufficient statistic $\left(\overline{X}_N, \hat{s}_N^2\right)$ and $\left(\overline{X}_N, \hat{s}_N\right)$ are also minimal and complete joint sufficient.

Example 5.13 Let $\mathbf{X}$ be an rv from $N(0, \sigma^2)$. Based on Example 5.12, a minimal and complete sufficient statistic for the population variance is $\sum_{i=1}^N x_i^2$. However, Example 5.10 shows that the data themselves are a sufficient statistic. Since the data are not a function of $\sum_{i=1}^N x_i^2$, they are not a minimal sufficient statistic.

Some additional examples, stated without derivation, include

1. If $\mathbf{X}_i \sim \text{unif}(x; a, b)$, then $\left(X_{(1)}, X_{(N)}\right)$ are complete joint sufficient statistics for (a, b);
2. If $\mathbf{X}_i \sim \text{unif}(x; a, b)$ with a (b) known, then $X_{(N)}\left(X_{(1)}\right)$ is a complete sufficient statistic for b (a);
3. If $\mathbf{X}_i \sim \exp\text{e}(x; \theta, \beta)$, where θ is a location parameter, then $\left(X_{(1)}, \overline{X}_N\right)$ are complete joint sufficient statistics for (θ, β); and
4. If $\mathbf{X}_i$ are standard Cauchy with a location parameter θ, then the order statistics are minimal sufficient statistics.

A statistic that does not depend on the parameters in the joint distribution is an *ancillary statistic*. By *Basu's theorem*, any complete sufficient statistic is independent of every ancillary statistic. Consequently, if a statistic $\hat{\lambda}$ is minimal sufficient and $\hat{\tau}$ is an ancillary statistic but $\hat{\tau} = g(\hat{\lambda})$ so that the two statistics are not independent, then $\hat{\lambda}$ is not complete. Conversely, independence between a minimally sufficient statistic and every ancillary statistic can be proved without deriving their joint distributions, which is often a difficult task.

Example 5.14 Two rvs $\mathbf{X}_1$ and $\mathbf{X}_2$ are drawn from $N(\mu, \sigma^2)$. Define the difference statistic $d = \mathbf{X}_1 - \mathbf{X}_2$ whose distribution is $N(0, 2\sigma^2)$. Since d does not depend on μ, it is an ancillary statistic for μ. However, if σ^2 is unknown, d is not ancillary for that parameter.

Let $\hat{\lambda}_k$ be an estimator for λ, and let $\hat{\theta}_k$ be sufficient for λ. Define $\hat{\phi}_k = \mathcal{E}\left[\hat{\lambda}_k | \hat{\theta}_k\right]$. Then $\mathcal{E}\left[\left(\hat{\phi}_k - \lambda\right)^2\right] \le \mathcal{E}\left[\left(\hat{\lambda}_k - \lambda\right)^2\right]$, with equality occurring only when $\hat{\lambda}_k = \hat{\phi}_k$. This is the

Rao-Blackwell theorem that was introduced independently by Rao (1945) and Blackwell (1947). The Rao-Blackwell estimator $\hat{\phi}_k$ is an improved estimator over the original one. The Rao-Blackwell theorem is the rationale for basing estimators on sufficient statistics because lower mse estimators than the sufficient statistic do not exist.

Example 5.15 A minimal sufficient statistic for the parameter p in a Bernoulli distribution is the sum of the data in a random sample. Take the mean of the first two data points as a trial estimator. Derive the Rao-Blackwell estimator, and compare its mse with the sufficient statistic.

Define the sufficient statistic $\hat{T} = \sum_{i=1}^{N} x_i$ so that

$$\mathcal{E}\left[\frac{(x_1 + x_2)/2}{\hat{T}}\right] = \mathcal{E}(x_1|\hat{T})/2 + \mathcal{E}(x_2|\hat{T})/2 = \mathcal{E}(x_1|\hat{T})$$

where the last term follows from exchangeability. The sample contains exactly $\hat{T}$ ones and $1 - \hat{T}$ zeroes conditional on $\hat{T}$, so $\mathcal{E}(x_1|\hat{T}) = \hat{T}/N$, and the Rao-Blackwell estimator is the sample mean.

The mses are given by the variances of the trial and Rao-Blackwell estimator because both are unbiased. The variance of a Bernoulli distribution is $p(1-p)$. Consequently, the mses of the trial and Rao-Blackwell estimators are $p(1-p)/2$ and $p(1-p)/N$, respectively. Presuming that $N > 2$, the mse of the Rao-Blackwell estimator is smaller than that of the trial estimator by a factor of $N/2$.

5.2.6 Statistical Decision Theory

The previous five sections have presented a nonexhaustive series of criteria or "beauty principles" that can be used to distinguish and categorize estimators. While it is not possible to simultaneously optimize all of them, a formal theory exists for comparing them that is called *decision theory*.

Consider a parameter λ and its estimator $\hat{\lambda}$. Define a *loss function* $\rho(\lambda, \hat{\lambda})$ that maps the parameter space onto the real line and measures the disparity between λ and $\hat{\lambda}$. Some examples of loss functions include

- 0–1 loss $\rho(\lambda, \hat{\lambda}) = \mathbf{1}(\lambda = \hat{\lambda})$;
- L_1 or absolute difference loss $\rho(\lambda, \hat{\lambda}) = |\lambda - \hat{\lambda}|$; and
- L_2 or squared difference loss $\rho(\lambda, \hat{\lambda}) = (\lambda - \hat{\lambda})^2$.

A loss function is an rv because the estimator $\hat{\lambda}$ depends on the random sample. Consequently, its expected value is called the *risk* of the estimator:

$$\mathcal{R}(\lambda, \hat{\lambda}) = \mathcal{E}\left[\rho(\lambda, \hat{\lambda})\right] = \int \rho\left[\lambda, \hat{\lambda}(x)\right] f(x; \lambda) dx \tag{5.18}$$

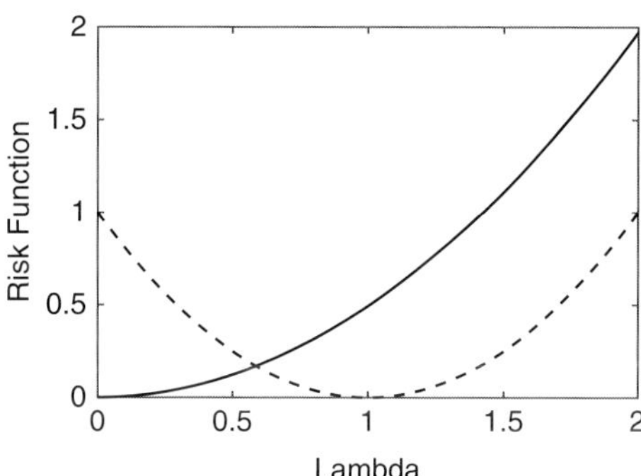

Figure 5.1 Risk functions for $\hat{\lambda}_1$ (solid line) and $\hat{\lambda}_2$ (dashed line).

For the L_2 loss function, the risk is identical to the mean squared error (5.3). Clearly, two estimators can be compared through their risk functions, but this does not necessarily provide an unambiguous result, as can be seen in the following example.

Example 5.16 Let an rv $\mathbf{X} \sim \text{rayl}(x; \lambda)$ using an L_2 loss function, and define two estimators $\hat{\lambda}_1 = \mathbf{X}$ and $\hat{\lambda}_2 = 1$. The risk functions corresponding to the two estimators are $\mathcal{R}(\lambda, \hat{\lambda}_1) = (3 - \sqrt{2\pi})\lambda^2$ and $\mathcal{R}(\lambda, \hat{\lambda}_2) = (1 - \lambda)^2$. Figure 5.1 compares the two, showing that neither dominates the other for all values of λ, so there is no clear decision between the two estimators.

What is needed is a summary statistic for the comparison of risk functions. An estimator $\hat{\lambda}$ is *admissible* if no other estimator $\hat{\lambda}'$ exists such that $\mathcal{R}(\lambda, \hat{\lambda}') \le \mathcal{R}(\lambda, \hat{\lambda})\ \forall\, \lambda$, where strict inequality exists for at least some values of the parameter. However, admissible estimators may be a large class.

A commonly used summary statistic utilizes minimization of the *maximum risk*:

$$\overline{\mathcal{R}}(\hat{\lambda}) = \sup_{\lambda}\left[\mathcal{R}(\lambda, \hat{\lambda})\right] \tag{5.19}$$

where the *supremum* is the least upper bound. For Example 5.16, the maximum risk is of little value if the Rayleigh distribution parameter is unbounded. However, suppose that it is known to lie on [1, 2]. In this case, the estimator $\hat{\lambda}_2$ is preferred because it has the smaller maximum risk. An estimator that minimizes the maximum risk for all possible estimators is called a *minimax rule*. The estimator $\hat{\lambda}$ is minimax if

$$\sup_{\lambda}\left[\mathcal{R}(\lambda, \hat{\lambda})\right] = \inf_{\tilde{\lambda}}\left\{\sup_{\lambda}\left[\mathcal{R}(\lambda, \tilde{\lambda})\right]\right\} \tag{5.20}$$

where the *infimum* is the greatest lower bound over all possible estimators $\tilde{\lambda}$. For standard Gaussian data, the sample mean is minimax for the population mean for any loss function that is convex and symmetric. No other estimator for the population mean is minimax, and consequently, the sample mean is the "best" estimator.

Example 5.16 For an L_2 loss function, consider the following five estimators for the population mean of a standardized Gaussian distribution given a set of N rvs:

1. Sample mean
2. Sample median
3. First value x_1
4. The constant 10
5. The weighted mean with weights w_i that sum to 1

The risk functions for these five estimators are, respectively, $1/N$, $\pi/(2N)$, 1, $(10-\mu)^2$, and $1/N + \sum_{i=1}^{N}(w_i - \overline{w})^2$. The second, third, and last of these are inadmissible because the first offers a lower risk for $N > 1$. The fourth is admissible because it is arbitrarily small when λ is near 10, but it is not minimax. Consequently, the minimax criterion favors the sample mean of these five choices of estimators.

5.3 Point Estimation: Method of Moments

The method of moments is the simplest way to estimate the parameters in a distribution from a set of iid observations and depends only on the law of large numbers. It requires that the population and sample moments be equated at successive orders until enough sample moments are available to compute the population moments. If there are p parameters, then this becomes a system of p equations in p unknowns that can be readily solved.

Example 5.18 For a set of rvs drawn from a normal population, $\mathcal{E}(\mathbf{X}) = \mu$ and $\mathcal{E}(\mathbf{X}^2) = \mu^2 + \sigma^2$, so the method of moments estimator is $\hat{\mu} = \overline{\mathrm{X}}_N$ and $\hat{\sigma}^2 = \overline{\mathrm{X}}_N^2 - \hat{\mu}^2$, where $\overline{\mathrm{X}}_N^2 = \frac{1}{N}\sum_{i=1}^{N} x_i^2$

Example 5.19 For a set of rvs drawn from the gamma distribution, $\mathcal{E}(\mathbf{X}) = \alpha/\beta$ and $\mathcal{E}(\mathbf{X}^2) = \alpha(\alpha+1)/\beta^2$. Then $\hat{\alpha} = \left(\overline{\mathrm{X}}_N\right)^2 / \left[\overline{\mathrm{X}}_N^2 - \left(\overline{\mathrm{X}}_N\right)^2\right]$ and $\hat{\beta} = \overline{\mathrm{X}}_N / \left[\overline{\mathrm{X}}_N^2 - \left(\overline{\mathrm{X}}_N\right)^2\right]$ are the method of moments estimators.

Method of moments estimators have the advantage of simplicity (typically, analytic simplicity) and can serve to initialize numerical procedures to compute statistical parameters in complex situations. Method of moments estimators are consistent but often biased, and are asymptotically Gaussian, but they lack the optimality property of efficiency and may or may not be sufficient statistics. In addition, they may yield estimates that lie outside the sample space, in which case they are unreliable. Further, they may not be unique; for example, the sample mean and sample variance are both method of moments estimators for the parameter in a Poisson distribution.

5.4 Point Estimation: Maximum Likelihood Estimator

It is natural to seek the value of the parameter λ for which the likelihood function $\mathcal{L}(\lambda; x_i)$ is maximized as the most probable value for the parameter because this maximizes the probability of observing the data that have been collected. In practice, it may be easier to maximize the log likelihood rather than the likelihood; both quantities have the same maximum. Equivalently, one can minimize the negative log likelihood, which often is the case for numerical optimization algorithms. Mathematically, this means seeking solutions of

$$\partial_\lambda \log \mathcal{L}(\lambda; x_i) = 0 \tag{5.21}$$

subject to $\partial_\lambda^2 \mathcal{L}(\lambda; x_i) < 0$. The solution $\hat{\lambda}_k$ to (5.21) is the *maximum likelihood estimator* (mle). In practice, the likelihood function may have multiple maxima, in which case the largest one is chosen, or it may have equal maxima, in which case the mle is not unique.

Example 5.20 Suppose that $\{\mathbf{X}_i\}$ are N independent rvs from a Poisson distribution with parameter λ. Derive the mle $\hat{\lambda}$ for λ.

The joint distribution is the product of the marginal pdfs because the data are iid

$$\text{pois}_N(x_i; \lambda) = e^{-N\lambda} \prod_{i=1}^{N} \frac{\lambda^{x_i}}{x_i!}$$

The log likelihood is

$$\log \mathcal{L}(\lambda; x_i) = -N\lambda + \sum_{i=1}^{N} (x_i \log \lambda - \log x_i!)$$

Setting the derivative with respect to λ to zero, the solution is the sample mean, and the second derivative with respect to λ is negative, so the sample mean is the mle.

Example 5.21 Suppose that $\{\mathbf{X}_i\}$ are independent rvs from $\mathrm{N}(\mu, \sigma^2)$ where neither μ nor σ^2 is known. Derive the mles for μ and σ^2.

The likelihood function is

$$\mathcal{L}(\mu, \sigma^2; x_i) = \frac{1}{(2\pi\sigma^2)^{N/2}} e^{(-1/2\sigma^2) \sum_{i=1}^{N} (x_i - \mu)^2}$$

The log likelihood function is

$$\log \mathcal{L}(\mu, \sigma^2; x_i) = -\frac{N}{2}\log 2\pi - \frac{N}{2}\log \sigma^2 - \frac{1}{2\sigma^2}\sum_{i=1}^{N}(x_i - \mu)^2$$

Taking the derivatives of $\log \mathcal{L}$ with respect to μ and σ^2 and setting them to zero yields $\hat{\mu} = \overline{\mathrm{X}}_N$ and $\hat{\sigma}^2 = \hat{s}_N^2$. Formally, it is necessary to also show that the second derivatives are negative, which is left as an exercise for the reader. It can also be shown that the sample mean is the mle for μ when σ^2 is known.

Example 5.22 The generalized extreme value distribution pdf (Section 3.4.9) is

$$\mathrm{gev}(x; \varsigma, \xi, \gamma) = \frac{1}{\gamma}\left(1 + \varsigma\frac{x - \xi}{\gamma}\right)^{-(1+1/\varsigma)} e^{-[1+\varsigma(x-\xi)/\gamma]^{-1/\varsigma}}$$

The likelihood and log likelihood functions are easily constructed from the distribution. Taking derivatives with respect to ς, ξ, and γ and setting them equal to zero gives three equations

$$\sum_{i=1}^{N}\left\{\frac{\varsigma + 1}{1 + \varsigma(x_i - \xi)/\gamma} - [1 + \varsigma(x_i - \xi)/\gamma]^{-(1+1/\varsigma)}\right\} = 0$$

$$\sum_{i=1}^{N}\log[1 + \varsigma(x_i - \xi)/\gamma] - \frac{\varsigma(1+\varsigma)}{\gamma}\sum_{i=1}^{N}\frac{x_i - \xi}{1 + \varsigma(x_i - \xi)/\gamma} + \sum_{i=1}^{N}N[1 + \varsigma(x_i - \xi)/\gamma]^{-1/\varsigma}$$
$$\times\left\{\frac{\log[1 + \varsigma(x_i - \xi)/\gamma]}{\varsigma} - \frac{x_i - \xi}{\varsigma[1 + \varsigma(x_i - \xi)/\gamma]}\right\} = 0$$

$$N - \sum_{i=1}^{N}\left\{(1 + 1/\varsigma)\frac{\varsigma(x_i - \xi)/\gamma}{1 + \varsigma(x_i - \xi)/\gamma} + \frac{x_i - \xi}{\gamma}[1 + \varsigma(x_i - \xi)/\gamma]^{-(1+1/\varsigma)}\right\} = 0$$

These are a coupled nonlinear set of equations that must be solved numerically. Further, it is necessary to show that the second derivatives are all negative, which is a real mess.

Example 5.23 In Example 4.4 it was stated that the parameters for the von Mises distribution could be estimated using the mean direction and the solution to a transcendental equation. Show that these are the mles.

The von Mises log likelihood function is given by

$$\log \mathcal{L}(\nu, \kappa|\theta_i) = \kappa\cos(\theta_i - \nu) - \log(2\pi) - \log I_0(\kappa)$$

Taking the derivative with respect to ν gives $\sum_{i=1}^{N}\sin(\theta_i - \nu) = 0$, whose solution $\hat{\nu}$ after expanding the sine function is the sample mean direction (4.25). Taking the second derivative with respect to v gives a result that is always negative, so the sample mean direction is the mle. Taking the derivative with respect to κ and using the fact that $\partial_\kappa I_0(\kappa) = I_1(\kappa)$ gives the transcendental equation $I_1(\kappa)/I_0(\kappa) = \overline{\mathrm{R}}$, which was solved numerically in Example 4.4. Since the second derivative of the log likelihood with respect to κ is negative, its solution is the mle.

Let $\Upsilon = \upsilon(\lambda)$ be a one-to-one (i.e., monotonically increasing or decreasing) mapping of λ. The likelihood function (or the log likelihood function) of υ (denoted $\hat{\upsilon}_k$) will be maximized at $\lambda = \hat{\lambda}_k$. This is the *equivariance property* of the mle: If $\hat{\lambda}_k$ is the mle for λ, then $\hat{\upsilon}_k = \upsilon(\hat{\lambda}_k)$ is the mle for Υ.

Example 5.24 It has been established that the mles for the mean and variance of a normal distribution are the sample mean and sample variance, respectively. Then the mle for the standard deviation is the square root of the sample variance.

Choosing the mle $\hat{\lambda}_k$ to maximize the likelihood function $\mathcal{L}$ is equivalent to choosing $\hat{\lambda}_k$ to maximize V in the Fisher-Neyman factorization theorem (5.16), and hence $\hat{\lambda}_k$ must be a function of only a sufficient statistic or must itself be a sufficient statistic. Further, $\hat{\lambda}_k$ will be unique and inherit all the optimality properties of sufficient statistics.

Example 5.25 The mle for the parameter in the Poisson distribution is the sample mean. Since it is an unbiased estimator, sufficiency establishes that it is the unique efficient estimator for the parameter.

Example 5.26 The sufficient statistic for the uniform distribution on $(0, \lambda)$ is the largest order statistic $x_{(N)}$. From inspection of the joint pdf given in Example 5.9, it is obvious that the mle is also the largest order statistic. However, this cannot be an unbiased estimator because $x_{(N)} \le \lambda$.

The mle has a number of desirable optimality properties. Under appropriate smoothness constraints on the likelihood function, the mle from an iid sample is an asymptotically consistent estimator. This applies whether there is a unique maximum for the likelihood function or not. The mle is also asymptotically normal with variance given by the inverse of the sample Fisher information (or for vector mles, a covariance matrix given by the inverse of the sample Fisher information matrix). Mathematically, this is

$$\frac{\hat{\lambda_N} - \lambda}{\sqrt{1/I_N\left(\hat{\lambda}_N\right)}} \xrightarrow{d} N(0,1) \tag{5.22}$$

Consequently, the standard error on the mle can be obtained from the Fisher information or alternately using the delta method of Section 4.7.4. The mle is asymptotically efficient, achieving the Cramér-Rao lower bound as the sample size approaches infinity. This means in particular that no asymptotically unbiased estimator has a lower asymptotic mse than the mle. The mle is also asymptotically minimax. However, note that all these properties are large sample ones and do not specify the finite sample properties. Finally, if the mle is unique, then it is a function of the minimal sufficient statistic. This is especially useful for the exponential family of distributions, where the minimal sufficient statistic is usually easy to find.

MATLAB simplifies the process of obtaining the mle for the distributions that it supports. For example, it provides the function *parmhat* = **gevfit**(*x*) that computes the three parameters in the generalized extreme value distribution given the data in *x*, where *parmhat* is a 3-vector containing the shape, scale, and location parameters. It provides similar estimators for the mle for most of the distributions described in Chapter 3. Alternately, invoking the **fitdist** function creates a distribution object with mle parameters. For example, *pd* = **fitdist**(*x*, 'Normal') creates a distribution object from the data in *x* with mle parameters for the mean and variance.

The same functionality can be obtained using the general mle function that is invoked using *parmhat* = **mle**(*x*, '*distribution*', *dist*), where *dist* is a character string that specifies the distribution. The list of distributions **mle** supports is more extensive than those directly implemented as pdf, cdf, and quantile functions.

The function **mle** also enables the user to define the distribution to be fit using *parmhat* = **mle**(*x*, '*pdf*', *pdf*, '*cdf*', *cdf*, '*start*', *start*), where *pdf* and *cdf* are function handles to custom pdf and cdf routines, and *start* is an initial guess for the parameters. It is also possible to substitute the negative log likelihood for the pdf and cdf using the keyword '*nloglf*'.

The function **mle** is a wrapper to call one of the two optimization functions in MATLAB, with the default being **fminsearch** and the alternative being **fmincon**. The former is an unconstrained nonlinear multivariable function minimizer based on the simplex method of Lagarias et al. (1998) that does not require numerical differentiation but is rather a direct search approach. The **fmincon** algorithm allows constraints to be applied to the minimization and, in addition, allows the gradient and Hessian matrix to be user provided for greater accuracy.

For complicated situations, the "Optimization Toolbox" provides additional function minimizers that may be more suitable. For example, the function **fminunc** is a multivariable nonlinear unconstrained function minimizer. Calling this routine with a user-supplied routine that provides the negative log likelihood function along with its gradient and Hessian typically gives better performance than **mle** with **fminsearch**. There are even more sophisticated algorithms in the MATLAB "Global Optimization Toolbox."

MATLAB provides the function **mlecov** that computes the asymptotic covariance matrix for the mle. In its simplest form, it is invoked using *acov* = **mlecov**(*parmhat*, *x*, '*pdf*', *pdf*), where *pdf* is a function handle. An alternative call based on the negative log likelihood function is also available.

Example 5.27 The file geyser.dat contains 299 measurements of the waiting time in minutes between eruptions of Old Faithful in Yellowstone National Park taken from Azzalini & Bowman (1990). A kernel density plot of the data shows that their distribution is bimodal. Fit a mixture Gaussian distribution to the data using the mle.

The mixture pdf for a single rv is given by

$$f\left(x;\ \mu_1, \mu_2, \sigma_1^2, \sigma_2^2, \alpha\right) = \alpha N\left(\mu_1, \sigma_1^2\right) + (1-\alpha) N\left(\mu_2, \sigma_2^2\right)$$

where $0 \le \alpha \le 1$. The log likelihood is given by

$$\log \mathcal{L}\left(\mu_1, \mu_2, \sigma_1^2, \sigma_2^2, \alpha; x\right) = \sum_{i=1}^{N} \log\left[\alpha \mathrm{N}\left(x_i; \mu_1, \sigma_1^2\right) + (1-\alpha)\mathrm{N}\left(x_i; \mu_2, \sigma_2^2\right)\right]$$

A MATLAB script to fit this distribution is

```
fun = @(lambda, data, cens, freq) - nansum(log(lambda(5)/
      (sqrt(2*pi)*lambda(3))* ...
      exp(-((data - lambda(1))/lambda(3)).^2/2) + ...
      (1 - lambda(5))/(sqrt(2*pi)*lambda(4))* ...
      exp(-((data - lambda(2))/lambda(4)).^2/2)));
start = [ 55 80 5 10 .4];
parm = mle(geyser, 'nloglf', fun, 'start', start)
```

Warning: Maximum likelihood estimation did not converge. Iteration limit exceeded.

```
  In stats/private/mlecustom at 241
  In mle at 229
parm =
   54.2026    80.3603    4.9520    7.5076    0.3076
```

Note the use of **nansum** in the negative log likelihood function to avoid overflow when the objective function is far from its minimum. The four arguments to the function handle must be provided even when the last two are not used. The start vector is an eyeball fit to the data plot in Figure 5.2, and the parameters are in the order μ_1, μ_2, σ_1, σ_2, α. The

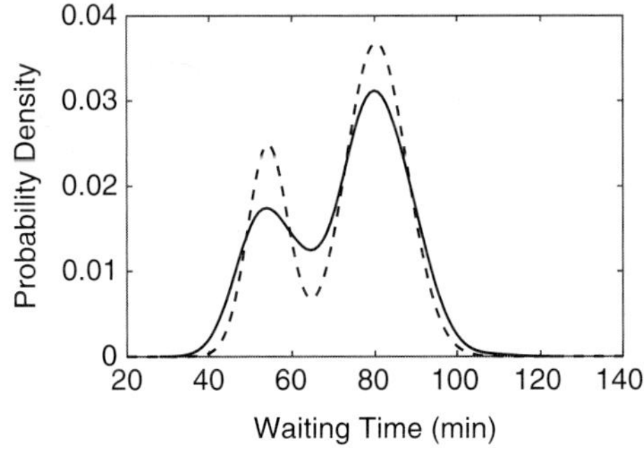

Figure 5.2 Kernel density estimator (solid line) and mixture Gaussian distribution with mle parameters (dashed line) for the Old Faithful waiting time data.

solution did not converge before the default limit to the number of iterations was reached. This can be set to a higher value using "options" and the **statset** function to change its value.

```
parm = mle(geyser, 'nloglf', fun, 'start', start, 'options',
statset('MaxIter', 300))
parm =
   54.2026    80.3603     4.9520     7.5076     0.3076
The result is identical to four significant figures.
m1 = parm(1);
m2 = parm(2);
s1 = parm(3);
s2 = parm(4);
a = parm(5);
ksdensity(geyser)
hold on
x = 20:.1:140;
plot(x, a/(sqrt(2*pi)*s1)*exp(-((x - m1)/s1).^2/2) + ...
     (1 - a)/(sqrt(2*pi)*s2)*exp(-((x - m2)/s2).^2/2))
```

The result is shown in Figure 5.2. The data are bimodal with approximate modes of 55 and 80 minutes. The fitted distribution matches the means in the kernel density estimator but is more sharply peaked, probably because a Gaussian distribution is not the appropriate model. A distribution with flatter peaks would be more appropriate.

5.5 Interval Estimation: Confidence and Tolerance Intervals

Confidence intervals are an alternative to point estimators for characterizing an unknown parameter. Instead of simply estimating the parameter, an interval on which it lies is constructed for which the probability may be stated a priori.

Assume that $\{\mathbf{X}_i\}$ are a random sample from a distribution with parameter λ. Let $\mathbf{Y}$ be an rv that is a function of $\{\mathbf{X}_i\}$ and λ with a continuous cdf $F(y)$ that does not explicitly depend on λ. There exists a number υ such that

$$F(\upsilon) = \Pr(y(x_i, \lambda) \le \upsilon) = 1 - \alpha \tag{5.23}$$

for which the probability inequality indicates a constraint on λ. The constraint may be written as $\xi_0 \le \lambda \le \xi_1$, where ξ_0 and ξ_1 are specific values of a statistic that does not depend on λ. Consequently,

$$F(\xi_1) - F(\xi_0) = \Pr(\xi_0 \le \lambda \le \xi_1) = 1 - \alpha \tag{5.24}$$

or, equivalently,

$$\int_{\xi_0}^{\xi_1} f(x)\,dx = 1-\alpha \tag{5.25}$$

The quantity $1-\alpha$ is the probability that the random interval (ξ_0, ξ_1) contains the parameter λ. The interval (ξ_0, ξ_1) is a *confidence interval* for λ, and ξ_0 and ξ_1 are the lower and upper *confidence limits*. The *confidence level* is $1-\alpha$.

Caution: A confidence interval is not a probability statement about λ because λ is a fixed parameter rather than an rv.

Example 5.28 Let $\{\mathbf{X}_i\}$ be a random sample from $N(\mu, \sigma^2)$ with unknown mean and variance. The statistic $\hat{t}_{N-1} = \sqrt{N-1}\left(\overline{\mathrm{X}}_N - \mu\right)/\hat{s}_N$, where $\hat{s}_N$ is the sample standard deviation given by the square root of (4.14), is distributed as Student's t (4.72) with $N-1$ degrees-of-freedom. As a result, the confidence interval on μ defined by

$$\Pr\left(\overline{\mathrm{X}}_N + \frac{\xi_0 \hat{s}_N}{\sqrt{N-1}} \le \mu \le \overline{\mathrm{X}}_N + \frac{\xi_1 \hat{s}_N}{\sqrt{N-1}}\right) = 1-\alpha$$

requires choosing a constant ξ such that $|\xi_0| = \xi_1 = \xi$, where

$$\int_{-\xi}^{\xi} \mathrm{tee}_{N-1}(x)\,dx = 1-\alpha$$

The value of ξ can be computed for given choices of α and N from the cdf $\mathrm{Tee}_{N-1}(x)$. Because of symmetry, $2\mathrm{Tee}_{N-1}(\xi) - 1 = 1-\alpha$, so $\mathrm{Tee}_{N-1}(\xi) = 1-\alpha/2$.

In Example 5.28, the confidence interval was obtained by assigning $\alpha/2$ of the probability to lie below $-\xi$ and above ξ in (5.24), respectively. This choice is called the *central confidence interval* but is in reality only one of an infinite number of possible choices. Confidence intervals with different amounts of probability below and above the lower and upper confidence limits are called *noncentral*. A noncentral confidence interval should be used (for example) when the penalty for overestimating a statistic is higher than that for underestimating it. In such an instance, the one-sided confidence interval obtained from $\Pr(\lambda \le \xi_1) = 1-\alpha$ would be appropriate. Central confidence limits typically have a shorter length than noncentral ones and are usually the most appropriate ones to use. Central confidence limits will be symmetric about the sample statistic only for symmetric distributions, such as Student's t.

Example 5.29 The central confidence interval for the binomial distribution may be obtained by solving

$$\sum_{k=0}^{j} \binom{N}{k} p_u^k (1 - p_u)^{N-k} = \alpha/2$$

and

$$\sum_{k=j}^{N} \binom{N}{k} p_l^k (1 - p_l)^{N-k} = \alpha/2$$

for p_l and p_u for a given choice of j and N. Let $N = 20$ and $j = 5$, so the mle for the distribution parameter $\hat{p} = 0.25$. An exact answer can be obtained by solving these two equations numerically, but an approximate answer ensues from using the binomial quantile function.

```
binoinv(.025, 20, .25)
ans =
   2
binoinv(.975, 20, .25)
ans =
   9
```

Note that the confidence interval (2, 9) is not symmetric about 5 because the binomial distribution is skewed when the probability $p \neq 0.5$.

The result in Example 5.29 is approximate because **binoinv** returns the least integer x such that the binomial cdf evaluated at x equals or exceeds the probabilities 0.025 or 0.975. This is called a *conservative confidence interval* and follows from the probability statement

$$\Pr(\xi_0 \leq \lambda \leq \xi_1) \geq 1 - \alpha \tag{5.26}$$

This confidence interval states that the parameter λ lies on (ξ_0, ξ_1) *at least* $100(1 - \alpha)\%$ of the time. The conservative confidence interval may be the only one that can be constructed, especially for discrete distributions.

Statistical problems often present with several parameters rather than just one, and hence the requirement for simultaneous confidence intervals on several parameters arises. One obvious approach would be simultaneous confidence interval assertions such as

$$\Pr(\xi_0 \leq \lambda_0 \leq \xi_1 \cap \zeta_0 \leq \lambda_1 \leq \zeta_1) = 1 - \alpha \tag{5.27}$$

This is rarely possible unless the parameters are independent, and even then it is usually impractical for more than two parameters.

With p parameters, the use of p univariate $1 - \alpha$ confidence intervals is inappropriate because the overall tail probability $p\alpha$ will exceed the desired value of α. The most widely used solution to this quandary is the Bonferroni confidence interval based on a family of probability inequalities, as formalized in Section 6.6. The general idea is to use the Student's t statistic as for the univariate confidence interval after dividing the significance level among the variables for which a simultaneous confidence interval is desired by using $1 - \alpha/p$ in the usual confidence interval expression. The Bonferroni approach does allow some variables to be given more significance than others, but generally they are treated equally. The result is conservative because the overall confidence interval on all the variables is no greater than $1 - \alpha$.

Example 5.30 Returning to the geyser data of Example 5.27, compute Bonferroni confidence intervals for the parameters.

The asymptotic covariance matrix is easily obtained by calling **mlecov**.

```
acov = mlecov(parm, geyser, 'nloglf', fun)
acov =
   0.4666    0.1545    0.1606   -0.1417    0.0064
   0.1545    0.4012    0.1176   -0.1235    0.0052
   0.1606    0.1176    0.2686   -0.1048    0.0050
  -0.1417   -0.1235   -0.1048    0.2571   -0.0049
   0.0064    0.0052    0.0050   -0.0049    0.0009
```

The off-diagonal terms are not vanishingly small but will be ignored for present purposes. The uncertainties on the parameters follow from (5.22) using the diagonal elements of the inverse of the Fisher information matrix (5.10) output by **mlecov**. Assuming that the total tail probability is 0.05, the normal p-quantile 2.3263 for a probability of $1 - 0.05/5 = 0.99$ is used to construct confidence intervals on the parameters. The MATLAB script is

```
xp = norminv(1 - .05/5);
sigma=sqrt(diag(acov));
[parm' - xp* sigma parm' + xp* sigma]
ans =
   52.6136   55.7917
   78.8868   81.8338
    3.7464    6.1576
    6.3280    8.6873
    0.2368    0.3784
```

Note that the confidence bounds on the means are much tighter than for the remaining parameters.

The concept of confidence limits extends to the order statistics in a straightforward manner and yields the remarkable result that the confidence intervals of the quantiles are distribution free that is due originally to Thompson (1936). Let the p-quantile be given by

$x_p = F^{-1}(p)$, where p is a probability level. The probability that x_p lies between two order statistics $X_{(r)}$ and $X_{(s)}$ (assuming that $r < s$) is (David 1981, sec. 2.5)

$$\begin{aligned}\Pr\left(X_{(r)} < x_p < X_{(s)}\right) &= \Pr\left(X_{(r)} \le x_p\right) - \Pr\left(X_{(s)} \le x_p\right) \\ &= \mathrm{Beta}(p; r, N - r + 1) - \mathrm{Beta}(p; s, N - s + 1)\end{aligned} \tag{5.28}$$

Equating (5.28) to the probability $1 - \alpha$ gives a result that does not depend on $F(x)$ or $f(x)$ and hence on the parent distribution. This result holds for any discrete or continuous distribution.

Example 5.31 Find the $\alpha = 0.05$ confidence interval on the sample median for sample sizes $N = 11$, 101, and 1001.

From symmetry considerations, the indices of the confidence interval on the sample median will be equal distances from that of the median; hence $s = N - r + 1$. The solution of

$$\mathrm{Beta}(0.5; i, N - i + 1) - \mathrm{Beta}(0.5; N - i + 1, i) \ge 0.95$$

for r given $N = 11$, 101, and 1001 can be obtained by direct evaluation starting with $i = 1$ and extending to $i = \lfloor N/2 \rfloor + 1$. For $N = 11$, this yields

```
n = 11;
for i = 1:floor(n/2) + 1
    [i betacdf(.5, i, n - i + 1) - betacdf(.5, n - i + 1, 1)]
end
1.0000     0.9990
2.0000     0.9932
3.0000     0.9653
4.0000     0.8828
5.0000     0.7178
6.0000     0.4844
```

so $X_{(3)} \le \tilde{X}_{11} \le X_{(9)}$. Repeating this for the remaining sample sizes gives $X_{(42)} \le \tilde{X}_{101} \le X_{(60)}$ and $X_{(474)} \le \tilde{X}_{1001} \le X_{(528)}$, respectively. These conservative confidence intervals apply for any distribution for the sample.

A closely related entity to the confidence interval is the *tolerance interval*, which is a confidence interval for a distribution rather than for a statistic. The distinction can be illustrated by considering a collection of rock samples, each of which has had its p-wave velocity measured in the laboratory. The upper confidence limit for the mean of the p-wave velocities is an upper limit to the mean of all the samples. The upper tolerance limit is an upper bound on the p-wave velocity that a specified fraction of the samples will not exceed.

A tolerance interval contains at least a proportion p of the population with probability α. Mathematically, the random interval (ξ_0, ξ_1) defined by

$$\Pr[F(\xi_1) - F(\xi_0) \geq p] = \alpha \tag{5.29}$$

where p is a free parameter, defines the tolerance interval at probability level α. The tolerance bound defined by (5.29) is independent of the usually unknown cdf $F(x)$ if and only if ξ_0 and ξ_1 are order statistics. Let $\{\mathbf{X}_{(i)}\}$ be the order statistics obtained by ranking a random sample $\{\mathbf{X}_i\}$, and let $\xi_0 = \mathbf{X}_{(r)}$ and $\xi_1 = \mathbf{X}_{(s)}$, where $1 \leq r < s \leq N$. As can be easily shown, the random interval $F(\mathbf{X}_{(s)}) - F(\mathbf{X}_{(r)}) \sim \text{Beta}(x, s - r, N - s + r + 1)$. Consequently, for a given choice of N, p, and α, the index difference $s - r$ can be computed using

$$\text{Beta}(1 - p, N - s + r + 1, s - r) = \alpha \tag{5.30}$$

where the symmetry property $\text{Beta}(x, a, b) = \text{Beta}(1 - x, b, a)$ has been applied. The outcome is a distribution-free method to estimate the tolerance interval for a random sample.

Example 5.32 Estimate the tolerance interval at $\alpha = 0.05$ for a sample of size 100 and $p =$ 0.5, 0.75, and 0.95.

Let $i = s - r$ so that $\text{Beta}(1 - p,\ 100 - i + 1,\ i) = 0.05$ is to be solved for i. By trial and error using MATLAB,

```
for i = 40:50
   [i betacdf (.5, 100 - i + 1, i) - 0.05]
end
ans =
40.0000   -0.0324
41.0000   -0.0216
42.0000   -0.0057
43.0000    0.0166
44.0000    0.0467
45.0000    0.0856
46.0000    0.1341
47.0000    0.1921
48.0000    0.2586
49.0000    0.3322
50.0000    0.4102
```

and hence $s - r = 42$ for $p = 0.5$. Any interval $(\mathbf{X}_{(r)}, \mathbf{X}_{(r+42)})$ for $r = 1, \ldots, 58$ serves as a tolerance interval for $p = 0.5$, $\alpha = 0.05$. Solving for the remaining two cases gives $s - r = 18$ and 2, respectively. In practice, it is usually the lower ($r = 0$) or upper ($s = N + 1$) tolerance intervals that are of interest. These limiting values establish the proportion of the population that exceed or fall below a given order statistic, respectively.

5.6 Ratio Estimators

Ratios are frequently of interest in the earth sciences and are a long-standing problem in statistics for which the estimation of a confidence interval is often difficult. There are two statistically distinct approaches: computing either the ratio of averages or the average of ratios. The latter often results in a distribution with infinite variance, and hence the median becomes the most useful measure of location; confidence intervals on it can be constructed from (5.28). It is more common to use a ratio of averages approach. If the numerator and denominator of a ratio are themselves averages, then each may be well behaved such that the classic central limit theorem pertains separately to the numerator and denominator, and hence either Fieller's theorem (described below) or the delta method of Section 4.7.4 may be used for statistical inference. However, it is difficult to quantify when the asymptotic limit is reached. Further, if the ratio has bounded support, there is no guarantee that a ratio of averages will maintain the bounds.

It is easy to show that the ratio estimator is always biased. Let an estimator $\hat{r}$ be obtained from the ratio of the sample means of the random variables **Y** and **X**. A first-order Taylor series expansion about the population parameters for a sample of size N yields

$$\mathcal{E}(\hat{r}) \approx r + \frac{1}{N\mu_X^2}\left(r\sigma_X^2 - cov[\mathbf{X},\mathbf{Y}]\right) \tag{5.31}$$

Consequently, the sample ratio estimate bias increases if the coefficient of variation (ratio of the standard deviation to the mean) of the denominator becomes large and is reduced when the covariance of the two variables has the same sign as r. A naive bias-corrected form follows by subtracting the second term of (5.31) from $\hat{r}$. However, the bias is $O(1/N)$ and hence is small when N is of moderate size.

The most general approach to confidence interval estimation for ratios is due to Fieller (1940, 1944, 1954). The construction of Fieller confidence limits for the ratio $r = \mu_y/\mu_x$ follows by noting that since linear combinations of Gaussian variables remain Gaussian, so is $\mu_y - r\mu_x$. Consequently, dividing the sample version of this term by an estimate of its standard deviation yields a quantity that is distributed as Student's t. That statistic is

$$\hat{T} = \frac{\overline{\mathrm{Y}}_N - \hat{r}\,\overline{\mathrm{X}}_N}{\sqrt{var\left(\overline{\mathrm{Y}}_N\right) - 2\hat{r}\,cov\left(\overline{\mathrm{Y}}_N,\overline{\mathrm{X}}_N\right) + \hat{r}^2 var\left(\overline{\mathrm{X}}_N\right)}} \tag{5.32}$$

$\hat{T}$ is a *pivot* because (5.32) has a distribution that does not depend on any of its constituent parameters. Let $t_{\alpha/2}$ be the $1-\alpha/2$ quantile of Student's t distribution with ν degrees-of-freedom. The Fieller confidence interval follows from the definition

$$1-\alpha = \Pr\left[-t_{\alpha/2} \le \hat{T} \le t_{\alpha/2}\right] = \Pr\left[\hat{\alpha}r^2 + \hat{\beta}r + \hat{\chi} = 0\right] \tag{5.33}$$

where

$$\hat{\alpha} = \left(\overline{X}_N\right)^2 - t_{\alpha/2}^2 var\left(\overline{X}_N\right)$$
$$\hat{\beta} = 2\left[t_{\alpha/2}^2 cov\left(\overline{Y}_N, \overline{X}_N\right) - \overline{Y}_N \overline{X}_N\right]$$
$$\hat{\chi} = \left(\overline{Y}_N\right)^2 - t_{\alpha/2}^2 var\left(\overline{Y}_N\right) \tag{5.34}$$

and corresponds to the set of r values for which $\hat{T}$ lies within the $1 - \alpha$ range of Student's t. The confidence interval is given by

$$\frac{-\hat{\beta} - \sqrt{\hat{\beta}^2 - 4\hat{\alpha}\hat{\chi}}}{2\hat{\alpha}} \leq r \leq \frac{-\hat{\beta} + \sqrt{\hat{\beta}^2 - 4\hat{\alpha}\hat{\chi}}}{2\hat{\alpha}} \tag{5.35}$$

provided that $\hat{\alpha} \geq 0$ and $\hat{\beta}^2 - 4\hat{\alpha}\hat{\chi} \geq 0$. Equation (5.35) is asymmetric about the sample estimate for r.

Equation (5.35) is the solution to (5.33) when the square of the denominator divided by its variance is significant, or $\left(\overline{X}_N\right)^2 / var\left(\overline{X}_N\right) > t_{\alpha/2}^2$. There are two other cases: when $\hat{\beta}^2 - 4\hat{\alpha}\hat{\chi} \geq 0$ and $\hat{\alpha} < 0$, the confidence interval is the complement of (5.35) given by

$$\left(-\infty, \frac{-\hat{\beta} + \sqrt{\hat{\beta}^2 - 4\hat{\alpha}\hat{\chi}}}{2\hat{\alpha}}\right] \leq r \leq \left[\frac{-\hat{\beta} - \sqrt{\hat{\beta}^2 - 4\hat{\alpha}\hat{\chi}}}{2\hat{\alpha}}, \infty\right) \tag{5.36}$$

whereas if $\hat{\beta}^2 - 4\hat{\alpha}\hat{\chi} < 0$ and $\hat{\alpha} < 0$, the confidence interval is $(-\infty, \infty)$. In both of these cases, the confidence interval is infinite, and little or nothing can be said about r.

The occurrence of infinite confidence intervals is a consequence of the denominator becoming arbitrarily close to zero, and Von Luxburg & Franz (2009) give an intuitive explanation for its effects. Equation (5.36) pertains when the denominator is not significantly different from zero, but the numerator is well defined. In that instance, as the numerator is divided by a number that is close to zero, the absolute value and sign of the result are uncontrolled, leading to a disjoint pair of infinite confidence intervals. If the numerator is also zero within the statistical uncertainty, then any result is possible because zero divided by zero is undefined. In either case, the statistic of interest is not well specified, and little can be concluded from the data.

A widely used alternative to Fieller's method is the delta method of Section 4.7.4. This can be applied directly to a ratio y/x by expanding that quantity in a two-variable first-order Taylor series and then taking the variance of the result. For the ratio of the sample means, this is

$$var\left(\frac{\overline{Y}_N}{\overline{X}_N}\right) \approx \frac{\left(\overline{Y}_N\right)^2}{\left(\overline{X}_N\right)^2}\left[\frac{var\left(\overline{Y}_N\right)}{\left(\overline{Y}_N\right)^2} - 2\frac{cov\left(\overline{Y}_N, \overline{X}_N\right)}{\overline{Y}_N \overline{X}_N} + \frac{var\left(\overline{X}_N\right)}{\left(\overline{X}_N\right)^2}\right] \tag{5.37}$$

It is common practice to neglect the middle covariance term in (5.37), although this can lead to large, undetectable errors.

Delta method confidence intervals using Student's t may be computed with the standard error obtained from (5.37). Such a confidence interval will always be symmetric about $\hat{r}$

and cannot be unbounded, in contrast to the Fieller confidence interval. The delta method will break down when the denominator of the ratio statistically approaches zero and will severely underestimate the confidence interval as that limit is approached. However, the delta method produces a reasonable approximation to the Fieller confidence interval when the denominator is sufficiently precise.

Hirschberg & Lye (2010) provide a lucid geometric representation and comparison of the Fieller and delta methods and suggest that the delta method produces a good approximation to the Fieller interval when the denominator of the ratio is precise and the signs of the ratio and the covariance of the numerator and denominator in (5.37) are the same, but the method is less accurate when the signs of the ratio and the covariance of the numerator and denominator in (5.37) differ. They also indicate that when their results differ, the Fieller interval provides better coverage and hence is preferred.

Gleser & Hwang (1987) proved that any statistical method that is not able to produce infinite confidence intervals for a ratio will lead to arbitrarily large errors or conversely, will have coverage probability that is arbitrarily small. For this reason, the Fieller approach is preferred to the delta method, and given that it does not require a substantial increase in computational burden, it is recommended for general use.

6 Hypothesis Testing

6.1 Introduction

In hypothesis testing, the goal is to make a decision to accept or reject some statement about the population based on a random sample. The starting point is a *statistical hypothesis* that is a conjecture about one or more populations. This may be phrased in working terms, or it may be stated mathematically. Some instances of the former include

- An engineer believes that there is a difference in accuracy between two sensors; and
- A climate scientist thinks that the mean surface temperature of Earth has increased over the past decade.

Statistical hypotheses are formulated in two parts under the Neyman-Pearson lemma (Neyman & Pearson 1933) that constitutes its theoretical basis. The Neyman-Pearson lemma will be described quantitatively in Section 6.5, but the basics are included at the outset. The first component is the *null hypothesis* denoted by H_0, which is the conjecture to be tested. Departures from the null hypothesis are usually of interest. The second part is the *alternate hypothesis* denoted by H_1. If H_0 is rejected, then H_1 is accepted. However, it can never be known with absolute certainty that a given hypothesis is true, and hence the statement is made that the null hypothesis is accepted (or rejected) at some probability level based on a statistical test. For example,

- The engineer formulates the null hypothesis that there is no difference in sensor accuracy against the alternate hypothesis that there is; and
- The climate scientist formulates the null hypothesis that there is no increase in surface temperature against the alternate hypothesis that there is.

This could be made more specific; for example, the test for the mean surface temperature μ could specify the hypothesis in the form

H_0: $\mu = 20°\text{C}$
H_1: $\mu > 20°\text{C}$

Hypotheses can either be simple or composite, with the latter predominating in the real world. This is best illustrated through examples. Let $\{\mathbf{X}_i\}$ be a set of random variables (rvs) from a Gaussian distribution having a known variance. A null hypothesis is formulated that specifies a mean of μ_1 against an alternate hypothesis that the mean is μ_2. Since σ^2 is known a priori, either hypothesis completely specifies the outcome. This is an example of a *simple hypothesis*.

Alternately, let $\{\mathbf{X}_i\}$ be a set of rvs from a discrete probability distribution. A null hypothesis is formulated that the rvs are drawn from a Poisson distribution against an alternate hypothesis that they are not Poisson. Neither hypothesis completely specifies the distribution (e.g., under H_0, the parameter λ for the Poisson distribution is not given), so this is an example of a *composite hypothesis*. Specifying λ in the null hypothesis would convert it to a simple hypothesis. In general, if a sampling distribution depends on k parameters and a hypothesis specifies m of them, the hypothesis is simple if $m = k$ and composite if $m < k$. Composite hypotheses involve $k - m$ *nuisance parameters* that must be accounted for in some way.

In summary, the steps in testing a hypothesis under the Neyman-Pearson lemma are as follows:

1. Formulate the null and alternate hypotheses using mathematical expressions. This typically makes a statement about a characteristic or a descriptive measure of a population.
2. Collect a random sample from the population of interest.
3. Calculate a statistic from the sample that provides information about the null hypothesis.
4. If the value of the statistic is consistent with the null hypothesis, then accept H_0. This requires knowledge of the probability distribution of the test statistic under the null hypothesis.
5. If the value of the statistic is not consistent with the null hypothesis, then reject H_0 and accept H_1. This is an application of the 0–1 decision rule of Section 5.2.6.

Example 6.1 Returning to the climate example, suppose that the scientist takes a random sample of 100 evenly distributed points on Earth and obtains historical temperature measurements at each over time. She uses the areal and temporal sample means to determine whether there is sufficient evidence to reject the null hypothesis and hence conclude that the mean temperature has increased. The sample mean that she obtains is 22°C, which is slightly larger than the 20°C value for the null hypothesis. However, the sample mean is an rv and has some variability associated with it; if the variance of the sample mean under the null hypothesis is large, then the difference between 22° and 20° might not be meaningful. If the scientist assumes that the surface temperature is normally distributed with a standard deviation of 6° based on prior experience, then she knows that the sample mean is normally distributed with mean 20°C and standard error $6°/\sqrt{100} = 0.6°C$ under the null hypothesis. Standardizing yields the test statistic $\hat{z} = (22 - 20)/0.6 = 3.33$. The test statistic is 3.33 standard deviations from the mean if the null hypothesis is true, and only ~0.3% of normally distributed rvs fall outside 3 standard deviations of the mean. Thus 22°C is not consistent with the null hypothesis except at an unreasonable probability level, and the alternate hypothesis is accepted.

This description illustrates the basic ideas behind hypothesis testing in statistics that pertain whether the approach is parametric (meaning that the distributions of the data and hence the test statistic are known a priori, as covered in this chapter), nonparametric (meaning that distributional assumptions are relaxed), or based on resampling methods (see Chapter 8). While there were earlier approaches to hypothesis testing, the Neyman-Pearson lemma constitutes the basis for testing statistical conjectures in modern works. Nearly concurrently with its presentation, Pitman (1937a) devised exact permutation methods based on the Neyman-Pearson approach that entailed only the assumption of exchangeability (see Section 2.6) but also required a level of computation that was not feasible until decades later. This led to the substitution of ranks for data in the 1940s and 1950s such that the permutation calculations need be completed only once and tabulated, resulting in many of the nonparametric methods described in Chapter 7. As computing power increased dramatically from the 1970s to present, the direct application of permutation methods became more widespread, and in addition, bootstrap resampling methods were invented by Efron (1979). These are covered in Chapter 8 and are typically preferred when dealing with real-world data.

6.2 Theory of Hypothesis Tests I

To formalize these statements in terms of set theory, hypothesis testing is a decision about whether a parameter λ lies in one subset of the parameter space $\mathcal{S}_0$ or in its complement $\mathcal{S}_1$. Consider a parameter λ whose value is unknown but that lies in the parameter space $\mathcal{S} = \mathcal{S}_0 \cup \mathcal{S}_1$. Assume that

$$\begin{aligned} \mathcal{S}_0 \cap \mathcal{S}_1 &= \varnothing \\ H_0 : \lambda &\in \mathcal{S}_0 \\ H_1 : \lambda &\in \mathcal{S}_1 \end{aligned} \tag{6.1}$$

The statistical algorithm to accept H_0 is called the *test procedure*, and the statistic used to make that decision is called the *test statistic*, denoted by $\hat{\lambda}$. The distribution of the test statistic under the null hypothesis is called the *null distribution*. The *critical region* (sometimes called the *rejection region*) is a region for the test statistic over which the null hypothesis would reject. The *critical value* is the locus for the test statistic that divides its domain into regions where H_0 will be rejected or accepted. The critical region depends on the distribution of the statistic under the null hypothesis, the alternate hypothesis, and the amount of error that can be tolerated. Typically, the critical region is in the tails of the distribution of the test statistic when H_0 is true. Consequently, there are three standard cases:

1. If a large value of the test statistic would reject the null hypothesis, then the critical region is in the upper tail of the null distribution. This is called an *upper tail test* and would be specified as H_1: $\lambda > \hat{\lambda}$.
2. If a small value of the test statistic would reject the null hypothesis, then the critical region is in the lower tail of the null distribution. This is called a *lower tail test* and would be specified as H_1: $\lambda < \hat{\lambda}$.
3. If either small or large values for the test statistic would reject the null hypothesis, then the critical region is in both the lower and upper tails. This is called a *two-tail test* and would be specified as H_1: $\lambda \neq \hat{\lambda}$.

Table 6.1 Outcomes from a Single Hypothesis Test

	Declared true	Declared false
True null	Correct $(1 - \alpha)$	Type 1 error (α)
False null	Type 2 error (β)	Correct $(1 - \beta)$

There are two kinds of error that can occur when a statistical hypothesis testing decision is made, and there is an inherent asymmetry between them. The first is called a *Type 1 error*, or a *false positive*, or *the error caused by rejecting the null hypothesis when it is true*. Its probability is denoted by α. Mathematically, this is the conditional probability statement

$$\alpha = \Pr\left(\hat{\lambda} \in \mathcal{S}_1 | H_0\right) \tag{6.2}$$

The second type of error is called a *Type 2 error*, or a *false negative*, or *the error caused by accepting the null hypothesis when it is false*. Its probability is denoted by β. Mathematically, this is

$$\beta = \Pr\left(\hat{\lambda} \in \mathcal{S}_0 | H_1\right) \tag{6.3}$$

Table 6.1 summarizes the possible outcomes and their probabilities from a single hypothesis test.

It is standard practice to search for significant evidence that the alternate hypothesis is true and not to reject the null hypothesis unless there is sufficient data-driven evidence to lead to that conclusion. The probability of making a Type 1 error is called the *significance level* of the test when the hypothesis is simple. If the null hypothesis is composite, the probability of a Type 1 error will depend on which particular part of the null hypothesis is true. In this case, the significance level is usually taken as the supremum of all the probabilities. The significance level is a free parameter chosen by the analyst and represents the maximum acceptable probability for a Type 1 error. Typical values for α lie between 0.01 and 0.05, although larger values might be used in special circumstances. The critical value is the quantile of the null distribution that gives a significance level of α.

Example 6.2 Returning to the climate example, the test statistic is $\hat{z} = 3.33$. A significance level α of 0.05 is chosen. Since the alternate hypothesis holds that the temperature has increased, a large value of the test statistic supports H_1. The critical value using the MATLAB function **norminv**(0.95) is 1.645. Thus, if $\hat{z} \geq 1.645$, H_0 can be rejected at the 0.05 probability level. This is an upper-tail test.

The probability of making a Type 2 error depends on β, which, in turn, depends on the sample size and the alternate hypothesis. The probability of not detecting a departure from the null hypothesis depends on the distribution of the test statistic under the alternate hypothesis, in contrast to Type 1 error, which depends on the distribution under the null hypothesis. However, there are many possible alternate hypotheses, so there are typically many distributions to consider. This makes consideration of Type 2 errors inherently more complicated. In most cases, the null distribution is one of the common sampling distributions of Section 4.9, whereas the alternate hypothesis is the noncentral form of the same distribution, with the noncentrality parameter describing the range of the alternate hypotheses that can be tested.

Rather than specifying or computing β, it is standard practice to determine the probability that H_0 is rejected when it is false, which is called the *power* of the test and is given by $\beta^C = 1 - \beta$. Mathematically, this is

$$\beta^C - \Pr\left(\hat{\lambda} \in S_1 | H_1\right) \tag{6.4}$$

The power is a measure of the ability of the hypothesis test to detect a false null hypothesis. The power depends on the magnitude of the difference between the population and hypothesized means, with power increasing as this rises. Since a larger value of α makes it easier to reject the null hypothesis, it increases the power of the test. In addition, a larger sample size increases the precision of the test statistic and hence increases the power of the test. It is standard practice in experimental design to choose α and β (typically, by placing a lower bound on it, such as requiring it to be at least 0.8) and determine the number of data required to achieve that significance level and power. Finally, for a given (α, β), a one-tail test is more powerful than a two-tail test and is preferred if it is appropriate for a given statistical problem.

Figure 6.1 illustrates the concepts of significance level and power. The solid line denotes the null distribution, and the dotted line shows an instance of the alternate distribution. The tail area for a two-sided hypothesis is the area beneath the gray portion of the null distribution, each part of which contains $\alpha/2$ of the tail probability. The probability of a Type 2 error β is the area under the alternate distribution extending from the lower $\alpha/2$ tail of the null distribution to its upper $\alpha/2$ tail. Consequently, the power is

$$\beta^C = 1 - F_{H1}\left(x_{1-\alpha/2}, \delta\right) + F_{H1}\left(x_{\alpha/2}, \delta\right) \tag{6.5}$$

where F_{H1} is the alternate cdf, x_p is the p-quantile of the null distribution, and δ is a measure of the distance between the test statistic and its hypothesized value. For an

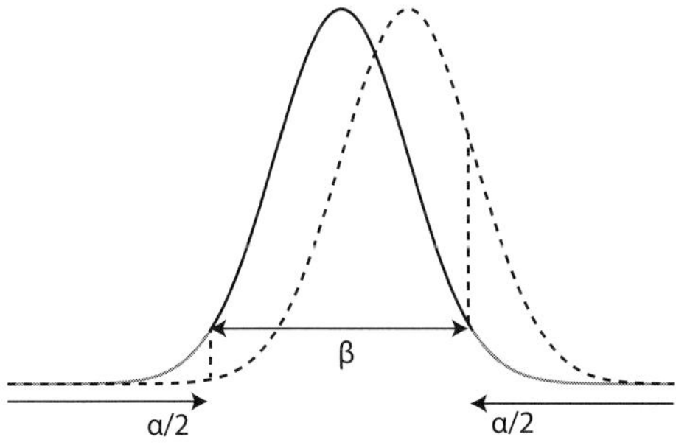

Figure 6.1 The null distribution (solid line) and an alternate distribution (dashed line) illustrating the role of α and β in hypothesis testing. The gray part of the null distribution denotes the tail area for a two-sided hypothesis, so a probability of $\alpha/2$ is given by the area beneath each gray segment. The Type 2 error probability β is the area beneath the alternate distribution extending between the $\alpha/2$ and $1 - \alpha/2$ quantiles of the null distribution.

upper-tail test, the last term in (6.5) is omitted, and $\alpha/2$ is replaced by α. A power curve can be constructed by computing (6.5) for a range of alternate hypotheses (i.e., by varying δ).

An ideal test would have $\alpha = \beta = 0$ or $\alpha = 0, \ \beta^C = 1$, but this is impossible to achieve in practice. Further, for a fixed sample size, in order to decrease α, β must be increased, or vice versa. Under the Neyman-Pearson paradigm, this conflict is resolved by accepting the asymmetry between Type 1 and Type 2 errors. The significance level α is a free parameter that is fixed in advance, and then a test and sample size are sought that maximize the power. The "best" test is defined to be the one that maximizes β^C for a given α out of all possible tests and is called *the most powerful test*. However, it is possible for a test to be most powerful for some but not all alternates. If a given test can be shown to be more powerful against all alternates, it is *uniformly most powerful* (or UMP). A test is *admissible* (Section 5.2.6) if it is UMP and no other test is more powerful against all alternates.

For a composite hypothesis, a test is *exact* if the probability of a Type 1 error is α for all the possibilities that make up the hypothesis. Exact tests are very desirable but often are difficult to devise. The permutation tests described in Section 8.3 are a notable exception. A test is *conservative* if the Type 1 error never exceeds α. An exact test is conservative, but the reverse may or may not hold. Finally, a test at a given significance level is *unbiased* if the power satisfies two conditions:

1. $\beta^C \le \alpha$ for all distributions that satisfy the hypothesis (i.e., β^C is conservative); and
2. $\beta^C \ge \alpha$ for every possible alternate hypothesis.

An unbiased test is more likely to reject a false hypothesis than a true one. A UMP test that is unbiased is called *UMPU*, and devising such a test is desirable but often unachievable, in which case a test that is most powerful against the most interesting alternates is typically sought.

Example 6.3 Returning yet again to the climate example to evaluate Type 2 errors, these values depend on the true mean, and hence the errors must be evaluated over a range of trial values. The following MATLAB script illustrates the approach:

```
mualt = 10:0.1:30;
cv = 1.645;
sigma = 0.6;
ct = cv + 20;
ctv = ct*ones(size(mualt)); %find the area under the curve to
the left of the critical value for each value of the true mean
beta = normcdf(ctv, mualt, sigma); %beta is the probability
of Type 2 error
power = 1 - beta;
p = 0.05*ones(size(mualt));
plot(mualt, power, mualt, p)
```

Figure 6.2 shows the result. As the true mean rises above the nominal value of 20, the power (which is the likelihood that the alternate hypothesis can be detected) rises rapidly.

Table 6.2 Cumulative Binomial Probabilities

	0	1	2	3	4	5	6	7	8	9	10
0.7	0.0000	0.0001	0.0016	0.0106	0.0473	0.1503	0.3504	0.6172	0.8507	0.9718	1.0000
0.6	0.0001	0.0017	0.0123	0.0548	0.1662	0.3669	0.6177	0.8327	0.9536	0.9940	1.0000
0.5	0.0010	0.0107	0.0547	0.1719	0.3770	0.6230	0.8281	0.9453	0.9893	0.9990	1.0000

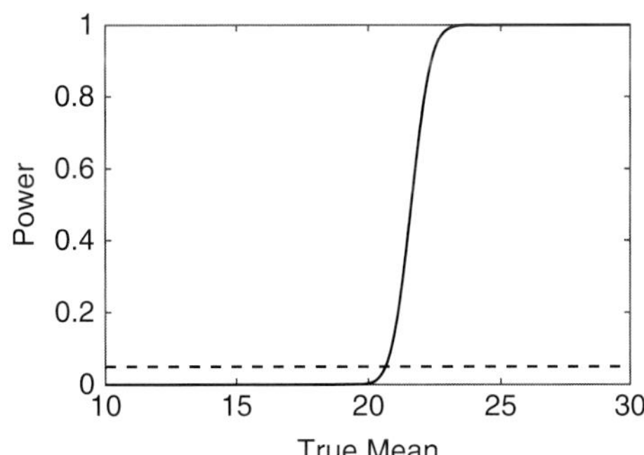

Figure 6.2 Power curve for the climate example. The horizontal dotted line denotes the 0.05 level.

At the $\beta^C = 0.05$ level, the test is unable to distinguish the nominal mean from a value below about 20.6 but then rapidly detects the alternate hypothesis. For a hypothesized mean of 22, the power is 0.723, so there is about a 72% chance that the null hypothesis will be rejected when it is false.

Example 6.4 Consider the problem of testing the value of the parameter p for a binomial rv with $N = 10$ trials. The hypotheses are H_0: $p = 0.5$ against H_1: $p > 0.5$. The number of successes $\hat{X}$ will be the test statistic, and the rejection region will be large values of $\hat{X}$ that are unlikely under H_0. A table of cumulative binomial probabilities $[\Pr(\mathbf{X} \leq N)]$ can be constructed using the MATLAB **binocdf**(0:10,10, p) function for $p = 0.5$ to 0.7.

The significance level of the test is the probability α of rejecting H_0 when it is true. Suppose that the rejection region consists of the values lying between 8 and 10. From the last row of Table 6.2 ($p = 0.5$), $\alpha = \Pr(\mathbf{X} > 7) = 1 - \Pr(\mathbf{X} \leq 7) = 0.0547$. If the rejection region consists of the values lying between 7 and 10, the significance level is 0.1719. In both cases, the null distribution is binomial with $p = 0.5$, $N = 10$.

Under the Neyman-Pearson paradigm, the value of α is chosen before a test is performed, and the test result is subsequently examined for Type 2 errors. Choose $\alpha = 0.0547$. If the true value of p is 0.6, the power of the test is $\Pr(\mathbf{X} \geq 8) = 0.1673$. If the true value of p is 0.7, then the power is 0.3828. Thus the power is a function of p, and it is obvious that it approaches α as $p \to 0.5$ and 1 as $p \to 1$.

Define the *p-value* to be the probability of observing a value of the test statistic as large as or larger than the one that is observed for a given experiment. A small number for the p-value is taken as evidence for the alternate hypothesis (or for rejection of the null hypothesis). The definition of *small* is subjective and up to the analyst. As a rough guide based largely on convention, a p-value that is (<0.01, 0.01–0.05, 0.05–0.10, >0.10) is (very strong, strong, weak, no) evidence for the alternate hypothesis (or rejection of the null hypothesis).

An important property of a p-value whose test statistic has a continuous distribution is that under the null hypothesis, the p-value has a uniform distribution on [0,1]. Consequently, if H_0 is rejected for a p-value below a value α, the probability of a Type 1 error is α.

Caution: Do not confuse the p-value with α. The latter is a parameter chosen before a given experiment is performed and hence applies to a range of experiments, whereas the p-value is a random variable that will be different for distinct experiments.

Caution: A large p-value is not necessarily strong evidence for the null hypothesis because this can also occur when the null hypothesis is false and the power of the test is low. Understanding the power of a given test and choosing tests with high power are important parts of hypothesis testing that are frequently neglected.

Caution: The p-value is the probability of observing data at least as far from the null value given that the null hypothesis is true. It cannot simultaneously be a statement about the probability that the null hypothesis is true given the observations.

The p-value concept is not without controversy, and the American Statistical Association recently issued a statement on its context, process, and purpose (Wasserstein & Lazar 2016). The use in this book is consistent with this statement, and it definitely deserves a read by all scientists who use hypothesis testing.

To summarize, a p-value hypothesis test proceeds as follows:

1. Determine the null and alternate hypotheses;
2. Find a test statistic $\hat{\lambda}$ that will provide evidence about H_0;
3. Obtain a random sample from the population of interest, and compute the test statistic;
4. Calculate the p-value (where F_{H0} is the cdf under the null hypothesis):
   a. Upper-tail test: $\hat{p} = 1 - F_{H0}(\hat{\lambda})$,
   b. Lower-tail test: $\hat{p} = F_{H0}(\hat{\lambda})$, and
   c. Two-tail test: $\hat{p} = 2 \times [1 - F_{H0}(|\hat{\lambda}|)]$ for a symmetric distribution and $\hat{p} = 2\times \min[1 - F_{H0}(\hat{\lambda}), F_{H0}(\hat{\lambda})]$ for an asymmetric distribution;
5. If the p-value is smaller than the specified Type 1 error α, then reject the null hypothesis., and do not take the p-value to be the Type 1 error; and
6. Compute the power of the test to determine the probability of rejecting the null hypothesis when it is false.

Example 6.5 For the climate example that has been beaten to a pulp,

```
mu = 20;
sig = 0.6;
xbar = 22;
```

```
zobs = (xbar - mu)/sig;
pval = 1 - normcdf(zobs, 0 , 1)
ans =
   4.34e-04
```

This is a very small number that very strongly rejects the null hypothesis.

The asymmetry in the Neyman-Pearson paradigm between the null and alternate hypotheses suggests that strategic choices of them can simplify the analysis and/or strengthen the conclusions. This is not a mathematical issue. Some typical guidelines include

- Make the null hypothesis the simpler of the two;
- If the consequences of incorrectly rejecting one hypothesis over the other are more serious, then this hypothesis should be made null because the probability of falsely rejecting it can be controlled by the choice of α; and
- In most scientific applications, the null hypothesis is a simple explanation that must be discredited to demonstrate the existence of some physical effect. This makes the choice natural.

6.3 Parametric Hypothesis Tests

6.3.1 The *z* Test

The *z test* pertains when the distribution of the test statistic under the null distribution is Gaussian. It is usually used to test for location (i.e., comparing the sample mean to a given value or comparing two sample means) in a situation where there are no nuisance parameters. In particular, the population variance must either be known a priori or be estimated with high precision. Because of the central limit theorem, for a sufficiently large sample (a rule of thumb is $N > 50$), many statistical tests can be cast as *z* tests. However, when the sample size is small and the variance is estimated from the data, the *t* test is preferred.

The most common application of the *z* test is comparing the mean estimated from data to a specified constant μ^* when the population variance σ^2 is known. The test statistic or *z-score* is

$$\hat{z} = \frac{\sqrt{N}\left(\overline{\mathrm{X}}_N - \mu^*\right)}{\sigma} \tag{6.6}$$

The null hypothesis is H_0: $\mu = \mu^*$ versus any of three possible alternate hypotheses:

1. If H_1 is two tailed (H_1:$\mu \neq \mu^*$), H_0 can be rejected if $|\hat{z}| \geq 2 \times \mathrm{N}^{-1}(1 - \alpha/2)$;
2. If H_1 is one tailed and predicts a population mean larger than that in the null hypothesis (upper-tail test), H_0 can be rejected if $\hat{z} \geq \mathrm{N}^{-1}(1 - \alpha)$; and
3. If H_1 is one tailed and predicts a population mean smaller than that in the null hypothesis (lower-tail test), H_0 can be rejected if $\hat{z} < 0$ and $|\hat{z}| \geq \mathrm{N}^{-1}(\alpha)$.

Example 6.6 Suppose that it is known that the population mean for the velocity of mantle rocks is 8 km/s with a population standard deviation of 2 km/s. The sample mean obtained from 30 rock specimens is 7.4 km/s. Can it be concluded that the sample came from a population with mean $\mu = 8$?

The hypotheses are H_0: $\mu = 8$ versus H_1: $\mu \neq 8$. The test statistic is $\hat{z} = -1.64$. Since $|\hat{z}| = 1.64$ and the two-tailed critical value is 1.96 [**norminv**(0.975)], the null hypothesis must be accepted at the 0.05 level. The p-value is $2 \times$ [1 − **normcdf**(1.64)], or 0.101, which is no evidence for the alternate hypothesis.

The power of the test for a hypothesized value of the mean μ_1 is

$$\beta^C = 1 - \Pr\left[\sqrt{N}(\mu^* - \mu_1)/\sigma - z_{\alpha/2} \le \hat{z}' \le \sqrt{N}(\mu^* - \mu_1)/\sigma + z_{\alpha/2}\right]$$

where $\hat{z}' = \sqrt{N}(\bar{X}_N - \mu_1)/\sigma$ is the hypothesized z-score, and $z_{\alpha/2} = 1.96$ for $\alpha = 0.05$. This expression may appear to be conflating the concepts of confidence intervals with hypothesis testing, but Section 6.4 will demonstrate their equivalence. Suppose that μ_1 is 7 km/s, so $\beta^C = 1 - \Pr(-0.8646 \le \hat{z}' \le 3.0554) = 1 -$ **normcdf**(3.0554) + **normcdf**(−0.8646) = 1 − 0.8052 = 0.1948, and there is a ~19% chance that the null hypothesis will be rejected when it is false. The power is low because the number of samples is small. Suppose that the power is required to be 0.9, and the number of samples needed to achieve this value is to be computed before the experiment is completed for a Type 1 error $\alpha = 0.05$. By trial and error, the result is $N \approx 260$.

Example 6.7 A meteorologist determines that during the current year there were 80 major storms in the northern hemisphere. He claims that this represents a significant increase over the long-term average. Based on data obtained over the past 100 years, it is known that there have been 70 ± 2 storms per year. Does 80 storms represent a significant increase from 70?

In this case, $N = 1$ for the single year with 80 storms. This gives $\hat{z} = 5$. The null and alternate hypotheses are H_0: $\mu = 70$ versus H_1: $\mu \neq 70$. Since $\hat{z} > 3.92$ using a two-sided test, the null hypothesis is rejected at the 0.05 level. The p-value is 5.7×10^{-7}.

MATLAB provides the function [h, p] = **ztest**(x, m, $sigma$) to perform a z test of the hypothesis that the data vector x comes from a normal distribution with mean m and standard deviation sigma (σ). It returns $h = 1$ to indicate that the null hypothesis is rejected at the 0.05 level and the p-value p. There are many additional options and, in particular, the keyword-value pair "Tail" with either "both" (default), "right," or "left."

6.3.2 The *t* Tests

The *t test* applies when the distribution for the test statistic follows Student's *t* distribution under the null hypothesis. It is undoubtedly the most widely used hypothesis test in statistics. There are a number of variants of the t test. The one-sample t test obtains when the mean is compared with a specified value, and both the population mean and the variance are estimated from the sample. The two-sample t test applies when the means of two independent populations are compared, and both the population means and the variances are estimated from the sample. It further requires that the variances of the two populations be identical, a condition that is called *homogeneity of variance* or *homoskedasticity*. In practice, this condition should always be verified using an F test (Section 6.3.4) or (preferably) Bartlett's M test for homogeneity of variance (Section 6.3.5), although the two-sample t test is remarkable insensitive to heteroskedasticity. Two-sample t tests may further be divided into unpaired and paired cases. The unpaired t test applies when iid samples are available from two populations. The paired t test is used in designed experiments where samples are matched in some way and is widely used in the biomedical fields.

The test statistic for the one-sample t test is

$$\hat{t} = \frac{\sqrt{N-1}\left(\bar{X}_N - \mu^*\right)}{\hat{s}_N} \tag{6.7}$$

where $\hat{s}_N$ is the sample standard error given by the square root of (4.14). Test evaluation is the same as for the single-sample z test, except that the critical value is taken from Student's t distribution and hence depends on the sample size.

Caution: The test statistic can also be expressed as

$$\hat{t} = \frac{\sqrt{N}\left(\bar{X}_N - \mu^*\right)}{\hat{s}'_N} \tag{6.8}$$

where $\hat{s'}_N$ is the unbiased standard deviation given by the square root of (4.15). Equations (6.7) and (6.8) are identical.

Example 6.8 A scientist claims that the average number of days per year that an oceanographer spends at sea is 30. In order to evaluate this statement, 20 oceanographers are randomly selected and state the following values: 54, 55, 60, 42, 48, 62, 24, 46, 48, 28, 18, 8, 0, 10, 60, 82, 90, 88, 2, and 54. Do the data support the claim?

The hypotheses are H_0: $\mu = 30$ versus H_1: $\mu \neq 30$. The sample mean is 43.95, and the sample standard deviation is 26.62. The test statistic is 2.28. The two-tailed critical value is obtained as **tinv**(0.975, 19) = 2.09 at the 0.05 level. The null hypothesis is rejected. The p-value is 2 × (1 – **tcdf**(2.28, 19)) = 0.0343, which is strong evidence for the alternate hypothesis. One-sided tests provide no evidence for the alternate hypothesis for the lower-tail version (p-value of 0.9828) but strongly accept it for the upper-tail test (p-value of 0.0172).

It is instructive to examine the power of the t test for this example. When the population mean is $\mu_1 \neq \mu^*$, then (6.7) has the noncentral t distribution with noncentrality parameter $\delta = \sqrt{N}(\mu_1 - \mu^*)/\sigma$, where σ is the population standard deviation that must be replaced with a sample estimate. From (6.5), the power of the single-sample t test is given by

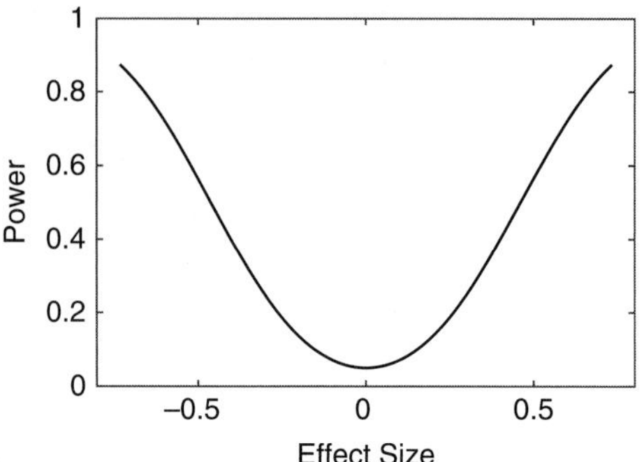

Figure 6.3 Power curve for the single-sample t test of Example 6.8.

$$\beta^C = 1 - \mathrm{Nctee}_{N-1}(\tau_{N-1}(1-\alpha/2), \delta) + \mathrm{Nctee}_{N-1}(-\tau_{N-1}(1-\alpha/2), \delta)$$

where $\tau_{N-1}(1-\alpha/2)$ is the $1-\alpha/2$ quantile of the central t distribution with $N-1$ degrees-of-freedom. A MATLAB script to evaluate the power is

```
x = [54 55 60  42 48 62 24 46 48 28 18 8 0 10 60 82 90 88 2 54];
mustar = 30;
sigma = std(x);
n = length(x);
tau = tinv(.975, n - 1);
mualt = 10:.1:50;
effect = (mualt - mustar)/sigma;
delta = sqrt(n)*effect;
power=1-nctcdf(tau, n-1, delta)+nctcdf(-tau, n-1, delta);
plot(effect, power)
```

Figure 6.3 shows the result plotted as effect size against power. The effect size is a measure of the difference between the population and postulated means normalized by the standard error. The power has a minimum value of 0.05 when $\mu_1 = \mu^*$ or effect size is zero and rises symmetrically on either side. A test whose power is no smaller than the significance level is unbiased, and in fact, the t test is a uniformly most powerful unbiased (UMPU) test. The power is about 0.58 for the observed effect size of 0.51, so there is an ~60% chance of rejecting the null hypothesis when it is false.

MATLAB gives a function [*h, p*] = **ttest**(*x*) to test the hypothesis that the data in *x* come from a normal distribution with mean zero and a Type 1 error of 0.05. The null hypothesis (mean is zero) is rejected if $h = 1$, where p is the p-value. The call [*h, p*] = **ttest**(*x*, 'Alpha', *alpha*) performs the test at the *alpha* probability level. The call [*h, p*] = **ttest**(*x*, 'Alpha', *alpha*, 'Tail', *tail*) allows the alternate hypotheses of "both" (default), "right" or "left" to be tested.

The t test can also be used to compare two paired samples. Pairing is frequently used in designed experiments in which a population is sorted into groups according to some characteristic and is especially important in biomedical testing. In this case, the t test evaluates the null hypothesis that the pairwise differences between two data sets have a mean of zero.

Pairing can be an effective experimental technique under some circumstances. To see this, consider N paired data $\{\mathbf{X}_i, \mathbf{Y}_i\}$, and assume that the means are $\{\mu_X, \mu_Y\}$ and the variances are $\{\sigma_X^2, \sigma_Y^2\}$. Further assume that distinct pairs are independent, and $\operatorname{cov}(\mathbf{X}_i, \mathbf{Y}_j) = \sigma_{XY}$. Consider the differences between the paired data $\mathbf{D}_i = \mathbf{X}_i - \mathbf{Y}_i$ that are independent with expected value $\mu_X - \mu_Y$ and variance $\sigma_X^2 + \sigma_Y^2 - 2\sigma_{XY}^2$. A natural estimator for $\mathbf{D}_i$ is the difference between the sample means of $\mathbf{X}_i$ and $\mathbf{Y}_i$ whose variance is $(\sigma_X^2 + \sigma_Y^2 - 2\sigma_{XY}^2)/N$. However, if $\mathbf{X}_i$ and $\mathbf{Y}_i$ were not paired but rather independent, then the variance of the difference would be $(\sigma_X^2 + \sigma_Y^2)/N$. The paired variance is smaller if the correlation between the pairs is positive, leading to a better experimental design.

Example 6.9 In his seminal 1908 paper, W. S. Gossett (aka Student) presented the paired results of yield from test plots of wheat, where half the plots were planted with air-dried seed and the remaining plots were planted with kiln-dried seed. The paired t test can be used to compare the two populations for a common mean.

```
gosset = importdata('gossett.dat');
[h, p, ci, stats] = ttest(gossett(:, 1), gossett(:, 2))
h =
    0
p =
    0.1218
ci =
    -78.1816
     10.7271
stats =
    tstat: -1.6905
       df: 10
       sd: 66.1711
```

The null hypothesis that the means are the same is accepted at the 0.05 level, and the p-value is 0.1218, which is no evidence for the alternate hypothesis.

The test statistic for the two-sample t test is

$$\hat{t} = \frac{\overline{X}_1 - \overline{X}_2}{\sqrt{\left(\frac{N_1\hat{s}_1^2 + N_2\hat{s}_2^2}{N_1 + N_2 - 2}\right)\left(\frac{1}{N_1} + \frac{1}{N_2}\right)}} \tag{6.9}$$

where the two samples have sizes N_1 and N_2, respectively, and $\overline{X}_i$ and $\hat{s}_i^2$ are the sample mean and variance for each. When $N_1 = N_2 = N$, the test statistic reduces to

$$\hat{t} = \frac{\sqrt{N-1}\left(\overline{X}_1 - \overline{X}_2\right)}{\sqrt{\hat{s}_1^2 + \hat{s}_2^2}} \tag{6.10}$$

Test evaluation is the same as for the single-sample case, except that the null and alternate hypotheses are H_0: $\mu_1 = \mu_2$ versus H_1: $\mu_1 \neq \mu_2$ (or H_1: $\mu_1 \geq \mu_2$ or ...), and the test degrees-of-freedom are $N_1 + N_2 - 2$. The reduction of two in the degrees-of-freedom reflects the computation of the sample means from the data, which are in fact a pair of linear constraints on them.

Example 6.10 A. A. Michelson made his famous measurements of the speed of light in 1879 and repeated a smaller number of measurements in 1882. These data are available in Stigler (1977, tables 6 and 7). Because the experiments were conducted over disjoint time intervals, they are presumed to be independent. Test the data to see if the sample means are different, and estimate the power of the test.

The data are in files michelson1.dat and michelson2.dat for the 1879 and 1882 experiments, respectively. The initial step in analyzing them is creating Gaussian q-q plots, as shown in Figure 6.4. These show two weak outliers for the 1879 data at the left side of the distribution but more systematic departures from normality for the 1882 data. The 1879 data show a stairstep pattern that is characteristic of numerical rounding that sometimes happens when data units are changed.

The null hypothesis to be tested is H_0: $\mu_1 = \mu_2$ versus H_1: $\mu_1 \neq \mu_2$, where μ_1 and μ_2 are the 1879 and 1882 means, respectively. There are 100 data in the 1879 data set and 24 data in the 1882 data set. Using MATLAB, $\overline{X}_1 = 852.40$, $\overline{X}_2 = 763.88$, $\hat{s}_1 = 78.61$, $\hat{s}_2 = 108.93$, and $\hat{t} = 4.53$. The critical value is **tinv**(0.975, 122) = 1.98, and the null hypothesis is rejected at the 0.05 level. The p-value is 1.40×10^{-5}, which is very strong evidence for the alternate hypothesis. Either the speed of light changed substantially over three years or systematic errors are present. However, the condition that the variances of the two populations are the same, as well as the test power, needs to be checked before drawing this conclusion too strongly.

Under the alternate hypothesis, (6.9) is distributed as the noncentral t distribution with noncentrality parameter

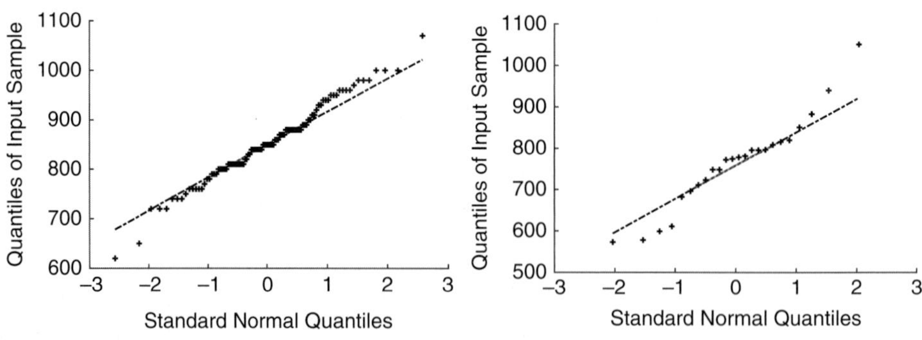

Figure 6.4 Quantile-quantile plots of the Michelson speed of light data from 1879 (left) and 1882 (right).

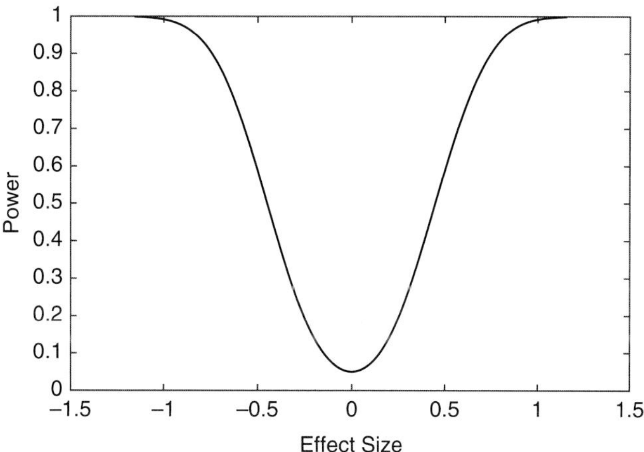

Figure 6.5 Power curve for the Michelson speed of light data.

$$\delta = \sqrt{N_1 N_2}(\mu_1 - \mu_2)/\sqrt{(N_1\sigma_1^2 + N_2\sigma_2^2)/(N_1 + N_2)}/\sqrt{N_1 + N_2}$$

A power curve for the Michelson data is easily constructed.

```
michelson1 = importdata('michelson1.dat');
michelson2 = importdata('michelson2.dat');
n1 = length(michelson1);
n2 = length(michelson2);
s1 = std(michelson1);
s2 = std(michelson2);
sp = sqrt((n1*s1^2 + n2*s2^2)/(n1 + n2 - 2));
tcrit = tinv(.975, n1 + n2 - 2);
mudiff = -100:.1:100;
effect = mudiff/sp;
delta = sqrt(n1*n2/(n1 + n2))*effect;
power = 1 - nctcdf(tcrit, n1 + n2 - 2, delta) + nctcdf
(-tcrit, n1 + n2 - 2, delta);
plot(effect, power)
```

Figure 6.5 shows the result. The power rises rapidly on either side of $\mu_1 - \mu_2 = 0$. For the observed difference of the sample means of about 88.5 (effect size of 1.02), the power is 0.99, so there is a 99% chance of rejecting the null hypothesis when it is false. This is a high value for the power, and hence there can be considerable confidence in the test, presuming that the variances of the two populations are comparable.

MATLAB includes the function $[h, p]$ = **ttest2**(x, y) that performs a two-sample t test on the data in vectors x and y. It returns $h = 1$ if the null hypothesis that the means are equal can be rejected at the 0.05 level, and p is the p-value. The call $[h, p]$ = **ttest2**(x, y, 'alpha',

alpha) gives the test at the *alpha* probability level. The call $[h, p]$ = **ttest2**(x, y, 'Alpha', *alpha*, 'Tail', *tail*) implements the test type when *tail* equals "both" (default), "right," or "left."

It is also possible to formulate an approximate two-sample t test when the variances of the two populations are not the same by using the test statistic

$$\hat{t} = \frac{\overline{X}_1 - \overline{X}_2}{\sqrt{\hat{s}_1^2/(N_1 - 1) + \hat{s}_2^2/(N_2 - 1)}} \tag{6.11}$$

and an approximation to the degrees-of-freedom. MATLAB uses the Satterthwaite (1946) value of

$$\nu = \frac{\left[\hat{s}_1^2/(N_1 - 1) + \hat{s}_2^2/(N_2 - 1)\right]^2}{\dfrac{\left[\hat{s}_1^2/(N_1 - 1)\right]^2}{N_1} + \dfrac{\left[\hat{s}_2^2/(N_2 - 1)\right]^2}{N_2}} - 2 \tag{6.12}$$

This is implemented in MATLAB as $[h, p]$ = **ttest2**(x, y, 'Vartype', 'unequal'). However, the t test for unequal population variances does not perform as well as some of the nonparametric tests described in Chapter 7 and hence is not recommended for general use.

Example 6.11 Apply the unequal variance version of the t test to the Michelson speed of light data.

```
[h, p] = ttest2(michelson1, michelson2, 'Vartype', 'unequal')
h =
    1
p =
    9.5099e-04
```

The conclusion is unchanged, and either there is not a disparity between the variances of the two data sets or the unequal variance t test does not do a good job of correction.

Example 6.12 In the 1950s and 1960s, experiments were carried out in several countries to see if cloud seeding would increase the amount of rain. A particular experiment randomly selected half of 52 clouds for seeding and measured the amount of rain from each cloud in acre-feet. These data are taken from Simpson, Olsen, & Eden (1975). The data are contained in the file cloudseed.dat, with the first and second columns corresponding to unseeded and seeded clouds.

As a start, Gaussian q-q plots of the data are produced and appear in Figure 6.6. The result shows marked departures from Gaussianity, being short tailed at the lower and long tailed at the upper ends of the distribution. Consequently, it is reasonable to question whether a t test will perform well.

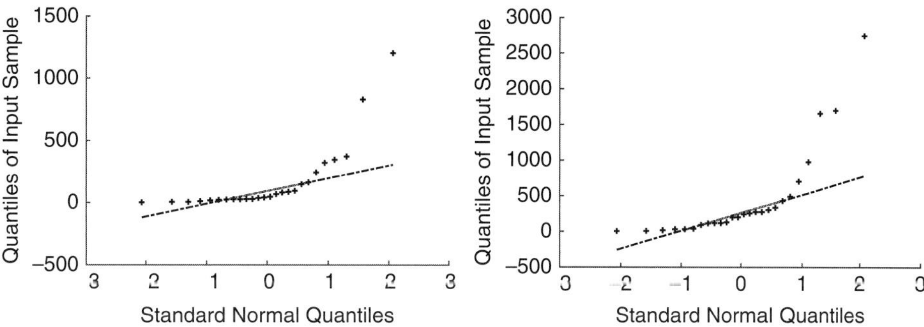

Figure 6.6 Quantile-quantile plots for the cloud seeding data. The unseeded data are shown on the left and the seeded data on the right.

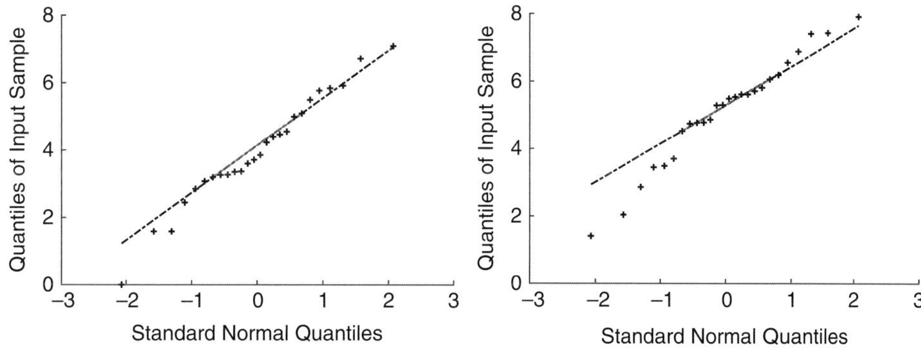

Figure 6.7 Quantile-quantile plots of the logs of the cloud seeding data. The unseeded data are shown on the left and the seeded data on the right.

A two-sample *t* test on the data yields

```
clouds = importdata('cloudseed.dat');
[h, p] = ttest2(clouds(:, 1), clouds(:, 2))
h =
   0
p =
   0.0511
```

The null hypothesis that the seeded and unseeded clouds yield different amounts of rain is weakly accepted. The test statistic is -1.9982 against a 0.05 critical value of 2.06, so the result is close to the critical region.

However, q-q plots of the logs of the data yield a result that is very close to being Gaussian (Figure 6.7). This suggests that a two-sample *t* test on the logs might yield a more accurate result.

```
[h, p] = ttest2(log(clouds(:, 1)),log(clouds(:, 2)))
h =
   1
```

```
p =
    0.0141
```

The null hypothesis is now strongly rejected by the t test. The lesson from this example is that the distribution of the data must be characterized before blindly applying a statistical test.

6.3.3 The χ^2 Test

The χ^2 test applies when the distribution for the test statistic follows the chi square distribution under the null hypothesis. In its simplest form, the χ^2 test is used to determine whether the variance computed from a sample has a specified value when the underlying distribution of the sample is Gaussian. Confusingly, there are other tests with the same name that are used in goodness-of-fit testing (Section 7.2.2) and other areas.

The test statistic for the χ^2 test is

$$\hat{\chi}^2 = \frac{N\hat{s}_N{}^2}{\sigma^2} \tag{6.13}$$

Test evaluation proceeds as follows:

1. If the alternate hypothesis is two tailed, H_0 can be rejected if $\hat{\chi}^2 \geq \mathrm{Chi}^{-1}(1-\alpha/2, N-1)$ or $\hat{\chi}^2 \leq \mathrm{Chi}^{-1}(\alpha/2, N-1)$;
2. If the alternate hypothesis is one tailed and predicts a population variance larger than in the null hypothesis, H_0 can be rejected if $\hat{\chi}^2 \geq \mathrm{Chi}^{-1}(1-\alpha, N-1)$; and
3. If the alternate hypothesis is one tailed and predicts a population variance smaller than that in the null hypothesis, H_0 can be rejected if $\hat{\chi}^2 \leq \mathrm{Chi}^{-1}(\alpha, N-1)$.

Example 6.13 A company claims that a certain type of battery used in an instrument has a lifetime of 9 months with a standard deviation of 2.24 months (or variance of 5 months squared). It is believed that the stated standard deviation is too low. To test this hypothesis, the lives of 30 batteries are measured, as contained in the file battery.dat. Do the data indicate that the standard deviation is other than 2.24 months?

The hypotheses to be compared are H_0: $\sigma^2 = 5$ versus H_1: $\sigma^2 \neq 5$. From the data, $\hat{s}_N^2 =$ 7.40, and $N = 30$. The test statistic is $\hat{\chi}^2 = 44.40$. From MATLAB, **chi2inv**(0.975, 29) = 45.72. Because $\hat{\chi}^2 < \mathrm{Chi}^{-1}(1-\alpha/2, N-1)$, the null hypothesis is accepted, although the test statistic is very close to the critical region. The p-value is 0.0674, so there is weak evidence for the alternate hypothesis. However, if H_1: $\sigma^2 > 5$, the null hypothesis is rejected because **chi2inv**(0.95, 9) = 42.56. The upper-tail p-value is 0.0337, which is strong evidence for the alternate hypothesis.

When the random variables used to compute the sample variance do not have zero expected value, the test statistic (6.13) has the noncentral chi square distribution with

$N-1$ degrees-of-freedom and noncentrality parameter $\delta = \sum_{i=1}^{N} \mu_i^2/\sigma^2$, where $\mu_i = \mathcal{E}(\mathbf{X}_i)$. A more reasonable way to measure the effect size is as the variance ratio difference $(\sigma_0^2 - \sigma^2)/\sigma^2$, where σ_0^2 is the target value of the variance, and $\delta = N(\sigma_0^2 - \sigma^2)/\sigma^2$. An effect size of zero corresponds to a variance ratio of unity and occurs when the power of the test is the Type 1 error α because the test is unbiased; the power rises asymmetrically on either side of this value due to the skewed form of the distribution. The power curve of the chi square test is

$$\beta^C = 1 - \text{Ncchi}(x_{1-\alpha/2}, N-1, \delta) + \text{Ncchi}(x_{\alpha/2}, N-1, \delta) \tag{6.14}$$

where Ncchi is the cdf for the noncentral chi square distribution.

Example 6.14 Compute a power curve for Example 6.13.

The observed value for the variance ratio is 1.48. The power calculation is easily implemented using MATLAB.

```
x = importdata('battery.dat');
sigma2 = 5;
n = length(x);
xlo = chi2inv(.025, n - 1);
xhi = chi2inv(.975, n - 1);
xhi1 = chi2inv(.95, n - 1);
varrat = 1:.01:2;
power = 1 - ncx2cdf(xhi, n - 1, n*(varrat - 1)) + ...
   ncx2cdf(xlo, n - 1, n*(varrat - 1));
power1 = 1 - ncx2cdf(xhi1, n - 1, n*(varrat - 1));
plot(varrat, power, varrat, power1, 'k--')
```

Figure 6.8 shows the power curves for the two-tailed and upper-tail tests against the effect size. At the observed variance ratio of 1.48, the powers for the two-tailed and upper-tail tests are only 0.37 and 0.48, respectively.

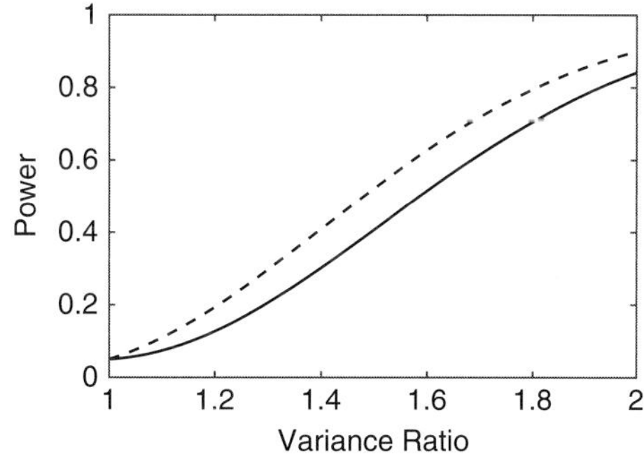

Figure 6.8 Power curves for a two-sided (solid line) and upper-tail (dashed line) test for the battery data.

MATLAB includes [*h*, *p*] = **vartest**(*x*, *v*), which performs a chi square test of the hypothesis that the data in *x* come from a normal distribution with variance *v*. It returns $h = 1$ if the null hypothesis can be rejected at the 0.05 level, and *p* is the *p*-value.

6.3.4 The *F* Test

The *F* test applies when the distribution for the test statistic follows the *F* distribution under the null hypothesis. In its simplest form, the *F* test is used to compare two variances computed from independent samples for which the underlying distributions are Gaussian. Consequently, it serves as a test for homogeneity of variance, although it is very sensitive to departures from normality and hence should be avoided for this purpose in favor of Bartlett's *M* test. The *F* test also finds extensive use in linear regression and modeling.

To test for homogeneity of variance, the test statistic is

$$\hat{F} = \frac{\hat{s}_{max}^2}{\hat{s}_{min}^2} \tag{6.15}$$

where $\hat{s}_{max}^2 = \max\left(\hat{s}_1^2, \hat{s}_2^2\right)$ and $\hat{s}_{min}^2 = \min\left(\hat{s}_1^2, \hat{s}_2^2\right)$, and $\hat{s}_1^2$ and $\hat{s}_2^2$ are the sample variances for the two populations. The test statistic $\hat{F}$ is assessed against critical values of the *F* distribution with $N_{max} - 1$, $N_{min} - 1$ degrees-of-freedom.

Example 6.15 For the Michelson speed of light data in Example 6.10, $\hat{F} = 11{,}866/6180.2 = 1.92$ and $F_{23,99}^{-1}(0.975) = 1.80$, so the null hypothesis H_0: $\sigma_1^2 = \sigma_2^2$ is rejected in favor of the alternate H_1: $\sigma_1^2 \neq \sigma_2^2$ at the 0.05 level. The *p*-value is 0.0292, which is strong evidence for the alternate hypothesis. This also holds for the upper-tail test H_1: $\sigma_1^2 > \sigma_2^2$, where $F_{23,99}^{-1}(0.95) = 1.64$. The *p*-value for the upper-tail test is 0.0146. Because the data fail a test for homogeneity of variance, the results from the *t* test in Section 6.3.2 are in doubt.

The power of the test is easily computed in a manner analogous to the *t* and χ^2 tests, with the result shown in Figure 6.9. At the observed variance ratio of 1.92, the power of the *F* test is only about 0.55 for the Michelson data, so one cannot place a large degree of confidence in the outcome.

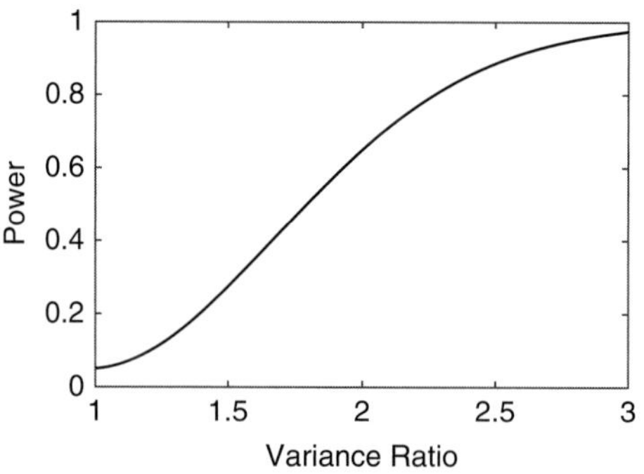

Figure 6.9 Power of the *F* test for variance homogeneity applied to the Michelson data.

```
xlo = finv(.025, 23, 99);
xhi = finv(.975, 23, 99);
varrat = 1:.01:3;
plot(varrat, 1 - ncfcdf(xhi, 23, 99, 24*(varrat - 1)) +
ncfcdf(xlo, 23, 99, 24*(varrat - 1)))
```

MATLAB includes the function $[h, p]$ = **vartest2**(x, y), which performs an F test for the hypothesis that x and y come from normal distributions with the same variance. It returns $h = 1$ if the null hypothesis that the variances are equal can be rejected at the 0.05 level, and the p-value is p.

6.3.5 Bartlett's *M* Test for Homogeneity of Variance

This test was introduced by Bartlett (1937) and is a more powerful test than the F test that applies to a population of independent variance estimates (i.e., to more than two estimates). It is also more robust to heteroskedasticity than the F test and is an unbiased test. The test statistic is

$$\hat{M} = \frac{(N-k)\log \hat{s}_p^2 - \sum_{i=1}^{k} N_i \log \hat{s'}_i^2}{1 + \frac{1}{3(k+1)}\left(\sum_{i=1}^{k} \frac{1}{N_i - 1} - \frac{1}{N-1}\right)} \tag{6.16}$$

where $N = \sum_{i=1}^{k} N_i$, and N_i is the degrees-of-freedom for the variance estimate $\hat{s'}_i^2$, and the pooled variance is

$$\hat{s}_p^2 = \frac{1}{N-k}\sum_{i=1}^{k}(N_i - 1)\hat{s'}_i^2 \tag{6.17}$$

For small samples, Bartlett showed that (6.16) is distributed according to χ_{k-1}^2 and can be tested for significance in the usual way. Because the null and alternate distributions cannot be fully characterized, power calculations are not straightforward. However, the asymptotic power can be obtained directly using the noncentral chi square distribution.

Example 6.16 For the speed of light data in Example 6.10, $\hat{M} = 4.97$ using (6.16). To test the null hypothesis H_0: $\sigma_1^2 = \sigma_2^2$ against the alternate H_1: $\sigma_1^2 \neq \sigma_2^2$, compute the p-value $2 \times \min(1 -$ **chi2cdf**(4.97, 1), **chi2cdf**(4.97, 1)) = 0.0516, which is weak evidence for the alternate hypothesis. This suggests that the data are nearly homoskedastic, in contrast to the F test.

MATLAB implements Bartlett's M test, although it is a bit difficult to use for comparing two data sets with different numbers of data. The function **vartestn**(x) computes Bartlett's test on the columns of x. Because it ignores NaN entries, it is necessary to add NaNs to the second Michelson data set to make it the same size as the first one. The following script accomplishes this:

```
michelson2 = [michelson2' NaN(1, 76)]';
x = [michelson1 michelson2];
[p, stats] = vartestn(x)
p =
   0.0265
stats =
   chisqstat: 4.9245
   df: 1
```

Note that the MATLAB function does not return the variable h, unlike the other tests described in this section. The difference between the directly calculated p-value in Example 6.16 and that from the MATLAB function is that the latter returns the upper-tail test, unlike the default for all of its other parametric tests, and no option is provided to change it, which is most unfortunate. However, it is unclear why the test statistics are slightly different because they do not depend on the nature of the test. The function **vartestn** also implements a series of variants on the M test that are more robust to heteroskedasticity.

6.3.6 The Correlation Coefficient

The correlation coefficient (sometimes called *Pearson's product moment correlation coefficient*) allows two populations drawn from a bivariate normal distribution to be compared. The test statistic is given by (4.79). The null hypothesis under test holds that there is no correlation between two data sets against the alternate hypothesis that there is correlation. The null distribution is given by (4.83), and the alternate distribution is given by (4.80).

The p-value can be computed by integrating (4.83) to get the null cdf. The result is

$$\mathrm{Corr}(x;0,N) = \frac{1}{2} + \frac{\Gamma[(N-1)/2]x\,{}_2F_1\left(\frac{1}{2},2-\frac{N}{2};\frac{3}{2};x^2\right)}{\sqrt{\pi}\;\Gamma(N/2-1)} \tag{6.18}$$

The hypergeometric series (4.81) terminates when either of the first two arguments are integers [or when the number of data is odd in (6.18)], and this must be accounted for in computing the hypergeometric function. The alternate cdf must be obtained by numerical integration of (4.80). These steps will be demonstrated by example.

The MATLAB **corr** function computes the correlation and the test p-value using a Student's t approximation rather than the exact result in (6.18). The function call is [*rho*, *p*] = **corr**(*x*1, *x*2), where *rho* is the correlation matrix, and *p* is the p-value.

Example 6.17 Data from a British government survey of household spending in the file alctobacc.dat taken from Moore & McCabe (1989) may be used to examine the relationship between household spending on tobacco products and alcoholic beverages.

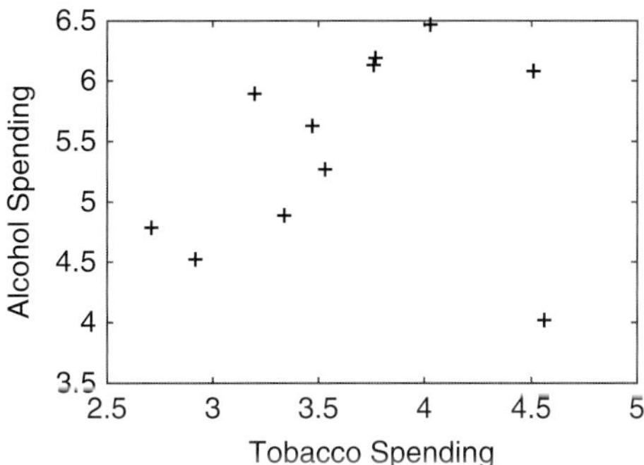

Figure 6.10 Alcohol and tobacco spending by British households in the 11 regions of the United Kingdom.

A scatterplot of spending on alcohol versus spending on tobacco in the 11 regions of Great Britain (Figure 6.10) shows an overall positive linear relationship, although there is a single outlying value at the lower right. The goal is assessing whether there is significant correlation between these two data sets and investigating whether the result is unduly influenced by a single datum. The results using the MATLAB function **corr** will be compared with those from (6.18).

```
alctobacc = importdata('alctobacc.dat')
[rhat, p] = corr(alctobacc(:, 1), alctobacc(:, 2))
rhat =
   0.2236
p =
   0.5087
N = length(alctobacc);
Corr = 0.5 + gamma((N - 1)/2)*rhat* ...
Hypergeometric2f1(1/2, 2 - N/2, 3/2, rhat^2)/(sqrt(pi)*gamma
(N/2 - 1));
pval = 2*min(Corr, 1 - Corr)
pval =
   0.5087
```

The null hypothesis that there is no correlation in these data is accepted, and the results from the MATLAB function **corr** and (6.18) are the same.

However, if the point at the lower right is eliminated (Northern Ireland, where they either smoke a lot and drink comparatively little or else fiddle with the numbers), the sample correlation coefficient changes to $\hat{r} = 0.7843$, the p-value changes to 0.0072, and the null hypothesis is rejected. The correlation between alcohol and tobacco consumption in Great Britain becomes real. To be strictly correct, after deleting the outlying value, the null distribution should be taken as the truncated form of (4.82), as described in Section 4.8.4. Due to its complexity, this will be neglected for pedagogical purposes.

6.3.7 Analysis of Variance

One-way analysis of variance or ANOVA is a method for comparing the means of two or more independent groups and is effectively an extension of the t test to more than two samples. The term *analysis of variance* may cause some confusion because the method uses variance to compare the means of the groups as opposed to comparing the variances directly. One-way ANOVA is balanced if each group has the same number of variables and otherwise is unbalanced.

Let M denote the number of groups to be compared. Let the population means be $\{\mu_1, \ldots, \mu_M\}$. ANOVA is a test of the null hypothesis $H_0 : \mu_1 = \mu_2 = \cdots = \mu_M$ against the alternate hypothesis H_1: not H_0, meaning that at least two of the population means are unequal. The assumptions for the test are

1. The distribution of the rvs for each of the M groups is Gaussian;
2. The data in each group are homoskedastic; and
3. The samples are independent.

ANOVA uses the *between* (sometimes called *hypothesis*) and *within* (sometimes called *error*) variances of the data. The between variance is a comparison of the sample mean of each group with the grand mean of all the data. The within variance is a comparison of the data in each group with its corresponding sample mean. The test statistic is the ratio of the between and within variances, and its null distribution is the F distribution. The p-value is the upper-tail probability that the test statistic exceeds the observed F value.

Let n_i denote the number of data in the ith group, $\overline{X}_i$ denote the corresponding sample mean, and $\overline{X}$ be the grand mean of all the data. The unbiased between variance is

$$\hat{s}'^2_b = \frac{1}{M-1}\sum_{i=1}^{M} n_i\left(\overline{X}_i - \overline{X}\right)^2 \tag{6.19}$$

Since (6.19) measures the variability among M means, it has $M - 1$ degrees-of-freedom.

In the balanced situation where all the n_i are identical, the data can be represented by the $N \times M$ matrix $\overleftrightarrow{\mathbf{X}}$, and the between sum of squares can be written succinctly using matrix notation as

$$w_b^2 = \frac{1}{N}\mathbf{j}_N \cdot \overleftrightarrow{\mathbf{X}} \cdot \left(\overleftrightarrow{\mathbf{I}}_M - \frac{1}{M}\overleftrightarrow{\mathbf{J}}_M\right) \cdot \overleftrightarrow{\mathbf{X}}^T \cdot \mathbf{j}_N^T \tag{6.20}$$

where $\mathbf{j}_N$ is a column N-vector of ones, $\overleftrightarrow{\mathbf{I}}_M$ is the $M \times M$ identity matrix, $\overleftrightarrow{\mathbf{J}}_M$ is an $M \times M$ matrix of ones, and $\hat{s}'^2_b = w_b^2/(M-1)$.

The within variance is

$$\hat{s}'^2_w = \frac{1}{N-M}\sum_{i=1}^{M}(n_i - 1)\hat{s}'^2_j \tag{6.21}$$

where $N = \sum_{j=1}^{m} n_i$,

$$\hat{s}'^2_j = \frac{1}{n_j - 1}\sum_{i=1}^{n_j}\left(x_{i,j} - \overline{X}_j\right)^2 \tag{6.22}$$

and $x_{i,j}$ denotes the ith datum in the jth group. In the balanced situation, the sum of squares can be written as

$$w_w^2 = \mathrm{tr}\left(\overleftrightarrow{\mathbf{X}} \cdot \overleftrightarrow{\mathbf{X}}^T\right) - \frac{1}{N}\mathbf{j}_N \cdot \overleftrightarrow{\mathbf{X}} \cdot \overleftrightarrow{\mathbf{X}}^T \cdot \mathbf{j}_N^T \qquad (6.23)$$

where tr denotes the matrix trace (sum of the diagonal elements), and $s'^2_w = w_w^2/[M(N-1)]$. The test statistic is

$$\hat{F} = \frac{\hat{s}'^2_b}{\hat{s}'^2_w} \qquad (6.24)$$

and is distributed as $F_{M-1,M(N-1)}$ under the null hypothesis. The results of an ANOVA are usually presented using a table, with the first and second rows denoting the between and within groups, respectively, and the last row presenting the total. The columns show the sum of squares, degrees-of-freedom, mean square, and F statistic.

MATLAB supports balanced one-way analysis of variance through the function [*p*, *anovatab*] = **anova1**(*x*), where *anovatab* is optional. The input variable *x* is a matrix whose columns are the data from each group. The returned variables are the *p*-value *p* and the ANOVA table *anovatab*. By default, **anova1** also produces a boxplot of the data. A more easily understood boxplot can be generated directly using the function **boxplot** (*x*, 'medianstyle', 'target') and is shown in Example 6.18. A boxplot consists of representations of the columns of *x* in which a target symbol is the median, the edges of the box are the 0.25 and 0.75 quantiles, whiskers extend to the limits of the data that the function considers to not be outliers, and outliers are marked with the symbol x. It is unclear how outliers are detected, and hence those markers should not be emphasized.

Example 6.18 Measurements have been made by seven laboratories of the content of a certain drug in tablet form in the file tablet.dat. There are 10 replicate measurements per laboratory. The goal is to determine whether there are systematic differences between the laboratory results.

```
[p, anovatab] = anova1(tablet);
```

The anova table is:

Source	SS	DF	MS	F
Between	0.15343	6	0.02556	11.9
Within	0.13536	63	0.00215	
Total	0.28879	69		

The *p*-value is 2 × **min**(1 - **fcdf**(*f*, 6, 63), **fcdf**(*f*, 6, 63)) = 1.40×10^{-8}, so the null hypothesis that the means are the same is rejected, which is also apparent in Figure 6.11.

Much of the terminology that underlies analysis of variance comes from devising designed experiments that are typically not feasible in the earth sciences but are common in the social and biological sciences. There is an extensive literature on designing experiments

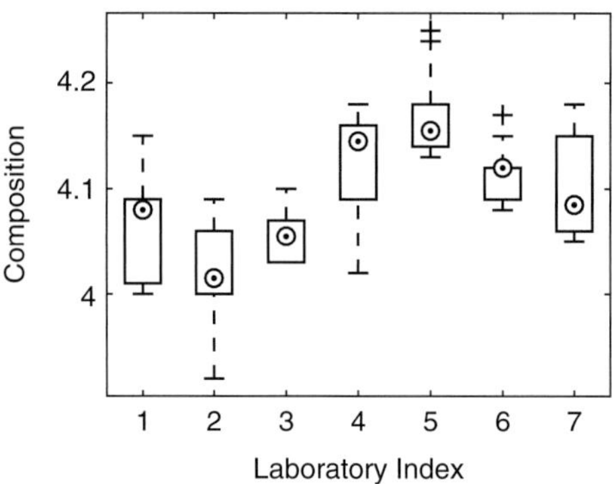

Figure 6.11 Boxplot for the tablet data from the seven laboratories of Example 6.17 shown on the *x*-axis with analysis results on the *y*-axis. The targets are the population medians, the solid boxes encompass the 0.25 to 0.75 quantiles, and the dotted whiskers show the full range of each data set. Two outliers in the fifth and sixth laboratory results are shown as + signs.

that is omitted in this book. The design of experiments typically leads to more complex situations where each group is divided into two or more categories, leading to two-way analysis of variance or a higher-order version. More information can be found in the classic text of Scheffé (1959) and in the more recent work of Hoaglin, Mosteller, & Tukey (1991). The power of an ANOVA test can be computed but is not as straightforward as in prior examples; see Murphy, Myors, & Wolach (2014) for details.

6.3.8 Sample Size and Power

There is an inherent tradeoff between the choice of the Type 1 error α, the power of a given statistical test β^C, and the number of data that are required to achieve a given (α, β^C). It is good statistical practice to estimate the required number of data to yield (α, β^C) prior to initiating data collection as an element of experimental design. Indeed, this is standard practice in the biomedical fields, where experimental designs are typically thorough and aimed at fulfilling regulatory requirements.

MATLAB provides a function n = **sampsizepwr**(testtype, $p0$, $p1$) that returns the number of values n required to achieve a power of 0.90 at a significance level of 0.05 for a two-sided test of types "z," "t," "var," and "p" for, respectively, the z, t, χ^2, and binomial proportion tests. The parameters $p0$ and $p1$ are the parameters under the null and alternate hypotheses that are test specific; for example, for the t test, $p0$ is $[\mu, \sigma]$ under the null hypothesis, and $p1$ is the mean under the alternate hypothesis. An alternate call *power* = **sampsizepwr**(testtype, $p0$, $p1$, [ ], n) returns the power that is realized for n data samples, and other variants are supported. It is possible to change the significance level and test type in the usual ways.

Caution: There are numerous power analysis applications available on the Internet; the most widely used appears to be G*Power, which is available at http://gpower.hhu.de.

However, many of these applications use approximations to the power whose accuracy can be limited unless the degrees-of-freedom are large. It seems that the MATLAB function **sampsizepwr** is not an exception. It is *strongly* recommended that the data analyst understand how to compute the power of a test directly and use that result instead of relying on black-box functions.

Example 6.19 For the number of days at sea data of Example 6.8, determine the number of data with the observed mean and standard deviation that would be required to achieve a power of 0.90 at a significance level of 0.05 for a two-sided test with population means of 30 and 40. Determine the power of a t test with the observed parameters for 10, 20, and 100 data and a population mean of 30.

```
x = [ 54 55 60  42 48 62 24 46 48 28 18 8 0 10 60 82 90 88 2 54];
sampsizepwr('t', [mean(x) std(x)], 30)
ans =
  43
sampsizepwr('t', [mean(x) std(x)], 40)
ans =
  505
```

Recalling that the mean of the data is 43.95 days, it is reasonable to expect that a much larger sample size will be required to discern a smaller effect size, as is observed.

```
sampsizepwr('t', [mean(x) std(x)], 30, [], 20)
ans =
  0.5976
sampsizepwr('t', [mean(x) std(x)], 30, [], 30)
ans =
  0.7714
sampsizepwr('t', [mean(x) std(x)], 30, [], 100)
ans =
  0.9993
```

Clearly, a much better hypothesis test would have ensued by asking 30 rather than 20 oceanographers for the number of days per year that they spend at sea because the power is nearly 0.8 for the larger sample.

6.4 Hypothesis Tests and Confidence Intervals

Let $\{\mathbf{X}_i\}$ be a random sample from a normal distribution having a known variance. Consider H_0: $\mu = \mu_0$ versus H_1: $\mu \neq \mu_0$. This is the z test described in Section 6.3.1, where the test statistic is $\hat{z} = \sqrt{N}(\bar{\mathrm{X}}_N - \mu_0)/\sigma$. The test accepts the null hypothesis when

$|\overline{X}_N - \mu_0| < \sigma z_{\alpha/2}/\sqrt{N}$, where $z_{\alpha/2}$ is the critical value from the Gaussian distribution. This is equivalent to $\overline{X}_N - \sigma z_{\alpha/2}/\sqrt{N} \le \mu_0 \le \overline{X}_N + \sigma z_{\alpha/2}/\sqrt{N}$, which is the $1 - \alpha$ confidence interval for μ_0. Comparing the acceptance region of the test with the confidence interval, it is clear that μ_0 lies within the confidence interval if and only if the hypothesis test accepts or, alternately, the confidence interval consists precisely of the values of μ_0 for which the null hypothesis is true. The example uses the two-tailed test, but equivalent statements can be made for the upper- or lower-tail test where the confidence intervals become $\overline{X}_N - \sigma z_\alpha/\sqrt{N} \le \mu_0 \le \infty$ or $-\infty \le \mu_0 \le \overline{X}_N + \sigma z_\alpha/\sqrt{N}$. In addition, the outcome applies generally for other types of hypothesis tests.

As a result, it is common practice to assess estimates of the mean (or another statistic) for different samples with their confidence intervals to evaluate the null hypothesis that they are the same, with the seemingly obvious conclusion that when there is no overlap of the confidence intervals, then the null hypothesis is rejected. However, this is not correct in general: rejection of the null hypothesis by examining overlap of the confidence intervals implies rejection by the hypothesis test, but failure to reject by examining overlap does not imply failure to reject by the hypothesis test. Consequently, the method of examining overlap is conservative in the sense that it rejects the null hypothesis less frequently. Schenker & Gentleman (2001) provide a detailed explanation, whereas Lanzante (2005) provides a simplified but lucid version. The fundamental problem is that comparison of the means of two samples uses their individual standard errors, whereas a two-sample test such as the two-sample t test of Section 6.3.2 uses the pooled standard error given by the square root of the sum of the squares of the individual standard errors. For equal-sized samples, it is not correct to evaluate $\overline{X}_N^1 - \overline{X}_N^2 \ge z_{\alpha/2}\hat{s}_1 + z_{\alpha/2}\hat{s}_2$, but it would be correct to evaluate $\overline{X}_N^1 - \overline{X}_N^2 \ge z_{\alpha/2}\sqrt{\hat{s}_1^2 + \hat{s}_2^2}$. However, the latter is not typically presented in plots of the means of different samples with their error bars. If $\hat{s}_1$ and $\hat{s}_2$ are comparable, then the discrepancy between the sum of the standard errors and the pooled standard error (hence the chance of drawing an incorrect conclusion) is large, whereas if one of the individual standard errors is much larger than the other, then the difference is negligible. It is much safer to use two sample tests to compare two populations than to rely on graphical comparisons.

6.5 Theory of Hypothesis Tests II

Section 6.2 provides a summary description of parametric hypothesis testing under the Neyman-Pearson lemma, and Section 6.3 contains a summary of the most commonly used parametric hypothesis tests. The underlying theory for these two sections is usually covered in more advanced statistical works, and only a summary of the key points is provided here. A very readable elementary treatment of this material is contained in Hogg & Craig (1995). A basic understanding of the principles underlying likelihood ratio tests has considerable value in applying them using modern resampling techniques.

6.5.1 Likelihood Ratio Tests for Simple Hypotheses

There are typically many tests at a given significance level with which the null hypothesis can be evaluated. It is desirable to select the "best" ones based on some set of objective criteria, where *best* means the test that has the correct significance level and is as or more powerful than any other test at that level.

Let Φ denote a subset of the sample space Ω. Φ is called the *best critical region of size* α for testing the simple hypotheses H_0: $\lambda = \lambda_0$ versus H_1: $\lambda = \lambda_1$ if, for every subset A of the sample space, the following properties hold:

1. $\Pr(\mathbf{X}_i \in \Phi | H_0) = \alpha$; and
2. $\Pr(\mathbf{X}_i \in \Phi | H_1) \geq \Pr(\mathbf{X}_i \in \mathrm{A} | H_1)$.

This definition states that if the null hypothesis holds, there will be a plurality of subsets A of Ω such that $\Pr(\mathbf{X}_i \in \mathrm{A} | H_0) = \alpha$. However, there will be one subset Φ such that when the alternate hypothesis is true, the power of the test is at least as great as the power of the test for any other subset A. Then Φ specifies the best critical region to test the null against the alternate hypothesis.

Consider null and alternate hypotheses for parameters λ_0 and λ_1 that are both simple. Suppose that H_0 specifies a null pdf $f_0(x)$ and H_1 specifies an alternate pdf $f_1(x)$. Given a specific random sample $\{\mathbf{X}_i\}$, the relative probabilities of H_0 and H_1 are measured by the *likelihood ratio* $\mathcal{L}_0(\lambda_0|x)/\mathcal{L}_1(\lambda_1|x)$, where $\mathcal{L}_i(\lambda_i|x)$ is the likelihood function for the null or alternate hypotheses. Let Φ be a subset of the sample space, and let κ be a positive number such that:

1. $\mathcal{L}_0(\lambda_0|x)/\mathcal{L}_1(\lambda_1|x) \leq \kappa$ for each point $x \in \Phi$;
2. $\mathcal{L}_0(\lambda_0|x)/\mathcal{L}_1(\lambda_1|x) \geq \kappa$ for each point $x \in \Phi^C$; and
3. $\Pr(\mathbf{X}_i \in \Phi | H_0) = \alpha$.

Then conditions 1–3 are necessary and sufficient for Φ to be a best critical region of size α for testing the two simple hypotheses. This is the Neyman-Pearson lemma of Neyman & Pearson (1933), which states that among all tests with a given Type 1 error probability α, the likelihood ratio test minimizes the probability of a Type 2 error β or else maximizes the power β^C.

Example 6.20 Let $\{\mathbf{X}_i\}$ be a random sample from a Gaussian distribution having unit variance. Consider H_0: $\mu = 0$ against H_1: $\mu = 1$. Let the significance level be α. By the Neyman-Pearson lemma, among all tests with significance α, the test that rejects for small values of the likelihood ratio is the most powerful test. The likelihood ratio can be easily computed because the distributions under the null and alternate hypotheses are both Gaussian

$$\hat{\Lambda} = \frac{\mathcal{L}_0(x \mid \mu = 0)}{\mathcal{L}_1(x \mid \mu = 1)} = \frac{\exp\left[-\sum_{i=1}^{N} x_i^2/2\right]}{\exp\left[-\sum_{i=1}^{N} (x_i - 1)^2/2\right]}$$

Taking the log and simplifying yields

$$\log \hat{\Lambda} = -\sum_{i=1}^{N} x_i^2/2 + \sum_{i=1}^{N} (x_i - 1)^2/2 = -\sum_{i=1}^{N} x_i + \frac{N}{2} \le \log \kappa$$

The best critical region is given by

$$\sum_{i=1}^{N} x_i \ge \frac{N}{2} - \log \kappa = \varsigma$$

and is equivalent to $\overline{\mathrm{X}}_N \ge \varsigma/N$, where ς is determined such that the size of the critical region is α. The distribution of the sample mean under the null hypothesis is $\mathrm{N}(0, 1/N)$. Consequently, ς can be estimated in the usual way from the Gaussian distribution. This shows that the z test of Section 6.3.1 is a likelihood ratio test.

Example 6.21 Let $\{\mathbf{X}_i\}$ be a random sample from a Cauchy distribution with location parameter λ. The hypotheses to be evaluated are H_0: $\lambda = 0$ against H_1: $\lambda = \lambda_1 \ne 0$. The null distribution is the Cauchy distribution with parameter $\lambda = 0$, whereas the alternate distribution is the same distribution with arbitrary nonzero location parameter. The likelihood ratio is

$$\hat{\Lambda} = \frac{\prod_{i=1}^{N} \frac{1}{1 + x_i^2}}{\prod_{i=1}^{N} \frac{1}{1 + (x_i - \lambda_1)^2}} = \prod_{i=1}^{N} \frac{1 + (x_i - \lambda_1)^2}{1 + x_i^2} \le \kappa(\alpha)$$

This is equivalent to a $2N$-order polynomial whose coefficients depend on κ and λ_1, and hence the form of the likelihood ratio test depends in a complex way on the choice of α.

6.5.2 Uniformly Most Powerful Tests

Consider the more complicated situation of a simple null hypothesis and a composite alternate hypothesis that is an ensemble of simple hypotheses, such as H_0: $\lambda = \lambda_0$ against H_1: $\lambda \ne \lambda_0$, where the significance level is α. The critical region will be different for each possible member of the ensemble of alternate hypotheses. However, it may be possible to establish that there is a best critical region for all possible alternate hypotheses, which is called the *uniformly most powerful critical region*, and the corresponding test is then *uniformly most powerful* (UMP) at a significance level α. UMP tests do not always exist, but if they do, the Neyman-Pearson lemma provides the formalism for their construction.

Example 6.22 Let $\{\mathbf{X}_i\}$ be a random sample from a Gaussian distribution having unit variance, and consider H_0: $\mu = \mu_0$ against H_1: $\mu > \mu_0$, where μ_0 is a constant. Let μ_1 be another constant that is different from μ_0. The Neyman-Pearson lemma gives

$$\log\hat{\Lambda} = -\sum_{i=1}^{N}(x_i-\mu_0)^2/2 + \sum_{i=1}^{N}(x_i-\mu_1)^2/2 = (\mu_0-\mu_1)\sum_{i=1}^{N}x_i + \frac{N}{2}\left(\mu_1^2-\mu_0^2\right) \le \log\kappa$$

where $\kappa > 0$. The inequality simplifies to

$$\sum_{i=1}^{N}x_i \ge \frac{N}{2}(\mu_0+\mu_1) - \frac{\log\kappa}{\mu_1-\mu_0}$$

provided that $\mu_1 > \mu_0$, and defines the critical region for the upper-tail test that is of interest. Setting the right side to a constant ς, it has been demonstrated that the upper-tail z test is UMP.

However, if the alternate hypothesis is changed to H_1: $\mu \ne \mu_0$, the log likelihood ratio reduces to

$$\sum_{i=1}^{N}x_i \le \frac{N}{2}(\mu_0+\mu_1) - \frac{\log\kappa}{\mu_1-\mu_0}$$

when $\mu_1 < \mu_0$, which is in conflict with the result when $\mu_1 > \mu_0$, and hence there is no UMP two-tailed test for the Gaussian distribution. In fact, two-tailed UMP tests do not exist in general.

If both the null and alternate hypotheses depend on a parameter λ, and there is a single sufficient statistic $\hat{\lambda}_k$ for λ, then it follows from the factorization theorem for sufficient statistics (5.16) that

$$\hat{\Lambda} = \frac{V\left(\hat{\lambda}_k,\lambda_0\right)}{V\left(\hat{\lambda}_k,\lambda_1\right)} \tag{6.25}$$

so the most powerful test is a function only of the sufficient statistic. This result further underlines the importance of sufficiency because tests using any statistic that is not sufficient will not be UMP.

Suppose that the random sample $\{\mathbf{X}_i\}$ is drawn from a member of the exponential family of distributions whose joint distribution is defined in (5.17). Let $k = 1$, so the joint distribution simplifies to

$$f(x_1,\ldots,x_N|\lambda) = e^{A(\lambda)B(x_i)+C(x_i)+D(\lambda)} \tag{6.26}$$

provided that $x_i \in \Omega$ and $\lambda \notin \Omega$. If $A(\lambda)$ is an increasing function of λ, then the likelihood ratio is

$$\hat{\Lambda} = \frac{\mathcal{L}_0(\lambda_0, x)}{\mathcal{L}_1(\lambda_1, x)} = \frac{e^{A(\lambda_0)\sum_{i=1}^{N} B(x_i) + \sum_{i=1}^{N} C(x_i) + N\,D(\lambda_0)}}{e^{A(\lambda_1)\sum_{i=1}^{N} B(x_i) + \sum_{i=1}^{N} C(x_i) + N\,D(\lambda_1)}} = e^{[A(\lambda_0) - A(\lambda_1)]\sum_{i=1}^{N} B(x_i) + N[D(\lambda_0) - D(\lambda_1)]} \tag{6.27}$$

If $\lambda_0 > \lambda_1$, the condition that $A(\lambda)$ is increasing means that the likelihood ratio is an increasing function of $\hat{\tau} = \sum_{i=1}^{N} B(x_i)$. Consequently, the likelihood ratio $\hat{\Lambda}$ is a monotone function of the sufficient statistic $\hat{\tau}$, and a test of H_0: $\lambda = \lambda_0$ versus H_1: $\lambda < \lambda_0$ using $\hat{\Lambda} \le \kappa$ reduces to one involving only the sufficient statistic $\hat{\tau} \le c$ for every $\lambda_1 < \lambda_0$ and is UMP. A similar statement can be applied to the upper-tail alternate hypothesis. This provides a method to devise UMP tests that are based on sufficient statistics for the exponential family of distributions.

6.5.3 Likelihood Ratio Tests for Composite Hypotheses

A general approach to likelihood ratio testing will be defined that does not result in a UMP test but that typically does yield a better test than can be constructed by other methods. Such tests play the same role in hypothesis testing that the mle does for parameter estimation. Suppose that there is a random sample $\{\mathbf{X}_i\}$ that has a likelihood function $\mathcal{L}(\boldsymbol{\lambda}|x_i)$ and that $\boldsymbol{\lambda}$ is a vector of parameters $(\boldsymbol{\lambda}_s, \boldsymbol{\lambda}_c)$, where $s + c = p$ is the total number of parameters. The null hypothesis is H_0: $\boldsymbol{\lambda}_s = \boldsymbol{\lambda}_s^*$ and is composite unless $c = 0$. The alternate hypothesis is H_1: $\boldsymbol{\lambda}_s \ne \boldsymbol{\lambda}_s^*$. The unconditional maximum of the likelihood function is obtained using the mles for $\boldsymbol{\lambda}_s$ and $\boldsymbol{\lambda}_c$, denoted by $\hat{\boldsymbol{\lambda}}_s$ and $\hat{\boldsymbol{\lambda}}_c$. The conditional maximum of the likelihood function is obtained using the mles for $\boldsymbol{\lambda}_c$ when H_0 holds and will in general differ from $\hat{\boldsymbol{\lambda}}_c$. Denote these mles by $\hat{\hat{\boldsymbol{\lambda}}}_c$. The likelihood ratio is

$$\hat{\Lambda} = \frac{\mathcal{L}_0\left(\boldsymbol{\lambda}_s^*, \hat{\hat{\boldsymbol{\lambda}}}_c|x_i\right)}{\mathcal{L}_1\left(\hat{\boldsymbol{\lambda}}_s, \hat{\boldsymbol{\lambda}}_c|x_i\right)} \tag{6.28}$$

Equation (6.28) is the maximum of the likelihood under the null hypothesis divided by the largest possible value for the likelihood, and $\hat{\Lambda} \to 1$ indicates that H_0 is acceptable. Thus there exists a critical value κ for $\hat{\Lambda}$ such that $\hat{\Lambda} \le \kappa$ rejects the null hypothesis. The critical value is determined from the distribution of $\hat{\Lambda}$ at a specified value of α if that distribution is available.

Example 6.23 Let $\{\mathbf{X}_i\}$ be a set of N random variables from $\mathrm{N}(\mu, \sigma^2)$, where μ and σ^2 are unknown. The composite null hypothesis H_0: $\mu = 1$, $\sigma^2 > 0$ will be tested against the composite alternate hypothesis H_1: $\mu \ne 1$, $\sigma^2 > 0$. The likelihood function for the unconditional alternate hypothesis is

$$\mathcal{L}_1\left(\mu, \sigma^2|x_i\right) = \left(2\pi\sigma^2\right)^{-N/2} e^{-\sum_{i=1}^{N}(x_i-\mu)^2/\left(2\sigma^2\right)}$$

whereas for the null hypothesis it is the same expression with $\mu = 1$. Under the null hypothesis, the mle for the variance is $\hat{s}_0^2 = (1/N)\sum_{i=1}^{N}(x_i - 1)^2$, whereas under the alternate hypothesis, the mles are $\hat{\mu} = \overline{X}_N$ and $\hat{s}_1^2 = (1/N)\sum_{i=1}^N \left(x_i - \overline{X}_N\right)^2 = \hat{s}_N^2$. Substituting into the likelihood functions for the numerator and denominator and simplifying yields

$$\hat{\Lambda} = \left[\frac{\sum_{i=1}^N \left(x_i - \overline{X}_N\right)^2}{\sum_{i=1}^N (x_i - 1)^2} \right]^{N/2} = \frac{1}{\left[1 + \hat{t}^2/(N-1)\right]^{N/2}}$$

where $\hat{t} = \sqrt{N-1}\left(\overline{X}_N - 1\right)/\hat{s}_N$ is the one-sample t test statistic (6.7) for a postulated mean of 1. The null hypothesis is rejected if $\hat{\Lambda} \le \kappa \le 1$ corresponding to $\hat{t} \ge \sqrt{(N-1)\left(\kappa^{-2/N} - 1\right)}$. Because $\hat{t}$ has Student's t distribution with $N - 1$ degrees-of-freedom, it has been shown that the likelihood ratio test is the one-sample t test. It rejects if the test statistic $\hat{t}$ exceeds a critical value for Student's t distribution.

Example 6.24 Let $\{\mathbf{X}_i\}$ and $\{\mathbf{Y}_j\}$ be sets of N and M independent rvs from Gaussian distributions with different means μ_1 and μ_2 and a common but unknown variance σ^2. A likelihood ratio test will be devised for H_0: $\mu_1 = \mu_2$, $\sigma^2 > 0$ versus H_1: $\mu_1 \ne \mu_2$, $\sigma^2 > 0$, where neither μ_1 nor μ_2 is specified. Both the null and alternate hypotheses are composite. The likelihood for the first set of samples is

$$\mathcal{L}\left(\mu_1, \sigma^2 | x\right) = \left(2\pi\sigma^2\right)^{-N/2} e^{-\sum_{i=1}^N (x_i - \mu_1)^2/\left(2\sigma^2\right)}$$

and similarly for the second set with M, y_i, and μ_2 substituting for N, x_i, and μ_1. The total likelihood is the product of the x and y likelihood functions. The total log likelihood is

$$\log \mathcal{L} = -\frac{N+M}{2}\log 2\pi - \frac{N+M}{2}\log\sigma^2 - \left[\sum_{i=1}^N (x_i - \mu_1)^2 + \sum_{i=1}^M (y_i - \mu_2)^2\right] / \left(2\sigma^2\right)$$

Under the null hypothesis, there exists a sample of size $M + N$ from a normal distribution with unknown mean and variance. The mles for the mean and variance under the null hypothesis are obtained by maximizing the likelihood ratio with $\mu_2 = \mu_1$, yielding

$$\hat{\mu}_0 = \left(N\overline{X}_N + M\overline{Y}_M\right)/(N+M)$$
$$\hat{s}_0^2 = \left[\sum_{i=1}^N (x_i - \hat{\mu}_0)^2 + \sum_{i=1}^M (y_i - \hat{\mu}_0)^2\right] / (N+M)$$

and a maximum for the likelihood of

$$\mathcal{L}_0 = \left(2\pi e \hat{s}_0^2\right)^{-(N+M)/2}$$

The unconditional log likelihood yields the unconstrained mles of $\hat{\mu}_1 = \overline{X}_N$, $\hat{\mu}_2 = \overline{Y}_M$, and

$$\hat{s}_1^2 = \left[\sum_{i=1}^{N}(x_i - \hat{\mu}_1)^2 + \sum_{i=1}^{M}(y_i - \hat{\mu}_2)^2\right] / (N+M)$$

The unconditional likelihood is given by

$$\mathcal{L}_1 = \left(2\pi e \hat{s}_1^2\right)^{-(N+M)/2}$$

The likelihood ratio is

$$\hat{\Lambda} = \frac{\mathcal{L}_0}{\mathcal{L}_1} = \left(\frac{\hat{s}_1^2}{\hat{s}_0^2}\right)^{N+M/2}$$

and rejects for $\hat{\Lambda} \leq \kappa$. Substituting for $\hat{s}_0^2$ and $\hat{s}_1^2$ and simplifying yields

$$\frac{\hat{s}_1^2}{\hat{s}_0^2} = \frac{1}{1 + \frac{NM}{N+M}\frac{\left(\overline{X}_N - \overline{Y}_M\right)^2}{\sum_{i=1}^{N}\left(x_i - \overline{X}_N\right)^2 + \sum_{i=1}^{M}\left(y_i - \overline{Y}_M\right)^2}} = \frac{1}{1 + \frac{\hat{t}^2}{N+M-2}}$$

where

$$\hat{t} = \sqrt{\frac{NM}{N+M}}\frac{\left|\overline{X}_N - \overline{Y}_M\right|}{\sqrt{\left(N\hat{s}_x^2 + M\hat{s}_y^2\right)/(N+M-2)}}$$

is the two-sample t statistic (6.9). Because the test rejects for large values of $\hat{t}^2$, it also rejects for large values of $\hat{t}$, and hence it has been shown that the likelihood ratio test is equivalent to the two-sample t test.

If the sampling distribution for $\hat{\Lambda}$ under H_0 is known, as in the preceding examples, then critical values and p-values can be obtained directly. However, in many instances, the sampling distribution is unknown, and recourse must be made to its asymptotic distribution. It can be shown that the asymptotic distribution of the likelihood ratio statistic (6.28) is noncentral chi square with noncentrality parameter $\delta = \left(\boldsymbol{\lambda}_s - \boldsymbol{\lambda}_s^o\right)^T \cdot \overleftrightarrow{\mathbf{I}}_F \cdot \left(\boldsymbol{\lambda}_s - \boldsymbol{\lambda}_s^o\right)$, where $\overleftrightarrow{\mathbf{I}}_F$ is the Fisher information matrix for the simple null hypothesis parameters. This reduces to the central chi square distribution when H_0 holds (i.e., when $\boldsymbol{\lambda}_s = \boldsymbol{\lambda}_s^*$). The distribution degrees-of-freedom is the difference between the number of free parameters under $H_0 \cup H_1$ minus the number of free parameters under H_0. This result is called *Wilks' theorem* and was introduced by Wilks (1938). It has an error that is typically $O(1/N)$. For (6.28), the number of free parameters under the null hypothesis is c because the data sample determines the parameters under the simple hypothesis. Under $H_0 \cup H_1$, there are $s + c$ free parameters, and consequently, the asymptotic distribution for (6.28) has s degrees-of-freedom. The asymptotic power of the likelihood ratio test follows directly from (6.5) using the noncentral chi square distribution.

Example 6.25 The mean monthly wind direction for a high-latitude region using the National Center for Environmental Prediction product is contained in the file wind.dat for calendar year 2004–5. The monthly means will average out periodic phenomena such as the tides to a substantial degree, simplifying the data statistics. The postulated mean over the 2-year interval is 240°. Derive the likelihood ratio test for the null hypothesis H_0: $\nu = \nu^*$, $\kappa > 0$ versus H_1: $\nu \neq \nu^*$, $\kappa > 0$ for the von Mises distribution, and apply it to the data.

Figure 6.12 is a rose diagram of the data showing a lot of scatter but with some tendency to cluster around ~140° and ~315°. The sample mean is 239.6°.

The mles for ν and κ that are needed for the unconstrained alternate hypothesis were derived in Example 5.23

$$\hat{\nu} = \tan^{-1}\left(\hat{S}/\hat{C}\right)$$
$$\frac{I_1(\hat{\kappa})}{I_0(\hat{\kappa})} = \frac{\sqrt{\hat{S}^2 + \hat{C}^2}}{N}$$

where $\hat{S} = \sum_{i=1}^N \sin\theta_i$ and $\hat{C} = \sum_{i=1}^N \cos\theta_i$. The second equation must be solved numerically. The denominator of the likelihood ratio is

$$\mathcal{L}_1 = \frac{e^{\hat{\kappa}\sum_{i=1}^N \cos(\theta_i - \hat{\nu})}}{(2\pi)^N I_0^N(\hat{\kappa})}$$

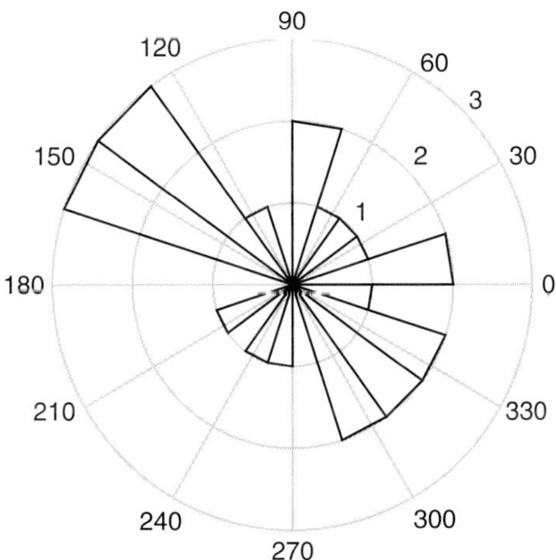

Figure 6.12 Rose diagram for the wind direction data.

The numerator of the likelihood ratio is

$$\mathcal{L}_0 = \frac{e^{\hat{\hat{\kappa}} \sum_{i=1}^{N} \cos(\theta_i - \nu*)}}{(2\pi)^N I_0^N\left(\hat{\hat{\kappa}}\right)}$$

where $\hat{\hat{\kappa}}$ satisfies

$$\frac{I_1\left(\hat{\hat{\kappa}}\right)}{I_0\left(\hat{\hat{\kappa}}\right)} = \frac{\hat{C}\cos\nu^* + \hat{S}\sin\nu^*}{N}$$

The test statistic is

$$-2\log\hat{\Lambda} = 2\hat{\kappa}\sum_{i=1}^{N}\cos(\theta_i - \hat{\nu}) + 2N\log I_0\left(\hat{\hat{\kappa}}\right) - 2\hat{\hat{\kappa}}\sum_{i=1}^{N}\cos(\theta_i - \nu^*) - 2N\log I_0(\hat{\kappa})$$

and is asymptotically distributed as chi square with one degree-of-freedom.

Solution of the equations for the precision parameters requires use of the MATLAB nonlinear equation solver **fzero**(x, *start*), where x is a function handle and *start* is an initial guess for the solution. A MATLAB script implementing the test is

```
wind = importdata('wind.dat');
s = sum(sind(wind));
c = sum(cosd(wind));
r = sqrt(s^2 + c^2);
nuhat = atan2d(s, c);
nustar = 240;
n = length(wind);
khat = fzero(@(x) besseli(1, x)/besseli(0, x) - r/n, 1);
khathat = fzero(@(x) besseli(1, x)/besseli(0, x) - (c*cosd
(nustar) + s*sind(nustar))/n, 1);
t1 = 2*khat*sum(cosd(wind - nuhat));
t2 = 2*n*log(besseli(0, khathat));
t3 = 2*khathat*sum(cosd(wind - nustar));
t4 = 2*n*log(besseli(0, khat));
test = t1 + t2 - t3 - t4
test =
    0.0024
1 - chi2cdf(test, 1)
ans =
   0.9609
```

The critical value is 5.02, and neither it nor the p-value rejects the null hypothesis. However, the data are not tightly clustered (see Figure 6.12) given that the mle for κ is only 3.1, so this is not surprising. If the test is repeated with a hypothesized mean of 150°, the test statistic is 41.6706, the p-value is 1.08e–10, and the null hypothesis is rejected.

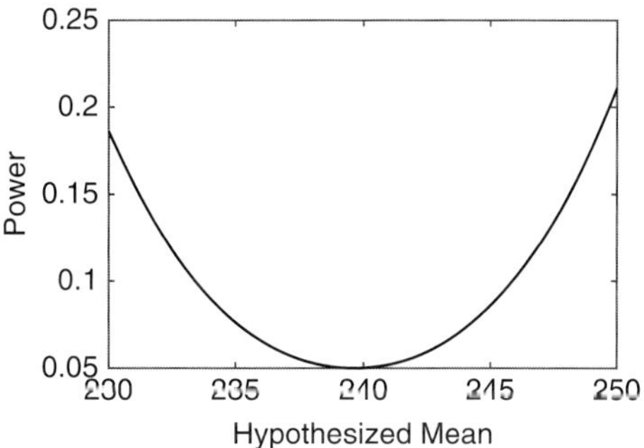

Figure 6.13 The power of the wind direction likelihood ratio test for the von Mises distribution.

The asymptotic power can be computed in the standard way. The variance can be estimated as $\hat{\sigma}^2 = 1/(\hat{R}\hat{\kappa})$, where $\hat{R} = \sqrt{\hat{S}^2 + \hat{C}^2}$, so $\delta = \sin^2(\hat{\nu} - \nu^*)/\hat{\sigma}^2$.

```
wind = importdata('wind.dat');
s = sum(sind(wind));
c = sum(cosd(wind));
nuhat = atan2d(s, c);
n = length(wind);
r = sqrt(c^2 + s^2);
khat = fzero(@(x) besseli(1, x)/besseli(0, x) - r/n, 1);
sigma2 = 1./(r*khat);
nustar = 230:.1:250;
delta = sind(nuhat - nustar).^2/sigma2;
xhi = chi2inv(.975, 1);
xlo = chi2inv(.025, 1);
power = 1 - ncx2cdf(xhi, 1, delta) + ncx2cdf(xlo, 1, delta);
plot(nustar, power)
```

The test has low power (Figure 6.13) and hence a low probability of rejecting the null hypothesis when it is false.

Likelihood ratio tests are usually carried out with respect to the general alternate hypothesis that the parameters are free. If a specific alternate hypothesis can be formulated, then better power generally can be obtained. For example, consider testing the null hypothesis that a set of rvs $\{\mathbf{X}_i\}$ is Poisson with parameter λ against the alternate hypothesis that the data are Poisson but with different parameters (i.e., rates) for each data point. Under the null hypothesis, the mle $\hat{\lambda}$ for λ is the sample mean of the data. Under the alternate hypothesis, the mles $\tilde{\lambda}_i$ are the corresponding data values x_i. The likelihood ratio is given by

$$\hat{\Lambda} = \frac{\prod_{i=1}^{N} \hat{\lambda}^{x_i} e^{-\hat{\lambda}} / x_i!}{\prod_{i=1}^{N} \tilde{\lambda}_i^{x_i} e^{-\tilde{\lambda}_i} / x_i!} = \prod_{i=1}^{N} \left(\frac{\bar{X}_N}{x_i} \right)^{x_i} e^{\left(x_i - \bar{X}_N \right)} \tag{6.29}$$

The statistic $-2 \log \hat{\Lambda} = 2 \sum_{i=1}^{N} x_i \log \left(x_i / \bar{X}_N \right)$ may be tested using the chi square approximation if N is large. There is one degree-of-freedom under the null hypothesis, and there are N degrees-of-freedom under the alternate, so the degrees-of-freedom for the test are $N - 1$. This test is called the *Poisson dispersion test*. It has high power against alternatives that are overdispersed compared with a Poisson process.

Example 6.26 A data set in the file tornados.dat contains the number of tornados in a single state from 1959 to 1990. Consider the null hypothesis that the data are Poisson against the alternate that they are too dispersed to be Poisson. The data are plotted in Figure 6.14.

There is some suggestion of an upward trend with time, although the data scatter is large, as would be expected for a Poisson process. Form the test

```
x = importdata('tornado.dat');
test = 2*sum(x.*log(x/mean(x)))
test =
    32.9726
```

There are 30 data, so the test has 29 degrees-of-freedom. The critical value is 45.72, and the p-value for the test is

```
1 - chi2cdf(test, 29)
ans =
   0.2788
```

The null hypothesis is accepted, and the data appear to be Poisson.

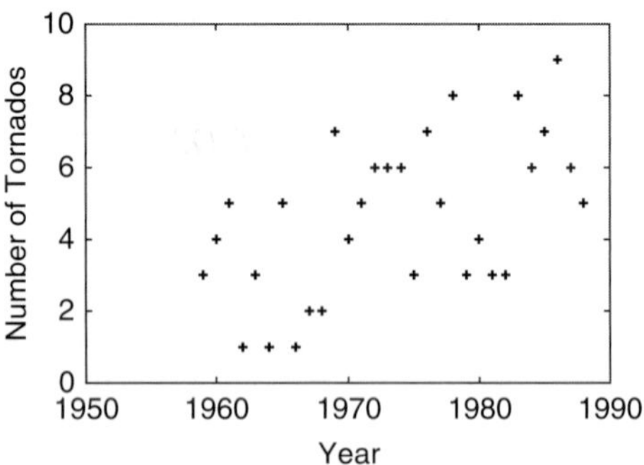

Figure 6.14 The number of tornados as a function of time.

6.5.4 The Wald Test

A simple approximation to the likelihood ratio test is the Wald test due to Wald (1943). This test is asymptotically efficient but often remains analytically approachable when the likelihood ratio is complicated. The Wald test is asymptotically equivalent to the likelihood ratio test but differs in performance with a finite sample.

Consider the likelihood ratio (6.28) for a single parameter λ under the null and alternate hypotheses H_0: $\lambda = \lambda^*$ versus $H_1 : \lambda \neq \lambda^*$, and expand $\log \mathcal{L}_0(\lambda^* | x_i)$ in a second-order Taylor series about $\log \mathcal{L}_1(\hat{\lambda} | x_i)$, yielding

$$-2 \log \hat{\Lambda} = \log \left[\frac{\mathcal{L}_0(\lambda^* | x_i)}{\mathcal{L}_1(\hat{\lambda} | x_i)} \right] \approx -2(\hat{\lambda} - \lambda^*) U(\hat{\lambda}) - (\hat{\lambda} - \lambda^*)^2 U'(\hat{\lambda}) \tag{6.30}$$

where $U(\hat{\lambda})$ is the score function (5.5) evaluated at $\hat{\lambda}$. The score function is zero at $\hat{\lambda}$, and consequently,

$$-2 \log \hat{\Lambda} \approx -N(\hat{\lambda} - \lambda^*)^2 \left[U'(\hat{\lambda})/N \right] \tag{6.31}$$

The estimator $\hat{\lambda}$ for λ under the null hypothesis is consistent, so $U'(\hat{\lambda})/N \to U'(\lambda^*)/N \to -\mathcal{I}(\lambda^*)/N$, where $\mathcal{I}(\lambda)$ is the Fisher information (5.7). Substituting into (6.31) gives

$$-2 \log \hat{\Lambda} \approx \hat{W} = (\hat{\lambda} - \lambda^*)^2 \mathcal{I}(\lambda^*) \tag{6.32}$$

where $\hat{W}$ is the Wald test statistic that is asymptotically distributed as χ_1^2.

The Wald test easily generalizes to multiple parameters and to composite hypotheses. Let $\hat{\boldsymbol{\lambda}}$ be a p-vector that is partitioned into s simple and c composite elements, where $s + c = p$, as in Section 6.5.3. For the composite hypotheses $H_0 : \hat{\boldsymbol{\lambda}}_s = \boldsymbol{\lambda}_s^*$ versus $H_1 : \hat{\boldsymbol{\lambda}}_s \neq \boldsymbol{\lambda}_s^*$, the Wald test statistic is

$$\hat{\mathrm{W}} = \left(\hat{\boldsymbol{\lambda}}_s - \boldsymbol{\lambda}_s^* \right)^T \cdot \left[\overleftrightarrow{\mathcal{I}}^{ss}(\hat{\boldsymbol{\lambda}}) \right]^{-1} \cdot \left(\hat{\boldsymbol{\lambda}}_s - \boldsymbol{\lambda}_s^* \right) \tag{6.33}$$

where $\hat{\boldsymbol{\lambda}}_s$ is the unconditional mle, and the Fisher information matrix is partitioned as

$$\overleftrightarrow{\mathcal{I}} \left(\hat{\boldsymbol{\lambda}} \right) = \begin{bmatrix} \overleftrightarrow{\mathcal{I}}_{ss} & \overleftrightarrow{\mathcal{I}}_{sc} \\ \overleftrightarrow{\mathcal{I}}_{cs} & \overleftrightarrow{\mathcal{I}}_{cc} \end{bmatrix} \tag{6.34}$$

and $\overleftrightarrow{\mathcal{I}}^{ss}(\hat{\boldsymbol{\lambda}})$ is the upper left element of the inverse of (6.34). By the block matrix inversion lemma:

$$\left[\overleftrightarrow{\mathcal{I}}^{ss}(\hat{\boldsymbol{\lambda}}) \right]^{-1} = \overleftrightarrow{\mathcal{I}}_{ss} - \overleftrightarrow{\mathcal{I}}_{sc} \cdot \overleftrightarrow{\mathcal{I}}_{cc}^{-1} \cdot \overleftrightarrow{\mathcal{I}}_{cs} \tag{6.35}$$

The Wald test statistic is asymptotically distributed as χ_s^2.

Example 6.27 Evaluate the Wald test for equivalence of the Michelson speed of light data in Example 6.10.

The test for equivalence of the means is the same as testing that the difference of the means is zero. The mles for the mean and standard deviation of the 1879 and 1881 data are given by the sample means and standard deviations, respectively. Consequently, an estimator for their difference is obtained by subtracting the sample means, yielding 88.52, and an approximate pooled variance is $\hat{s}_1^2/N_2 + \hat{s}_2^2/N_1 = 376.14$. The Wald test statistic (6.32) is the square of the difference of the means divided by the pooled variance, or 20.83. The critical value from the χ_1^2 distribution for a two-tail test is 5.02, and hence the null hypothesis that the difference of the means is zero is rejected. The asymptotic p-value is 1.00×10^{-5}. This result is very close to that from the two-sample t test in Example 6.10.

6.5.5 The Score Test

The score test is an alternative to the Wald and likelihood ratio tests and was introduced by Rao (1947). All three of these tests are asymptotically equivalent but will behave differently for a finite-size sample.

The derivation of the score test closely parallels that of the Wald test. Consider a single parameter λ and a simple hypothesis under the null and alternate hypotheses H_0: $\lambda = \lambda^*$ versus H_1: $\lambda \neq \lambda^*$. Compute the score function $U(\hat{\lambda})$ as in (6.30), and let $\mathcal{I}(\hat{\lambda})$ be the Fisher information. The score test statistic is

$$\hat{\mathrm{W}}' = \frac{U^2(\lambda^*)}{\mathcal{I}(\lambda^*)} \tag{6.36}$$

and is asymptotically distributed as χ_1^2. The score test statistic depends only on the null value for the test statistic, unlike the Wald statistic (6.32).

The score test easily generalizes to multiple parameters and composite hypotheses. Let $\hat{\boldsymbol{\lambda}}$ be a p-vector that is partitioned into s simple and c composite elements, where $s + c = p$ as in Section 6.5.4. For the composite hypotheses H_0: $\hat{\boldsymbol{\lambda}}_s = \boldsymbol{\lambda}_s^*$ versus H_1: $\hat{\boldsymbol{\lambda}}_s \neq \boldsymbol{\lambda}_s^*$, the score test statistic is

$$\hat{\mathrm{W}}' = \mathbf{U}^T(\hat{\lambda}_s) \cdot \overleftrightarrow{\mathcal{I}}^{-1}(\hat{\lambda}_s) \cdot \mathbf{U}(\hat{\lambda}_s) \tag{6.37}$$

where $\hat{\lambda}_s$ is the mle under the null hypothesis. The score test statistic (6.37) is distributed as χ_s^2 when the null hypothesis is true.

Score tests are frequently used in directional statistics. The simplest example is the Rayleigh test for uniformity, which is the score test for uniformity under a von Mises model. The log likelihood for the von Mises distribution is

$$\begin{aligned}\log L(\nu, \kappa|\theta) &= \kappa \sum_{i=1}^{N} \cos(\theta_i - \nu) - N \log I_0(\kappa) \\ &= (\kappa \cos \nu, \kappa \sin \nu) \cdot \begin{pmatrix} \sum_{i=1}^{N} \cos \theta_i \\ \sum_{i=1}^{N} \sin \theta_i \end{pmatrix} - N \log I_0(\kappa)\end{aligned} \tag{6.38}$$

The von Mises parameters can be taken as $(\kappa \cos \nu, \ \kappa \sin \nu)$ instead of $(\nu, \ \kappa)$. The score function is

$$\mathbf{U} = \begin{pmatrix} \sum_{i=1}^{N} \cos \theta_i \\ \sum_{i=1}^{N} \sin \theta_i \end{pmatrix} - N \frac{I_1(\kappa)}{I_0(\kappa)} \begin{pmatrix} \cos \nu \\ \sin \nu \end{pmatrix} \tag{6.39}$$

In the uniform limit when $\kappa \to 0$, the score becomes the first term in (6.39). From the central limit theorem for directional data in Section 4.7.3, the asymptotic covariance matrix (4.44) in the uniform limit is $\overleftrightarrow{\mathbf{I}}_2/2$. The score statistic is

$$\hat{\mathrm{W}}' = 2N\bar{\mathrm{R}}^2 \tag{6.40}$$

where $\bar{\mathrm{R}}$ is the mean resultant length. The score statistic (6.40) is asymptotically distributed as χ_2^2. The Rayleigh test is most powerful against von Mises alternatives, and because $\bar{\mathrm{R}}$ is invariant under rotation and reflection of data, the test shares these properties.

Example 6.28 Apply the Rayleigh test to the wind direction data of Example 6.25.

```
wind = importdata('wind.dat');
s = sum(sind(wind));
c = sum(cosd(wind));
rbar = sqrt(s^2 + c^2)/length(wind);
rayl = 2*length(wind)*rbar^2
rayl -
   32.0933
1 - chi2cdf(rayl, 2)
ans =
  1.0740e-07
```

The null hypothesis that the data are uniformly distributed is rejected, which is no surprise in light of Figure 6.12.

6.6 Multiple Hypothesis Tests

There are situations where it is necessary to compute many simultaneous independent hypothesis tests. This topic has become prominent because of recent advances in fields such as genomics, where microarrays make it possible to measure thousands to many hundreds of thousands of genes at the same time, or in image processing, where many pixels in a scene must be tested together. In geophysics, an example occurs in the inversion of multiple data sets from distinct sites. It is customary to lump all the misfit across all data sets and sites into a single misfit measure such as the chi square statistic of Section 6.3.3 and then test the result using just the single statistic. This approach loses information about the misfit at each site and for each data set type. Instead, it would make more sense to test each data set and site separately and merge the results into a multiple hypothesis test.

However, there is a problem in multiple hypothesis testing because if a particular test is carried out at a significance level $\alpha = 0.05$ for rejecting the null hypothesis when it is true, then for an ensemble of tests, about one of twenty will be false positives. The problem is that there is no control on the false positive rate for the ensemble.

It is easy to compute the probability of observing a false positive for N independent tests, each of which has a significance level α. The distribution of false positives is binomial with probability α and number of trials N. The probability of observing n false positives is $\text{bin}(n; \alpha, N)$. Consequently, it is straightforward to assess whether the observed number of significant tests in an ensemble is reasonable or not.

Example 6.29 For 100 simultaneous independent tests with each test having a significance level of 0.05, what is the probability of observing 1, 10, or 20 significant tests?

The probability is $\sum_{i=n}^{N} \text{bin}(i; \alpha, N) = 1 - \text{Bin}(n - 1; \alpha, N)$ with $N = 100$. Using the MATLAB **binocdf** function, the answers are 0.994, 0.028, and 1.05e–07. For 10 or 20 observations, there clearly must be some real positives. The remaining issue is which ones are they?

The simplest way to address this issue is the Bonferroni method introduced in Section 5.5, which will now be formalized. Let the ensemble false positive probability be A. In the statistics literature, A is often called the *family-wide error rate* (FWER), or the probability that at least one Type 1 error occurs, and the significance level for a particular test α is the *comparison-wise error rate* (CWER). The probability of not making a Type 1 error in a particular test is $1 - \alpha$, and for N such independent tests, the probability becomes $(1 - \alpha)^N$. Consequently, the probability of at least one false positive is

$$A = 1 - (1 - \alpha)^N \tag{6.41}$$

Equation (6.41) can be easily inverted to yield the CWER required to achieve a specified FWER for a given number of tests:

Table 6.3 Outcomes from N Simultaneous Hypothesis Tests

	Significant	Not significant	Total
True null	v	u	N_0
False null	s	t	N_1
Total	r	w	N

$$\alpha = 1 - (1 - A)^{1/N} \tag{6.42}$$

A simpler version obtains by approximating $(1 - A)^{1/N} \approx 1 - A/N$, yielding the Bonferroni CWER $\alpha = A/N$. The probability of the false rejection of any of the null hypotheses is no more than α. However, the result is very conservative because either (6.42) or the Bonferroni approximation to it makes it unlikely that even a single null hypothesis will be falsely rejected. In addition, the power of each test is reduced by the use of very small significance levels.

Table 6.3 summarizes the outcomes from simultaneously performing N independent hypothesis tests, whereas Table 6.1 covered the outcomes from a single test. Each of the N hypotheses is presumed to have a p-value p_i, and the ith hypothesis is deemed significant if the corresponding null hypothesis is rejected through a small p-value. The unobservable random variables v, s, u, and t are, respectively, the number of false rejections, the number of true rejections, the number of true acceptances, and the number of false rejections. The observable random variables $r = s + v$ and $w = u + t$ are, respectively, the number of significant and insignificant tests. Similarly, the total number of true and false null hypotheses N_0 and N_1 are unknown but fixed. Finally, the proportion of false rejections is v/r for $r > 0$, and the proportion of false acceptances is t/w for $w > 0$. Consequently, the FWER is simply $\Pr(v \geq 1)$.

Neither (6.42) nor the Bonferroni approximation that controls the FWER accounts for the fact that rejection of a given hypothesis reduces the number of hypothesis tests for the ensemble. This led to the development of a number of sequential approaches to multiple hypothesis testing, culminating in the seminal work of Benjamini & Hochberg (1995), who introduced the concept of *false discovery rate* (FDR), which is the expected proportion of false rejections of the null hypothesis. From Table 6.3, the FDR is $\mathcal{E}(v/r)$, with the understanding that this quantity is zero when $r = 0$. The motivation for FDR is that multiple hypothesis testing is usually a screening step in a data analysis algorithm, and the data underlying tests that accept will often be subjected to further analysis. In this case, it is more important to ensure that all the true tests are included, accepting that a fraction of false positives will be included but controlling their quantity.

The Benjamini-Hochberg one-stage step-up algorithm is as follows:

1. Compute the order statistics $p_{(k)}$ from the p-values obtained by applying N individual hypothesis tests;
2. For independent tests, define $\beta_k = kq/N$, where $k = 1, \ldots, N$, and q is an a priori probability level (typically 0.05). For dependent tests, divide β_k by $\sum_{k=1}^{N}(1/k)$;

3. Define the rejection threshold $R = \max\left(p_{(k)} < \beta_k\right)$; and
4. If R is nonzero, reject all null hypotheses for which $p_k \leq R$, and otherwise retain all hypotheses.

Benjamini & Hochberg (1995) proved that the FDR from this procedure is always less than or equal to $N_0 q/N$, and more generally, they proved that the algorithm is a solution to the constrained maximization problem of choosing α that maximizes the number of rejections subject to the constraint $\alpha N/(N - N_0) \leq q$.

The following MATLAB script implements the Benjamini-Hochberg algorithm:

```
function [RejTh, Index] = Benjamini(pval, varargin)
%computes the rejection threshold ReJTh and the index Index
of p-values
%pval that reject the null hypothesis using the Benjamini-
Hochberg method
%input variables
%pval - vector of pvalues (required)
%q - FWER probability level (optional), default is 0.05
%cm = turn on dependence factor (optional), default is
1 for independence
[n p] = size(pval);
switch nargin
   case 1
      q = 0.05;
      cm = 1;
   case 2
      q = varargin{1} ;
      cm = 1;
   case 3
      q = varargin{1} ;
      i = 1:max(n, p);
      cm = sum(1./i);
end
if n < p
   pval1 = pval';
else
   pval1 = pval;
end
beta = (1:n)'* q/(cm* n);
pvalo = sort(pval1);
i = find(pvalo - beta <= 0);
if isempty(i)
   RejTh = 0;
else
   RejTh = pvalo(i(length(i)));
```

```
end
Index = find(pvall <= RejTh);
End
```

Example 6.30 Suppose that ten independent hypothesis tests are computed, leading to the following set of ordered p-values:

```
0.00013 0.00333 0.00579 0.00956 0.01345 0.26489 0.35487
0.55698 0.65345 0.99111
```

Apply a Bonferroni and Benjamini-Hochberg multiple hypothesis test at the 0.05 level.

The Bonferroni threshold with $\alpha = 0.05$ is 0.005, and hence the first two tests are rejected. For the Benjamini-Hochberg method, β_i is given by the following:

```
0.0050    0.0100    0.0150    0.0200    0.0250    0.0300
0.0350    0.0400    0.0450    0.0500
```

Comparing the two shows that $R = 0.01345$ and $r = 5$. The first five tests are rejected. The FDR has an upper bound of 0.025. Note that the data suggest the presence of two populations, with the first five having small and the second five much larger p-values.

Multiple hypothesis testing continues to be an area of active research. Genovese & Wasserman (2002) showed that the Benjamini-Hochberg algorithm minimizes the expected proportion of unrejected hypotheses for which the alternate holds, or in other words, it minimizes the false nondiscovery rate. Benjamini, Krieger, & Yekutieli (2006) introduced a two-stage multiple hypothesis estimator in which the initial step is the original algorithm that is used to estimate R, and the second stage uses a corrected value of q based on that estimate. The result is a more powerful multiple hypothesis testing procedure.

7 Nonparametric Methods

7.1 Overview

The hypothesis tests discussed to this point have been based on the premise that the data come from a known distribution in which some or all of the parameters are unknown. These are called *parametric tests.*

This chapter covers nonparametric tests of two classes: goodness-of-fit tests and rank-based hypothesis tests. The former begins with the little-used likelihood ratio test for a multinomial distribution that is more general than its widely used asymptotic approximation, the Pearson's χ^2 test, and deserves greater attention. The Kolmogorov-Smirnov and Anderson-Darling tests for data distribution assessment are then reviewed, and the Jarque-Bera test for Gaussianity is summarized.

Rank-based nonparametric tests are typically permutation tests (see Section 8.3) applied to the data ranks and were devised during the 1940s and 1950s before computers were powerful enough to apply permutation methods directly to data. Instead, it made sense to tabulate their results for data ranks. Such tests are less sensitive to distributional assumptions and work better than parametric tests with small samples. They typically lack power in comparison with parametric tests but are more robust to departures from parametric test assumptions. The sign, signed rank, and rank sum tests are described. The Ansari-Bradley test for dispersion is then summarized. A pair of nonparametric tests for correlation, the Spearman rank correlation and Kendall's τ, are then described. The chapter closes with a description of meta-analysis that allows p-values from separate studies to be combined.

7.2 Goodness-of-Fit Tests

Let $\{\mathbf{X}_i\}$ be a random sample from an unknown cdf $F(x)$. The null hypothesis is H_0: $F(x) = F^*(x)$, where $F^*(x)$ is a specified target distribution that is to be tested against a suitable alternate hypothesis, which is often H_1: $F(x) \neq F^*(x)$. This is called a *goodness-of-fit test.* It is a simple hypothesis if all the parameters in the target distribution are specified and a composite hypothesis if they are not.

7.2.1 Likelihood Ratio Test for the Multinomial Distribution

Partition the sample space $\mathcal{S}$ for a random variable $\mathbf{X}$ into k subspaces $\{\mathcal{S}_i\}$ that exhaust $\mathcal{S}$. Assume that N independent observations of the random variable (rv) are available, and let $\{\mathbf{Y}_i\}$ be the number of rvs in $\mathcal{S}_i$. The joint distribution of $\{\mathbf{Y}_i\}$ is multinomial with parameters N and $\{\pi_i\}$, where $\pi_i = \Pr(\mathbf{X} \in \mathcal{S}_i)$. A likelihood ratio test for the goodness-of-fit to a multinomial model can be devised in the usual way. The hypothesis under test is H_0: $\pi_i = p_i$, for $i = 1, \ldots, k$, versus H_1: $\pi_i \neq p_i$, for at least one value of i where p_i is known.

The likelihood ratio will be formulated relative to an alternate hypothesis where the k class probabilities are free but subject to the constraint that they sum to 1, so the alternate hypothesis has $k-1$ degrees-of-freedom. If m additional parameters are estimated from the data for the null hypothesis, the test degrees-of-freedom are reduced to $k-m-1$. Typically, at least one parameter will be needed for the null hypothesis, so $m \geq 1$.

The numerator of the likelihood ratio is

$$\mathcal{L}(N_1, \ldots, N_k, p_1, \ldots, p_k | y_i) = \frac{N!}{y_1! \cdots y_k!} p_1(\lambda)^{y_1} \cdots p_k(\lambda)^{y_k} \tag{7.1}$$

and is maximized when λ is the multinomial maximum likelihood estimator (mle) $\hat{\lambda}$. The corresponding probabilities will be denoted as $p_i\left(\hat{\lambda}\right)$. The denominator of the likelihood ratio is the multinomial pdf with p_i set to the multinomial mle $\hat{p}_i = y_i/N$. The likelihood ratio is

$$\hat{\Lambda} = \frac{\frac{N!}{y_1! \cdots y_k!} p_1\left(\hat{\lambda}\right)^{y_1} \cdots p_k\left(\hat{\lambda}\right)^{y_k}}{\frac{N!}{y_1! \cdots y_k!} \hat{p}_1^{y_1} \cdots \hat{p}_k^{y_k}} = \prod_{i=1}^{k} \left[\frac{p_i\left(\hat{\lambda}\right)}{\hat{p}_i}\right]^{y_i} = \prod_{i=1}^{k} \left(\frac{Np_i}{y_i}\right)^{y_i} \tag{7.2}$$

Taking logs and using $y_i = N\hat{p}_i$ yields

$$-2\log\hat{\Lambda} = -2N\sum_{i=1}^{k} \hat{p}_i \log\left[\frac{p_i\left(\hat{\lambda}\right)}{\hat{p}_i}\right] = 2\sum_{i=1}^{k} o_i \log\left(\frac{o_i}{e_i}\right) \tag{7.3}$$

where the observed number of occurrences is $o_i = N\hat{p}_i = y_i$, and the expected number of occurrences is $e_i = Np_i\left(\hat{\lambda}\right)$. The null hypothesis is rejected if $\hat{\Lambda}$ or its logarithm is small enough. The exact distribution for $\hat{\Lambda}$ is unknown, and hence the asymptotic χ^2_{k-m-1} approximation must be used.

A significant weakness of the multinomial likelihood ratio test (as well as the Pearson chi square test covered in the next section) is that the subdivision of the sample space $\mathcal{S}$, and hence choice of the boundaries between the k classes, is arbitrary, yet different parameterizations will result in different values for the likelihood ratio $\hat{\Lambda}$. Further, if the class boundaries are chosen from the data, they become rvs and hence should be incorporated into the model. A general rule of thumb is to choose the classes such that they have equal probability under the null hypothesis and choose $k \sim 10$, although often neither criterion is fulfilled in practice.

The multinomial likelihood ratio test is consistent (and hence asymptotically unbiased). It is unbiased if the classes are chosen so that their corresponding probabilities are approximately equal and will have the best power when the classes are approximately the same size. The test is somewhat robust to the presence of a small number of extreme values because of the logarithm in (7.3).

Example 7.1 Rutherford, Geiger, & Bateman (1910) present some of the first statistical data for radioactive decay. Their experiments measured α decay in polonium over 2608 disjoint 72-second intervals. The columns in the Table 7.1 give the number of counts and the number of occurrences.

Based on physics, the data should be Poisson, and the sample mean constitutes the mle for the parameter in the distribution.

```
rutherford = [ 57 203 383 525 532 408 273 139 45 27 10 4 1 1];
c = [ 0 1 2 3 4 5 6 7 8 9 10 11 13 14];
param = sum(rutherford.* c)/sum(rutherford)
param =
   3.8715
n = sum(rutherford);
p = poisspdf(c, param);
loglambda = 2* sum(rutherford.* log(rutherford./(n* p)))
loglambda =
   20.0336
1 - chi2cdf(loglambda, length(c) - 2)
ans =
   0.0665
```

The null hypothesis that the data are Poisson is accepted, although not strongly.

Table 7.1 Radioactive Decay Data

Counts	Number
0	57
1	203
2	383
3	525
4	532
5	408
6	273
7	139
8	45
9	27
10	10
11	4
13	1
14	1

Table 7.2 V-1 Bomb Hits in London

Hits	Number
0	229
1	211
2	93
3	35
4	7
5 and over	1

Source: Clarke (1946).

Example 7.2 Table 7.2 presents the number of V-1 flying bomb hits in a 36 square kilometer part of South London during World War II. The area was divided into 0.25-km squares, and the number of hits per square is contained in Table 7.2. Assess whether the hits were random or not.

If the hits were random, then a Poisson distribution should fit the data. The observed occurrences are given in the second column of the table. The expected occurrences are given by the Poisson probabilities scaled by the total number of hits.

```
bomb = importdata('bomb.dat');
o = bomb(:, 2);
e = poisspdf(bomb(:, 1), sum(bomb(:, 1).*bomb(:, 2))/sum
     (bomb(:, 2)))* ...
sum(bomb(:, 2));
loglambda = 2*sum(o.*log(o./e))
loglambda =
   1.4995
1 - chi2cdf(loglambda, length(bomb) - 2)
ans =
   0.8267
```

The null hypothesis is accepted.

Example 7.3 A data set in the file quakes.dat consists of 62 measurements of the time in days between successive earthquakes taken from Hand et al. (1994). Assess the goodness-of-fit of the data to the exponential distribution.

```
quakes = importdata('quakes.dat');
[f, x] = ecdf(quakes);
plot(x, f, x, expcdf(x,mean(quakes)))
```

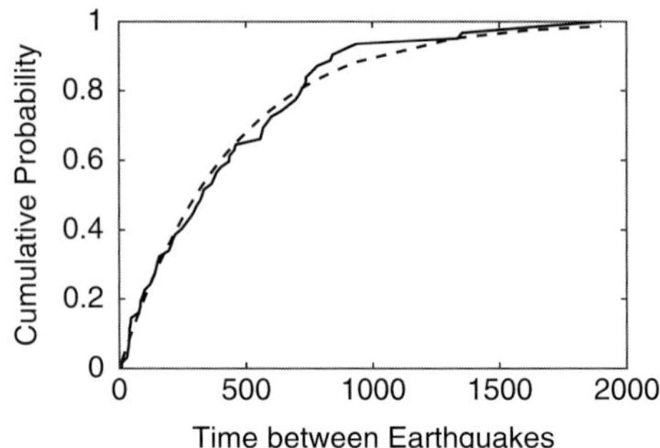

Figure 7.1 The empirical (solid line) and exponential (dashed line) cdfs for the earthquake interval time data.

The empirical and exponential cdfs with the mle for the distribution parameter are compared in Figure 7.1.

This problem is most easily solved by using the **histogram** function to compute the counts in each of a given number of bins, then integrate the pdf between the histogram bin boundaries to get the theoretical probabilities, and then convert to theoretical counts. The difficulty with an exponentially distributed variate is that the counts in some of the bins may be zero and will not likely be approximately equal without carefully choosing the bin edges.

```
h = histogram(quakes, 7, 'Normalization', 'count');
o = h.Values;
p = [];
for i = 1:7
   p(i) = integral(@(x) exppdf(x, mean(quakes)), h.BinEdges
            (i), h.BinEdges(i) + h.BinWidth);
end
e = p*length(quakes);
loglambda = 2*sum(o.*log(o./e))
loglambda =
   5.9478
1 - chi2cdf(loglambda, 5)
ans =
     0.3113
```

The null hypothesis that the data are exponential is accepted. However, the number of data per bin is far from even. They will be adjusted to make that more true at the cost of fewer bins.

```
edges = [0 75 200 400 600 1000 1960];
h = histogram(quakes, edges)
o = h.Values;
p = [];
for i = 1:6
```

```
    p(i) = integral(@(x) exppdf(x, mean(quakes)), h.BinEdges
           (i), h.BinEdges(i + 1));
end
e = p*length(quakes);
loglambda = 2*sum(o.*log(o./e))
loglambda =
    3.3545
1 - chi2cdf(loglambda, 4)
ans =
    0.5003
```

The conclusion is unchanged.

7.2.2 Pearson's χ^2 Test for Goodness-of-Fit

The Pearson (1900) goodness-of-fit test differs from the multinomial likelihood ratio (LR) test in that the former uses the asymptotic normal approximation rather than the exact form of the multinomial distribution. This is one of the most widely used statistical tests in the earth sciences and perhaps in science overall. The LR and Pearson tests are asymptotically equivalent but will give different results for small samples. Cochran (1952) provides a thorough introductory exposition on the chi square test that is as relevant today as it was over sixty years ago and should be required reading for entering graduate students in geophysics.

The Pearson test of goodness-of-fit can be derived from (7.3) using a Taylor series approximation. If the null hypothesis is true and the number of data is large, $o_i \approx e_i$. The Taylor series about e_i is

$$o_i \log\left(\frac{o_i}{e_i}\right) = (o_i - e_i) + \frac{1}{2e_i}(o_i - e_i)^2 + \cdots \tag{7.4}$$

so that

$$-2\log\hat{\Lambda} \approx 2\sum_{i=1}^{k}(o_l - e_i) + \sum_{i=1}^{k}\frac{(o_i - e_i)^2}{e_i} \tag{7.5}$$

The first term on the right is zero, and hence the LR test reduces to the Pearson test. However, because it is an approximation, the Pearson test will typically be less accurate than the multinomial LR test.

Consequently, the Pearson test for H_0: $\pi_i = p_i, \quad i = 1, \ldots, k$, has the test statistic

$$\hat{Q}_N = \sum_{i=1}^{k}\frac{(y_i - Np_i)^2}{Np_i} = \sum_{i=1}^{k}\frac{(o_i - e_i)^2}{e_i} \tag{7.6}$$

If the null hypothesis is true, then $\hat{Q}_N$ is asymptotically χ^2_{k-1} for large samples. If m parameters are estimated from the data, the degrees-of-freedom decrease to $k - m - 1$. The null hypothesis rejects when $\hat{Q}_N > \kappa$, where $\Pr(\hat{Q}_N > \kappa) = \alpha$ is chosen from the chi square distribution.

While they are asymptotically equivalent, for finite numbers of data, the multinomial LR test offers better performance if for no other reason than that fewer approximations and assumptions are needed to derive it. It is typically more robust because the logarithm of a ratio replaces the square of a difference. However, the Pearson test is far more widely used, presumably because historically the use of logarithms required a slide rule or table lookup rather than simple arithmetic. This is hardly a valid reason today, and the multinomial LR test is recommended for standard use over the Pearson test.

Computing the power of either the multinomial LR or the Pearson test is complicated because there are a large number of possible alternates and is typically not of much interest because the tests are usually used to assess fit in the absence of a well-defined alternate. Guenther (1977) gives an example of a power calculation for a single specific alternate involving loaded die. The simple case where the class probabilities are completely specified by the null hypothesis in the large sample limit was treated by Patnaik (1949). Power computation becomes much more complicated when the class probabilities are estimated from the data using, for example, maximum likelihood; see Mitra (1958) and Broffitt & Randles (1977) for elaboration.

Example 7.4 Returning to the Rutherford α decay data of Example 7.1, the test statistic is

```
q = sum((rutherford - n*p).^2./(n*p))
q =
    21.4973
1 - chi2cdf(q, length(c) - 2)
ans =
   0.0436
```

The null hypothesis that the data are Poisson is rejected.

However, as an illustration of the effect that class boundaries can have on the result, lump the last four together and repeat the test.

```
c = [ c(1:10) 10];
rutherford = [ rutherford(1:10) 16];
param = sum(rutherford.*c)/sum(rutherford)
param =
    3.8673
n = sum(rutherford);
p = poisspdf(c, param);
q = sum((rutherford - n*p).^2./(n*p))
q =
    14.8620
```

```
1 - chi2cdf(q, length(c) - 2)
ans =
   0.0948
```

The null hypothesis that the data are Poisson is accepted. A similar result is obtained when the last six classes are combined.

Example 7.5 Returning to the V-1 flying bomb data of Example 7.2

```
bomb = importdata('bomb.dat');
o = bomb(:, 2);
e = poisspdf(bomb(:,1), sum(bomb(:,1).*bomb(:,2))/sum(bomb
    (:,2)))*sum(bomb(:,2));
q = sum((o - e).^2./e)
q =
    1.0543
1 - chi2cdf(q, length(bomb) - 2)
ans =
   0.9014
```

This result is consistent with the one obtained using the multinomial LR test. However, the classes have a widely divergent numbers of hits. Merging the last four classes yields a more even distribution of hits, although it reduces the degrees-of-freedom for the test to 1. The result is a test statistic of 28.4423 and a p-value of 9.65×10^{-8}. The null hypothesis that the data are Poisson is rejected with this class structure. The problem with this data set is that it was not structured for statistical analysis because the bin sizes are too small to give adequate sampling.

In MATLAB, $[h, p]$ = **chi2gof**(x) performs a Pearson chi square test of the data in x to a normal distribution with mean and variance estimated from the data. If $h = 1$, the null hypothesis that x is a random sample from a normal distribution can be rejected at the 5% level, and p contains the p-value. For other distributions, the cdf must be provided as a function handle, as shown by the following example.

Example 7.6 Apply the Pearson chi square test to the earthquake interval data in Example 7.3.

```
[h, p, stats] = chi2gof(quakes, 'cdf',{@expcdf, expfit(quakes)})
h =
   0
p =
   0.3206
chi2stat: 3.5014
```

```
df: 3
edges: [ 9.0000 198.2000 387.4000 576.6000 765.8000 1.9010e+03]
O: [ 21 14 8 10 9]
E: [ 22.5985 13.8407 8.9788 5.8248 10.7572]
```

The null hypothesis that the data are drawn from an exponential distribution with a mean given by that of the data is accepted at the 0.05 level. However, the test only has three degrees-of-freedom. The degrees-of-freedom can be increased by writing a MATLAB script, as in Example 7.3.

7.2.3 Kolmogorov-Smirnov Test

This is not an exotic drink made from vodka and prune juice but rather a nonparametric test for the fit of a set of data to a specified distribution. Consider a random sample of size N drawn from some unknown continuous distribution. An empirical cdf $\hat{F}_N(x)$ can be constructed from the order statistics, as shown in Section 4.8.2. This is essentially what the MATLAB functions **ecdf** and **histogram** produce. The result is a monotonically increasing set of step functions with step size $1/N$ assuming the data are unique and larger steps if they are not. Recall from Section 4.8.2 that $\hat{F}_N(x) \xrightarrow{p} F(x)$, where $F(x)$ is the true cdf for the sample, and further, by the Glivenko-Cantelli theorem, $\sup_x\left|\hat{F}_N(x) - F(x)\right| \xrightarrow{as} 0$.

Define the Kolmogorov-Smirnov (K-S) test statistic (Kolmogorov 1933; English translation is given in Kotz & Johnson 1992)

$$\hat{D}_N = \sup_x\left|\hat{F}_N(x) - F(x)\right| \tag{7.7}$$

By the Glivenko-Cantelli theorem, $\lim_{N\to\infty} \hat{D}_N = 0$. The statistic $\hat{D}_N$ is used to evaluate the hypotheses

$$\begin{aligned} H_0 &: \quad F(x) = F^*(x) \\ H_1 &: \quad F(x) \neq F^*(x) \end{aligned} \tag{7.8}$$

The K-S test also can be applied to the two one-sided alternate hypotheses $F(x) \leq F^*(x)$ and $F(x) \geq F^*(x)$. The K-S statistic for these two alternates may be easily derived from the order statistics as

$$\hat{D}_N^- = \max\left[\max_i\left(F^*\left(x_{(i)} - (i-1)/N\right), 0\right)\right] \tag{7.9}$$

$$\hat{D}_N^+ = \max\left[\max_i\left(i/N - F^*\left(x_{(i)}\right), 0\right)\right] \tag{7.10}$$

where

$$\hat{D}_N = \max\left(\hat{D}_N^-, \hat{D}_N^+\right) \tag{7.11}$$

The distribution of $\hat{D}_N$ when H_0 is true is available in asymptotic form (Feller 1948, 1950)

$$\lim_{N\to\infty} \Pr\left(\hat{D}_N \le \frac{z}{\sqrt{N}}\right) = 1 - 2\sum_{i=1}^{\infty}(-1)^{i-1}e^{-2i^2z^2} \tag{7.12}$$

Two approximate critical values of the asymptotic distribution are $c_{0.05} = 1.3581$ and $c_{0.01} = 1.6726$ for the two-sided alternate. These are reasonably accurate for $N > 75$. Stephens (1970) gives approximations to the K-S statistic that are suitable for computation.

Another version of the K-S test can be used to compare two empirical cdfs (Smirnov 1939). Let $\{\mathbf{X}_i\}$ and $\{\mathbf{Y}_j\}$ be N and M independent rvs from two different populations, respectively, and let their empirical cdfs be $\hat{F}_N(x)$ and $\hat{G}_M(x)$. Testing the null hypothesis that the population cdfs are the same against the alternate that they are not uses a K-S statistic in the form

$$\hat{D}_{MN} = \max_i\left(\left|\hat{F}_N(t_i) - \hat{G}_M(t_i)\right|\right) \tag{7.13}$$

where t_i is an element of the merged rvs $\{\mathbf{X}_i, \mathbf{Y}_j\}$. The asymptotic distribution is given by (7.12) if N is replaced by $MN/(M+N)$ under some continuity conditions (Feller 1948).

The K-S test offers some advantages over the multinomial LR test or Pearson test. The K-S test does not depend on the target cdf or on data grouping, and it is typically more powerful. However, there are a few caveats. First, the K-S test applies only to continuous distributions. Second, the K-S test is relatively insensitive to the tails of the distribution, yet it is in the tails that differences are typically largest (but also where outliers are a problem). The K-S test is most sensitive near the distribution mode. Third, if the data are used to estimate the distribution parameters, then the asymptotic distribution is not fully valid, and the p-value will be biased. This is typically not an issue for large numbers of data but may be for a limited sample size and can be overcome using the Lilliefors test (described later) or resampling (as described in Section 8.2.5).

In MATLAB, the function [*h*, *p*, *ksstat*, *cv*] = **kstest**(*x*, *cdf*, *alpha*, *type*) rejects the null hypothesis at the 0.05 level if $h = 1$, where x is a data vector, p is the asymptotic p-value, and the target cdf is the standardized normal if cdf is not supplied. The parameters *alpha* and *type* are the significance level and the test type, respectively. The returned variables *ksstat* and *cv* are the test statistic and the critical value.

Caution: Test rejection is based on a comparison of the asymptotic p-value with *alpha*. It is strongly recommended that the actual critical value *cv* and the test statistic in *ksstat* be compared as well.

The function **kstest2**($x1, x2$) performs the two-sample K-S test to evaluate whether two data samples are drawn from the same distribution. The remaining parameters are as for **kstest**.

A modification of the K-S test is the Lilliefors test (Lilliefors 1967), which pertains to a target distribution where the mean and variance are derived from the data. The disadvantage of the Lilliefors test is that it depends on the target cdf, unlike the K-S test. In MATLAB, this test is [*h*, *p*, *kstat*, *cv*] = **lillietest**(*x, alpha, dist*). The parameters are as for **kstest**, but dist may only be "*norm*," "*exp*," or "*ev*," where the latter refers to the extreme value distribution.

Example 7.7 Estimate the fit of the earthquake data of Example 7.3 to a Weibull distribution using a two-sided K-S test.

This requires first computing the parameters in the Weibull cdf

$$\mathrm{Weib}(x|\alpha, c) = 1 - e^{-(x/\alpha)^c}$$

The mles for α and c are obtained by solving

$$\hat{\alpha} = \left(\frac{1}{N}\sum_{i=1}^{N} x_i^{\hat{c}}\right)^{1/\hat{c}}$$

$$\hat{c} = \left[\left(\sum_{i=1}^{N} x_i^{\hat{c}} \log x_i\right)\left(\sum_{i=1}^{N} x_i^{\hat{c}}\right)^{-1} - \frac{1}{N}\sum_{i=1}^{N} \log x_i\right]^{-1}$$

The second equation may be easily solved iteratively for $\hat{c}$ using **fzero** with 1 as a starting value, and then the first equation may be solved for $\hat{\alpha}$. However, MATLAB makes this even simpler because the **wblfit** function solves for the two mle parameters directly.

```
parmhat = wblfit(quakes)
parmhat =
   450.2379      1.0798
quakes = sort(quakes);
cdf = [quakes' wblcdf(quakes', parmhat(1), parmhat(2))];
[h, p, ksstat, cv] = kstest(quakes, cdf)
h =
   0
p =
   0.9006
ksstat =
   0.0700
cv =
   0.1696
```

The null hypothesis that the data are drawn from a Weibull distribution with the computed parameters is accepted at the 0.05 level, and the p-value is large. Figure 7.2 compares the empirical and model cdfs and is substantially the same as Figure 7.1.

Example 7.8 The K-S test will be applied to a set of random draws from a standardized Gaussian distribution before and after it is contaminated with four outliers at the upper end of the distribution.

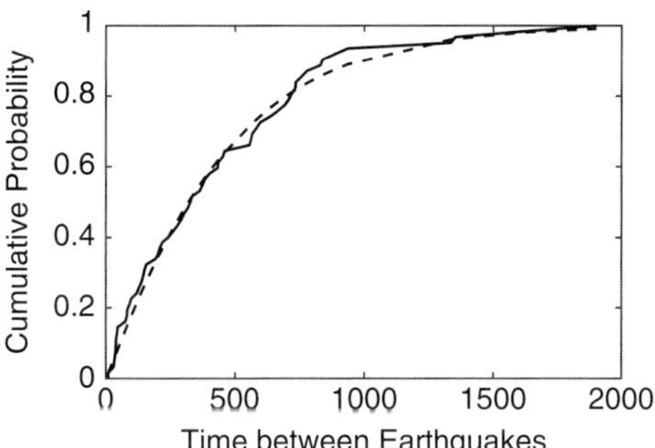

Figure 7.2 Empirical and Weibull cdfs for the earthquake interval data.

```
rng default;
data = normrnd(0, 1, 1000, 1);
[h, p, kstat, cv] = kstest(data)
h =
   0
p =
   0.3975
kstat =
   0.0282
cv =
   0.0428
data(1000) = 6.5;
data(999) = 6.0;
data(998) = 5.5;
data(997) = 5;
qqplot(data)
[h, p, kstat, cv] = kstest(data)
h =
   0
p =
   0.5415
kstat =
   0.0252
cv =
   0.0428
```

The K-S test result is almost unchanged when the outliers are present, and in fact, the *p*-value has increased slightly. This illustrates the insensitivity of the K-S test in the tails of the distribution. Figure 7.3 is a q-q plot that shows the four outliers that depart substantially from a Gaussian expectation.

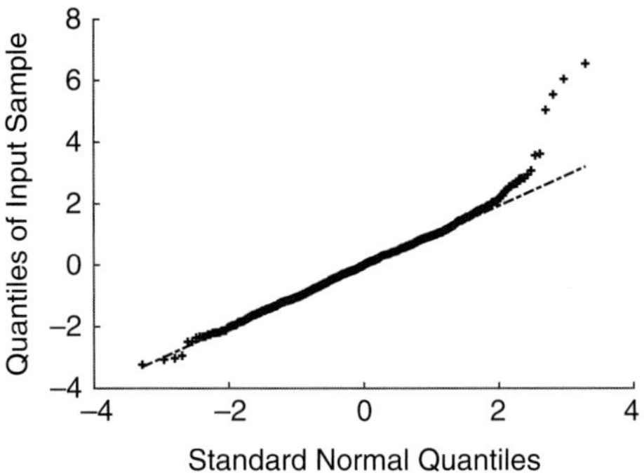

Figure 7.3 Quantile-quantile plot for the contaminated Gaussian data.

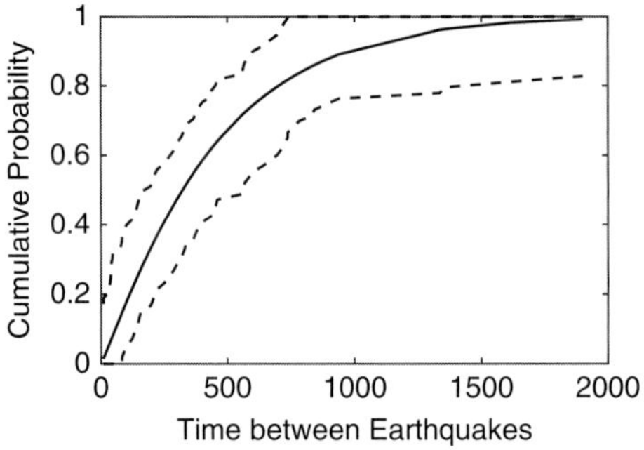

Figure 7.4 The 95% confidence interval for the earthquake interval data compared with the Weibull cdf.

Because the absolute difference between the empirical and target cdf is measured, the K-S statistic also provides a way to place confidence limits on the entire cdf and on probability plots. Regardless of what the true cdf $F(x)$ is,

$$\Pr\left[\hat{D}_N = \sup_x\left|\hat{F}_N(x) - F(x)\right| > \frac{c_\alpha}{\sqrt{N}}\right] = \alpha \tag{7.14}$$

which may be inverted to yield

$$\Pr\left[\hat{F}_N(x) - \frac{c_\alpha}{\sqrt{N}} \le F(x) \le \hat{F}_N(x) + \frac{c_\alpha}{\sqrt{N}}\right] = 1 - \alpha \tag{7.15}$$

There is a probability $1 - \alpha$ that the true cdf $F(x)$ lies entirely within the band $\pm c_\alpha/\sqrt{N}$ around the empirical cdf $\hat{F}_N(x)$.

Example 7.9 Place a 95% confidence band on the Weibull cdf obtained for the earthquake data in Example 7.7.

Figure 7.4 shows the result. Bounds have been placed at 0 and 1 to keep the probabilities meaningful.

```
cv = 1.3581;
plot(x, wblcdf(x, parmhat(1), parmhat(2)), x, max(0, f -
cv/sqrt(length(quakes))), ...
    x, min(1, f + cv/sqrt(length(quakes))))
```

The K-S critical value also may be used to put a $1 - \alpha$ confidence interval on p-p, variance-stabilized p-p, and q-q plots (see Sections 4.8.6 and 4.8.7), as described in Michael (1983). Let c_α denote the critical value for the K-S test at probability level α, and assume that the order statistics have been standardized to zero mean and unit variance. Then the $1 - \alpha$ confidence band is obtained by plotting the abscissa for each probability plot type against (p-p plot) $u_i \pm c_\alpha$, where $u_i = (i - 0.5)/N$ is a uniform quantile (variance-stabilized p-p plot) $2\sin^{-1}\left[\sqrt{\sin^2(r_i) \pm c_\alpha}\right]/\pi$, where $r_i = 2\sin^{-1}(\sqrt{u_i})/\pi$, and (q-q plot) $F^{*-1}(u_i \pm c_\alpha)$, where $F^*(x)$ is the target cdf.

Example 7.10 Assess the fit of the Gaussian distribution for the contaminated data of Example 7.7, and place confidence bounds on variance-stabilized p-p and q-q plots of the data.

```
u = ((1:1000) - .5)/1000;
r = 2/pi*asin(sqrt(u));
plot(r, 2/pi*asin(sqrt(normcdf(sort(data), mean(data), std
(data)))), 'r+')
hold on
plot(r,2/pi*asin(sqrt(min(1,sin(pi*r/2).^2+cv))),r,2/pi*...
       asin(sqrt(max(0, sin(pi*r/2).^2 - cv))))
hold off
x = norminv(u, mean(data), std(data));
plot(x, sort((data - mean(data))/std(data)), 'r+')
hold on
plot(x, norminv(u + cv, mean(data),std(data)), x, norminv
(u - cv, mean(data), std(data)))
```

The outliers appear as a slight downturn at the top of the distribution in the p-p plot of Figure 7.5, and the confidence band is too wide to enable detection of the outliers. The outliers are much more apparent in the q-q plot of Figure 7.6, but the widening of the confidence band in the distribution tails prevents their direct detection. However, the slope of the q-q plot, which gives the standard deviation of the data, has been torqued sufficiently that the q-q plot crosses the lower confidence bound at around 1 on the x-axis, and the null hypothesis that the data are Gaussian would be rejected, or at least suspicion about its validity would be raised.

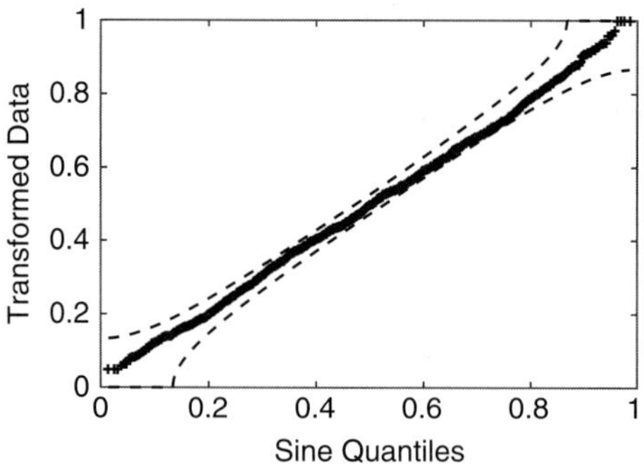

Figure 7.5 Variance-stabilized p-p plot for the contaminated Gaussian data of Example 7.7, along with the 95% confidence band derived from the K-S critical value.

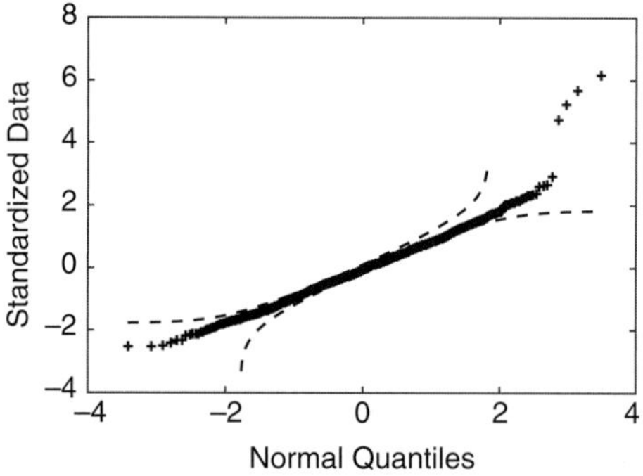

Figure 7.6 Quantile-quantile plot for the contaminated Gaussian data of Example 7.7, along with the 95% confidence band derived from the K-S critical value.

7.2.4 Cramér–von Mises Tests

Cramér–von Mises tests provide a more general quadratic criterion for evaluating goodness-of-fit. Let $\hat{F}_N(x)$ and $F^*(x)$ be the empirical cdf for a set of N rvs and the target distribution, respectively. The generalized Cramér–von Mises statistic is given by

$$\hat{\omega}^2 = \int_{-\infty}^{\infty} \left[\hat{F}_N(x) - F^*(x)\right]^2 w(x) f^*(x)dx \tag{7.16}$$

where $w(x)$ is a weight function. When $w(x) = 1$, (7.16) is the Cramér–von Mises statistic, for which the exact distribution is presented by Csörgő & Faraway (1996). When $w(x) = \{F^*(x)[1 - F^*(x)]\}^{-1}$, (7.16) is the Anderson-Darling (A-D) test first proposed by Anderson & Darling (1952, 1954), which is more powerful than the K-S test. The A-D test is most sensitive in the distribution tails but has the disadvantage that the critical values depend on the target distribution. The critical values also depend on whether the distribution parameters are known or have been estimated because they are used in standardizing the data.

The A-D test statistic is

$$\hat{A}^2 = -N - \frac{1}{N}\sum_{i=1}^{N}(2i-1)\left\{\log F^*\left(x_{(i)}\right) + \log\left[1 - F^*\left(x_{(N-i+1)}\right)\right]\right\} \tag{7.17}$$

For the Gaussian distribution, Stephens (1976) gives asymptotic two-sided test critical values at $\alpha = 0.05$ (0.01) of 2.492 (3.857) when μ and σ^2 are known, 1.088 (1.541) when only σ^2 is known, 2.315 (3.682) when only μ is known, and 0.751 (1.029) when both μ and σ^2 must be estimated, where the mles are used for the unknown parameters. For the exponential distribution, the critical values are 1.326 (1.943) when the parameters are unknown. Other critical points have been calculated and appear in the statistics literature.

MATLAB implements the A-D test as [*h*, *p*, *adstat*, *cv*] = **adtest**(*x*) and supports the Gaussian, exponential, extreme value (Gumbel), lognormal, and Weibull distributions. It defaults to the Gaussian, and alternate distributions may be selected using the keyword value pair "Distribution" followed by "norm," "exp," "ev," "logn," or "weibull." It also allows control of the level of significance and how the A-D critical value is calculated.

Example 7.11 For the earthquake data of Example 7.3, the A-D test for the null hypothesis that the data are exponential against the alternate that they are not may be tested with the following MATLAB script:

```
quakes = importdata('quakes.dat');
quakes = sort(quakes/mean(quakes));
n = length(quakes);
i - 1:n;
u1 = expcdf(quakes);
u2 = 1 - expcdf(quakes(n - i + 1));
a2 = -n - sum((2*i - 1).*(log(u1) + log(u2)))/n
a2 =
   0.3666
```

This is much smaller than the critical value of 1.326 at the 0.05 level, and hence the null hypothesis is accepted. Alternately, this can be done directly with the MATLAB function

```
[h, p, adstat, cv] = adtest(quakes, 'Distribution', 'exp')
h =
   0
p =
   0.7066
adstat =
   0.3666
```

```
cv =
   1.3130
```

The p-value is 0.7066, and hence the null hypothesis is strongly supported, and the data are exponential.

Example 7.12 Returning to the contaminated Gaussian example of Example 7.7, the A-D test is applied to the data after adding four outliers.

```
[h, p, adstat, cv] = adtest(data)
h =
   1
p =
   0.0027
adstat =
   1.2665
cv =
   0.7513
```

The test statistic is 1.27, which substantially exceeds the critical value, and the null hypothesis is rejected. This illustrates the sensitivity of the A-D test to the distribution tails, in contrast to the K-S test.

7.2.5 Jarque-Bera Test

The Jarque-Bera (J-B) or skewness-kurtosis score test for normality was introduced by Jarque & Bera (1987). The test statistic is

$$\hat{\tau} = \frac{N}{6}\left[(\hat{s}_3)^2 + (\hat{s}_4 - 3)^2/4\right] \tag{7.18}$$

and is a test of the joint null hypothesis that the skewness and the excess kurtosis are zero against the alternate that they are not. For Gaussian data, the J-B statistic is asymptotically chi square with two degrees-of-freedom and can be tested in the usual way, whereas Monte-Carlo simulation can also be used to get more accurate p-values. MATLAB implements the Jarque-Bera test as [*h*, *p*, *jbstat*, *cv*] = **jbtest**(*x*).

Example 7.13 Reanalyze the Gaussian simulation of Example 7.7 using the J-B test without and with the four upper outliers.

```
rng default
data = normrnd(0, 1, 1000, 1);
```

```
[h, p] = jbtest(data)
h =
   0
p =
   0.0850
data(1000) = 6.5;
data(999) = 6.0;
data(998) = 5.5;
data(997) = 5;
[h, p] = jbtest(data)
Warning: p is less than the smallest tabulated value,
returning 0.001.
h =
   1
p =
   1.0000e-03
```

The test accepts the null hypothesis for the Gaussian random samples, although the p-value of 0.0850 means that the support is weak. It easily detected the presence of the outliers.

7.3 Tests Based on Ranks

7.3.1 Properties of Ranks

Section 6.3 presented methods for testing single samples (e.g., the z test, single-sample t test, and chi square test) and for comparing the properties of two samples (e.g., two-sample t test, F test, and correlation coefficient). There are a wide variety of nonparametric tests for similar situations that replace the data with their ranks, yielding results that are invariant under any monotone transformation of the data and moderating the influence of unusual data because the most extreme values are pulled toward the center. However, this can be a disadvantage at the same time because closely spaced observations are spread apart. An excellent introduction to rank-based statistical tests that includes MATLAB code is given by Kvam & Vidakovic (2007), and this section relies on that book. A more advanced treatment of the theory is given by Hettmansperger & McKean (1998).

The ranks of an N-fold random sample $\{\mathbf{X}_i\}$ are easily obtained from its order statistics

$$\text{rank}\left(X_{(i)}\right) = i \tag{7.19}$$

or directly from the data values as

$$\text{rank}(x_i) = \sum_{j=i}^{N} \mathbf{1}\left(x_i \geq x_j\right) \tag{7.20}$$

Assuming that a random sample is exchangeable, it follows that $\{\mathbf{X}_i\} \xrightarrow{d} \{\mathbf{X}_{\varsigma_i}\}$, where ς_i represents a finite permutation of the indices. As a result,

$$\Pr\left[\text{rank}\left(\mathbf{X}_{(i)}\right) = i\right] = \frac{1}{N} \tag{7.21}$$

meaning that the ranks are distributed as discrete uniform rvs. Let $\Upsilon_i = \text{rank}\left(\mathbf{X}_{(i)}\right)$ denote the ranks of the rvs $\{\mathbf{X}_i\}$. It follows that

$$\begin{aligned} \mathcal{E}(\Upsilon_i) &= \sum_{i=1}^{N} \frac{i}{N} = \frac{N+1}{2} \\ var(\Upsilon_i) &= \sum_{i=1}^{N} \frac{i^2}{N} - \frac{(N+1)^2}{4} = \frac{N^2-1}{12} \end{aligned} \tag{7.22}$$

MATLAB supports ranks through the function r = **tiedranks**(x). This returns the ranks in r, and if any values in x are tied, it gives their average rank.

7.3.2 Sign Test

The sign test is a simple nonparametric procedure that pertains to continuous distributions for the null hypothesis H_0: $\tilde{\mu} = \tilde{\mu}^*$, where $\tilde{\mu}$ is the population median, and $\tilde{\mu}^*$ is a specified value, against either one- or two-sided alternates. It substantially predates most of statistical theory, originating in Arbuthnott (1710), who used a sign test to show that male births exceeded female births for most of the prior century. For a random sample $\{\mathbf{X}_i\}$, assign different categorical variables (such as + or −) when $x_i > \tilde{\mu}^*$ and when $x_i < \tilde{\mu}^*$. If the null hypothesis holds, then $\Pr(\mathbf{X}_i > \tilde{\mu}^*) = \Pr(\mathbf{X}_i < \tilde{\mu}^*) = 1/2$ by the definition of the median. Let the test statistic $\hat{\Xi}$ denote the total number of samples where $x_i > \tilde{\mu}^*$ given by

$$\hat{\Xi} = \sum_{i=1}^{N} \mathbf{1}(x_i > \tilde{\mu}^*) \tag{7.23}$$

Under the null hypothesis, the test statistic is a binomial variable with probability 1/2, or $\hat{\Xi} \sim \text{bin}(N, 1/2)$.

At a significance level α, if the test is upper tail or lower tail, then the critical values are given, respectively, by integers that are larger than or equal to ξ_α or smaller than or equal to ξ'_α. These are given, respectively, by the smallest or largest integer for which

$$\Pr\left(\hat{\Xi} \geq \xi_\alpha | H_0\right) = \frac{1}{2^N} \sum_{i=\xi_\alpha}^{N} \binom{N}{i} < \alpha \tag{7.24}$$

$$\Pr\left(\hat{\Xi} \leq \xi'_\alpha | H_0\right) = \frac{1}{2^N} \sum_{i=0}^{\xi'_\alpha} \binom{N}{i} < \alpha \tag{7.25}$$

If the test is two tailed, then the critical values are integers either less than or equal to $\xi'_{\alpha/2}$ or greater than or equal to $\xi_{\alpha/2}$ and may be obtained as the union of (7.24) and (7.25) after replacing α with $\alpha/2$.

Similarly, for an observed value of the test statistic $\hat{\Xi}$, the p-values for an upper- or lower-tail test are given, respectively, by

$$p = \frac{1}{2^N}\sum_{i=\hat{\Xi}}^{N}\binom{N}{i} = \frac{1}{2^N}\sum_{i=0}^{N-\hat{\Xi}}\binom{N}{i} \tag{7.26}$$

and

$$p = \frac{1}{2^N}\sum_{i=0}^{\Xi}\binom{N}{i} \tag{7.27}$$

The two-tailed p-value follows by taking $\hat{\Xi}' = \min\left(\hat{\Xi}, N - \hat{\Xi}\right)$ and computing

$$p = \frac{1}{2^{N-1}}\sum_{i=0}^{\hat{\Xi}'}\binom{N}{i} \tag{7.28}$$

MATLAB implements the two-sided sign test by default as the function [*p*, *h*, *stats*] = **signtest**(*x*, *mustar*), where p is the p-value, h is a logical variable (1 for rejection and 0 for acceptance), and *stats* is a structure that contains details about the test. Note that MATLAB reverses the order of the p-value and accept/reject flag for the nonparametric tests compared with the parametric tests, which is an unnecessary source of confusion. A tail probability of 0.05 is assumed. The variable *mustar* is the postulated median that defaults to zero if omitted. One-sided versions of the test may be computed using the keyword "tail" with either the value "left" or "right."

Example 7.14 Returning to Example 6.8, the sign test for the median (rather than the mean) number of days an oceanographer spends at sea per year being 30 is

```
x = [ 54 55 60  42 48 62 24 46 48 28 18 8 0 10 60 82 90 88 2 54];
[p, h, stats] = signtest(x, 30)
p =
    0.2632
h =
    0
stats =
    zval: NaN
    sign: 13
```

The null hypothesis that the median number of days a scientist spends at sea per year is 30 is accepted, in contrast to the t test on the mean value. This result uses the default exact statistical distribution that returns a z-value of NaN due to the small size of the sample. The value of the test statistic is given by stats.sign. The sign test also accepts both single-sided versions of the test, in contrast to the t test in Example 6.8, which rejects for the upper-tail test.

In some experimental designs (especially in the biomedical fields), the data are paired. For example, subjects might be grouped according to their age or weight or medical condition and then randomly assigned to test and control groups. The pairing causes the samples to be dependent. However, the sign test can be applied in this situation, in which case the null hypothesis is that the median of the difference between the two groups is zero, which is not the same as stating that their medians are identical.

Example 7.15 A data set consisting of the remission times for 42 leukemia patients is contained in the file remiss.dat and is taken from Hand et al. (1994). The patients were divided into two equal-sized groups: a test group that was treated with a new drug and a control group that was not, with the latter contained in the first 21 rows of the data. Test the null hypothesis that the median value of the differences between the test and control groups is zero.

```
remiss = importdata('remiss.dat');
x = remiss(1:21, 2);
y = remiss(22:42, 2);
[p, h, stats] = signtest(x, y)
p =
    0.0015
h =
    1
stats =
    zval: NaN
    sign: 3
```

The null hypothesis is strongly rejected. Figure 7.7 shows the two data sets against patient number, clearly indicating that the test group typically is in remission for much longer than the control group.

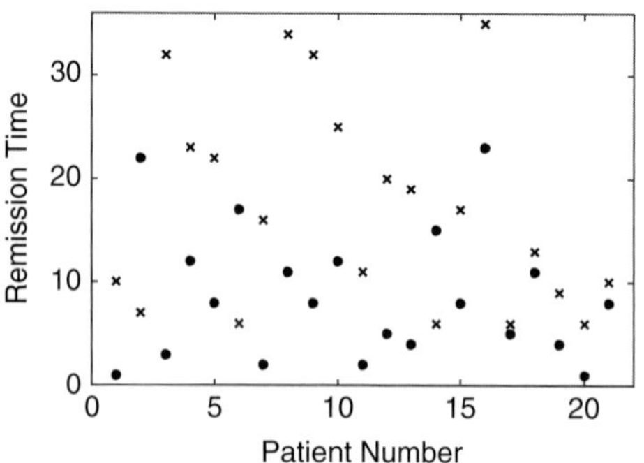

Figure 7.7 The remission time for the test group (x) and control group (filled circles) in Example 7.15.

Example 7.16 Returning to Gossett's wheat data from Example 6.9, use the sign test to evaluate the null hypothesis that the difference in yield between air- and kiln-dried wheat seeds has a median value of zero.

```
gossett = importdata('gossett.dat');
[p, h, stats] = signtest(gossett(:, 1), gossett(:, 2))
p =
   0.5488
h =
   0
stats =
   zval: NaN
   sign: 4
```

The null hypothesis is accepted at the 0.05 level, which also pertains to the t test in Example 6.9.

7.3.3 Signed Rank Test

A significant limitation of the sign test is that only the sign of the difference between a sample and the postulated median enters it. This led Wilcoxon (1945) to suggest that the absolute difference should be included with the sign in the test, and that turned into the test that bears his name. Let $\hat{D}_i = x_i - \tilde{\mu}^*$ denote the difference between a specific element in the random sample and a postulated value for the median $\tilde{\mu}^*$. The key assumption behind the Wilcoxon signed rank test is that the $\{\hat{D}_i\}$ are symmetric about zero if the null hypothesis is true, so the probability of observing a positive or negative D_i is about the same. This is equivalent to assuming that the underlying distribution is symmetric about its median. Hodges & Lehmann (1956) showed that the efficiency of the signed rank test is $3/\pi \approx 0.95$ for Gaussian data and never lower than 0.864 for any location distribution alternate. Consequently, the penalty for using this nonparametric test over a parametric test is not large. Further, including the absolute difference in the test raises the efficiency over that of the sign test.

The Wilcoxon signed rank test statistic is calculated by:

1. Ranking the absolute values of the differences $|\hat{D}_i|$;
2. Restoring the signs of the differences to the ranks, yielding the signed ranks; and
3. Calculating the sum of those ranks that have positive signs.

In effect, the ranked absolute differences are added to the sign test as weights.

Under the null hypothesis H_0: $\tilde{\mu} = \tilde{\mu}^*$, the expected value of the sum of the positive and negative differences should be the same. Following on (7.23), the test statistics are

$$\hat{\Xi}^{+} = \sum_{i=1}^{N} \text{rank}\left(\left|\hat{D}_i\right|\right) \mathbf{1}\left(\hat{D}_i > 0\right) \tag{7.29}$$

$$\hat{\Xi}^{-} = \sum_{i=1}^{N} \text{rank}\left(\left|\hat{D}_i\right|\right) \left[1 - \mathbf{1}\left(\hat{D}_i > 0\right)\right] \tag{7.30}$$

Under the null hypothesis, the signs $\left\{\mathbf{1}\left(\hat{D}_i > 0\right)\right\}$ are iid Bernoulli variables with probability ½ and are independent of the $\left\{\left|\hat{D}_i\right|\right\}$. Consequently, the expected value of $\hat{\Xi}^{+}$ is $N(N+1)/4$, and its quantiles can be computed, and similarly for $\hat{\Xi}^{-}$. These results are exact and are pertinent for small samples.

It follows from (7.29) to (7.30) that

$$\hat{\Xi}^{+} + \hat{\Xi}^{-} = \sum_{i=1}^{N} i = \frac{N(N+1)}{2} \tag{7.31}$$

and hence

$$\hat{\Xi} = \hat{\Xi}^{+} - \hat{\Xi}^{-} = 2\sum_{i=1}^{N} \text{rank}\left(\left|\hat{D}_i\right|\right) \mathbf{1}\left(\hat{D}_i > 0\right) - \frac{N(N+1)}{2} \tag{7.32}$$

For large samples, a Gaussian limiting form for $\hat{\Xi}$ is appropriate, and hence the usual Gaussian quantiles pertain.

MATLAB contains the function [*p*, *h*, *stats*] = **signrank**(*x*, *mustar*), which performs a two-sided test of the hypothesis that the sample in *x* comes from a distribution whose median is *mustar*. The structure *stats* contains the signed rank statistic. The function **signrank** uses the exact distribution or the normal approximation depending on the size of the sample.

Example 7.17 Returning to Examples 6.8 and 7.14, use the Wilcoxon signed rank test to evaluate the null hypothesis that the number of days per year that an oceanographer spends at sea is 30.

```
x = [ 54 55 60  42 48 62 24 46 48 28 18 8 0 10 60 82 90 88 2 54];
[p, h, stats] = signrank(x, 30)
p =
    0.0457
h =
    0
stats =
    zval: 1.9985
    signedrank: 158.5
```

The null hypothesis is weakly rejected at the 0.05 level, as was also the case for the *t* test, but not for the sign test. However, an upper-tail test fails weakly (*p*-value of 0.0239), and a lower-tail test accepts (*p*-value of 0.9782), which is very similar to the *t* test result in Example 6.8.

The Wilcoxon signed rank test also can be applied to the differences between two paired samples, as shown by Wilcoxon (1946). In this case, the test determines whether the median of the difference is zero and not whether their medians are the same.

Example 7.18 Returning to the leukemia data from Example 7.15, test the null hypothesis that the median of the difference between the test and control groups is zero.

```
remiss = importdata('remiss.dat');
x = remiss(1:21, 2);
y = remiss(22:42, 2);
[p, h, stats] = signrank(x, y)
p =
   0.0051
h =
   1
stats =
   zval: -2.8018
   signedrank: 35
```

The null hypothesis is rejected at the 0.05 level, as it was for the sign test.

Example 7.19 Returning to Student's wheat data from Examples 6.9 and 7.16, use the sign rank test to evaluate the null hypothesis that the difference in yield between air- and kiln-dried wheat seeds has a median value of zero.

```
gossett = importdata('gossett.dat');
[p, h, stats] = signrank(gossett(:, 1), gossett(:, 2))
p =
   0.1230
h =
   0
stats =
   signedrank: 15
```

The null hypothesis is accepted. Note that the exact distribution is used for this small sample, whereas the asymptotic normal approximation was used in Examples 7.17 and 7.18. The keyword value pair "method," "exact" could have been used to override the defaults for those two examples.

7.3.4 Rank Sum Test

The Wilcoxon (1945) rank sum test (which was introduced along with the signed rank test in a single paper) and the Mann & Whitney (1947) U test are equivalent, although they were derived from different assumptions and were long thought to be distinct. They are sometimes known as the *Wilcoxon-Mann-Whitney test*, but the term *rank sum test* is simpler. They are the nonparametric equivalent of the two-sample t test, but they do not require the assumption of a Gaussian population. The rank sum test also can be used with ordinal data, in contrast to the t test. Two key distinctions between the rank sum and the sign or signed rank tests are that the former directly tests whether the medians of two data sets are the same rather than testing the median of the difference, and there is no pairing requirement for the rank sum test, so different sized samples can be accommodated. Lehmann (1953) investigated the power of two-sample rank tests, which is a complex subject unless specific alternates are considered.

Suppose that $\{\mathbf{X}_i\}$ are N samples from a distribution $f(x)$ and $\{\mathbf{Y}_j\}$ are M samples from a distribution $g(y)$, and further suppose that $\{\mathbf{X}_i\}$ and $\{\mathbf{Y}_j\}$ are independent. The null hypothesis is that the two distributions are identical against a suitable alternate. A test statistic can be obtained by grouping all $M+N$ observations together, ranking them and then calculating the sum of the ranks of those samples that actually come from $\{\mathbf{X}_i\}$, yielding

$$\hat{\Omega} = \sum_{i=1}^{M+N} i\ \mathbf{1}(\mathbf{X}_i) \tag{7.33}$$

Equation (7.33) is the Wilcoxon rank sum statistic. Its exact distribution can be computed through combinatoric arguments that are not simple. A Gaussian approximation can be used for large samples. Additional corrections are needed in the event of ties in the ranks, but the details will be omitted.

An alternate test statistic can be estimated by comparing all N values in $\{\mathbf{X}_i\}$ to all M values in $\{\mathbf{Y}_j\}$ and then computing the proportion of the comparison for which $x_i > y_j$, giving

$$\hat{U} = \sum_{i=1}^{N}\sum_{j=1}^{M} \mathbf{1}\left(\mathbf{X}_i > \mathbf{Y}_j\right) \tag{7.34}$$

Equation (7.34) is the Mann-Whitney test statistic. Its exact distribution is the same as for the Wilcoxon rank sum statistic.

MATLAB provides the function [*p*, *h*, *stats*] = **ranksum**(*x*, *y*) that takes the original data in the vectors x and y, computes the test statistic, and returns the two-sided p-value in p, a 0–1 value indicating acceptance or rejection in h and a statistical structure in *stats*. Additional keyword value pairs can be added to change the test type and the method for computing the p-value. The function **ranksum** uses the exact distribution rather than the normal approximation for a small sample.

Example 7.20 Return to the Michelson speed of light data from Example 6.10.

```
michelson1 = importdata('michelson1.dat');
michelson2 = importdata('michelson2.dat');
[p, h, stats] = ranksum(michelson1, michelson2)
p =
   4.4407e-05
```

```
h =
   1
stats =
    zval: 4.0833
    ranksum: 6.8955e+03
```

The null hypothesis that the medians of the two populations are the same is rejected, and the speed of light has evidently changed between 1879 and 1882 or else systematic errors that were not accounted for are present. This result was computed using the normal approximation to the distribution of the test statistic. The exact p-value can be obtained, although this takes several minutes of computation

```
[p, h, stats] =ranksum(michelson1, michelson2, 'method', 'exact')
p =
   2.3681e-05
h =
   1
stats =
   ranksum: 6.8955e+03
```

The conclusion is unchanged and is consistent with the two-sample t test result in Example 6.10.

Example 7.21: Returning to the cloud seeding data from Example 6.12, compute the p-value for the rank sum test to assess the null hypothesis that the medians of the rainfall from unseeded and seeded clouds are the same.

```
clouds = importdata('cloudseed.dat');
[p, h] = ranksum(clouds(:, 1), clouds(:, 2))
p =
   0.0138
h =
   1
```

The null hypothesis is rejected. By contrast, the two-sample t test in Example 6.12 weakly accepted the null hypothesis that the means of unseeded and seeded clouds were the same.

Carrying out the rank sum test on the logs of the data

```
[ p, h] = ranksum(log(clouds(:, 1)), log(clouds(:, 2)))
p =
   0.0138
h =
   1
```

The p-value is unchanged from that with the original data because the ranks are not affected by a monotone transformation such as the logarithm. The p-value is very close to the two-sample t test result on the logs of the data. This example illustrates the robustness to departures from normality for rank-based tests.

7.3.5 Ansari-Bradley Test

This test was proposed by Ansari & Bradley (1960) and is a two-sample nonparametric test of dispersion for continuous populations having the same median and shape. The alternate hypothesis holds that the two populations have the same median and shape but different dispersions. Let two sets of rvs $\{\mathbf{X}_i, \mathbf{Y}_i\}$ have M and N values and be drawn from distributions with cdfs F and G, respectively. Under the null hypothesis, the two distributions differ at most by a scale factor on the argument, so $F(u) = G(\lambda u)$. The Ansari-Bradley (A-B) test evaluates H_0: $\lambda = 1$ versus H_1: $\lambda \neq 1$, or one-tailed alternates, based on ranks.

The two data sets are merged and sorted into ascending values, and ranks are assigned from both ends working toward the center, beginning with unity and incrementing by one each time, so that for $M + N$ an even number, the array of ranks is

$$1, 2, ..., (M + N)/2, (M + N)/2, ..., 2, 1 \tag{7.35}$$

whereas if the number of data is odd, the center value is unique. The test statistic is the sum of the ranks associated with the x data and can be written as

$$\hat{w} = \sum_{i=1}^{(M+N+1)/2} i\ \mathbf{1}(\mathbf{X}_i) + \sum_{i=(M+N+1)/2}^{M+N} (M + N + 1 - i)\ \mathbf{1}(\mathbf{X}_i) \tag{7.36}$$

Small values of the test statistic indicate a larger dispersion for the x data, and large values indicate the reverse. The null distribution can be determined exactly based on combinatoric arguments and is asymptotically Gaussian.

MATLAB implements the A-B test as [*h*, *p*, *stats*] = **ansaribradley**(*x*, *y*) (note that the order of p and h is reversed compared with all other nonparametric tests) and defaults to a two-tailed test, with the computation method for the null distribution determined from the number of data. One-tailed tests and the computation method can be controlled in the usual way. The exact test can take considerable computation with a large number of data.

Example 7.22 Returning to the Michelson speed of light data from Examples 6.10 and 6.15, test the null hypothesis that the dispersion of the data sets from different years is the same.

```
[h, p] = ansaribradley(michelson1, michelson2)
h =
    1
p =
    0.0109
```

The test rejects the null hypothesis. However, the medians are 850 and 776 for the 1879 and 1882 data. After adjusting so that the medians are identical, the result is

```
michelson1 = michelson1 - median(michelson1);
michelson2 = michelson2 - median(michelson2);
[h, p] = ansaribradley(michelson1, michelson2)
h =
   0
p =
   0.5764
```

The test now accepts the null hypothesis with a p-value of 0.58, which is no evidence for the alternate hypothesis. This should be contrasted with the F test in Section 6.3.4 that rejected the null hypothesis and the Bartlett's M test result of Section 6.3.5 that barely accepted it.

7.3.6 Spearman Rank Correlation Test

Spearman (1904) proposed the rank correlation statistic for bivariate data that bears his name long before most of the other work in nonparametric statistics was published in the 1940s and beyond. Suppose that $\{\mathbf{X}_i\}$ and $\{\mathbf{Y}_i\}$ are N random samples, and let $\{r_i\}$ and $\{s_i\}$ denote their ranks obtained from (7.20). The Spearman correlation coefficient is just the Pearson correlation coefficient from Section 6.3.6 computed using the ranks:

$$\hat{r}_s = \frac{\sum_{i=1}^{N} (r_i - \bar{r})(s_i - \bar{s})}{\sqrt{\sum_{i=1}^{N} (r_i - \bar{r})^2 \sum_{i=1}^{N} (s_i - \bar{s})^2}} \tag{7.37}$$

This expression can be simplified considerably by recognizing that $\bar{r} = \bar{s} = (N+1)/2$ and that the two terms in the denominator are each equal to $N\left(N^2 - 1\right)/12$. Parameterizing the expression in terms of the differences between the ranks $\hat{D}_i = r_i - s_i$ yields

$$\hat{r}_s = 1 - \frac{6\sum_{i=1}^{N} \hat{D}_i^2}{N\left(N^2 - 1\right)} \tag{7.38}$$

The exact permutation distribution for the Spearman statistic has been derived and can be approximated as a Gaussian for large data sets. These are used to test the null hypothesis that the data are uncorrelated against the alternate hypothesis that they are not

MATLAB implements the Spearman correlation coefficient within the function **corr** that also provides the Pearson correlation coefficient, although this requires the input of additional parameters. The command [*rho*, *p*] = **corr**(*x*, *y*, 'type', 'spearman') returns the correlation coefficient *rho* and p-value p given data in x and y.

Example 7.23 Returning to the tobacco and alcohol consumption data from Example 6.17:

```
alctobacc = importdata('alctobacc.dat');
[rhat,p] =corr(alctobacc(:,1),alctobacc(:,2),'type','spearman')
rhat =
   0.3727
p =
   0.2606
```

The null hypothesis that the data are uncorrelated is accepted at the 0.05 level. For comparison, the results in Example 6.17 were $\hat{r} = 0.2236$ and a p-value of 0.5087. The effect of the Northern Ireland outlier is still present, although the correlation has risen and the p-value has fallen. After removing the Northern Ireland data point, the correlation rises to 0.8303 and the p-value drops to 0.0056, resulting in rejection of the null hypothesis, just as happened in Example 6.17.

7.3.7 Kendall's Tau

This statistic uses the probability of concordance and discordance:

$$\begin{aligned} \rho_c &= \Pr\left[\left(\mathbf{X}_i - \mathbf{X}_j\right)\left(\mathbf{Y}_i - \mathbf{Y}_j\right) > 0\right] \\ \rho_d &= \Pr\left[\left(\mathbf{X}_i - \mathbf{X}_j\right)\left(\mathbf{Y}_i - \mathbf{Y}_j\right) < 0\right] \end{aligned} \tag{7.39}$$

to define the Kendall τ statistic

$$\tau = \rho_c - \rho_d \tag{7.40}$$

When the marginal distributions of $\mathbf{X}$ and $\mathbf{Y}$ are continuous, then $\rho_c + \rho_d = 1$, and hence $\tau = 2\rho_c - 1 = 1 - 2\rho_d$, so

$$\frac{\rho_c}{\rho_d} = \frac{1+\tau}{1-\tau} \tag{7.41}$$

To apply the Kendall τ estimator to data, it is necessary to use estimates for the probability of concordance and discordance based on relative frequency. For N data $\{x_i, y_i\}$, define

$$\begin{aligned} \hat{\upsilon}_{ij} &= \operatorname{sgn}\left(x_i - x_j\right) \\ \hat{\omega}_{ij} &= \operatorname{sgn}\left(y_i - y_j\right) \\ \hat{\alpha}_{ij} &= \hat{\upsilon}_{ij}\hat{\omega}_{ij} \end{aligned} \tag{7.42}$$

where $\operatorname{sgn}(x)$ is -1, 0, 1 as x is negative, zero, or positive. It is obvious that $\Pr\left(\hat{\alpha}_{ij} = 1\right) = \rho_c$ and $\Pr\left(\hat{\alpha}_{ij} = -1\right) = \rho_d$, and hence $\mathcal{E}\left(\hat{\alpha}_{ij}\right) = \tau$. An estimator for τ is

$$\hat{\tau}_N = \frac{1}{\binom{N}{2}} \sum_{i=1}^{N} \sum_{j=i+1}^{N} \hat{\alpha}_{ij} \tag{7.43}$$

and is unbiased. Equation (7.43) is Kendall's sample τ. It can easily be shown that (7.43) is equivalent to

$$\hat{\tau}_N = \frac{\sum_{i=1}^{N}\sum_{j=1}^{N} \hat{\upsilon}_{ij}\hat{\omega}_{ij}}{\sqrt{\left(\sum_{i=1}^{N}\sum_{j=1}^{N} \hat{\upsilon}_{ij}^2\right)\left(\sum_{i=1}^{N}\sum_{j=1}^{N} \hat{\omega}_{ij}^2\right)}} \tag{7.44}$$

which is the Pearson correlation coefficient with the data replaced by signed differences. A test of the null hypothesis that the two sets of rvs have zero correlation follows from the fact that $\tau = 0$ in that instance, and hence the test should reject for large values of $|\hat{\tau}_N|$. The null distribution can be derived based on combinatorics and is asymptotically Gaussian.

MATLAB implements the Kendall τ within the **corr** function as a different value for the keyword "type."

Example 7.24 Returning to the tobacco and alcohol consumption data from Examples 6.17 and 7.23:

```
alctobacc = importdata('alctobacc.dat');
[rho, p] =corr(alctobacc(:,1),alctobacc(:,2),'type','kendall')
rho =
   0.3455
p =
   0.1646
```

This is comparable to the result using the Spearman statistic in Example 7.23. The Northern Ireland outlier is still producing zero correlation. After removing that data point, the correlation rises to 0.6444 and the p-value is 0.0091, resulting in rejection of the null hypothesis that the data are uncorrelated.

7.3.8 Nonparametric ANOVA

Conventional ANOVA (Section 6.3.7) will fail in the presence of heteroskedasticity or non-Gaussian variates, just as the t test or (especially) the F test breaks down under these conditions. Nonparametric multivariate tests have been developed to address these problems. For one-way ANOVA, the Kruskal-Wallis test requires only that the samples be independent and that their underlying distributions be identically shaped and scaled. The hypotheses tested by the Kruskal-Wallis test are H_0: $\tilde{\mu}_1 = \tilde{\mu}_2 = \cdots = \tilde{\mu}_M$, where $\tilde{\mu}_i$ is the population median of the ith group, versus H_1: not H_0, meaning that at least two of the medians are not the same. The Kruskal-Wallis test procedure is:

1. Rank all the data, disregarding their grouping and assigning the average of the ranks that would have been received to any tied values.
2. Form the test statistic

$$\hat{K} = (N-1)\frac{\sum_{i=1}^{M} n_i(\bar{r}_i - \bar{r})^2}{\sum_{i=1}^{M}\sum_{j=1}^{n_i}\left(r_{ij} - \bar{r}\right)^2} \tag{7.45}$$

where r_{ij} is the global rank of the jth observation from the ith group, $\bar{r} = (N+1)/2$ and $\bar{r}_i = (1/n_i)\sum_{j=1}^{n_i} r_{ij}$.

3. Since the denominator of (7.45) is $(N-1)N(N+1)/12$, the test statistic reduces to

$$\hat{K} = \frac{12}{N(N+1)}\sum_{i=1}^{M} n_i\bar{r}_i - 3(N+1) \tag{7.46}$$

4. The test statistic is corrected for ties by dividing (7.46) by $1 - \sum_{i=1}^{T}\left(t_i^3 - t_i\right)/N\left(N^2 - 1\right)$, where T is the number of groupings of different tied ranks, and t_i is the number of tied data with a particular value in the ith group. The correction is unimportant unless the number of ties is very large.
5. Assess the test statistic against the Kruskal-Wallis distribution or, for a sufficiently large sample, the chi square approximation $\chi^2_{M-1}(\alpha)$, rejecting if the test statistic is larger than the critical value (or computing the chi square p-value in the usual way).

MATLAB supports the Kruskal-Wallis test through the function [*p*, *anovatab*] = **kruskalwallis**(*x*), where the argument and returned values are identical to those for **anova1**.

Example 7.25 Return to the laboratory tablet data in Example 6.17.

```
[p, anovatab] = kruskalwallis(tablet)
```

The anova table is

Source	SS	DF	MS	chi-sq
Between	14756.4	6	2459.4	35.76
Within	13713.6	63	217.68	
Total	28470	69		

The p-value is 3.06×10^{-6}, so the null hypothesis that the medians are the same is rejected at the 0.05 level. This is the same conclusion obtained with ANOVA in Example 6.17.

7.4 Meta-analysis

Meta-analysis provides ways to combine the inferences from a number of studies conducted under similar conditions. This is a common procedure in biomedical fields, where p-values are always reported from trials, and might prove useful in the earth sciences once reporting of p-values becomes more routine. The key remaining issue is the phrase "under similar conditions," which has to be approached with caution.

While parametric meta-analysis methods exist, these typically require additional information that is not usually provided in the scientific literature. In many instances, all that is available are the p-values from individual studies, in which case nonparametric methods must be applied. Birnbaum (1954) summarizes the issues.

The simplest technique for combining p-values is the inverse chi square method of Fisher (1932). Under the null hypothesis, the p-value is uniformly distributed on [0, 1] so that $-2\log p_i \sim \chi_2^2$. By the additivity property of the chi square distribution, it follows for m p-values that

$$-2\sum_{i=1}^{m}\log p_i \sim \chi_{2m}^2 \tag{7.47}$$

and hence the combined p-value satisfies

$$p = 1 - \mathrm{Chi}\left(-2\sum_{i=1}^{m}\log p_i, 2m\right) \tag{7.48}$$

The combined p-value can be assessed in standard ways.

An alternative approach uses the probability integral transformation of Section 4.8.1. Implementing this for the Gaussian distribution, note that if $\{z_i\}$ are m standardized Gaussian variables, then $\sum_{i=1}^{m} z_i/\sqrt{m}$ is also $\mathrm{N}(0, 1)$. Consequently, the combined p-value can be computed as

$$p = 1 - \Phi\left[\frac{1}{\sqrt{m}}\sum_{i=1}^{m}\Phi^{-1}(1 - p_i)\right] \tag{7.49}$$

This method is most suitable for upper-tail tests.

Example 7.26 Professors A and B are independently trying to evaluate a theory for which they have formulated a hypothesis test based on Bernoulli trials. The hypotheses to be tested are H_0: $p = 0.25$ versus H_1: $p > 0.25$. Professor A teaches large classes each semester and uses the attendees to conduct his experiments. He carries out two trials and obtains p-values of 0.04 and 0.009. Professor B teaches small classes each quarter and carries out ten experiments, yielding p-values of 0.22, 0.15, 0.23, 0.17, 0.20, 0.18, 0.14, 0.08, 0.31, and 0.21. None of Professor B's experiments rejected the null hypothesis at the 0.05 level, whereas both of Professor A's do. However, the combined *p*-values must be computed to compare their respective outcomes. This example is taken from Kvam & Vidakovic (2007, page 108).

```
pvala = [ .04 .009];
pvalb = [ .22 .15 .23 .17 .20 .18 .14 .08 .31 .21];
1 - chi2cdf(-2* sum(log(pvala)), 4)
ans =
   0.0032
1 - normcdf(sum(norminv(1 - pvala))/sqrt(2))
ans =
   0.0018
1 - chi2cdf(-2* sum(log(pvalb)),20)
ans =
   0.0235
1 - normcdf(sum(norminv(1 - pvalb))/sqrt(10))
ans =
   0.0021
```

Despite the fact that none of her individual experiments rejected the null hypothesis, the combination of Professor B's ten experimental *p*-values does so at the 0.05 level. By contrast, Professor A's raw and combined *p*-values both reject the null hypothesis.

8 Resampling Methods

8.1 Overview

A variety of parametric estimators for the mean and variance (among other statistics) for iid data and for testing hypotheses about them have been established in Chapters 4 through 6. However, in many instances the data analyst will not have information on or want to make assumptions about the data distribution and hence may not be certain of the applicability of parametric methods. In the real world, it is often true that departures from expected conditions occur, including:

- Mixtures of distributions,
- Outliers,
- Complex distributions for realistic problems (e.g., multivariate forms),
- Noncentrality, and
- Unknown degrees-of-freedom due to correlation and heteroskedasticity

This has led to the development of estimators that are based on resampling of the available data that are covered in this chapter. The most widely used of these is the bootstrap described in the next section, which was first proposed by Efron (1979) and has undergone considerable evolution since that time. In recent years, permutation hypothesis tests that typically are superior to their bootstrap analogue have become more widely used and are described in Section 8.3. A linear approximation to the bootstrap called the *jackknife* is described in Section 8.4 and offers computational efficiency at a slight cost in accuracy, but it is becoming less important as the power of computers grows.

8.2 The Bootstrap

8.2.1 The Bootstrap Distribution

Instead of assuming a particular parametric form for the sampling distribution, such as Student's *t*, the empirical distribution based on the available sample plays the central role in the bootstrap method. The bootstrap is based on sampling with replacement, and as a result, the bootstrap is typically not exact. Davison & Hinkley (1997) and Efron & Tibshirani (1998) are very readable treatises on the bootstrap.

Example 8.1 The file geyser.dat contains 299 measurements of the waiting time in minutes between eruptions of Old Faithful at Yellowstone National Park and was used to illustrate the maximum likelihood estimator (mle) in Example 5.27. A kernel density estimator for the data is shown in Figure 5.2. Characterize the empirical pdf of the data, and then resample from it with replacement 30, 100, 300, 1000, 3000, and 10,000 times. Finally, compute the sample mean for each replicate. Plot the resulting bootstrap distributions for the sample mean. Repeat for the kurtosis.

Figure 5.2 shows that the data are bimodal, peaking at around 55 and 80 minutes. Figure 8.1 shows a Gaussian q-q plot for the data, indicating that they are shorter tailed than Gaussian data at the top and bottom of the distribution but with an excess of probability near the distribution mode.

```
n = length(geyser);
b = [30 100 300 1000 3000 10000];
bins = [10 20 30 30 50 50];
rng default %initialize random number seed so result will
be reproducible
for i=1:6
    ind = unidrnd(n, n, b(i)); %use unidrnd to get indices to
    sample the data with replacement
    boot = geyser(ind); %resample the data
    meanb = mean(boot);
    mean(meanb)
    subplot(3, 2, i)
    histogram(meanb, bins(i), 'Normalization', 'pdf')
end
ans =
    72.5831
ans =
    72.2098
ans =
    72.3521
ans =
    72.2766
ans =
    72.3384
ans =
    72.3201
```

For comparison purposes, the sample mean of the geyser data is 72.3144. Figure 8.2 shows the results. For a limited number of bootstrap replicates (in this case, under 300), the resulting distribution does not resemble the $N(\mu, \sigma^2/N)$ sampling distribution of the mean. As the number of replicates increases to 10,000, the bootstrap distribution looks increasingly Gaussian and, in addition, is consistent as the spread of the distribution drops.

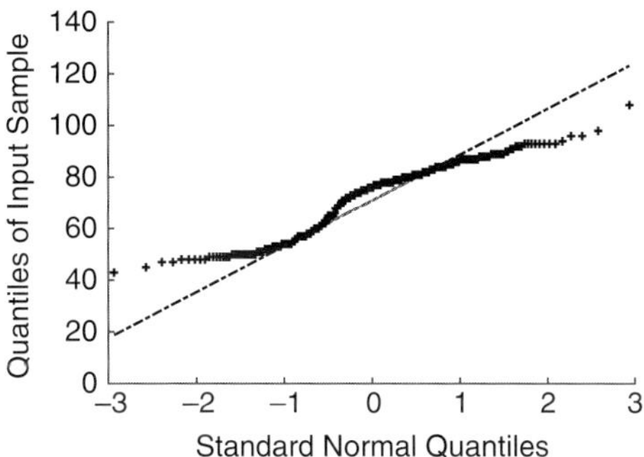

Figure 8.1 Gaussian q-q plot for the Old Faithful geyser data.

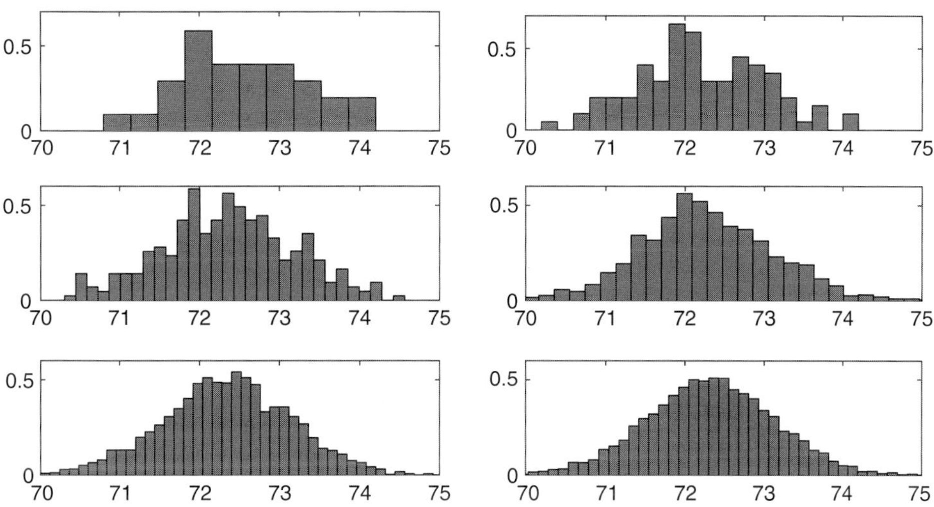

Figure 8.2 Bootstrap distributions for the sample mean of the Old Faithful geyser data computed using (top row) 30 and 100, (middle row) 300 and 1000, and (bottom row) 3000 and 10,000 bootstrap replicates from the data.

However, the bootstrap distribution is centered on the sample mean rather than the unknown population mean.

Figure 8.3 shows the bootstrap distributions for the kurtosis of the Old Faithful data whose parametric sampling distribution for Gaussian data is known but complicated. However, Figure 8.1 shows that the geyser data are far from Gaussian. For reference, the sample kurtosis of the geyser data is 1.9916. As with the bootstrap distributions for the mean, the result is far from well defined when the number of bootstrap replicates is small but is increasingly clear as the number of replicates rises beyond 1000 or so. The result appears to be slightly asymmetric, with a longer right than left tail.

It might seem as though something illegitimate is going on because resampling appears to create data out of nothing. However, the resampled data are not used as if they are new

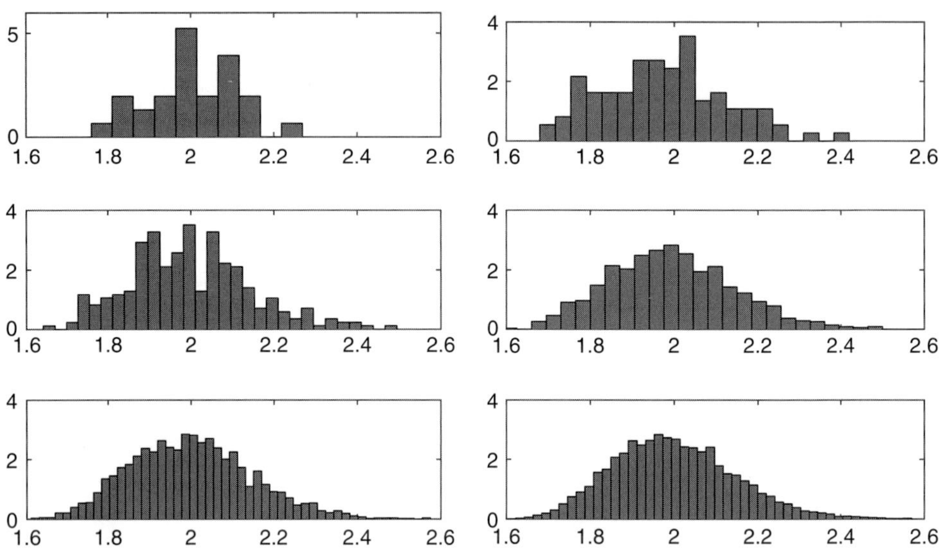

Figure 8.3 Bootstrap distributions for the sample kurtosis of the Old Faithful geyser data computed using (top row) 30 and 100, (middle row) 300 and 1000, and (bottom row) 3000 and 10,000 bootstrap samples from the data.

data. Rather, the bootstrap distribution of, for example, the sample mean is only used to determine how the sample mean of the actual data varies through random sampling. It is quite reasonable to use the same random sample to estimate a parameter such as the sample mean and also to compute its variability. Indeed, this is what is done parametrically when the sample mean is used to determine the population mean and the sample variance is used to determine the dispersion about the sample mean. The difference with the bootstrap is that the bootstrap distribution of the sample mean is used to determine the sample mean and its variance. The other thing that distinguishes the bootstrap is that direct appeal to the classical central limit theorem is not applied to determine whether a sampling distribution is approximately Gaussian. It is what it is, as seen in Figures 8.2 and 8.3.

The approach of this section is a *nonparametric bootstrap* because no assumptions were made about the distribution of the data being resampled, and replicates were simply drawn at random to compute a statistic of interest. The only requirement that this entails is an iid random sample for the data.

An alternate approach is the *parametric bootstrap*, in which there is belief that the random sample is drawn from a particular distribution, and hence the model distribution is fit to the data using an estimator such as the mle. Subsequently, resampling from the fitted distribution is used to estimate statistical properties of the data set that typically are too complicated to compute directly from the sample. This serves as an alternative to the delta method of Section 4.7.4 that does not require a first-order Taylor series approximation.

There are two types of error with the bootstrap: statistical and simulation errors. The first is due to the difference between the population cdf F and the empirical cdf $\hat{F}$ that decreases with the size of the data set and based on the choice of the statistic λ_N^*. Whereas the former cannot be controlled because F is unknown, the latter can be minimized by choosing λ_N^* so that its empirical distribution is approximately Gaussian. This can sometimes be achieved through transformation, as described later. However, just as parametric estimators often deliver poor performance with small samples, bootstrap estimators also suffer when the sample size is small; there is no substitute for a large random sample. Simulation error is caused by sampling from the empirical rather than population distribution and typically can be minimized by using a large number of replicates. Given the power of modern computers, simulation error ought not to be a significant limitation in using the bootstrap any more.

8.2.2 Bootstrap Parameter Estimation

Suppose that the statistic of interest is $\hat{\lambda}_N$, along with derived quantities such as its standard error. Obtain B bootstrap samples from the original data by resampling with replacement. The notation $\mathbf{X}_k^*$ will be used to denote the kth bootstrap sample, each of which contains N data. Apply the estimator for $\hat{\lambda}_N$ to each bootstrap sample to get a set of bootstrap replicates $\hat{\lambda}_N^*(\mathbf{X}_k^*)$ for $k = 1, \ldots, B$, from which estimates of derived statistics may be obtained. For example, the standard error of $\hat{\lambda}_N$ follows from

$$\mathrm{SE}_B\left(\hat{\lambda}_N\right) = \left\{\frac{1}{B}\sum_{k=1}^{B}\left[\hat{\lambda}_N^*(\mathbf{X}_k^*) - \bar{\lambda}^*\right]^2\right\}^{1/2} \tag{8.1}$$

where the sample mean of the bootstrap replicates is

$$\bar{\lambda}^* = \frac{1}{B}\sum_{k=1}^{B}\hat{\lambda}_N^*(\mathbf{X}_k^*) \tag{8.2}$$

While statistical lore holds that 50–200 replicates suffice to estimate the standard error on a statistic, the results in Section 8.2.1 suggest that more may be required to properly simulate the sampling distribution, and it is always good practice to examine the bootstrap distribution for accuracy.

A bootstrap estimate for the bias can also be obtained. Recall from Section 5.2.2 that the bias is defined as the difference between the expected and population values of a statistic. The bootstrap analogue is

$$\hat{B}\left(\hat{\lambda}_N\right) = \bar{\lambda}^* - \hat{\lambda}_N \tag{8.3}$$

from which the bias-corrected estimate obtains

$$\widehat{\lambda}_N = \hat{\lambda}_N - \hat{B} = 2\hat{\lambda}_N - \bar{\lambda}^* \tag{8.4}$$

Example 8.2 For the Old Faithful data set in Examples 5.27 and 8.1, estimate the skewness along with its bias (8.3) and standard error (8.1) using the bootstrap.

```
n = length(geyser);
rng default
b = 1000; %number of bootstrap replicates
theta = skewness(geyser)
```

```
theta =
   -0.3392
ind = unidrnd(n, n, b);
xboot = geyser(ind); %resample the data
thetab = skewness(xboot);
mean(thetab) %bootstrap mean
ans =
   -0.3432
mean(thetab) - theta %bootstrap bias
ans =
   -0.0040
std(thetab) %bootstrap standard error
ans =
   0.1026
```

Note that slightly different answers are obtained from the data and from the replicates. Further, if the script is run a second time without issuing the "**rng** default" command, another slightly different answer results. The variability represents the effect of random resampling.

MATLAB includes a function called *boot* = **bootstrp(***nboot*, *bootfun*, *data*) that obtains *nboot* replicates using the function handle *bootfun* and data in *data* and returns the function applied to the data in each row of *boot*. With 10,000 bootstrap replicates using the geyser data, this gives

```
rng default
bmat = bootstrp(b, @skewness, geyser);
mean(bmat)
ans =
   -0.3421
mean(bmat) - skewness(geyser)
ans =
   -0.0029
std(bmat)
ans =
   0.0991
```

The function **bootstrp** will parallelize if a cluster is available, such as occurs after executing the command **parpool**('local'). However, due to the vagaries of process scheduling on multiple cores, the "**rng** default" command will no longer remove variability from the answer.

Example 8.3 Evaluate the effect of changing the number of bootstrap replicates for the skewness statistic using the geyser data.

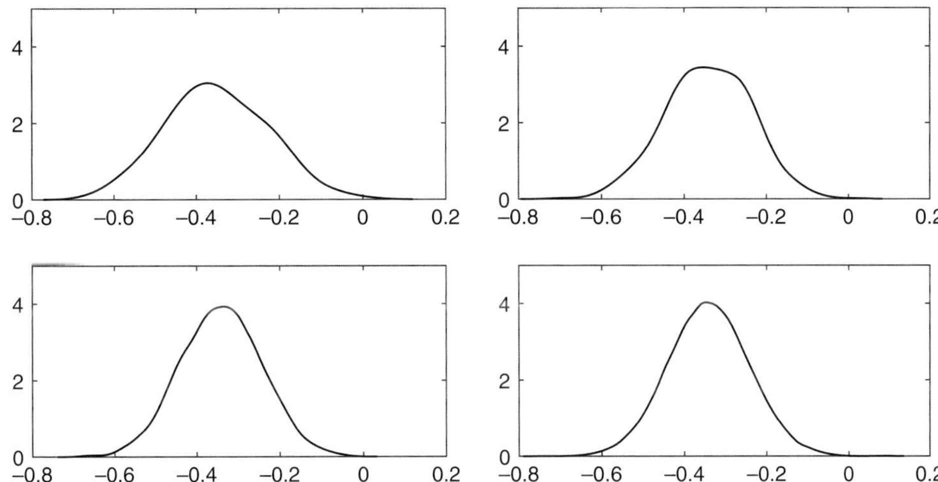

Figure 8.4 Kernel density pdfs of the bootstrap skewness for the Old Faithful data with 50 (upper left), 500 (upper right), 5000 (lower left), and 50,000 (lower right) replicates.

This can be achieved by examining the empirical distribution of the bootstrap skewness estimates for 50, 500, 5000, and 50,000 replicates, as shown in Figure 8.4. The results indicate that a large number of replicates results in a smoother, more symmetric distribution.

```
rng default
b = [50 500 5000 50000];
for i=1:4
    bmat = bootstrp(b(i), @skewness, geyser);
    subplot(2, 2, i)
    ksdensity(bmat)
end
```

Example 8.4 Apply the bootstrap to the interquartile range for the geyser data. Explain the result.

The bootstrap can be used with the interquartile range simply by substituting **iqr** for **skewness** in the preceding script. The mean of the interquartile range and its standard deviation are (23.755, 1.7154), (23.7775, 1.7462), (23.6853, 1.7084), and (23.6642, 1.6943) for 50, 500, 5000, and 50,000 bootstrap replicates compared with the sample interquartile applied to the data of 24.0000. Whereas the mean of the bootstrap replicates for the interquartile range appears to be settling down as the number of samples increases, the standard deviation is somewhat variable. Figure 8.5 shows kernel density estimates for the four bootstrap simulation sizes. As the number of bootstrap replicates increases, multiple peaks in the distribution are resolved. This occurs because the order statistics are concentrated on the sample values that are quite

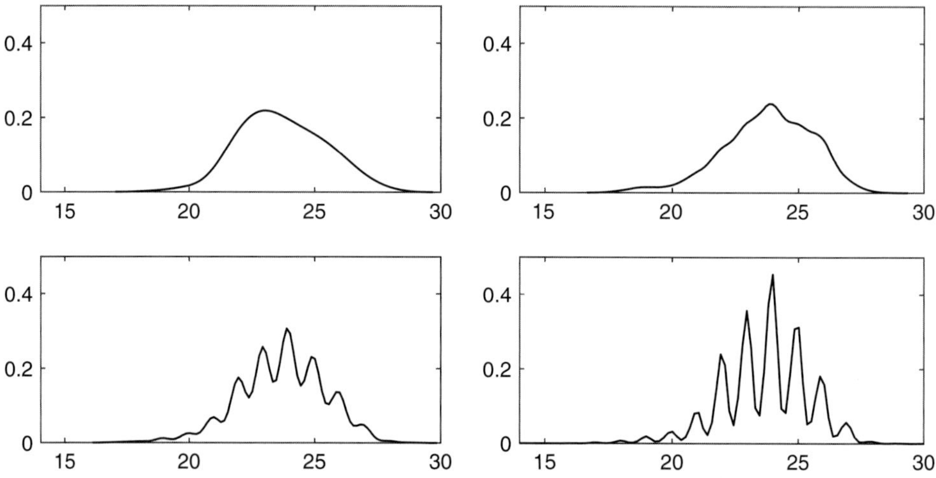

Figure 8.5 Kernel density pdfs of the bootstrap interquartile range for the Old Faithful data with 50 (upper left), 500 (upper right), 5000 (lower left), and 50,000 (lower right) replicates.

discrete unless the data set is very large. As a consequence, the bootstrap distribution becomes increasingly discrete as the number of replicates rises, and a naive nonparametric bootstrap approach fails.

The issue seen in Example 8.4 can be addressed using the *smoothed bootstrap*, which replaces the discrete jumps in the empirical cdf from which the bootstrap replicates are obtained with smoothed transitions. The smoothed bootstrap is easily implemented by first taking a bootstrap sample in the usual way and then adding random draws from the Gaussian distribution with a variance h^2 to each, where h is the smoothing bandwidth.

Example 8.5 Apply the smoothed bootstrap to the interquartile range of the geyser data.

This will be illustrated with the 50,000 bootstrap replicate example shown in the lower right panel of Figure 8.5. The smoothing bandwidth must be chosen by trial and error.

```
rng default
h = [.707 .866 1 1.414];
for i = 1:4
    x = bmat + normrnd(bmat, h(i).^2*ones(size(bmat)));
    subplot(2,2,i)
    ksdensity(x)
end
```

The result is shown in Figure 8.6. A smoothing bandwidth of about 1 appears appropriate.

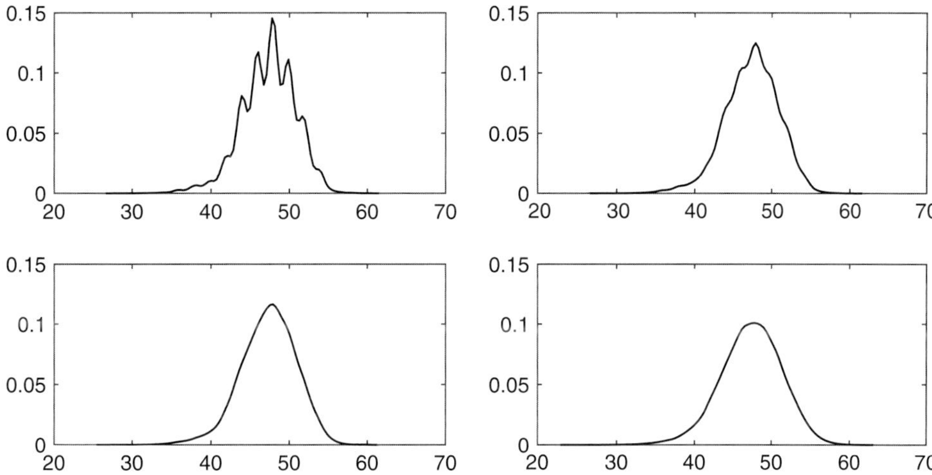

Figure 8.6 Kernel density pdfs of 50,000 bootstrap replicates for the interquartile range for the Old Faithful data smoothed by a Gaussian kernel with a bandwidth of (top row) 0.707 and 0.866 and (bottom row) 1.0 and 1.414.

8.2.3 Bootstrap Confidence Intervals

The bootstrap is widely used to obtain confidence intervals when the statistic of interest is complicated so that its sampling distribution is hard to obtain or work with. *Standard bootstrap confidence intervals* can be found in the usual way by substituting the bootstrap estimates for the statistic (or its bias-corrected version) and the standard deviation for the sample mean and sample standard error in the usual t confidence interval formula, yielding

$$\Pr\left[\bar{\lambda}^* - t_{N-1}(1-\alpha/2)\ \mathrm{SE_B}\left(\hat{\lambda}_N\right) \le \lambda \le \bar{\lambda}^* + t_{N-1}(1-\alpha/2)\ \mathrm{SE_B}\left(\hat{\lambda}_N\right)\right] = 1-\alpha \qquad (8.5)$$

Where $\bar{\lambda}^*$ and $\mathrm{SE_B}\left(\hat{\lambda}_N\right)$ are given by (8.2) and (8.1), respectively.

Several enhancements to this straightforward approach have been proposed. The *bootstrap-t confidence interval* is based on the normalized statistic

$$z_k^* = \frac{\hat{\lambda}_N^* - \bar{\lambda}^*}{\mathrm{SE_B}\left(\hat{\lambda}_N\right)} \qquad (8.6)$$

If a formula for the standard error of the kth bootstrap replicate exists, then it can be used to estimate z_k^*, but otherwise the bootstrap may be applied (in other words, bootstrap the bootstrap replicates). The $\left\{z_k^*\right\}$ are then ranked, and the $\alpha/2$ and $1-\alpha/2$ quantiles are located in the set for use in the standard formula in place of the t_{N-1} quantile. To achieve reasonable accuracy, the number of bootstrap replicates needs to be large, with a minimum of 1000 recommended and preferably more.

The *bootstrap percentile method* directly uses the $\alpha/2$ and $1-\alpha/2$ quantiles of the bootstrap distribution as the confidence interval and has better stability and convergence properties than either the standard or bootstrap-*t* confidence interval.

The *bias-corrected and accelerated* (BCa) *method* adjusts the confidence limits obtained from the percentile method to correct for bias and skewness. It is obtained as the proportion of the bootstrap estimates $\hat{\lambda}_k^*$ that are less than $\hat{\lambda}_N$ (or $\bar{\lambda}^*$). The bias factor is

$$z_0 = \Phi^{-1}(p_0) \tag{8.7}$$

$$p_0 = \frac{1}{B}\sum_{i=1}^{B} \mathbf{1}\left(\hat{\lambda}_k^* < \hat{\lambda}_N\right) \tag{8.8}$$

Define the acceleration factor that measures the rate of change in the bootstrap standard error to be

$$a_0 = \frac{\sum_{i=1}^{B}\left(\bar{\lambda}^* - \hat{\lambda}_k^*\right)^3}{6\left[\sum_{i=1}^{B}\left(\bar{\lambda}^* - \hat{\lambda}_k^*\right)^2\right]^{3/2}} \tag{8.9}$$

The BCa confidence interval is obtained using the quantiles

$$q_1 = \Phi\left\{z_0 + \left(z_0 + z_{\alpha/2}\right)/\left[1 - a_0\left(z_0 + z_{\alpha/2}\right)\right]\right\} \tag{8.10}$$

$$q_2 = \Phi\left\{z_0 + \left(z_0 + z_{1-\alpha/2}\right)/\left[1 - a_0\left(z_0 + z_{1-\alpha/2}\right)\right]\right\} \tag{8.11}$$

This reduces to the bootstrap percentile interval when $z_0 = a_0 = 0$.

Example 8.6 Returning to the Old Faithful data of Example 5.27, apply the bootstrap to get the double-sided 95% confidence limits on skewness and kurtosis using the first three methods just described.

```
rng default
b = 10000;
bmat = bootstrp(b, @skewness, geyser);
xm = mean(bmat);
xs = std(bmat, 1);
t = tinv(.975, length(geyser) - 1);
[xm - t*xs xm + t*xs]
zhat = (bmat - xm)./xs;
zhat = sort(zhat);
[xm + zhat(round(.025*b))*xs xm + zhat(round(.975*b))*xs]
bmats = sort(bmat);
[bmats(round(b*.025)) bmats(round(b*.975))]
```

For the skewness, the results are

```
-0.5370   -0.1472
-0.5371   -0.1484
-0.5371   -0.1484
```

For the kurtosis, the results are

```
1.7117   2.2879
1.7433   2.3120
1.7433   2.3120
```

MATLAB provides the function *ci* = **bootci**(*nboot*, {*bootfun*, *data*}, 'type', 'xxx') that returns the confidence interval in *ci* using *nboot* replicates and bootstrap function handle *bootfun*. The parameter after "type" specifies the method used to get the confidence interval and may be "norm" for the normal interval with bootstrapped bias and standard error, "per" for the basic percentile method, "cper" for the bias-corrected percentile method, "bca" for the bias-corrected and accelerated percentile method (default), and "stud" for the studentized confidence interval. As for **bootstrp**, **bootci** will parallelize if a cluster pool is available.

Example 8.7 Apply the BCa method to the geyser data.

```
rng default
ci = bootci(10000, @skewness, geyser)
ci =
   -0.5219
   -0.1334
ci = bootci(10000, @kurtosis, geyser)
ci -
   1.7575
   2.3443
```

The results in Examples 8.12 and 8.13 are comparable because the bootstrap distributions for the skewness and kurtosis of these data are fairly symmetric. In general, the standard and bootstrap-t intervals are obsolete and should not be used. The percentile method works well when the number of bootstrap replicates is large, and the BCa is the most consistent estimator for general applications.

Example 8.8 Returning once again to the earthquake interval data from Example 7.3, compute the skewness and kurtosis along with bootstrap percentile and BCa confidence intervals.

```
skewness(quakes)
ans =
    1.4992
kurtosis(quakes)
ans =
   5.5230
rng default
bootci(10000, {@skewness, quakes}, 'type', 'per')
ans =
   0.6487
   2.0201
bootci(10000, @skewness, quakes)
ans =
   0.9836
   2.3748
bootci(10000, {@kurtosis, quakes}, 'type', 'per')
ans =
   2.7588
   8.4796
bootci(10000, @kurtosis, quakes)
ans =
   3.5865
   10.5179
```

The differences between the two types of bootstrap estimators is substantial and cannot be accounted for by the fact that the BCa confidence interval is bias corrected while the percentile one is not. The cause can be found by examining the bootstrap distributions for the skewness and kurtosis (Figure 8.7), both of which are skewed either to the left

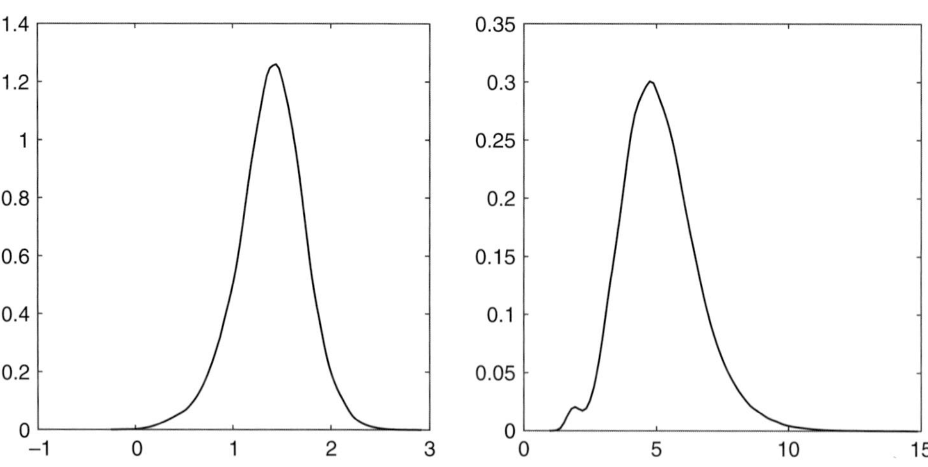

Figure 8.7 Kernel density pdfs for the bootstrap distributions using 10,000 replicates for skewness (left) and kurtosis (right) computed using the earthquake interval data.

(skewness) or to the right (kurtosis). In this circumstance, the percentile confidence interval will be biased, whereas the BCa interval will offer better performance.

8.2.4 Bootstrap Hypothesis Tests

Let an upper-tail hypothesis test on the statistic λ that is estimated by the estimator $\hat{\lambda}$ be evaluated using the p-value $p = 1 - F(\hat{\lambda})$. If the cdf $F(x)$ is known, then this is straightforward. More typically, the cdf is not known, and recourse is made to an asymptotic result whose accuracy is unclear. The alternative approaches are to use a nonparametric method (Section 7.3), a permutation test, or a bootstrap hypothesis test. The latter differs from the permutation test in that sampling is done with replacement. Implementing the bootstrap hypothesis test is straightforward: generate B bootstrap replicates X_k^* of the random sample being tested, and then compute a *bootstrap test statistic* $\hat{\lambda}_N^*(\mathrm{X}_k^*)$ from each using the same procedure as for $\hat{\lambda}$. The bootstrap upper-tail p-value is obtained by comparing the bootstrap test statistics with the test statistic based on all the data

$$\hat{p} = 1 - \hat{F}(\hat{\lambda}) = \frac{1}{B}\sum_{i=1}^{B} \mathbf{1}\left[\hat{\lambda}_N^*(\mathrm{X}_k^*) > \hat{\lambda}\right] \tag{8.12}$$

A lower-tail test can be implemented by reversing the argument in (8.12), whereas a two-tail test uses the equal-tail bootstrap p-value

$$\hat{p} = \frac{2}{B}\min\left\{\sum_{i=1}^{B} \mathbf{1}\left[\hat{\lambda}_N^*(\mathrm{X}_k^*) \geq \hat{\lambda}\right], \sum_{i=1}^{B} \mathbf{1}\left[\hat{\lambda}_N^*(\mathrm{X}_k^*) < \hat{\lambda}\right]\right\} \tag{8.13}$$

In any instance, the bootstrap p-value is the proportion of the bootstrap test statistics that are more extreme than the observed test statistic based on the entire sample.

The bootstrap test procedure outlined here is exact when the population test statistic is a pivot (i.e., does not depend on any unknown parameters), so $\hat{\lambda}$ and $\hat{\lambda}_N^*$ have the same distribution if the null hypothesis holds, and the number of bootstrap samples B is such that $\alpha(B+1)$ is an integer. The result is called a *Monte Carlo test* and actually predates introduction of the bootstrap by two decades. When these conditions do not hold, the resulting bootstrap hypothesis test will not be exact. However, it is good practice to enforce the number of samples for a Monte Carlo test on all bootstrap hypothesis tests.

It is important to ensure that the random samples are obtained from the null distribution for the test, which is not the same as the empirical cdf of the data in most cases. The appropriate approach for a one-sample location test is as follows:

1. Obtain B random draws from the empirical cdf of the data;
2. Compute a location parameter $\hat{\lambda}$ based on the entire sample;
3. Compute bootstrap test statistics that are proportional to $\hat{\lambda}_N^* - \hat{\lambda}$; and
4. Compute the p-value by comparing the bootstrap test statistics against a sample test statistic that is proportional to $\hat{\lambda} - \mu^*$, such as the one-sample t statistic, where μ^* is the postulated location parameter.

Example 8.9 Returning to the number of days an oceanographer spends at sea from Example 6.8, evaluate the null hypothesis that it is 30 days against the alternate hypothesis that it is not using 99, 999, 9999, and 99,999 bootstrap replicates.

```
rng default
b = [99 999 9999 99999] ;
data = [54 55 60  42 48 62 24 46 48 28 18 8 0 10 60 82 90 88 2 54] ;
that = (mean(data) - 30)./std(data, 1); %sample test statistic
       against the hypothesized value
for i=1:4
    bmat = bootstrp(b(i), @(t)(mean(t) - mean(data))./std
    (t, 1), data);
    pval = 2*min(sum(bmat >= that), sum(bmat < that))./b(i)
    %two tail p-value
end
pval =
   0.0404
pval =
   0.0320
pval =
   0.0402
pval =
   0.0378
```

The test rejects the null hypothesis that an oceanographer spends 30 days per year at sea, although not strongly. The distribution of the bootstrap test statistic is shown in Figure 8.8 for 99,999 replicates. The value of the test statistic based on the original data is 0.5240.

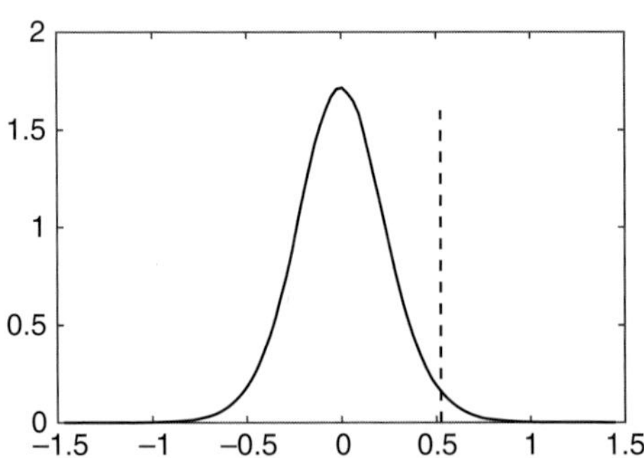

Figure 8.8 Kernel density pdf for 99,999 bootstrap replicates in a test of the hypothesis that an oceanographer spends 30 days per year at sea. The vertical dashed line is the test statistic based on all the data.

For a two-sample test for the equality of means by comparing M data $\{\mathbf{X}_i\}$ and N values $\{\mathbf{Y}_i\}$, the following algorithm pertains:

1. Center the two data sets around their grand mean as $\tilde{\mathbf{X}}_i = \mathbf{X}_i - \overline{\mathrm{X}}_N + \left(\overline{\mathrm{X}}_N + \overline{\mathrm{Y}}_N\right)/2$ and similarly for $\mathbf{Y}_i$;
2. Sample separately with replacement from each centered data set;
3. Evaluate the two-sample t statistic for each sample; and
4. Compute the p-value by comparing the bootstrap statistics with the sample value of the test statistic.

Example 8.10 Returning to the cloud seeding data of Example 6.12, compute a bootstrap test for the null hypothesis that there is no effect from cloud seeding against the alternate that there is.

```
rng default
b = [99 999 9999 99999] ;
that = (mean(clouds(:, 1)) - mean(clouds(:, 2)))./ ...
       sqrt(var(clouds(:, 1)) + var(clouds(:, 2))); %two
       sample t statistic
gmean = (mean(clouds(:, 1)) + mean(clouds(:, 2)))/2; %grand
       mean of the data
for i=1:4
    bmat1 = bootstrp(b(i), @(x)[ mean(x) var(x, 1)], clouds
         (:, 1) - mean(clouds(:, 1)) + gmean);
    bmat2 = bootstrp(b(i), @(x)[ mean(x) var(x,1)], clouds(:,
         2) - mean(clouds(:, 2)) + gmean);
    boot = (bmat1(:, 1) - bmat2(:, 1))./sqrt(bmat1(:, 2) +
         bmat2(:, 2));
    pval = 2*min(sum(boot > that), sum(boot <= that))./b(i)
end
pval =
   0.0606
pval =
   0.0160
pval =
   0.0174
pval =
   0.0182
```

The null hypothesis is rejected, and there is a difference in rainfall between seeded and unseeded clouds. This is in agreement with the rank sum test but in contrast with the two-sample t test. Figure 8.9 shows the bootstrap distribution that is asymmetric due to the logarithmic nature of the data. The sample test statistic is -0.3919. A nearly identical result

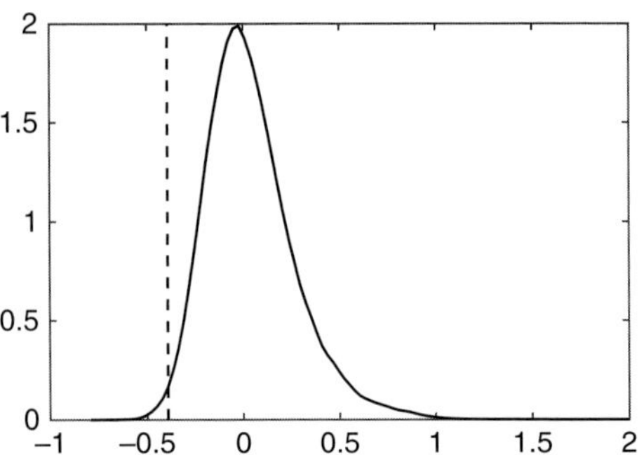

Figure 8.9 Kernel density pdf of the bootstrap distribution of the test statistic for the cloud seeding data. The vertical dashed line is the test statistic based on all the data.

is obtained after taking logs of the data. This example demonstrates the robustness of bootstrap tests against departures from Gaussianity.

To test whether two samples are drawn from the same distribution, use the following algorithm:

1. Obtain B random draws from the data set obtained by pooling $\{\mathbf{X}_i\}$ and $\{\mathbf{Y}_i\}$. The first M observations in the bootstrap sample become $\mathbf{X}_i^*$, and the last N become $\mathbf{Y}_i^*$.
2. Compute a bootstrap test statistic that measures the difference between the two samples. This could be as simple as the difference of their means.
3. Compute the p-value by comparing the bootstrap test statistics with the sample test statistic.

Example 8.11 Returning to the earthquake interval data from Example 7.3, evaluate the null hypothesis that the data are exponential and Rayleigh by testing against random draws with mle parameters for each.

```
rng default
rnd = exprnd(mean(quakes), size(quakes));
data = [quakes rnd]';
b = [99 999 9999 99999];
that = mean(quakes) - mean(rnd);
for i = 1:4
    bmat = bootstrp(b(i), @(x) (mean(x(1:length(quakes))) -
            mean(x(length(quakes)+1:2*length(quakes)))),data);
    pval = 2*min(sum(bmat > that),sum(bmat <= that))./b(i)
end
```

```
pval =
   0.4848
pval =
   0.4725
pval =
   0.4992
pval =
   0.4937
```

The result is quite unequivocal. The test statistic is 121.873. Figure 8.10 shows the bootstrap test distribution for 99,999 replicates.

The test will be repeated using the difference of the medians as the test statistic. The result is

```
pval =
   0.1616
pval =
   0.1441
pval =
   0.1414
pval =
   0.1376
```

The null hypothesis that the data are exponential is accepted, and the conclusion is unchanged. Figure 8.11 shows the bootstrap distribution for 99,999 replicates, which is somewhat long tailed and shows evidence for clustering around the order statistics, as in Example 8.4. The test statistic is 135.8992.

Repeating using the Rayleigh distribution and the difference of the means as the metric gives

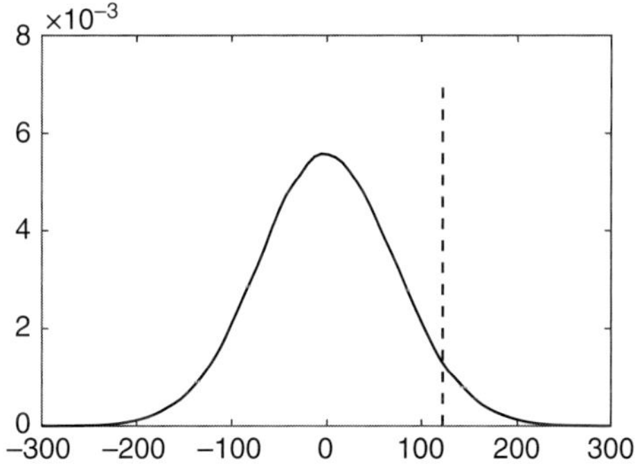

Figure 8.10 Kernel density pdf of the bootstrap distribution for the earthquake data using the difference between the means as the bootstrap statistic and an exponential distribution as the target statistic. The vertical dashed line is the test statistic based on all the data.

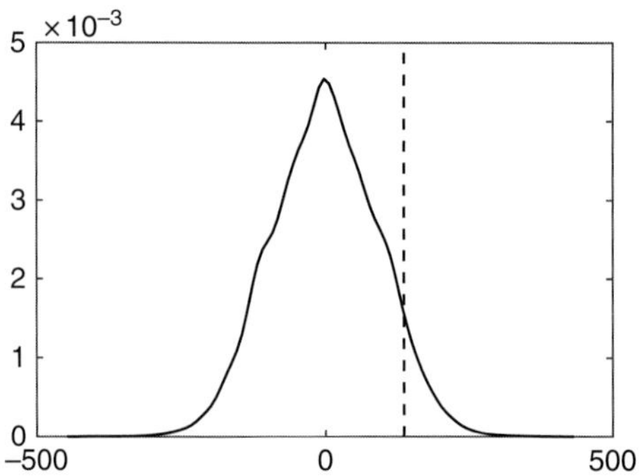

Figure 8.11 Kernel density pdf of the bootstrap distribution for the earthquake data using the difference of the medians as the bootstrap statistic and an exponential distribution as the target statistic. The vertical dashed line is the test statistic based on all the data.

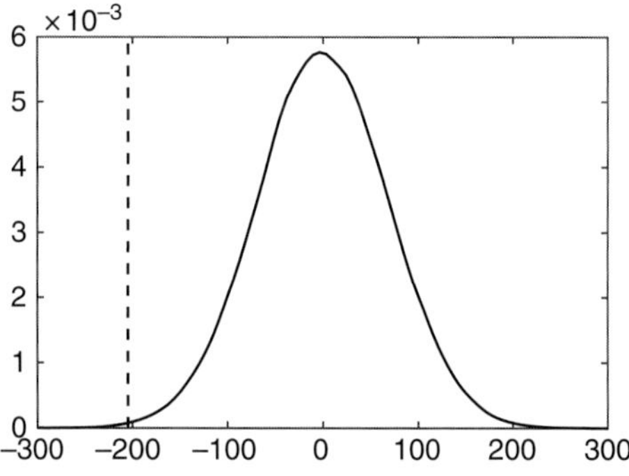

Figure 8.12 Kernel density pdf of the bootstrap distribution for the earthquake data using the difference of the medians as the bootstrap statistic and a Rayleigh distribution as the target statistic. The vertical dashed line is the test statistic based on all the data.

```
pval =
   0
pval =
   0.0040
pval =
   0.0034
pval =
   0.0030
```

The null hypothesis that the earthquake data are Rayleigh is strongly rejected. Figure 8.12 shows the bootstrap distribution, which is long tailed. The test statistic is −204.6115.

Bootstrap hypothesis tests are typically not exact because there is always a difference between the true and empirical cdfs, and the best results obtain when the empirical cdf is as close as possible to the true cdf in the vicinity of the critical value, although that is difficult to ensure in practice without extensive simulation. However, when the test statistic is asymptotically pivotal, a bootstrap hypothesis test has a lower probability of error in terms of sample size N than any asymptotic test using the same test statistic. The reverse is true when the test statistic is not asymptotically pivotal. Further, the power of a bootstrap test is comparable to that of an asymptotic test when the test statistic is asymptotically pivotal.

8.2.5 Bias Correction for Goodness-of-Fit Tests

It is well known that goodness-of-fit tests such as the Kolmogorov-Smirnov (K-S) and Anderson-Darling (A-D) procedures require that the parameters in the target distribution be known a priori and hence will yield a biased test if the parameters are estimated from the set of observations. This issue can be obviated by using the Lilliefors test, presuming that the data distribution is known, which rarely holds in practice. However, it is possible to remove the bias from goodness-of-fit tests using a Monte Carlo approach. The steps in the procedure are as follows:

1. Obtain the K-S test statistic for the N observations against a target distribution whose parameters are estimated using the mle;
2. Obtain N random draws with replacement from the target distribution using the same mle parameters;
3. Compute the K-S statistic for the random draw;
4. Repeat steps 2 and 3 a large number of times; and
5. Compute the p-value.

Example 8.12 Remove the bias from the Weibull fit to earthquake data in Example 7.7.

```
parmhat = wblfit(quakes);
cdf = [sort(quakes)' wblcdf(sort(quakes)', parmhat(1),
parmhat(2))];
[h, p ,ksstat] = kstest(quakes, cdf);
n = length(quakes);
b = 999999;
sps = [];
parfor i = 1:b;
   draw = sort(wblrnd(parmhat(1), parmhat(2), 1, n));
   cdf = [draw' wblcdf(draw', parmhat(1), parmhat(2))];
   [h, p, ks1] = kstest(draw, cdf);
   sps = [sps ks1];
end
pval=2*min(sum(sps >= ksstat) + 1, sum(sps < ksstat) + 1)/(b+1)
pval =
   0.1995
```

The p-value used in this script differs slightly from the one usually used with bootstrap hypothesis tests and is further discussed in Section 8.3.1. The sample p-value is 0.9006 and

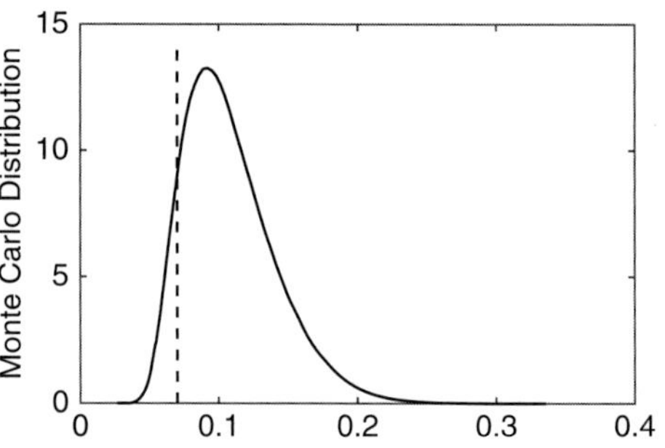

Figure 8.13 Monte Carlo distribution of the earthquake K-S statistic (solid line), along with the test statistic (dashed line).

is clearly upward biased, although the correction does not change the outcome of the hypothesis test. The simulated distribution (Figure 8.13) is asymmetric, as would be expected because it represents that of the K-S statistic.

Example 8.13. Remove the bias from the K-S test applied to contaminated Gaussian data in Example 7.8.

```
data = normrnd(0, 1, 1000, 1);
data(1000) = 6.5;
data(999) = 6.0;
data(998) = 5.5;
data(997) = 5;
[h, p, kstat] = kstest(data);
par1 = mean(data);
par2 = std(data, 1);
b = 99999;
sps = [];
parfor i = 1:b;
   draw = sort(normrnd(par1, par2, 1000, 1));
   cdf = [draw normcdf(draw, par1, par2)];
   [h, p, ks1] = kstest(draw, cdf);
   sps = [sps ks1];
end
pval = 2*min(sum(sps >= kstat) + 1, sum(sps < kstat) + 1)/(b + 1)
pval =
   0.9192
```

This compares with the sample p-value of 0.5415. The conclusion is unchanged, but downward bias of the sample K-S statistic is evident.

8.3 Permutation Tests

8.3.1 Principles

Permutation tests were introduced by Pitman (1937a; 1937b; 1938) but were quite impractical for statistical purposes until modern computers became available. An overview is provided by Good (2000, 2005), and a more advanced treatise is that of Pesarin & Salmaso (2010) or Berry, Mielke, & Johnston (2016). Permutation tests have wide applicability and can be used to analyze data that are Gaussian, close to Gaussian, or markedly non-Gaussian. A permutation test is typically at least as powerful as the alternatives for large samples (Hoeffding 1952). It is also nearly distribution free, requiring only very simple assumptions about the population. Further, a key characteristic of the permutation test is its accuracy. If the null hypothesis is true, then there is nearly exactly a probability α that the permutation p-value will be less than α, and under particular circumstances, a permutation test becomes exact. These are advantages to the permutation test over the bootstrap test described in Section 8.2.4 because the latter is neither exact nor conservative and is typically less powerful than a permutation test.

The key steps comprising the application of a permutation test are:

1. Choose a test statistic λ;
2. Compute the value of the test statistic for the original set(s) of data;
3. Combine the data into a pooled data set;
4. Obtain the permutation distribution of λ by randomly permuting the pooled data and recomputing the test statistic;
5. Repeat step 4 many times; and
6. Accept (or not) the null hypothesis according to the p-value for the original test statistic compared with the permutation distribution.

For a two sample test, if the null hypothesis holds, then a sample taken from the pooled data set ensuing from the merger of $\mathbf{X}$ and $\mathbf{Y}$ will be same as that from either sample taken alone. Consequently, if a sample is taken from the pooled data set, the first M values may be used to represent the first data set, and the last N values may be used to represent the second data set. If the difference between the sample means (or any other statistic that measures the difference in location) of the original data does not lie within the middle 95% of the permutation distribution, then a double-sided permutation test rejects the null hypothesis. Similar arguments pertain for upper- and lower-tail tests.

To express this more formally, let $\mathbf{Z}_{(i)}$ denote the order statistics of the merged data set, and define the indicator function $\mathbf{1}\left(\mathbf{Z}_{(i)} \in \mathbf{X}\right)$ that determines whether the ith merged order statistic came from the $\mathbf{X}_i$. The combination of the order statistics and indicator function contains all the information present in the original two samples. The vector $\boldsymbol{\Psi}$ that results from applying the indicator function to all the merged data consists of M ones and N zeros, and by combinatoric arguments, there are $(N + M)!/(M!N!)$ unique ways of dividing the

$M + N$ elements into two subsets of size M and N. Consequently, each realization of $\mathbf{\Psi}$ is a permutation that has a probability of occurrence of $M!N!/(M+N)!$ under the null hypothesis. Define a test statistic $\hat{\lambda}(\mathbf{Z}_{(i)}, \mathbf{\Psi})$ that measures the difference between the sampled populations. Let $\mathbf{\Psi}_k^*$ denote one of the possible permutations, and let the *permutation replication* of the test statistic be $\hat{\lambda}_k^*(\mathbf{Z}_{(i)}, \mathbf{\Psi}_k^*)$. The distribution that results from a set of permutation replications is the permutation distribution. The two-sided equal-tail permutation p-value is

$$p_{\text{perm}} = 2 \times \min\left\{\frac{1}{B+1}\left[\sum_{k=1}^{B} \mathbf{1}\left(\hat{\lambda}_k^* \geq \hat{\lambda}\right) + 1\right], \frac{1}{B+1}\left[\sum_{k=1}^{B} \mathbf{1}\left(\hat{\lambda}_k^* < \hat{\lambda}\right) + 1\right]\right\} \tag{8.14}$$

where B is the number of permutations and is evaluated in the usual way. One-tailed tests also may be used. The choice of test statistic is usually not critical, although in some circumstances a poor choice may result in reduced power.

Permutation tests are based on sampling without replacement, in contrast to a bootstrap hypothesis test that uses sampling with replacement. A sufficient condition for a permutation test to be exact and unbiased is exchangeability of the observations in a combined sample.

The permutation p-value in (8.14) differs from

$$\hat{p} = 2 \times \min\left[\frac{1}{B}\sum_{k=1}^{B} \mathbf{1}\left(\hat{\lambda}_k^* \geq \hat{\lambda}\right), \frac{1}{B}\sum_{k=1}^{B} \mathbf{1}\left(\hat{\lambda}_k^* < \hat{\lambda}\right)\right] \tag{8.15}$$

used in Section 8.2.4. Equation (8.15) is unbiased, but its Type 1 error rate is given by (Phipson & Smyth 2010)

$$\Pr(\hat{p} \leq \alpha) = \frac{\lfloor B\alpha \rfloor + 1}{B+1} \tag{8.16}$$

and may be larger or smaller than α as that parameter is near 0 or 1, respectively. Equation (8.15) also can yield a zero p-value. Neither of these issues is likely to be important for single hypothesis tests, but both can lead to problems when multiple hypotheses are simultaneously tested, as described in Section 6.6. By contrast, while (8.14) is biased, it yields an exact p-value presuming that there are no duplicate permutations or permutations equal to the original data in the simulation, and cannot be smaller than $1/(B+1)$. Consequently, it is preferred in practice. For sampling with replacement, (8.14) gives a conservative upper bound on the p-value.

8.3.2 One-Sample Test for a Location Parameter

While permutation methods are usually applied to two or more samples, they can be used for some types of one-sample tests. If the underlying distribution for a data set is symmetric about the location parameter λ, then

$$\Pr(\mathbf{X} \leq \lambda - x) = F(\lambda - x) = \Pr(\mathbf{X} \geq \lambda + x) = 1 - F(\lambda + x)$$

or

$$F(\lambda - x) + F(\lambda + x) = 1$$

For a test for a particular value of the location parameter λ_0, a suitable test statistic is the sum of the deviations of the data about λ_0, which should be close to zero if the null hypothesis holds. Under the alternate hypothesis, randomizing the signs of the deviations of the data about λ_0 will make the result either smaller or larger than zero. Consequently, permuting the signs of the deviations and reattaching the results to their absolute value, and repeating the process many times, will define a permutation distribution to which the sum of the deviations of the original data can be compared to define a p-value. This works because for a symmetric distribution the absolute values of the deviations are a sufficient statistic for the sample.

It is important to examine the permutation distribution for symmetry, which is most easily accomplished by comparing it with a standard distribution such as the Gaussian. This does *not* imply that the permutation distribution should correspond to some parametric form. Romano (1990) proved that a permutation test for a location parameter is asymptotically exact if the data distribution has finite variance. If the permutation distribution is almost symmetric, the test will be almost exact even for small numbers of data. Further, the permutation test has nearly the power of a Student t test for Gaussian data and large samples.

Example 8.14 Returning to Example 6.8, devise a permutation test for the null hypothesis that the number of days per year that an oceanographer spends at sea is 30 against an upper-tail alternate. The MATLAB function **randperm**(n) produces a random permutation of the integers from 1 to n and will be used to obtain the permutations.

```
x = [54 55 60  42 48 62 24 46 48 28 18 8 0 10 60 82 90 88 2 54];
x1 = sum(x - 30); %test statistic
n = length(x);
s = sign(x - 30); %signs of the differences
sps = [];
b = 10000;
perm = zeros(b, n);
m = [];
parfor i = 1:b
   perm(i, :) = randperm(n);
   if perm(i, :) == 1:n;
       m = [m i];
   end
end
if ~isempty(m)
    perm(m', :) = []; % ensure original data are not included
end
perm = unique(perm, 'rows'); %ensure sampling without replacement
```

```
b = length(perm)
b =
   10000
parfor i = 1:b
   sp = s(perm(i, :)); % permute the signs
   x2 = sum(sp.*abs(x - 30)); % permutation test statistic
   sps = [sps x2];
end
pval = (sum(sps >= x1) + 1)/(b +1) %exact p-value
pval =
   0.0281
```

The null hypothesis is rejected, but only weakly, and was also weakly rejected using a t test in Example 6.8. The p-values for 100,000 and 1 million permutations are 0.0267 and 0.0262, respectively. In each case, there were no duplicate permutations in the array *perm*. The permutation distribution for 1 million permutations is shown in Figure 8.14 and appears symmetric.

8.3.3 Two-Sample Test for a Location Parameter

For a two sample test, a statistic is needed to measure the difference between two sets of random variables, such as the difference of their sample means or, equivalently, the sum of the data from one of the treatment groups. To demonstrate the validity of the latter as a test statistic, write out the difference of the means of two sets of rvs:

$$\bar{\mathbf{X}} - \bar{\mathbf{Y}} = \frac{M+N}{MN}\sum_{i=1}^{M} x_i - \frac{1}{N}\left(\sum_{i=1}^{M} x_i + \sum_{i=1}^{M} y_i\right) \tag{8.17}$$

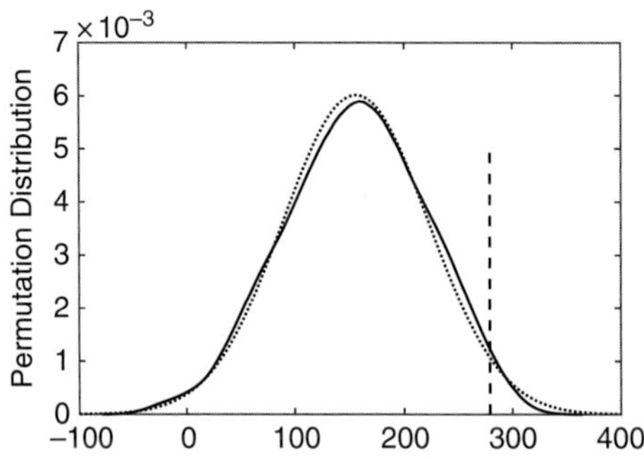

Figure 8.14 Kernel density permutation distribution (solid line) for the oceanographer days at sea example compared with a Gaussian pdf with mle parameters (dotted line). For comparison, the test statistic (dashed line) based on the original data is 279. The rejection region is to its right for an upper-tail test.

The second term in (8.17) is a constant, so the difference of means is a monotone function of the sum of the data in the first treatment group. Provided that the distributions of **X** and **Y** have finite second moments and the sample sizes are the same, the permutation test for two location parameters based on the sum of the observations in **X** is asymptotically exact (Romano 1990).

Example 8.15 Use a permutation test on the Michelson speed of light data to assess whether the location parameters in each year are the same.

```
michelson1 = importdata('michelson1.dat');
michelson2 = importdata('michelson2.dat');
n1 = length(michelson1);
n2 = length(michelson2);
n = n1 + n2;
michelson = [michelson1' michelson2'] ';
s = sum(michelson1); %measure of the null hypothesis
sps = [];
b = 10000;
perm = zeros(b, n);
m = [];
parfor i = 1:b
  perm(i, :) = randperm(n);
  if perm(i, :) == 1:n;
     m = [m i];
  end
end
if ~isempty(m)
   perm(m', :) = []; % ensure original data are not included
end
perm=unique(perm, 'rows'); %ensure sampling without replacement
b = length(perm)
b =
  10000
parfor i = 1:b
  michelsonp = michelson(perm(i, :)); %permuted version of
  the merged data
  sp = sum(michelsonp(1:n1));
  sps = [sps sp];
end
pval = 2*min(sum(sps >= s) + 1, sum(sps < s) + 1)/(b + 1) %
permutation p-value
pval =
      1.9998e-04
```

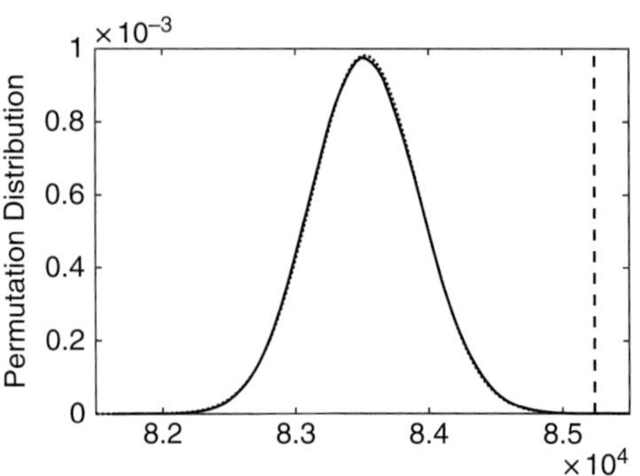

Figure 8.15 The permutation null distribution with 1 million samples (solid line) for the Michelson speed of light data compared with a standard Gaussian distribution (dotted line) with mle parameters. The vertical dashed line shows the value of the test statistic.

The result with 1 million permutation replicates is 1.8000×10^{-5}. The two-sample t test gives a two-sided p-value of 1.4×10^{-5} for comparison. Note that only a small fraction of the possible permutations has been sampled even with 1 million replicates because the total number of permutations is 26,010,968,307,696,038,491,182,501, so the sampling is effectively without replacement even in the absence of enforcing that condition. The empirical null distribution for 1 million permutations compared with a standardized Gaussian (Figure 8.15) shows a strong resemblance. The Gaussian distribution is used as an exemplar symmetric distribution and does not imply that the permutation distribution should be Gaussian.

However, there are a significant number of duplicate values in the 1879 data that are presumably due to rounding during unit conversion and are apparent in the stairstep pattern in the q-q plot of Figure 6.4. The duplicates can be removed using the MATLAB **unique** function. This leaves only 30 of 100 values in the 1879 data set and 22 of 24 values in the 1882 data set. Repeating the permutation test yields a p-value of 0.0057, so the conclusion is unchanged by removing the duplicates.

Example 8.16 Apply a permutation test to evaluate the null hypothesis that seeded and unseeded clouds produce the same amount of rain using the data from Example 6.12.

```
data = importdata('cloudseed.dat');
n = length(data);
cloud = [data(:, 1)' data(:, 2)']';
s = mean(data(:, 1)) - mean(data(:, 2));
sps = [];
b = 1000000;
perm = zeros(b, 2*n);
```

```
m = [];
parfor i = 1:b
   perm(i, :) = randperm(2*n);
   if perm(i, :) == 1:2*n;
       m = [m i];
   end
end
if ~isempty(m)
    perm(m', :) = []; % ensure original data are not included
end
perm=unique(perm, 'rows'); %ensure sampling without replacement
b = length(perm)
b =
   1000000
parfor i = 1:b
   cloudp = cloud(perm(i, :));
   sp = mean(cloudp(1:n)) - mean(cloudp(n + 1:2*n);
   sps = [sps sp];
end
pval = 2*min(sum(sps >= s) + 1, sum(sps < s) + 1)/(b +1)
pval =
   0.0441
```

The null hypothesis that the seeded and unseeded clouds produce the same amount of rain is rejected, although only weakly. A two-sample t test in Example 6.12 produced weak acceptance of the alternate hypothesis with a p-value of 0.0511. The permutation distribution is approximately symmetric (Figure 8.16) but is platykurtic compared with a Gaussian. The test statistic based on the original data is -277.3962.

Repeating after taking the logarithms of the data gives a p-value of 0.0143, so the conclusion is unchanged, in contrast to the parametric test in Example 6.12. This example shows how permutation tests are not dependent on distributional assumptions.

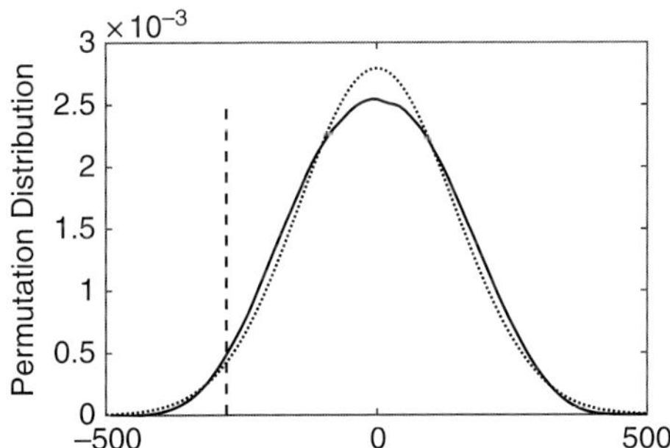

Figure 8.16 Permutation distribution for 1,000,000 replicates of the cloud seeding data (solid) compared with a Gaussian distribution with mle parameters (dotted line). The vertical dashed line shows the value of the test statistic.

8.3.4 Two-Sample Test for Paired Data

Rather than being based on random sampling from a population as in Section 8.3.3, a permutation test applied to paired data requires the assumption that the assignment to treatment groups is random. However, the permutation approach is very similar. The permutation distribution (often called the *randomization distribution* in this application) is then constructed to determine whether this statistic is large enough to establish that there is a significant difference between the treatments. Ernst (2004) describes estimation of the additive treatment effect by inverting the permutation test.

Example 8.17 Returning to the leukemia remission data from Example 7.15, use a permutation approach to assess whether a new drug improves survival.

```
remiss = importdata('remiss.dat');
x = remiss(1:21, 2);
y = remiss(22:42, 2);
s = sum(x);
data = [x' y'];
n = length(data);
sps = [];
b = 1000000;
perm = zeros(b, n);
m = [];
parfor i = 1:b
   perm(i, :) = randperm(n);
   if perm(i, :) == 1:n
        m = [ m i];
   end
end
if ~isempty(m)
    perm(m', :) = []; % ensure original data are not included
end
perm=unique(perm, 'rows'); %ensure sampling without replacement
b = length(perm)
b =
   1000000
parfor i = 1:b
   datap = data(perm(i, :));
   sp = sum(datap(1:n/2);
   sps = [sps sp];
end
pval = 2*min(sum(sps >= s) + 1,sum(sps < s) + 1)/(b + 1)
pval =
   0.0021
```

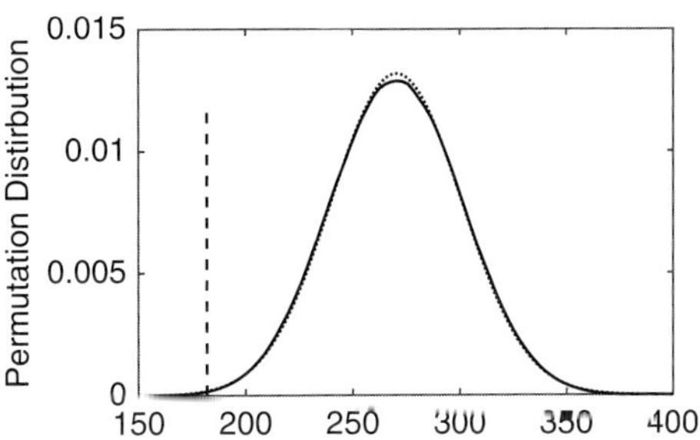

Figure 8.17 Permutation distribution for 1,000,000 replicates of the leukemia remission data (solid) compared with a Gaussian distribution with mle parameters (dotted line). The vertical dashed line shows the value of the test statistic.

This compares favorably with the sign test result in Example 7.15 and the signed rank test result in Example 7.18. Figure 8.17 shows the permutation distribution for this example.

8.3.5 Two-Sample Test for Dispersion

On initial reflection, it would seem reasonable to compare the variances of two populations by using the same procedure as for location parameters but using the squares of the observations instead of the observations themselves. However, this will fail unless the means of the two populations are known or are known to be the same. This follows directly because $\mathcal{E}(\mathbf{X}^2) = \sigma^2 + \mu^2$.

The solution is to center each data set about its median and then use the difference of the sum of the squares (or the absolute values) of the deviations from the median as a test statistic. In essence, this centers each distribution about its middle. If the number of samples is odd, then one of these values will be zero and hence not informative. It should be discarded. In practice, this issue is important only for small sample sizes or when data have undergone unit conversions with truncation.

Example 8.18 Use a permutation approach to test that the dispersion in the two Michelson speed of light data sets are the same after removing duplicates.

```
x1 = importdata('michelson1.dat');
x2 = importdata('michelson2.dat');
x1 = unique(x1);
x1 = x1 - median(x1);
x2 = unique(x2);
x2 = x2 - median(x2);
x = [x1' x2']';
n1 = length(x1);
n2 = length(x2);
```

```
s = sum(x1.^2)/n1 - sum(x2.^2)/n2
sps = [];
b = 100000;
perm = zeros(b, n1 +n2 );
m = [];
parfor i = 1:b
   perm(i, :) = randperm(n1 + n2);
   if perm(i, :) == 1:n1 + n2;
       m = [m i];
   end
end
if ~isempty(m)
    perm(m', :) = []; % ensure original data are not included
end
perm=unique(perm, 'rows'); %ensure sampling without replacement
b = length(perm)
b =
   1000000
parfor i = 1:b
   xp = x(perm(i, :));
   sp = sum(xp(1:n1).^2)/n1 - sum(xp(n1 + 1:n1 + n2).^2)/n2;
   sps = [sps sp];
end
pval=2*min(sum(sps>=s) + 1,sum(sps<s) + 1)/(b + 1)
pval =
   0.5809
```

The null hypothesis is accepted. The permutation distribution is slightly asymmetric (Figure 8.18). The test statistic computed from the original data is −2606.4 and is located in the lower half of the permutation distribution.

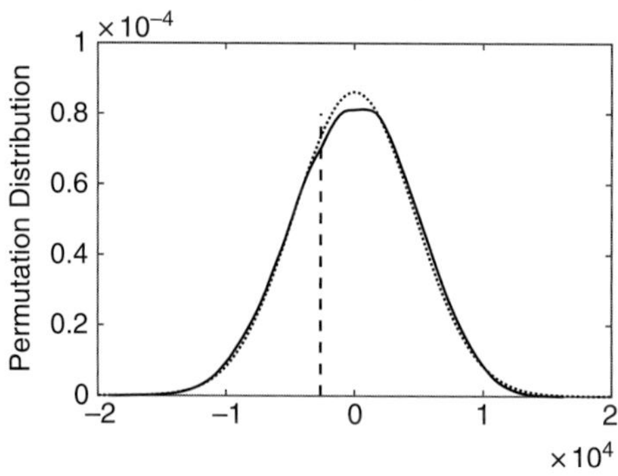

Figure 8.18 Permutation distribution for the Michelson speed of light data dispersion test using the squared deviations of the observations from their median as a test statistic (solid line) compared with a Gaussian distribution with mle parameters (dotted line). There are 1 million replicates in the sample. The vertical dashed line is the test statistic.

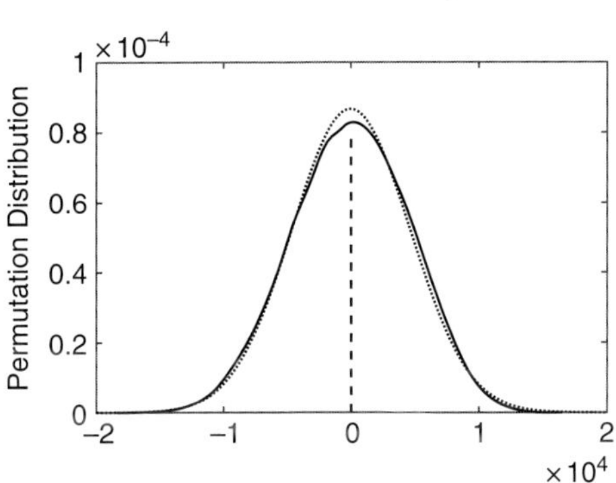

Figure 8.19 Permutation distribution for the Michelson speed of light dispersion test using the absolute values of the deviations of the observations from their median as a test statistic (solid line) compared with a Gaussian pdf with mle parameters (dotted line). The vertical dashed line shows the test statistic. There are 1 million replicates in the sample.

An alternative measure of dispersion that is more robust is the difference of the sum of the absolute values of the deviations from the median. Repeating the test using this metric gives a p-value of 0.9956 for 1 million permutations, strongly accepting the null hypothesis that the dispersion is the same in the two data sets. The test statistic based on the original data is -0.9848. The permutation distribution is more symmetric (Figure 8.19), although the statistical outcome is not changed.

8.4 The Jackknife

The jackknife is a linear approximation to the bootstrap that is computationally simple and efficient, making it suitable for large data sets where the bootstrap would be prohibitive, although as computational power grows, this argument for the jackknife gets weaker. Thomson & Chave (1991) provide a review of the jackknife. Let $\{\mathbf{X}_i\}$, $i = 1, \ldots, N$, be independent samples drawn from an unknown distribution or mixture of distributions, and let λ be some parameter that is to be estimated using the estimator $\hat{\lambda}$. Let $\hat{\lambda}_N$ be the estimate of λ computed using all the data. Subdivide the data into N groups of size $N-1$ by sampling with replacement. Denote the estimate of λ obtained from the ith subset as $\hat{\lambda}_{\backslash i}(\mathbf{X}_1, \ldots \mathbf{X}_{i-1}, \mathbf{X}_{i+1}, \ldots \mathbf{X}_N)$, where the subscript $\backslash i$ denotes the delete the ith datum estimate. The jackknife literature usually involves substitute data called *pseudo-values* given by

$$p_i = N\hat{\lambda}_N - (N-1)\hat{\lambda}_{\backslash i} \tag{8.18}$$

The jackknife estimate of λ is the mean of (8.18)

$$\begin{aligned}\tilde{\lambda} &= \frac{1}{N}\sum_{i=1}^{N} p_i \\ &= N\hat{\lambda}_N - \frac{N-1}{N}\sum_{i=1}^{N}\hat{\lambda}_{\backslash i}\end{aligned} \tag{8.19}$$

If $\mathcal{E}(\hat{\lambda}_N) = \lambda + (a/N) + O(1/N^2)$, where a is a constant, it can shown that the jackknife estimator eliminates the $1/N$ term and hence reduces the bias. It was for this purpose that it was originally introduced in the 1950s.

For a linear statistic such as the sample mean, where $\hat{\lambda}_N = \overline{X}_N$

$$\begin{aligned}\tilde{\lambda} &= \sum_{I=1}^{N} x_i - \frac{N-1}{N}\left(\sum_{j=1}^{n}\frac{1}{N-1}\sum_{\substack{i=1\\ i\neq j}}^{N} x_i\right) \\ &= \sum_{j=1}^{N}\left(x_j - \frac{1}{N}\sum_{\substack{i=1\\ i\neq j}}^{N} x_i\right) \\ &= \sum_{j=1}^{N}\left(\frac{N+1}{N}x_j \quad - \quad \frac{1}{N}\sum_{i=1}^{N} x_i\right) \\ &= (N+1)\overline{X}_N - N\overline{X}_N \\ &= \overline{X}_N\end{aligned} \tag{8.20}$$

which is the same as $\hat{\lambda}_N$. Since the mean is a linear statistic, the jackknife will not improve its calculation. However, when the estimator is not linear in the data, this will not generally be true.

The most important application of the jackknife occurs in estimating the variance of $\hat{\lambda}_N$. Recalling that the standard or mle estimator is the unbiased sample variance divided by N when $\hat{\lambda}_N = \overline{X}_N$, the jackknife formula follows from the definition of the pseudovalues in (8.18).

$$\begin{aligned}\hat{s}_J^2 &= \frac{1}{N(N-1)}\sum_{i=1}^{N}\left(p_i - \tilde{\lambda}\right)^2 \\ &= \frac{N-1}{N}\sum_{i=1}^{N}\left(\hat{\lambda}_{\backslash i} - \frac{1}{N}\sum_{j=1}^{N}\hat{\lambda}_{\backslash i}\right)^2 \\ &= \frac{N-1}{N}\sum_{i=1}^{N}\left(\hat{\lambda}_{\backslash i} - \bar{\lambda}\right)^2\end{aligned} \tag{8.21}$$

where $\bar{\lambda}$ is the mean of the delete-one estimates. This is computationally simple: compute the delete-one estimates of the statistic, compute their mean, and compute the variance. While one might expect that $\hat{s}_J^2 = var(\tilde{\lambda})$, small sample simulations show that $\hat{s}_J^2 = var(\hat{\lambda}_N)$.

An important property of the jackknife is (Efron & Stein 1981)

$$\mathcal{E}\left(\hat{s}_J^2\right) > \frac{\sigma^2}{N} \tag{8.22}$$

This holds even for data that are not identically distributed. Thus the jackknife yields a conservative estimate for the variance.

Under general conditions, it can be shown that $(\tilde{\lambda}-\lambda)/\hat{s}_j \sim t_{N-1}$ and $(\hat{\lambda}_N-\lambda)/\hat{s}_j \sim t_{N-1}$. This allows the construction of approximate confidence intervals using the jackknife in the usual way:

$$\Pr\left[\hat{\lambda}_N - t_{N-1}(1-\alpha/2)\ \hat{s}_J \leq \lambda \leq \hat{\lambda}_N + t_{N-1}(1-\alpha/2)\ \hat{s}_J\right] = 1-\alpha \tag{8.23}$$

This holds for any statistic, not just the mean.

However, the jackknife should not be used blindly on statistics that are markedly non-Gaussian. Based on simulations, the Student t model breaks down in that case unless something is done to make the statistic more Gaussian. Two good examples are the bounded statistics variance and correlation coefficient, which must be nonnegative or lie on $[-1, 1]$, respectively. The solution for the variance is to jackknife logs rather than raw estimates. Let $\{\hat{s}_k^2\}$ be a set of raw delete-one variance estimates, so that

$$\log \hat{s}_{\backslash i}^2 = \log\left(\frac{1}{N}\sum_{\substack{k=1\\ i\neq k}}^{N} \hat{s}_k^2\right) \tag{8.24}$$

$$\log \bar{s}^2 = \frac{1}{N}\sum_{j=1}^{N} \log \hat{s}_{\backslash j}^2 \tag{8.25}$$

$$\hat{s}_J^2 = \frac{N-1}{N}\sum_{i=1}^{N}\left(\log \hat{s}_{\backslash i}^2 - \log \bar{s}^2\right)^2 \tag{8.26}$$

Because of the log transformation, $\left(\log \hat{s}_{\backslash i}^2 - \log \bar{s}^2\right)/\hat{s}_J \sim t_{N-1}$, and approximate confidence intervals can be constructed in the usual way:

$$\Pr\left(\bar{s}\, e^{-t_{N-1}(1-\alpha/2)\hat{s}_J} \leq \lambda \leq \bar{s}\, e^{t_{N-1}(1-\alpha/2)\hat{s}_J}\right) = 1-\alpha \tag{8.27}$$

To jackknife the correlation coefficient, an inverse hyperbolic tangent transformation is made before proceeding in a like manner.

Example 8.19 Use the jackknife to estimate the skewness and its standard error for the Old Faithful data set.

```
n = length(geyser);
skew = skewness(geyser)
skew =
   -0.3392
skewt = zeros(1, n);
for i =1:n
    geyt = geyser;
```

```
    geyt(i) = []; % leave i-th point out
    skewt(i) = skewness(geyt);
end
mean(skewt)
ans =
  -0.3392
sqrt((n - 1)/n*sum((skewt - mean(skewt)).^2))
ans =
  0.1004
```

Note that the same result is obtained by directly computing the skewness from the data and by averaging the delete-one estimates, and that a slightly different standard deviation as compared with the bootstrap ensues.

MATLAB provides the function *jackstat* = **jackknife**(*jackfun*, *x*) to compute the jackknife from the $N \times p$ array *x* using the function handle *jackfun* and returns the $N \times p$ array of jackknife statistics *jackstat*. Example 8.19 can be obtained more simply as

```
jackstat = jackknife(@skewness, geyser);
sqrt((n - 1)/n*sum((jackstat - mean(jackstat)).^2))
ans =
  0.1004
```

Example 8.20 Compute the median and its standard error for the geyser data using the jackknife and the bootstrap. The sample median of the data is 76.

Use the same scripts as in Example 8.8 and 8.19 but change the **skewness** function to the **median** function. The result for the jackknife is zero, which does not make much sense on initial glance. The delete-one estimates must be identical to account for a zero standard error. The result with the bootstrap is 76.2390. The problem is that the median is not a smooth statistic, and the jackknife breaks down for estimators based on order statistics. However, the bootstrap continues to function satisfactorily.

---

9 Linear Regression

9.1 Motivating Example

Suppose that the salinity of seawater is measured as a function of its temperature T, confining pressure p, and electrical conductivity σ, as implemented in the widely used CTD instrument. Further suppose that a quantitative relation between salinity and one or more of these entities is wanted for prediction purposes. One might expect a relation of the general form $s = S(T, p, \sigma)$, where the function S might:

1. Be simple and dependent on only one variable, as in $s = \beta_0 + \beta_1 T$;
2. Be multivariate and first order, as in $s = \beta_0 + \beta_1 T + \beta_2 p + \beta_3 \sigma$;
3. Depend on higher-order terms, as in $s = \beta_0 + \beta_1 T + \beta_2 T^2 + \cdots$; or
4. Involve more complicated functions, as in $s = \beta_0 + \beta_1 p e^{-\beta_2 T} + 1/(1 + \beta_4 \sigma)$,

where the $\{\beta_i\}$ are constants to be estimated from measurements. Solving the problem requires both determining the form of the model (i.e., the nature of the functional relation, the number of terms, etc.), and computation and assessment of the statistical fit of data to the model.

A general method that originated from the work of Gauss (1823) exists to address this type of problem when the model is linear in the parameters $\{\beta_i\}$. Examples 1 to 3 above are linear in the parameters, but example 4 is not. However, $s = \beta_0 e^{-\beta_1 T}$ is linear when logs are taken, so $\log s = \log \beta_0 - \beta_1 T$, but the error structure from solving the second equation will not be the same as for the first form.

Consider first just the salinity-temperature data. One method for constructing a straight line $s = \beta_0 + \beta_1 T$ is to minimize the sum of the squared vertical deviations of the data points from the line. This implicitly assumes (in this case) that temperature is measured without error or at least has relative errors (i.e., some estimate of the error divided by the data value) in temperature that are small compared with those of the salinity data. One might think that a more rational approach would be to minimize the sum of the squared orthogonal deviations of the data points from the line (i.e., the squared minimum distance of each data point from the line). This would implicitly assume that the relative errors on both data were the same for each data pair $\{s_i, T_i\}$, but this is rarely the case.

More generally, if both the data pairs $\{s_i, T_i\}$ and some estimate of the errors on each, as well as their covariance, are known, a more exact way of solving for the best-fitting line can be devised. Such information is almost never available, so this approach is almost never used.

When $T = T_i$, the corresponding predicted salinity value is $s_i = \beta_0 + \beta_1 T_i$, and the vertical distance between $\{s_i, T_i\}$ and the line is $|s_i - (\beta_0 + \beta_1 T_i)|$. Suppose that there are N data, and let Q denote the sum of the squares of the vertical distances over the N data points

$$Q = \sum_{i=1}^{N} [s_i - (\beta_0 + \beta_1 T_i)]^2 \tag{9.1}$$

The method of least squares specifies that β_0 and β_1 should be chosen to minimize Q. This is obtained by setting the derivatives of Q with respect to β_0 and β_1 to zero. In fact, to be formal, it must further be determined that the solution is a minimum rather than a maximum by showing that the second derivatives of Q with respect to the two parameters are upward convex, although this is often omitted. The result is

$$\partial_{\beta_0} Q = -2\sum_{i=1}^{N} [s_i - (\beta_0 + \beta_1 T_i)] = 0 \tag{9.2}$$

$$\partial_{\beta_1} Q = -2\sum_{i=1}^{N} T_i[s_i - (\beta_0 + \beta_1 T_i)] = 0 \tag{9.3}$$

and reduces to the following pair of equations:

$$\hat{\beta}_0 N + \hat{\beta}_1 \sum_{i=1}^{N} T_i = \sum_{i=1}^{N} s_i \tag{9.4}$$

$$\hat{\beta}_0 \sum_{i=1}^{N} T_i + \hat{\beta}_1 \sum_{i=1}^{N} T_i^2 = \sum_{i=1}^{N} s_i T_i \tag{9.5}$$

These are called the *normal equations* for $\hat{\beta}_0$ and $\hat{\beta}_1$, and their solution is the least squares estimator for the line. Solving this pair of equations for the unknown parameters gives

$$\hat{\beta}_0 = \bar{s}_N - \hat{\beta}_1 \bar{T}_N \tag{9.6}$$

$$\hat{\beta}_1 = \frac{\sum_{i=1}^{N} s_i T_i - N \bar{s}_N \bar{T}_N}{\sum_{i=1}^{N} T_i^2 - N\left(\bar{T}_N\right)^2} \tag{9.7}$$

This formalism can be easily extended to include higher-order polynomial terms, and an arbitrary kth order polynomial can be easily accommodated. In this case, the constants $\{\beta_i\}$ are chosen to minimize

$$Q = \sum_{i=1}^{N} \left[s_i - \left(\beta_0 + \beta_1 T_i + \cdots + \beta_k T_i^k\right)\right]^2 \tag{9.8}$$

If the $k + 1$ partial derivatives of Q with respect to β_i are computed and set to zero, the normal equations are obtained as $k + 1$ equations in $k + 1$ unknowns. These will have a solution if and only if the determinant of the left-hand side is nonzero. Note also that there must be at least as many data N as there are parameters or else the problem is under-determined with no unique solution.

The formalism easily generalizes to the multiple regression case where there is more than one predictor variable (i.e., T, p, and σ in the salinity example) for either a first-order (i.e., straight line) or higher-order model. This is most easily accomplished using the notation of matrix algebra. Let $\mathbf{y}$ be an $N \times 1$ vector of *response variables* (such as the salinity data in the earlier example). An $N \times 1$ vector has N rows and 1 column. Let $\overleftrightarrow{\mathbf{X}}$ be an $N \times p$ matrix of *predictor variables* [such as T, p, and σ in the earlier example, where the model may be allowed to have k powers for each variable so that $p = 3 \times (k + 1)$]. The $p \times 1$ vector of parameters $\boldsymbol{\beta}$ that satisfies $\min \left\| \mathbf{y} - \overleftrightarrow{\mathbf{X}} \cdot \boldsymbol{\beta} \right\|^2$, where $\| \|^2$ denotes the L_2 or least squares norm, is

$$\overleftrightarrow{\mathbf{X}}^T \cdot \left(\mathbf{y} - \overleftrightarrow{\mathbf{X}} \cdot \boldsymbol{\beta} \right) = 0 \tag{9.9}$$

where $\cdot$ denotes the inner product. Rearranging terms, the matrix form of the normal equation estimator is

$$\hat{\beta} = \left(\overleftrightarrow{\mathbf{X}}^T \cdot \overleftrightarrow{\mathbf{X}} \right)^{-1} \cdot \overleftrightarrow{\mathbf{X}}^T \cdot \mathbf{y} \tag{9.10}$$

It can be easily shown that for any other vector γ,

$$\left\| \mathbf{y} - \overleftrightarrow{\mathbf{X}} \cdot \hat{\boldsymbol{\beta}} \right\|^2 \leq \left\| \mathbf{y} - \overleftrightarrow{\mathbf{X}} \cdot \gamma \right\|^2 \tag{9.11}$$

so the least squares solution produces the minimum sum of squared residuals.

9.2 Statistical Basis for Linear Regression

Consider the *linear model* for p parameters given N data, where $p \leq N$ is assumed

$$\mathbf{y} = \overleftrightarrow{\mathbf{X}} \cdot \boldsymbol{\beta} + \boldsymbol{\varepsilon} \tag{9.12}$$

where $\boldsymbol{\varepsilon}$ is an $N \times 1$ vector of unobservable random errors. The following four assumptions will be made:

1. The model is linear in the parameters. This is implied by (9.12).
2. The error structure is additive. This is also implied by (9.12).
3. The random errors have zero mean and equal variance while being mutually uncorrelated. The first condition can be forced approximately (and exactly so asymptotically) by adding a column of ones to $\overleftrightarrow{\mathbf{X}}$. The second condition can be forced by weighting the rows of the regression problem by some measure of data quality. These conditions are equivalent to requiring that all of the data have equal influence on the result.
4. $\text{Rank}\left(\overleftrightarrow{\mathbf{X}}\right) = p$ (note that this is the rank of a matrix, which is not the same as the ranks of data defined in Section 7.3.1), so $\overleftrightarrow{\mathbf{X}}^T \cdot \overleftrightarrow{\mathbf{X}}$ is positive definite and hence invertible. The rank is the number of independent columns in $\overleftrightarrow{\mathbf{X}}$. If $\text{rank}\left(\overleftrightarrow{\mathbf{X}}\right) < p$, then the model

is flawed because some of the parameters are not independent, and the solution will not be unique.

In elementary statistical textbooks, $\overleftrightarrow{\mathbf{X}}$ is usually taken to have no measurement error, in which case the statistical model corresponding to (9.12) is

$$\begin{aligned}\mathcal{E}(\mathbf{y}) &= \overleftrightarrow{\mathbf{X}} \cdot \boldsymbol{\beta} \\ cov(\mathbf{y}) &= \sigma^2 \overleftrightarrow{\mathbf{I}}_N\end{aligned} \tag{9.13}$$

where $\boldsymbol{\beta}$ and σ^2 are the population values of the parameters and variance, and $\overleftrightarrow{\mathbf{I}}_N$ is the $N \times N$ identity matrix. In reality, least squares is usually applied for the case where $\overleftrightarrow{\mathbf{X}}$ contains random variables (rvs). This can be accommodated in the statistical model using the conditional expectation and covariance

$$\begin{aligned}\mathcal{E}\left(\mathbf{y}|\overleftrightarrow{\mathbf{X}}\right) &= \overleftrightarrow{\mathbf{X}} \cdot \boldsymbol{\beta} \\ cov\left(\mathbf{y}|\overleftrightarrow{\mathbf{X}}\right) &= \sigma^2 \overleftrightarrow{\mathbf{I}}_N\end{aligned} \tag{9.14}$$

The theory of least squares applies equally well when $\overleftrightarrow{\mathbf{X}}$ contains rvs under very general conditions (Shaffer 1991).

The least squares estimator for (9.12) is

$$\hat{\boldsymbol{\beta}} = \left(\overleftrightarrow{\mathbf{X}}^T \cdot \overleftrightarrow{\mathbf{X}}\right)^{-1} \cdot \overleftrightarrow{\mathbf{X}}^T \cdot \mathbf{y} \tag{9.15}$$

The following properties and definitions apply to the least squares solution:

1. The conditional expected value of $\hat{\boldsymbol{\beta}}$ is

$$\mathcal{E}\left(\hat{\beta}|\,\overleftrightarrow{\mathbf{X}}\right) = \beta + \left(\overleftrightarrow{\mathbf{X}}^T \cdot \overleftrightarrow{\mathbf{X}}\right)^{-1} \cdot \overleftrightarrow{\mathbf{X}}^T \cdot \boldsymbol{\varepsilon} \tag{9.16}$$

   using (9.15) and (9.12). The unconditional expected value is

$$\mathcal{E}(\hat{\boldsymbol{\beta}}) = \mathcal{E}\left[\mathcal{E}\left(\hat{\boldsymbol{\beta}}|\,\overleftrightarrow{\mathbf{X}}\right)\right] = \boldsymbol{\beta} \tag{9.17}$$

   using assumption 3 above. Consequently, the least squares estimator is unconditionally unbiased.

2. The conditional covariance matrix for the parameters is

$$\begin{aligned}cov\left(\hat{\boldsymbol{\beta}}|\,\overleftrightarrow{\mathbf{X}}\right) &= cov\left[\left(\overleftrightarrow{\mathbf{X}}^T \cdot \overleftrightarrow{\mathbf{X}}\right)^{-1} \cdot \overleftrightarrow{\mathbf{X}}^T \cdot \left(\overleftrightarrow{\mathbf{X}} \cdot \boldsymbol{\beta} + \boldsymbol{\varepsilon}\right)|\,\overleftrightarrow{\mathbf{X}}\right] \\ &= cov\left[\left(\overleftrightarrow{\mathbf{X}}^T \cdot \overleftrightarrow{\mathbf{X}}\right)^{-1} \cdot \overleftrightarrow{\mathbf{X}}^T \cdot \boldsymbol{\varepsilon}|\,\overleftrightarrow{\mathbf{X}}\right] \\ &= \mathcal{E}\left[\left(\overleftrightarrow{\mathbf{X}}^T \cdot \overleftrightarrow{\mathbf{X}}\right)^{-1} \cdot \overleftrightarrow{\mathbf{X}}^T \cdot \boldsymbol{\varepsilon} \cdot \boldsymbol{\varepsilon}^T \cdot \overleftrightarrow{\mathbf{X}} \cdot \left(\overleftrightarrow{\mathbf{X}}^T \cdot \overleftrightarrow{\mathbf{X}}\right)^{-1} |\,\overleftrightarrow{\mathbf{X}}\right] - \\ &\quad \mathcal{E}\left[\left(\overleftrightarrow{\mathbf{X}}^T \cdot \overleftrightarrow{\mathbf{X}}\right)^{-1} \cdot \overleftrightarrow{\mathbf{X}}^T \cdot \boldsymbol{\varepsilon}|\,\overleftrightarrow{\mathbf{X}}\right] \cdot \mathcal{E}\left[\boldsymbol{\varepsilon}^T \cdot \overleftrightarrow{\mathbf{X}} \cdot \left(\overleftrightarrow{\mathbf{X}}^T \cdot \overleftrightarrow{\mathbf{X}}\right)^{-1} |\,\overleftrightarrow{\mathbf{X}}\right] \\ &= \sigma^2 \left(\overleftrightarrow{\mathbf{X}}^T \cdot \overleftrightarrow{\mathbf{X}}\right)^{-1}\end{aligned} \tag{9.18}$$

where the second line follows because $\boldsymbol{\beta}$ is a parameter and the last line follows from assumption 3 above. By the law of total covariance, the unconditional covariance matrix of the parameters is

$$\begin{aligned} cov(\hat{\boldsymbol{\beta}}) &= \mathcal{E}\left[cov\left(\hat{\beta}|\overleftrightarrow{\mathbf{X}}\right)\right] + cov\left[\mathcal{E}\left(\hat{\beta}|\overleftrightarrow{\mathbf{X}}\right)\right] \\ &= \mathcal{E}\left[cov\left(\hat{\beta}|\overleftrightarrow{\mathbf{X}}\right)\right] + \mathcal{E}\left[\mathcal{E}\left(\hat{\beta}|\overleftrightarrow{\mathbf{X}}\right) \cdot \mathcal{E}\left(\hat{\beta}|\overleftrightarrow{\mathbf{X}}\right)^T\right] - \\ &\quad \mathcal{E}\left[\mathcal{E}\left(\hat{\beta}|\overleftrightarrow{\mathbf{X}}\right)\right] \cdot \left\{\mathcal{E}\left[\mathcal{E}\left(\hat{\beta}|\overleftrightarrow{\mathbf{X}}\right)\right]\right\}^T \\ &= \sigma^2\mathcal{E}\left[\left(\overleftrightarrow{\mathbf{X}}^T \cdot \overleftrightarrow{\mathbf{X}}\right)^{-1}\right] \equiv \sigma^2\overleftrightarrow{\mathbf{Z}}^{-1} \end{aligned} \tag{9.19}$$

3. The N-vector of *predicted values* for the response variable is

$$\hat{\mathbf{y}} = \overleftrightarrow{\mathbf{X}} \cdot \hat{\boldsymbol{\beta}} = \overleftrightarrow{\mathbf{X}} \cdot \left(\overleftrightarrow{\mathbf{X}}^T \cdot \overleftrightarrow{\mathbf{X}}\right)^{-1} \cdot \overleftrightarrow{\mathbf{X}}^T \cdot \mathbf{y} \equiv \overleftrightarrow{\mathbf{H}} \cdot \mathbf{y} \tag{9.20}$$

where $\overleftrightarrow{\mathbf{H}}$ is the $N \times N$ *hat matrix*. The predicted values have the following characteristics:

$$\mathcal{E}\left(\hat{\mathbf{y}}|\overleftrightarrow{\mathbf{X}}\right) = \overleftrightarrow{\mathbf{X}} \cdot \boldsymbol{\beta} \tag{9.21}$$

$$cov\left(\hat{\mathbf{y}}|\overleftrightarrow{\mathbf{X}}\right) = \sigma^2\,\overleftrightarrow{\mathbf{H}} \tag{9.22}$$

The hat matrix has the following properties:

a. Because of the relation between the predicted and observed response variables, $\overleftrightarrow{\mathbf{H}}$ is a projection matrix. As a result, $\overleftrightarrow{\mathbf{H}}$ must be symmetric (its entries are identical above and below the main diagonal), idempotent $\left(\overleftrightarrow{\mathbf{H}} \cdot \overleftrightarrow{\mathbf{H}} = \overleftrightarrow{\mathbf{H}}\right)$, and have eigenvalues of either 0 or 1.
b. The entries on the main diagonal (denoted by $\hat{h}_{ii}$) are always real and satisfy $0 \le \hat{h}_{ii} \le 1$.
c. If $\hat{h}_{ii} = 0$, then the corresponding entry in $\hat{\mathbf{y}}$ is zero and hence is not affected by that entry in $\mathbf{y}$. If $\hat{h}_{ll} - 1$, then the corresponding entries in $\hat{\mathbf{y}}$ and $\mathbf{y}$ are identical, and the model fits the datum exactly. In this instance, only the ith row of $\overleftrightarrow{\mathbf{X}}$ has any influence on the regression problem, leading to the term high (or extreme) *leverage* to describe that point.
d. Chave & Thomson (2003) showed that when the rows of $\overleftrightarrow{\mathbf{X}}$ are complex multivariate normal, the distribution of the hat matrix diagonal is beta $\left(\hat{h}_{ii}, p, N-p\right)$. This relationship holds for real predictors as a special case. From the properties of the beta distribution, the expected value of $\hat{h}_{ii}$ is p/N. Knowledge about the distribution of $\hat{h}_{ii}$ will be useful in model assessment.

4. The regression *residuals* are given by

$$\hat{\mathbf{r}} = \mathbf{y} - \overleftrightarrow{\mathbf{X}} \cdot \hat{\boldsymbol{\beta}} = \left(\overleftrightarrow{\mathbf{I}}_N - \overleftrightarrow{\mathbf{H}}\right) \cdot \mathbf{y} \tag{9.23}$$

5. The *sum of squared residuals* is given by

$$\hat{\mathbf{r}}^T \cdot \hat{\mathbf{r}} = \mathbf{y}^T \cdot \left(\overleftrightarrow{\mathbf{I}}_N - \overleftrightarrow{\mathbf{H}}\right) \cdot \mathbf{y} \tag{9.24}$$

using the idempotence property of $\overleftrightarrow{\mathbf{I}}_N$ and $\overleftrightarrow{\mathbf{H}}$.

6. An unbiased estimate for the variance σ^2 is

$$\hat{\sigma}^2 = \frac{\hat{\mathbf{r}}^T \cdot \hat{\mathbf{r}}}{N - p} \tag{9.25}$$

and an estimator for $\overleftrightarrow{\mathbf{Z}}$ in (9.19) is

$$\hat{\overleftrightarrow{\zeta}} = \overleftrightarrow{\mathbf{X}}^T \cdot \overleftrightarrow{\mathbf{X}} \tag{9.26}$$

7. The residuals $\hat{\mathbf{r}}$ are uncorrelated with the predicted values $\hat{\mathbf{y}}$.
8. The parameter vector $\hat{\boldsymbol{\beta}}$ is the best linear unbiased estimator (BLUE) for $\boldsymbol{\beta}$. This statement is the *Gauss-Markov theorem* that underlies least squares.
9. While statements 1 through 8 do not depend on distributional assumptions, if $\boldsymbol{\varepsilon} \sim N_N(0,\ \sigma^2)$, meaning that the random errors are N-variate Gaussian with zero mean and common variance σ^2, then $\hat{\boldsymbol{\beta}}$ is also the maximum likelihood estimate and hence is asymptotically Gaussian

$$\frac{\hat{\boldsymbol{\beta}} - \boldsymbol{\beta}}{\hat{\sigma}} \xrightarrow{d} N(0, 1) \tag{9.27}$$

The regression parameters are then distributed according to

$$\hat{\boldsymbol{\beta}} \sim N_p\left(\boldsymbol{\beta},\ \sigma^2 \hat{\overleftrightarrow{\zeta}}^{-1}\right) \tag{9.28}$$

$$\hat{\mathbf{y}} \sim N_N\left(\overleftrightarrow{\mathbf{X}} \cdot \boldsymbol{\beta},\ \sigma^2\, \overleftrightarrow{\mathbf{H}}\right) \tag{9.29}$$

$$(N - p)\hat{\sigma}^2/\sigma^2 \sim \chi^2_{N-p} \tag{9.30}$$

Further, the regression estimator is consistent $(\hat{\boldsymbol{\beta}} \xrightarrow{p} \boldsymbol{\beta})$, and $\hat{\boldsymbol{\beta}}$ and $\hat{\sigma}^2$ are independent, serving as sufficient statistics for estimating $\boldsymbol{\beta}$ and σ^2.

It is very important to assess the compatibility of a given set of data with the conditions implied by statements 1–9. Considerable emphasis will be given to this topic in what follows. If the conditions are violated, it is very possible (one might say, it is guaranteed) that the parameters, and especially estimates of their variances or confidence limits, will be biased.

9.3 Numerical Considerations

The normal equations can be solved in MATLAB using a single line of code

```
beta = (x' * x)\ x' * y
```

where the backslash operator \ or, equivalently, the function **mldivide** solves a set of linear equations, generally using an LDL or Cholesky decomposition for the normal equations. However, problems can occur when $\overleftrightarrow{\mathbf{X}}$ has entries that cover a large range and 32-bit arithmetic is used. Consider the following example

$$\overleftrightarrow{\mathbf{X}} = \begin{bmatrix} 1 & 1 & 1 \\ \varepsilon & 0 & 0 \\ 0 & \varepsilon & 0 \\ 0 & 0 & \varepsilon \end{bmatrix}$$

$$\mathbf{y} = \begin{bmatrix} 1 \\ 0 \\ 0 \\ 0 \end{bmatrix} \tag{9.31}$$

where ε is a small number. Equation (9.31) corresponds to the situation where a quadratic polynomial is being fit to a data set when the sum $\beta_0 + \beta_1 + \beta_2$ can be measured accurately, but the three terms in the polynomial cannot be measured well individually. The terms in the normal equations are

$$\overleftrightarrow{\mathbf{X}}^T \cdot \overleftrightarrow{\mathbf{X}} = \begin{bmatrix} 1+\varepsilon^2 & 1 & 1 \\ 1 & 1+\varepsilon^2 & 1 \\ 1 & 1 & 1+\varepsilon^2 \end{bmatrix}$$

$$\overleftrightarrow{\mathbf{X}}^T \cdot \mathbf{y} = \begin{bmatrix} 1 \\ 1 \\ 1 \end{bmatrix}$$

$$\hat{\boldsymbol{\beta}} = \frac{1}{3+\varepsilon^2} \begin{bmatrix} 1 \\ 1 \\ 1 \end{bmatrix} \tag{9.32}$$

If $\varepsilon = 10^{-4}$, $1 + \varepsilon^2 = 1.00000001$ and $\overleftrightarrow{\mathbf{X}}^T \cdot \overleftrightarrow{\mathbf{X}}$ will be nearly singular. It will not be invertible using single-precision (32-bit) arithmetic. Double-precision (64-bit) arithmetic can solve the problem, but if ε gets smaller, there are still going to be inaccuracies. A test of the invertibility of $\overleftrightarrow{\mathbf{X}}^T \cdot \overleftrightarrow{\mathbf{X}}$ can be made by examining the condition number using the MATLAB function **cond**. If the result is close to 1, then things are fine, but if it is large, there will be inaccuracies.

One solution to this dilemma is to restate the model so that the basis functions (i.e., the polynomial terms in the quadratic expansion used in this example) are orthogonal. In this case, the first three Chebyshev polynomials could be used instead of a simple quadratic. This makes setting up the model a bit more difficult but can help with numerical roundoff. In fact, this was standard practice at the advent of computers when ongoing attention to numerical issues was essential. It is rarely used today.

A better and more general approach is to solve the least squares problem without ever forming the normal equations. The $\overleftrightarrow{\mathbf{X}}$ matrix can always be factored into the product of two matrices

$$\overleftrightarrow{\mathbf{X}} = \overleftrightarrow{\mathbf{Q}} \cdot \overleftrightarrow{\mathbf{R}} \tag{9.33}$$

where $\overleftrightarrow{\mathbf{Q}}$ is $N \times p$, $\overleftrightarrow{\mathbf{Q}}^T \cdot \overleftrightarrow{\mathbf{Q}} = \overleftrightarrow{\mathbf{I}}_p$, and $\overleftrightarrow{\mathbf{I}}_p$ is the $p \times p$ identity matrix. A matrix with the properties of $\overleftrightarrow{\mathbf{Q}}$ is called *orthonormal*. The second matrix $\overleftrightarrow{\mathbf{R}}$ is $p \times p$ and *upper triangular*: all the elements to the left of the main diagonal are zero. Solve the least squares problem using the qr decomposition:

$$\mathbf{y} = \overleftrightarrow{\mathbf{X}} \cdot \boldsymbol{\beta} = \overleftrightarrow{\mathbf{Q}} \cdot \overleftrightarrow{\mathbf{R}} \cdot \boldsymbol{\beta} \tag{9.34}$$

so that

$$\overleftrightarrow{\mathbf{Q}}^T \cdot \mathbf{y} = \overleftrightarrow{\mathbf{R}} \cdot \boldsymbol{\beta} \tag{9.35}$$

The left side of (9.35) is $p \times 1$. The last (pth) row of $\boldsymbol{\beta}$ multiplied by the (p, p) element of $\overleftrightarrow{\mathbf{R}}$ is equal to the last element of the left-hand side. Stepping up one row, there is only one unknown because the pth element of $\boldsymbol{\beta}$ is known. This process of stepping back row by row to get the elements of $\boldsymbol{\beta}$ is called *back substitution*. The least squares problem has been solved without ever computing $\overleftrightarrow{\mathbf{X}}^T \cdot \overleftrightarrow{\mathbf{X}}$, so numerical roundoff issues have been minimized.

In MATLAB, qr decomposition can be used explicitly to solve a least squares problem in two steps.

```
[q, r] = qr(x);
beta = r\q'*y
```

The whole process is numerically stable, unlike the normal equations. Rather than this two-step process, the overdetermined least squares problem (9.12) can be solved using qr decomposition with the single line

```
beta = x\y
```

This approach is highly recommended over invoking the normal equations in practice.

A MATLAB alternative is the function [*b*, *bint*, *r*] = **regress**(*y*, *x*), which takes response *y* and predictor variables *x* and returns the *p* parameters *b*, a $p \times 2$ matrix *bint* of confidence intervals on *b*, and residuals *r*. Two additional returned variables may be added on the left: an $n \times 2$ matrix *rint* that can be used to detect outliers (not recommended because the methods described below are superior) and a block of four statistics in *stats*. The latter contains $\hat{R}^2$, the *F* statistic, its *p*-value, and the error variance.

Caution: The parameters returned in *stats* may not be accurate unless the predictor matrix includes a column of ones.

A second useful matrix decomposition is the *singular value decomposition* (svd). Let $\overleftrightarrow{\mathbf{X}}$ be $N \times p$ with rank $k \le p$. There exists an $N \times N$ orthogonal matrix $\overleftrightarrow{\mathbf{U}}$, a $p \times p$ orthogonal matrix $\overleftrightarrow{\mathbf{V}}$, and an $N \times p$ diagonal matrix $\overleftrightarrow{\mathbf{S}}$ such that

$$\overleftrightarrow{\mathbf{X}} = \overleftrightarrow{\mathbf{U}} \cdot \overleftrightarrow{\mathbf{S}} \cdot \overleftrightarrow{\mathbf{V}}^T \tag{9.36}$$

The entries in $\overleftrightarrow{\mathbf{S}}$ are arranged so that they are nonincreasing; all of them are nonnegative, and exactly k are strictly positive. The svd solution is closely related to the eigenvalue-eigenvector decomposition of $\overleftrightarrow{\mathbf{X}}^T \cdot \overleftrightarrow{\mathbf{X}}$ and $\overleftrightarrow{\mathbf{X}} \cdot \overleftrightarrow{\mathbf{X}}^T$ because

$$\left(\overleftrightarrow{\mathbf{X}}^T \cdot \overleftrightarrow{\mathbf{X}}\right) \cdot \overleftrightarrow{\mathbf{V}} = \overleftrightarrow{\mathbf{V}} \cdot \overleftrightarrow{\mathbf{S}}^2$$
$$\left(\overleftrightarrow{\mathbf{X}} \cdot \overleftrightarrow{\mathbf{X}}^T\right) \cdot \overleftrightarrow{\mathbf{U}} = \overleftrightarrow{\mathbf{U}} \cdot \overleftrightarrow{\mathbf{S}}^2 \tag{9.37}$$

Thus the right singular vectors in the columns of $\overleftrightarrow{\mathbf{V}}$ are the eigenvectors of $\overleftrightarrow{\mathbf{X}}^T \cdot \overleftrightarrow{\mathbf{X}}$, and the left singular vectors in the columns of $\overleftrightarrow{\mathbf{U}}$ are the eigenvectors of $\overleftrightarrow{\mathbf{X}} \cdot \overleftrightarrow{\mathbf{X}}^T$. The squares of the singular values in the diagonal of $\overleftrightarrow{\mathbf{S}}$ are the corresponding eigenvalues. However, the svd (9.36) exists when $\overleftrightarrow{\mathbf{X}}$ is singular, whereas (9.37) cannot be solved in that case.

A solution to the least squares problem $\mathbf{y} = \overleftrightarrow{\mathbf{X}} \cdot \boldsymbol{\beta}$ using the svd is

$$\boldsymbol{\beta} = \overleftrightarrow{\mathbf{V}} \cdot \overleftrightarrow{\mathbf{S}}^{-1} \cdot \overleftrightarrow{\mathbf{U}}^T \cdot \mathbf{y} \tag{9.38}$$

where $\overleftrightarrow{\mathbf{S}}^{-1}$ is just $\mathrm{diag}(1/s_i)$. Additional information is obtained for free with the svd approach. First, the condition number of $\overleftrightarrow{\mathbf{X}}$ is the ratio of the largest to the smallest eigenvalue in $\overleftrightarrow{\mathbf{S}}$. If this is large (i.e., approaching floating-point infinity), then the problem is ill posed, and care is needed in estimating a solution. Second, the rank of $\overleftrightarrow{\mathbf{X}}$ can be determined from the number of nonzero singular values. If the rank of $\overleftrightarrow{\mathbf{X}}$ is smaller than p, then the least squares model is ill posed and hence is deficient.

Suppose that $\mathbf{y} = \overleftrightarrow{\mathbf{X}} \cdot \boldsymbol{\beta}$ is to be solved subject to the equality constraints $\mathbf{G} \cdot \boldsymbol{\beta} = \mathbf{a}$. While there are more elegant ways to solve this constrained least squares problem, a simple approach is to consider the augmented problem $\overleftrightarrow{\mathbf{W}} \cdot \boldsymbol{\beta} = \mathbf{z}$, where $\mathbf{z}$ is just $\mathbf{a}$ stacked over $\mathbf{y}$, and $\overleftrightarrow{\mathbf{W}}$ is $\overleftrightarrow{\mathbf{G}}$ stacked over $\overleftrightarrow{\mathbf{X}}$. Then apply weights of 1 to the rows involving $\overleftrightarrow{\mathbf{G}}$ and weights of τ to the rows involving $\overleftrightarrow{\mathbf{X}}$, where τ is a small number. This will yield $\boldsymbol{\beta}$ subject to the equality constraints.

It is also possible to solve problems like $\mathbf{y} = \overleftrightarrow{\mathbf{X}} \cdot \boldsymbol{\beta}$ subject to the constraint $\boldsymbol{\beta} > 0$. For example, it may be known that the parameters are always positive on physical grounds, yet variability in the data may result in negative values in the solution to the linear regression problem. The MATLAB function **lsqnonneg** implements nonnegative least squares. The MATLAB function **lsqlin** implements a wider range of constrained least squares problems with matrix equality and inequality constraints. Least squares solutions subject to constraints will always have a larger residual sum of squares than for an unconstrained solution.

9.4 Statistical Inference in Linear Regression

Once the solution to the least squares problem (9.12) is obtained, the statistical properties of the fit as described in Section 9.2 and the uncertainties on the parameters must be evaluated. This section outlines a series of statistical tests that should be applied in a regression context.

9.4.1 Analysis of Variance

Analysis of variance or ANOVA was defined in Section 6.3.7, and only the specifics for its application to linear regression will be covered here. Define the total sum of squares (TSS) of the response variable

Table 9.1 Generic Analysis of Variance Table for Regression

Source	*df*	Sum of squares	Mean squares	*F*	*p*-Value
Due to regression (explained)	$p-1$	ESS	$M_E = \text{ESS}/(p-1)$	$\hat{F} = M_E/M_R$	$1 - \text{Eff}(\hat{F}, p-1, N-p)$
Due to error (residual)	$N-p$	RSS	$M_R = \text{RSS}/(N-p)$		
Total	$N-1$	TSS			

$$\text{TSS} = \left(\mathbf{y} - \overline{\text{Y}}_N\right)^T \cdot \left(\mathbf{y} - \overline{\text{Y}}_N\right) \tag{9.39}$$

The TSS may be partitioned into two components:

$$\text{TSS} = (\mathbf{y} - \hat{\mathbf{y}})^T \cdot (\mathbf{y} - \hat{\mathbf{y}}) + \left[\hat{\mathbf{y}}^T \cdot \hat{\mathbf{y}} - N\left(\overline{\text{Y}}_N\right)^2\right] \tag{9.40}$$

The first term is the residual or between sum of squares (RSS $=\hat{\mathbf{r}}^T \cdot \hat{\mathbf{r}}$), and the second term is the explained or within sum of squares (ESS).

A test of H_0: $\boldsymbol{\beta} = 0$ against H_1: $\boldsymbol{\beta} \neq 0$ (i.e., not all the elements of $\boldsymbol{\beta}$ are zero) is given by the statistic

$$\hat{F} = \frac{\text{ESS}/(p-1)}{\text{RSS}/(N-p)} \tag{9.41}$$

which is distributed as $F_{p-1,N-p}$ if the null hypothesis is true. Note that this is a fairly blunt test because one is usually interested in seeing if a particular regression parameter is required, not whether all of them are required at once. It is effectively a test of the applicability of regression. The ANOVA results are usually presented in the form of an ANOVA table, as shown generically in Table 9.1.

It is also common practice to present the RSS divided by the TSS

$$\hat{Q}^2 = \frac{\hat{\mathbf{r}}^T \cdot \hat{\mathbf{r}}}{\left(\mathbf{y} - \overline{\mathbf{Y}}_N\right)^T \cdot \left(\mathbf{y} - \overline{\mathbf{Y}}_N\right)} \tag{9.42}$$

where $\hat{Q}^2$ is the proportion of the variation in the data $\mathbf{y}$ that is not explained by the regression. Conversely, $\hat{R}^2 = 1 - \hat{Q}^2$ (i.e., ESS divided by TSS) is the proportion of the variation that is explained by regression. Both $\hat{Q}^2$ and $\hat{R}^2$ lie between 0 and 1. However, either of these parameters is at best a distraction and at worst a fraud, and should generally be avoided in assessing a linear regression.

The following MATLAB function computes the ANOVA table for a regression problem:

```
function [a] = Anovar(y, x, w)
%y - response variable
%x - predictor matrix
%w -- vector of weights lying between 0 and 1
```

```
%one - set to zero or one as x does not/does contain a
column of ones
p = size(x, 2);
one = 0;
for i = p:-1:1
   ii = find(x(:, i) ~= 1);
   if isempty(ii)
        one = 1;
        break
   end
end
i = find(w ~= 0);
b = (repmat(w(i), 1, p).*x(i, :))\(w(i).*y(i));
n = length(i);
wy = w(i).*y(i);
wyhat = w(i).*yhat;
tss = (wy - mean(wy))'*(wy - mean(wy));
rss = (wy - wyhat)'*(wy - wyhat);
ess = wyhat'*wyhat - n*mean(wy)^2;
me = ess/(p - one);
mr = rss/(n - p);
f = me/mr;
fpval = 1 - fcdf(f, p - one, n - p);
a = [p - one ess me f fpval;
n - p rss mr 0 0;
n - one tss 0 0 0];
```

9.4.2 Hypothesis Testing on the Regression Estimates

Let β_j^* be a particular value for the jth entry in the parameter vector. It is useful to test the hypotheses

$$\begin{aligned} H_0 &: \hat{\beta}_j = \beta_j^* \\ H_1 &: \hat{\beta}_j \neq \beta_j^* \end{aligned} \tag{9.43}$$

Because $\hat{\sigma}^2/\sigma^2 \cdot (N \quad p)\chi^2_{N-p}$ independent of $\hat{\boldsymbol{\beta}}$ when Gaussianity holds, the statistic

$$\hat{t}_i = \frac{\sqrt{\hat{\zeta}_{jj}}\left(\hat{\beta}_j - \beta_j^*\right)}{\hat{\sigma}} \tag{9.44}$$

is distributed according to t_{N-p}, where $\hat{\zeta}_{jj}$ is a diagonal element of (9.26). The null hypothesis is rejected if the p-value for the test statistic is below the Type 1 error level α.

Because the case $\beta_j^* = 0$ is usually of interest as a test of whether a given regression coefficient is required or not, it is common practice in regression to compute $\hat{\beta}_j$, its standard error from (9.19) with the sample statistics (9.25) and (9.26), the test statistic $\hat{t}_j$ from (9.44),

Table 9.2 Generic t Table for Regression

Predictor	Coefficient	Standard error	t-Statistic	p-Value
X_1				
X_2				
$\hat{\sigma} =$ $\hat{R}^2 =$				

and the relevant p-value from Student's t distribution and present these together in the form of a table, as shown in Table 9.2. There are as many rows in the table as there are predictors. However, it is important to note that each row of the table is not simply a test that $\hat{\beta}_j = 0$ but rather that $\hat{\beta}_j = 0$ holds in a linear model that includes all the remaining predictors and nothing else. It is not an absolute test of $\hat{\beta}_j = 0$.

More generally, if the null hypothesis that a linear combination of the elements of $\hat{\boldsymbol{\beta}}$ have a given value against the alternate that they do not is under test, the test statistic

$$\hat{t} = \frac{\mathbf{c}^T \cdot \left(\hat{\boldsymbol{\beta}} - \boldsymbol{\beta}^*\right)}{\sqrt{\hat{\sigma}^2 \mathbf{c}^T \cdot \left(\mathbf{X}^T \cdot \mathbf{X}\right)^{-1} \cdot \mathbf{c}}} \tag{9.45}$$

is formed, where $\mathbf{c}$ is a matrix of constants. This is assessed against t_{N-p}.

The following MATLAB script computes the t table for a linear regression:

```
function [ b, bse, t, pval] =Treg(y, x, w)
p = size(x, 2);
i = find(w ~= 0);
b = (repmat(w(i), 1, p).*x(i, :))\(w(i).*y(i));
yhat = x(i, :)*b;
n = length(i);
zeta = inv((repmat(w(i), 1, p)'.*x(i, :)')*(repmat(w(i), 1,
p).*x(i, :)));
sigma2 = (w(i)'.*(y(i) - yhat)')*(w(i).*(y(i) - yhat))/(n - p);
bse = sqrt(sigma2*diag(zeta));
t = b./bse;
pval = 2*(1 - tcdf(abs(t), n - p));
```

9.4.3 Confidence Intervals

The $1 - \alpha$ confidence interval on $\hat{\beta}_j$ is just the set of all values $\left\{\beta_j^*\right\}$ for which H_0: $\hat{\beta}_j = \beta_j^*$ is acceptable at a level of significance α. The set of all $\left\{\hat{\beta}_j\right\}$ such that $\left|\hat{t}_j\right| < c$ forms a confidence interval on β_j at the $1 - \alpha$ confidence level. This is equivalent to the statement

$$\hat{\beta}_j - t_{N-p}(1-\alpha/2)\sqrt{\hat{\mathbf{r}}^{\mathrm{T}} \cdot \hat{\mathbf{r}} / \left[(N-p)\hat{\zeta}_{jj}\right]} \leq \beta_j \leq \hat{\beta}_j + t_{N-p}(1-\alpha/2)\sqrt{\hat{\mathbf{r}}^{\mathrm{T}} \cdot \hat{\mathbf{r}} / \left[(N-p)\hat{\zeta}_{jj}\right]} \quad (9.46)$$

With more than one regression coefficient, use of the Bonferroni approach, where α is replaced by α/p, is recommended to avoid underestimation of the confidence intervals.

Jackknifing regression solutions introduces a new twist because the problem is unbalanced: there is one variable on the left-hand side, but there are p variables on the right-hand side. This has led to the suggestion (based on simulations) to use a modified pseudovalue to that defined in Section 8.4. The pseudovalue is now a vector rather than a scalar. The modified version takes the form

$$\hat{\mathrm{P}}_i = \left[N\left(1 - \hat{h}_{ii}\right) + 1\right]\hat{\boldsymbol{\beta}} - N\left(1 - \hat{h}_{ii}\right)\hat{\boldsymbol{\beta}}_{\setminus i} \quad (9.47)$$

where $\hat{\boldsymbol{\beta}}_{\setminus i}$ is the solution to the regression when the ith row is deleted from $\mathbf{y}$ and $\overleftrightarrow{\mathbf{X}}$, and $\hat{h}_{ii}$ is the corresponding diagonal element of the hat matrix (9.20). The jackknifed covariance matrix is given by

$$\overleftrightarrow{\boldsymbol{\Sigma}}_J = \frac{N}{N-p}\sum_{i=1}^{N}\left(1 - \hat{h}_{ii}\right)^2 (\hat{\boldsymbol{\beta}} - \hat{\boldsymbol{\beta}}_{\setminus i}) \cdot (\hat{\boldsymbol{\beta}} - \hat{\boldsymbol{\beta}}_{\setminus i})^T \quad (9.48)$$

The standard error on the elements of $\hat{\boldsymbol{\beta}}$ is given by the square root of the diagonal elements of this quantity and should be used in the usual t confidence bound (9.46) in place of $\sqrt{\hat{\mathbf{r}}^{\mathrm{T}} \cdot \hat{\mathbf{r}} / \left[(N-p)\hat{\zeta}_{jj}\right]}$.

The bootstrap may be applied to regression problems to estimate the regression parameters and their confidence bounds and to carry out hypothesis tests on the regression estimates. Obtain B bootstrap samples from the original data by resampling with replacement from both the response variable $\mathbf{y}$ and the predictor variables $\overleftrightarrow{\mathbf{X}}$. Apply the least squares estimator to each bootstrap sample to get a set of bootstrap replicates for the regression paramcters $\left\{\widehat{\boldsymbol{\beta}}_k^*\right\}$ in the form of a $B \times p$ matrix. The bootstrap estimate of the regression coefficients is a p-vector given by

$$\overline{\boldsymbol{\beta}}^* = \frac{1}{B}\sum_{i=1}^{B}\widehat{\boldsymbol{\beta}}_k^* \quad (9.49)$$

The replicates also define the distribution of $\hat{\boldsymbol{\beta}}$ or $\overline{\boldsymbol{\beta}}^*$, and estimates for statistics may be derived from them. For example, the unbiased covariance matrix of $\hat{\boldsymbol{\beta}}$ or $\overline{\boldsymbol{\beta}}^*$ follows from

$$\overleftrightarrow{\boldsymbol{\Sigma}}_B(\hat{\boldsymbol{\beta}}) = \frac{1}{B-p}\sum_{k=1}^{B}\left(\widehat{\boldsymbol{\beta}}_k^* - \overline{\boldsymbol{\beta}}^*\right) \cdot \left(\widehat{\boldsymbol{\beta}}_k^* - \overline{\boldsymbol{\beta}}^*\right)^T \quad (9.50)$$

Confidence intervals may be obtained using the bootstrap percentile method by sorting the columns of $\widehat{\boldsymbol{\beta}}_k^*$ and taking the $\lfloor B\alpha/(2p)\rfloor$ and $\lfloor B(1-\alpha)/(2p)\rfloor$ entries as the confidence interval for each. The BCa method is applied in an analogous manner.

The bootstrap can also be used to assess the significance of the regression coefficient estimates using a double bootstrap procedure. The inner bootstrap produces an estimate for the regression coefficients by sampling with replacement from the data. The outer bootstrap computes a one-sample t statistic. If the sample estimate for the t statistic is subtracted from the bootstrap t statistic and then compared with the sample t statistic, the bootstrap p-value can be computed as in Section 8.2.4 and used to assess the corresponding regression coefficient estimate. This will be illustrated in Section 9.5.2.

9.4.4 The Runs and Durbin-Watson Tests

It is important to assess the regression residuals for randomness (hence lack of correlation) to evaluate consistency with assumption 3 in Section 9.2. The nonparametric runs test (sometimes called the *Wald-Wolfowitz runs test*) is a simple way to achieve this. Kvam & Vidakovic (2007, sec. 6.5) summarize the approach.

A sequence is defined to be 0 or 1 as the regression residuals corresponding to a given datum are below or above the mean. The sequence should be well mixed between the two values if it is random, and clustering of values suggests nonrandomness. As a measure of clustering, define the test statistic $\hat{R}$ to be the number of homogeneous runs in a sequence of 0s and 1s, including the first one. For example, the 15-term sequence 001110001011001 contains $\hat{R} = 8$ runs. If $\hat{R}$ is too large, then the sequence is showing a degree of anticorrelation, whereas if $\hat{R}$ is too small, then groupings or trends exist in the sequence. The null hypothesis holds that $\hat{R}$ is consistent with randomness, whereas the alternate hypothesis pertains if $\hat{R}$ is either too large or too small.

The distribution for R can be obtained using combinatorial arguments. Let the number of 0s and 1s be n and m, respectively, where $n + m = N$ is the total number of data. The null distribution for R is given by

$$
\begin{aligned}
f_R(r) &= \frac{2\binom{n-1}{r/2-1}\binom{m-1}{r/2-1}}{\binom{N}{n}} \qquad r \text{ even} \\
f_R(r) &= \frac{\binom{n-1}{(r-1)/2}\binom{m-1}{(r-3)/2} + \binom{n-1}{(r-3)/2}\binom{m-1}{(r-1)/2}}{\binom{N}{n}} \qquad r \text{ odd}
\end{aligned}
\tag{9.51}
$$

for $r = 2, \ldots, N$. The mean and variance of R are given by

$$\mu_R = 1 + \frac{2nm}{N} \tag{9.52}$$

$$\sigma_R^2 = \frac{2nm(2nm - N)}{N^2(N-1)} \tag{9.53}$$

A normal approximation exists for n, $m > 15$. The null hypothesis is accepted if $\mu_R - z_{\alpha/2}\sigma_R - 0.5 < \hat{R} < \mu_R + z_{\alpha/2}\sigma_R + 0.5$. Alternatively, exact p-values can be computed from the distribution.

MATLAB implements the runs test as $[h, p] =$ **runstest**(x), where h is a 0–1 decision rule that indicates acceptance or rejection of the null hypothesis, and p is the p-value. An additional structure can be included on the left side that contains N, m, n, and r. The default approach is assessing r against the mean of x, but a second argument on the right [such as **median**(x)] will change this. Additional name-value pairs may be used to change the method for computing p-values and the type of alternate hypothesis (the default is a two-tailed test).

The Durbin-Watson test evaluates the presence of autocorrelation in the residuals from a regression. It was introduced by Durbin & Watson (1950, 1951, 1971). The presence of positive autocorrelation in the residuals will result in upward bias in the ANOVA F test and downward bias in the parameter standard errors, hence increasing the false rejection rate on the parameters themselves.

The Durbin-Watson test statistic is given by

$$\hat{d} = \frac{\sum_{i=2}^{N} (\hat{r}_i - \hat{r}_{i-1})^2}{\sum_{i=1}^{N} \widehat{r}_i^2} \tag{9.54}$$

Since $\hat{d} \approx 2(1 - \hat{\rho})$, where $\hat{\rho}$ is the Pearson correlation coefficient, a value $\hat{d} = 2$ indicates no autocorrelation, and $0 \leq \hat{d} \leq 4$. A small value for $\hat{d}$ indicates positive autocorrelation, and vice versa. The distribution for $\hat{d}$ is complicated, but if the predictor matrix $\overleftrightarrow{\mathbf{X}}$ is available, then the null distribution can be exactly computed.

MATLAB implements the Durbin-Watson test as $p =$ **dwtest**(r, x), where p is the two-tail p-value by default, r is the regression residuals, and x is the predictor matrix. A second argument on the left returns the test statistic. Additional name-value pairs on the right control the method for computing the p-value and the type of alternate hypothesis.

Caution: dwtest will not return reliable results unless the predictor matrix contains a column of ones.

9.5 Linear Regression in Practice

9.5.1 Assessing the Results

Influential data is a catch-all term for data that are in some sense unusual or that exert an excessive amount of control on the outcome of a linear regression. Influential data are usually classified as outliers and leverage points when they are manifest in the residuals and hat matrix diagonal, respectively.

A flowchart is presented in Figure 9.1 that outlines the steps toward fitting and assessing a linear regression model. The steps will be explored in turn using data. In practice, most scientists only perform the first, second, and last steps and ignore model assessment and

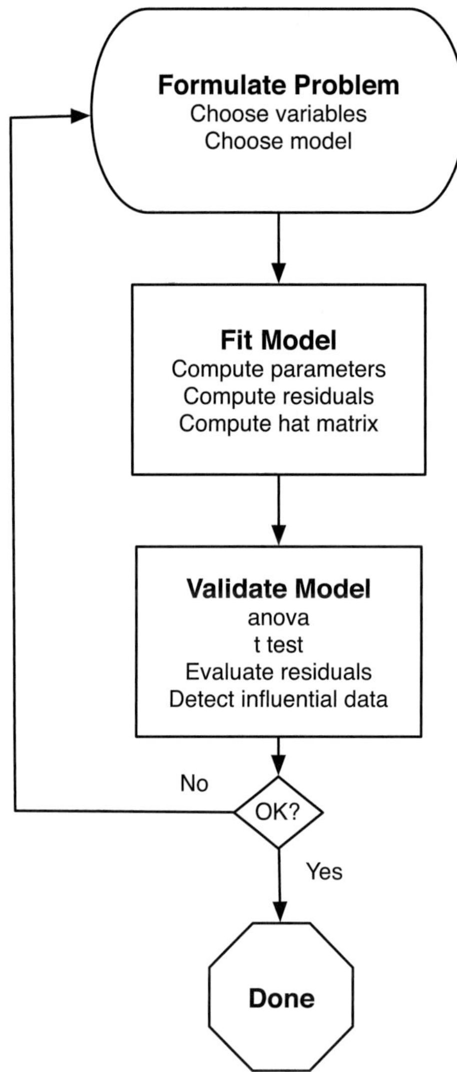

Figure 9.1 Flowchart for analyzing data using linear regression.

goodness-of-fit tests. This can lead to erroneous conclusions despite a seemingly self-consistent least squares result.

There is no totally satisfactory approach to model assessment. In particular, there are no reliable statistical tests that can be applied to the parameters or other derived quantities (i.e., the residuals or the hat matrix diagonal) to test for outliers or influential data. It is possible to take two approaches: (1) plotting of various derived quantities to detect bad data and (2) robust estimators that automatically reduce the influence of unusual data, as described in Section 9.6. The first of these is largely useful for small data sets, whereas the latter are essential for large data sets.

The detection of problems in regression can be carried out in many ways:

1. A test for the null hypothesis that the residuals are random versus the alternate that they are not, as described in Section 9.4.4. In addition, plotting $\hat{\mathbf{r}}$ against the time at which the data were collected (if this information is available) can be used to detect clustering or temporal effects.
2. A test for the null hypothesis that the residuals are uncorrelated against the alternate that they are not is provided by the Durbin-Watson test, as described in Section 9.4.4.
3. Plot the residuals $\hat{\mathbf{r}}$ against each of the predictor variables in $\overleftrightarrow{\mathbf{X}}$. These should be scatter diagrams. If there is discernible correlation, then the parameters in $\hat{\boldsymbol{\beta}}$ and especially their variances may be biased. However, this is inherently a qualitative process and quickly becomes cumbersome as the number of predictors increases.
4. Plot the residuals $\hat{\mathbf{r}}$ against the predicted values $\hat{\mathbf{y}}$. These two quantities should be independent and uncorrelated. Typical breakdowns include curvature that suggests the presence of nonlinearity (and hence the possibility that the model is not correct) and a change in the residual variance with fitted value, which suggests heteroskedasticity (i.e., residuals that do not share a common variance, indicating outliers or a mixture of populations).
5. A statistical rule of thumb holds that any datum for which $\hat{h}_{ii} > 2p/N$ is suspect as a leverage point, and hence ought to be investigated thoroughly and possibly removed. One way to do this is by plotting the hat matrix diagonal normalized by its expected value $\left(N\hat{h}_{ii}/p\right)$ against observation number. If any of the values exceeds 2, then investigate it.
6. Examine the distribution of the residuals using a quantile-quantile plot. A plot of normal quantiles against ranked and ordered residuals will be a straight line if the residuals are Gaussian. Systematic departures from a straight line suggest that the residuals may not be Gaussian, in which case parametric statistical inference about the least squares parameters may be biased. The presence of a few extreme residuals will be apparent as sharp departures from linearity at the upper and lower extremes. These are outliers, and will bias the regression procedures. The consistency of the residuals with the normal distribution can be quantified using the likelihood ratio or chi square tests of goodness-of-fit, or the Kolmogorov-Smirnov or Anderson-Darling tests.
7. Examine the distribution of the hat matrix diagonal elements against the beta distribution using a quantile-quantile plot. The presence of extreme values at the upper end is suggestive of leverage points that will have undue influence. Note that there is not a requirement in linear regression for the predictors to be Gaussian, so systematic departures from a beta distribution are not necessarily a matter for concern.
8. Plot the residuals against the hat matrix diagonal elements (leverage-regression plot). This allows one to determine whether outliers and leverage points occur more or less independently or are correlated. The latter is a matter for special concern.

Overall, this process is of necessity somewhat subjective. In addition, it is possible for a single outlier or leverage point to mask the presence of other ones. As a result, an iterative procedure must be applied. An initial least squares solution, along with the residuals and hat matrix diagonal, is obtained. The model is assessed using a subset of the eight tests and plots just listed. If problems are detected, then either the approach needs to be reevaluated or influential data are present. Influential data need to be removed (this process is called *censoring* in statistics), and the entire process needs to be repeated until no further problems are detected. It is not unusual to have to reject 10% of a data set in the earth sciences (and there are examples where this approaches 50%).

Note also that one cannot eliminate some elements in a data set by censoring and continue to use the original distribution [either Gaussian or beta$(p, N-p)$] for reference. If this is done, the result will appear unduly short tailed, and there is risk of missing influential data. The process of censoring means that the target distribution becomes a truncated form of the original one, as discussed in Section 4.8.4.

9.5.2 Examples

Example 9.1 In 1929, Edwin Hubble investigated the relationship between the distance of a galaxy from Earth and the velocity with which it is receding. Galaxies appear to be moving away from Earth isotropically (i.e., independent of the direction in which one looks). This is thought to be the result of the "Big Bang" origin of the universe. The data Hubble collected are distances (megaparsecs, where 1 parsec = 3.26 light-years) to 24 galaxies and their recession velocities (km/s). Hubble's law is

$$\text{Recession velocity} = H \times \text{distance}$$

where H is Hubble's constant, which is thought to be about 75 km/s/Mp. By tracking backward in time, the galaxies appear to meet in the same place. Consequently, $1/H$ can be used to estimate the time since the Big Bang and hence is a measure of the age of the universe. A regression analysis on these data will be performed, but there is no constant term in Hubble's law, and hence the requirement for no constant term must be verified first. These data are taken from Hubble (1929).

First, a regression analysis will be performed on the data, including a constant term, and the results will be evaluated using the ANOVA, t, runs, and Durbin-Watson tests.

```
hubble = importdata('hubble.dat');
y = hubble(:, 2);
x = [ hubble(:, 1) ones(size(y))] ;
w = ones(size(y));
[a] = Anovar(y, x, w);
[b, bse, t, tprob] = Treg(y, x, w);
r = y - x*b;
r2 = a(1, 2)/a(3, 2)
r2 =
   0.6235
[h, p] = runstest(r)
h =
   0
```

```
p =
  1
p = dwtest(r, x)
p =
  0.8353
```

The runs and Durbin-Watson tests indicate that the regression residuals are random and uncorrelated. The remaining results are used to fill out the ANOVA and t tables in Tables 9.3 and 9.4.

The correlation coefficient is $\hat{R}^2 = 0.6235$. The null hypothesis that no regression is required is rejected, although data scatter reduces $\hat{R}^2$.

The slope predictor is strongly required, but the p-value for the intercept accepts the null hypothesis that it is zero, and hence the intercept may be neglected for further analysis.

```
x = hubble(:, 1);
[a] = Anovar(y, x, w);
[b, bse, t, tprob] = Treg(y, x, w);
r = y - x*b;
r2 = a(1, 2)/a(3, 2)
r2 =
   0.6194
[h, p] = runstest(r)
h =
   0
p =
   0.5264
p = dwtest(r, x)
p =
   0.9473
```

Table 9.3 ANOVA Table for Hubble Data with an Intercept

Source	*df*	Sum of squares	Mean squares	*F*	*p*-Value
Due to regression	1	1.9766e+06	1.9766e+06	36.4377	4.4775e−06
Due to error	22	1.1934e+06	0.3188		
Total	23	3.1701e+06			

Table 9.4 Regression t Table for Hubble Data with an Intercept

Predictor	Coefficient	Standard error	t-Statistic	p-Value
X_1	454.1584	75.2371	6.0364	4.4775e−06
X_2	−40.7836	83.4389	−0.4888	0.6298

Table 9.5 ANOVA Table for Hubble Data without an Intercept

Source	*df*	Sum of squares	Mean squares	*F*	*p*-Value
Due to regression	1	1.9637e+06	1.9637e+06	37.4376	3.0554e−06
Due to error	23	1.2064e+06	5.2452e+04		
Total	24	3.1701e+06			

Table 9.6 Regression *t* Table for Hubble Data without an Intercept

Predictor	Coefficient	Standard error	*t*-Statistic	*p*-Value
X_1	423.9373	42.1541	10.0568	6.8686e−10

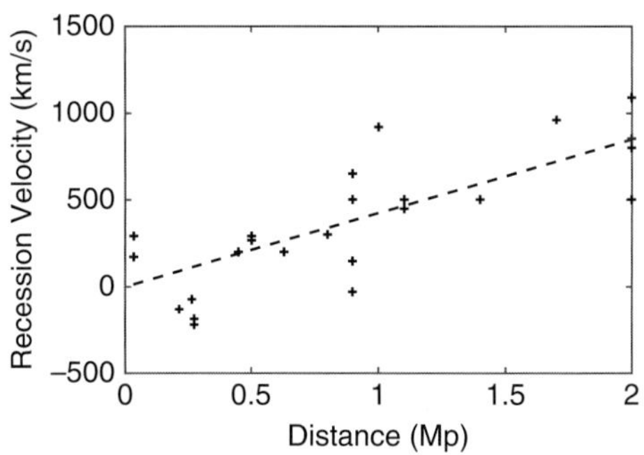

Figure 9.2 The Hubble data (plus signs) with the best-fit linear regression line (dashed line).

The runs and Durbin-Watson tests continue to show that the residuals are random and uncorrelated. The ANOVA and *t* tables are shown in Tables 9.5 and 9.6.

Regression continues to be required. The correlation coefficient is reduced slightly to 0.6194.

The null hypothesis that the slope coefficient is zero is strongly rejected. Figure 9.2 shows the data and the line fitted to them.

```
plot(hubble(:, 1), hubble(:, 2), '+', sort(hubble(:, 1 )),
b*sort(hubble(:, 1)))
```

Quantile-quantile plots for the regression residuals and the hat matrix diagonal can be produced to detect influential data.

```
i = 1:length(hubble);
p = (i - .5)/length(hubble);
plot(norminv(p), sort(r), '+')
```

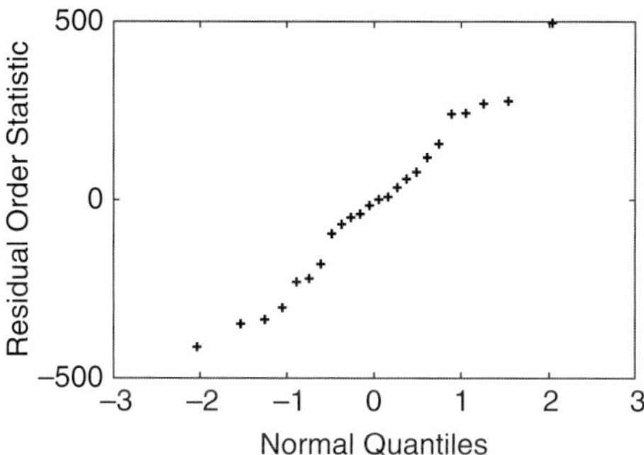

Figure 9.3 Quantile-quantile plot of the Hubble residuals against Gaussian quantiles.

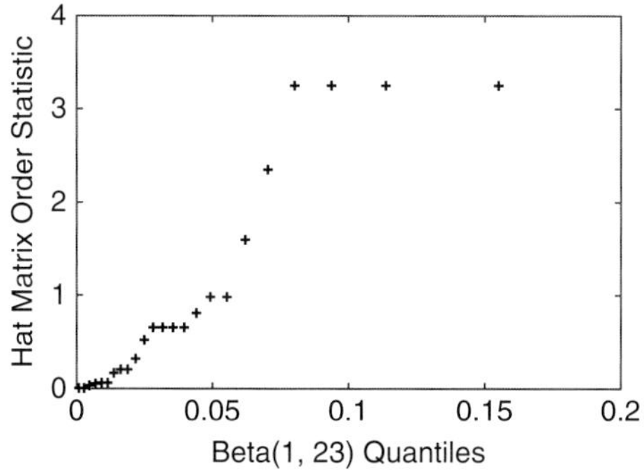

Figure 9.4 Quantile-quantile plot of the hat matrix diagonal normalized to its expected value against beta(1,23) quantiles.

The residual q-q plot in Figure 9.3 shows no outliers.

```
h = x*inv(x'*x)*x';
hat = diag(h)*length(hubble);
plot(betainv(p, 1, length(hubble) - 1), sort(hat), '+')
```

The hat matrix diagonal q-q plot in Figure 9.4 does not display major leverage points, although the flattening at the top of the distribution is strange. It suggests that beyond some value it is not possible to measure the distance, and hence a limiting value was used. Overall, the fit is consistent and acceptable.

Example 9.2 Measurements have been made of the logarithms of the effective surface temperature and light intensity of stars from the cluster Cygnus OB1. A linear relationship will be evaluated for these data, and the fit will be assessed to see whether it is significant. These data were presented in Rousseeuw & Leroy (1987, p. 27).

A least squares line is first fit to the data after adding a column of ones to the predictor matrix to allow for a nonzero intercept, and then the result is plotted in Figure 9.5. The MATLAB script is

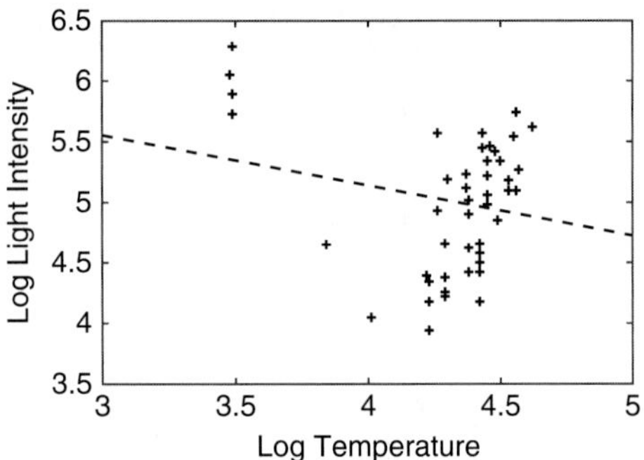

Figure 9.5 The star data (+s) and the best-fit line using all the data (dashed line).

```
star = importdata('star.dat');
y = star(:, 2);
x = [star(:, 1)  ones(size(y))];
b = x\y
b =
  -0.4133
   6.7935
x1 = 3.0:.1:5.0;
y1 = b(1)*x1 + b(2);
plot(x(:, 1), y, '+', x1, y1, '-')
r = y - x*b;
[h, p] = runstest(r)
h =
   0
p =
   0.0982
p = dwtest(r, x)
p =
   0.1929
```

While the runs and Durbin-Watson tests show that the residuals are random and uncorrelated, the result is counterintuitive because the line that fits the majority of the data should have a positive slope and pass through the cloud of data at the right side of the plot. Evidently, the four points at the upper left are "pulling" the fit. ANOVA and t-statistic analyses of the regression parameters are shown in Tables 9.7 and 9.8.

```
w = ones(size(y));
a = Anovar(y, x, w);
[b bse t tprob] = Treg(y, x, w);
r2 = a(1,2)/a(3,2)
```

Table 9.7 ANOVA Table for the Star Data

Source	*df*	Sum of squares	Mean squares	*F*	*p*-Value
Due to regression	1	0.6646	0.6646	2.0846	0.1557
Due to error	45	14.3464	0.3188		
Total	46	15.0110			

Table 9.8 Regression *t* Table for the Star Data

Predictor	Coefficient	Standard error	*t*-Statistic	*p*-Value
X_1	−0.4133	0.2863	−1.4438	0.1557
X_2	6.7935	1.2365	5.4940	0.0000

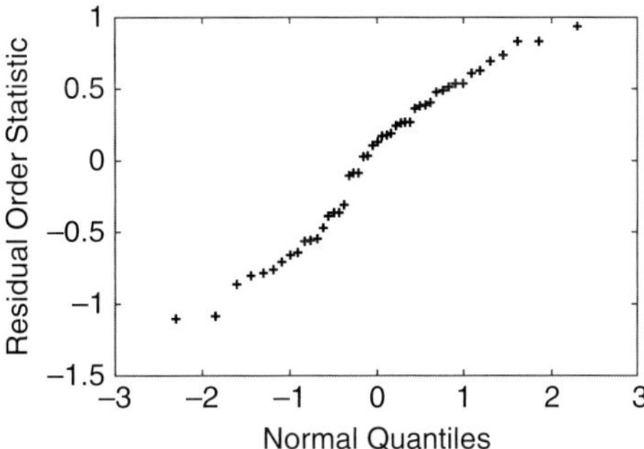

Figure 9.6 Quantile-quantile plot for the star residuals against Gaussian quantiles.

The correlation coefficient is $\hat{R}^2 = 0.0443$. The regression is fitting nothing very well, and the ANOVA *p*-value rejects the need for regression. Table 9.8 shows the *t* table.

The slope coefficient is not distinguishable from zero, but the intercept is. Overall, the result is a very complicated way to estimate the sample mean.

Quantile-quantile plots of the regression residuals and hat matrix diagonal elements against the appropriate normal and beta distributions are examined next.

```
i = 1:length(y);
p = (i - .5)/length(y);
[r1, i1] = sort(r);
plot(norminv(p), r1, '+')
```

The residual q-q plot (Figure 9.6) is approximately straight and free of outliers, although there is some weak systematic curvature.

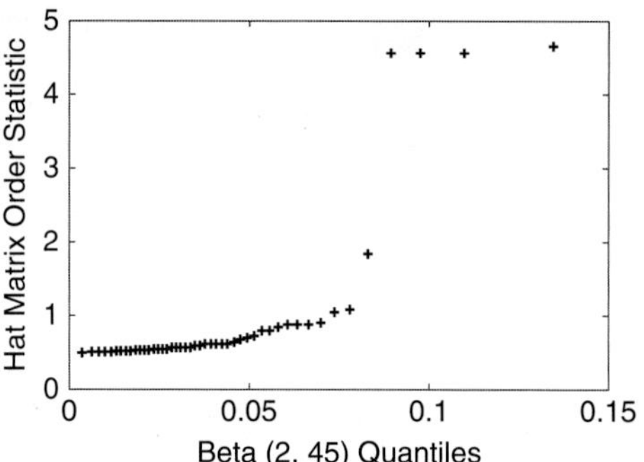

Figure 9.7 Quantile-quantile plot for the hat matrix diagonal normalized by its expected value against beta(2,45) quantiles for the star data.

```
h = x* inv(x'* x)* x';
hii = diag(h);
[h1, i1] = sort(hii);
plot(betainv(p, 2, length(y) - 2), length(y)* h1/2, '+')
```

The hat matrix q-q plot (Figure 9.7) shows the presence of four high-leverage points and possibly a fifth. By examining the indices in the variable $i1$, the four extreme values can be identified as the four stars in the upper left corner of Figure 9.5. These stars are red giants and hence differ from the majority of the stars in the cluster. Their presence is exerting strong leverage on the regression line.

The four anomalous stars are eliminated, and the regression procedure is redone.

```
w(34) = 0;
w(30) = 0;
w(20) = 0;
w(11) = 0;
a = Anovar(y, x, w);
[b bse t tprob] = Treg(y, x, w);
r2 = a(1,2)/a(3,2);
i = find(w ~= 0);
r = y(i) - x(i, :)* b;
[h, p] = runstest(r)
h =
   0
p =
   0.0668
x1 = x(i, :);
p = dwtest(r, x1)
p =
   2.3707e-05
```

Table 9.9 ANOVA Table for the Star Data Less Four Leverage Points

Source	*df*	Sum of squares	Mean squares	*F*	*p*-Value
Due to regression	1	3.9072	3.9072	23.7264	1.6971e−05
Due to error	41	6.7518	0.1647		
Total	42	10.6590			

Table 9.10 Regression *t* Table for the Star Data Less Four Leverage Points

Predictor	Coefficient	Standard error	*t*-Statistic	*p*-Value
X_1	2.0467	0.4202	4.8710	0.0000
X_2	−4.0565	1.8441	−2.1997	0.0335

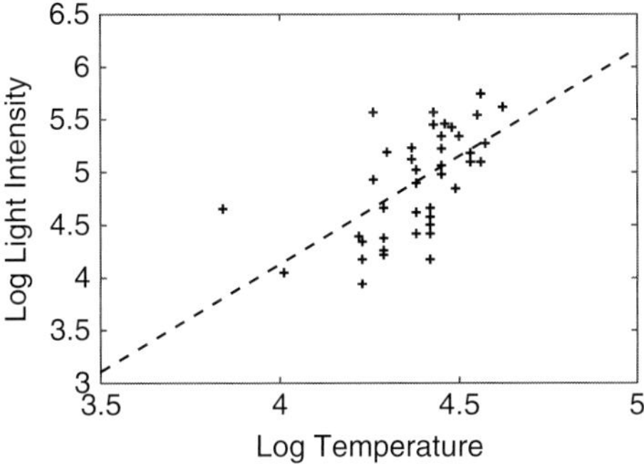

Figure 9.8 The star data (+) and the best-fit linear regression line (dashed line) for the star data after eliminating the four largest leverage points.

The runs test accepts the null hypothesis that the residuals are random, but the Durbin-Watson test rejects the null hypothesis that they are uncorrelated. Investigation of this issue will be deferred.

The correlation coefficient is $\hat{R}^2 = 0.3666$. The null hypothesis that all the regression coefficients are zero is rejected. Table 9.10 shows the regression *t* table. Both coefficients are significant.

```
x1 = 3.0:.1:5.0;
y1 - b(1)*x1 + b(2);
plot(x(i, 1), y(i), '+', x1, y1, '-')
```

A plot of the result (Figure 9.8) shows a reasonable fit to the data, but the slope has changed to a positive one.

Quantile-quantile plots of the regression residuals and hat matrix diagonal may be examined using the appropriate truncated forms of the distributions.

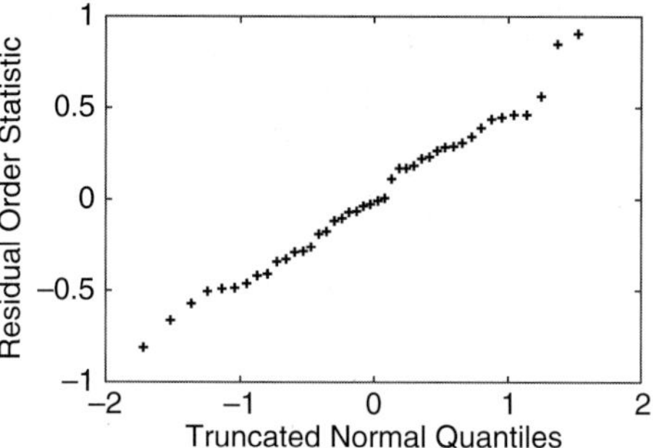

Figure 9.9 Quantile-quantile plot for the residuals from fitting the star data after removing four leverage points against truncated Gaussian quantiles.

```
n = length(y);
m = length(i);
ii = 1:m;
%assume 2 values truncated from bottom and 2 from top of
normal distribution
mr1 = 2;
mr2 = 2;
p = (normcdf(norminv((n - mr2 - .5)/n)) - normcdf(norminv
((mr1 - .5)/n)))*(ii - .5)/m + normcdf(norminv((mr1 - .5)/
n));
[r1, i1] = sort(r);
plot(norminv(p), r1, '+')
h = x(i, :)*inv(x(i, :)'*x(i, :))*x(i, :)';
hii = m/2*diag(h);
[h1, i1]=sort(hii);
%truncation of beta distribution is only at top
mh2 = 4;
p = betacdf(betainv((n - mh2 - .5)/n, 2, n - 2), 2, n - 2)*
(ii - .5)/m;
plot(betainv(p, 2, length(y) - 2), h1, '+')
```

The residual q-q plot (Figure 9.9) is much straighter and free of outliers, but leverage issues still exist (Figure 9.10). One more point (number 7) should be eliminated. The runs and Durbin-Watson p-values are 0.2757 and 3.81e−04, respectively, so once again the residuals are random but correlated. Repeating the procedure increases $\hat{F}$ to 34.8208 and $\hat{R}^2$ to 0.4654 and yields the t table in Table 9.11.

The intercept coefficient has changed by a large amount as a result of eliminating one more point. The residual q-q plot (Figure 9.11) shows a weak upper outlier as a result of the censoring, but it is not significant given the shape of the uncertainty band on a

Table 9.11 Regression t Table for the Star Data Less Five Leverage Points

Predictor	Coefficient	Standard error	t-Statistic	p-Value
X_1	2.8028	0.4750	5.9009	0.0000
X_2	−7.4035	2.0905	−3.5415	0.0010

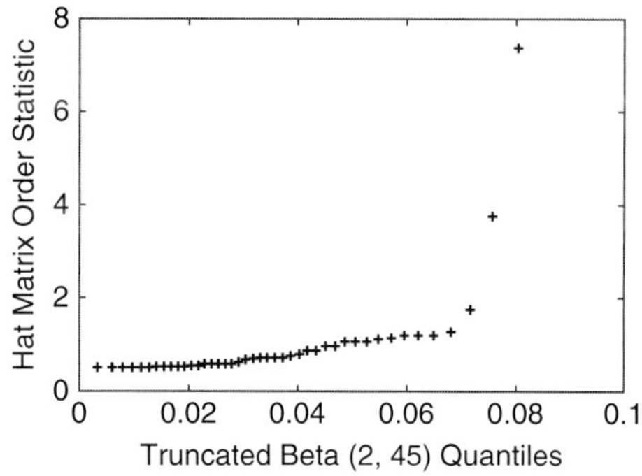

Figure 9.10 Quantile-quantile plot for the hat matrix diagonal normalized by its expected value against truncated beta(2,45) quantiles for the star data after removing four leverage points.

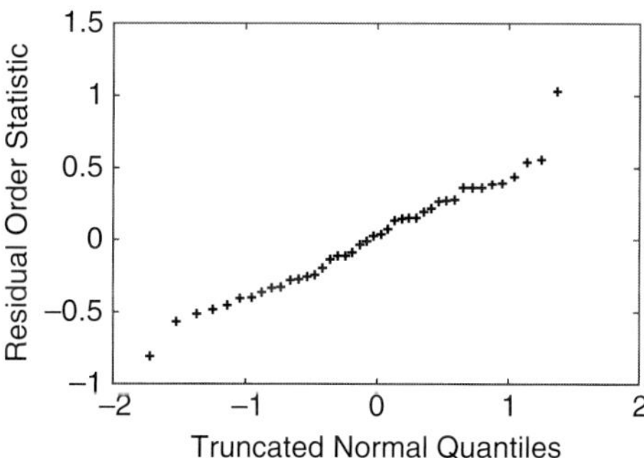

Figure 9.11 Quantile-quantile plot for the residuals from fitting the star data after removing five leverage points against truncated Gaussian quantiles.

q-q plot (Section 7.1.3), and the hat matrix q-q plot (Figure 9.12) shows that point 14 is a remaining leverage point.

Eliminating point 14 and repeating gives runs and Durbin-Watson test p-values of 0.3464 and 0.0013, respectively. The F statistic and the correlation coefficient are nearly unchanged at 29.2656, and 0.4287. The t table is only slightly changed (Table 9.12).

The residual and hat matrix q-q plots are free of outliers. A plot of the residuals against datum number (Figure 9.13) shows some clustering of the residuals for a small data index, which is probably the reason the Durbin-Watson test rejected.

Table 9.12 Regression *t* Table for the Star Data Less Six Leverage Points

Predictor	Coefficient	Standard error	*t*-Statistic	*p*-Value
X_1	2.9840	0.5516	5.4098	0.0000
X_2	−8.2077	2.4328	−3.3738	0.0017

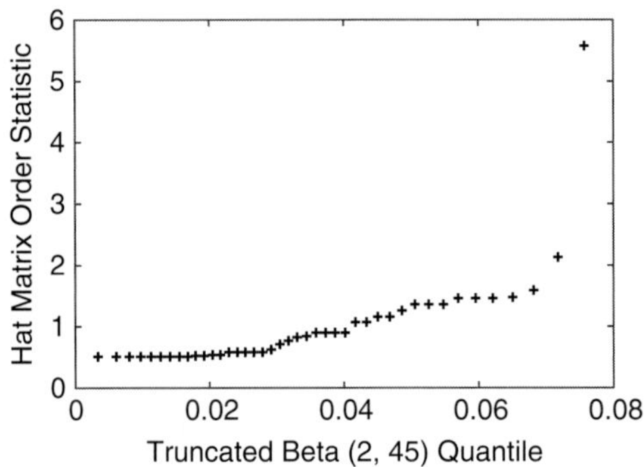

Figure 9.12 Quantile-quantile plot for the hat matrix diagonal normalized by its expected value against truncated beta(2,45) quantiles for the star data after removing five leverage points.

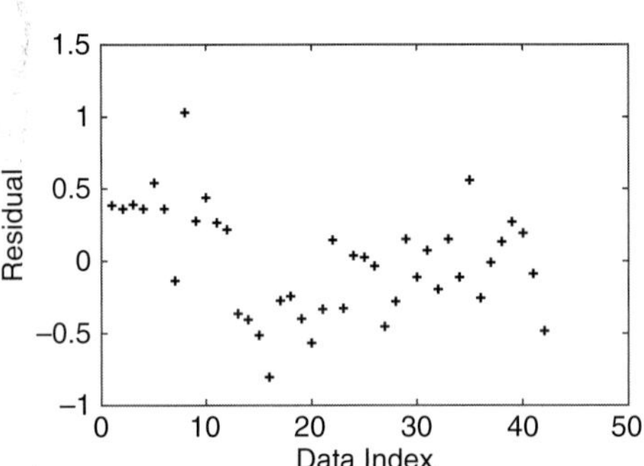

Figure 9.13 Scatter plot of the residuals from the final star data linear regression agains the datum number.

Example 9.3 Repeat the analysis of the star data using the bootstrap to estimate the coefficients and test them for significance.

```
star = importdata('star.dat');
y = star(:, 2);
x = [star(:, 1)  ones(size(y))];
rng default
```

```
boot = bootstrp(99999, @(x, y) x\y, x, y);
mean(boot)
ans =
 -0.2951    6.2706
```

The bootstrap distributions of the regression parameters (Figure 9.14) are skewed by the leverage points, with a long upper tail for the slope and a long lower tail for the intercept.

To implement the double bootstrap, the inner bootstrap and computation of the t statistic will be done using a function placed in the default directory as follows

```
function [t] = func(x, y, b)
opt = statset('UseParallel', true); %parallelize the bootstrap
bmat = bootstrp(999, @(x, y) x\y, x, y, 'Options', opt);
sd = std(bmat, 1);
t = (mean(bmat) - b)./sd;
end
b = x\y;
r = y - x*b;
sd = sqrt(r'*r/(length(y) - 2)*diag(inv(x'*x)));
that = b'./sd';
boot = bootstrp(999, @(x, y) func(x, y, b'), x, y);\
for i = 1:2
  pval(i) = 2*min(sum(boot(:, i) > that(i)), sum(boot(:, i)
  <= that(i)))/999;
end
pval
pval =
  0.0801        0
that
that =
   -1.4438     5.4940
```

The slope parameter is not significant, but the intercept is required. Delete the four leverage points and repeat.

```
y(11) = [];
y(19) = [];
y(28) = [];
y(31) = [];
x(11, :) = [];
x(19, :) = [];
x(28, :) = [];
x(31, :) = [];
b = x\y;
r = y - x*b;
```

```
sd = sqrt(r'*r/(length(y)-2)*diag(inv(x'*x)));
that = b'./sd';
rng default
boot = bootstrp(9999, @(x, y)(x'*x)\x'*y, x, y);
mean(boot)
ans =
   2.1827    -4.6602
```

The bootstrap distributions for the slope and intercept parameters (Figure 9.16) are still skewed in the same sense as in Figure 9.14, but the long tails are gone. Continue with the significance tests.

```
pval
pval =
   0.0280       0.3023
that
that =
   4.8710       -2.1997
```

The p-value for the slope is very small, but that for the intercept is large such that the latter is not required. The bootstrap distributions for the test statistic are skewed in the same sense as in Figure 9.15, but with substantially shorter tails and nearly bimodal form. Rejection of the intercept term is due to the long left tail in Figure 9.17 and is suspect. Eliminating the seventh and fourteenth data gives a bootstrap estimate for the regression coefficients of 2.8456 and −7.5848 for the slope and intercept. This is substantially different from that without the four largest leverage points, especially for the intercept. The bootstrap distributions for the regression coefficients are now fairly symmetric (Figure 9.18), in contrast with those in Figures 9.14 and 9.16. However, the width of the

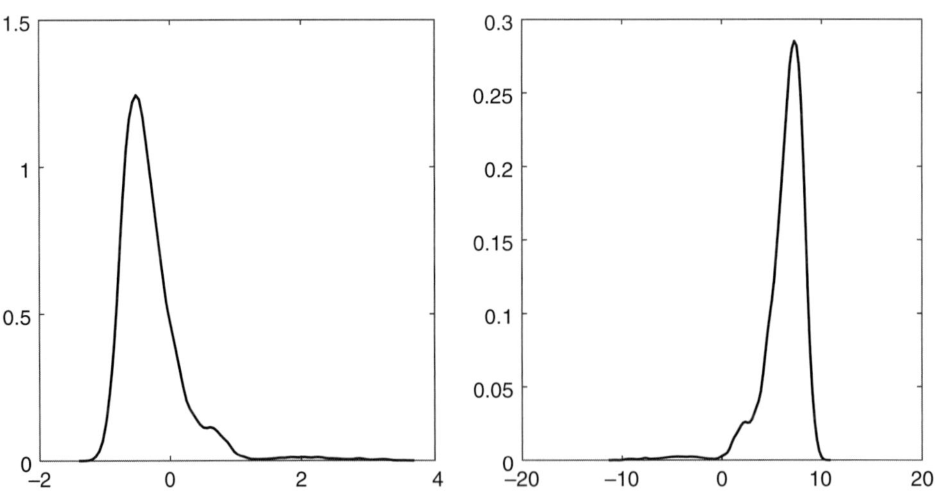

Figure 9.14 Bootstrap distributions for the slope (left) and intercept (right) regression coefficients for the star data.

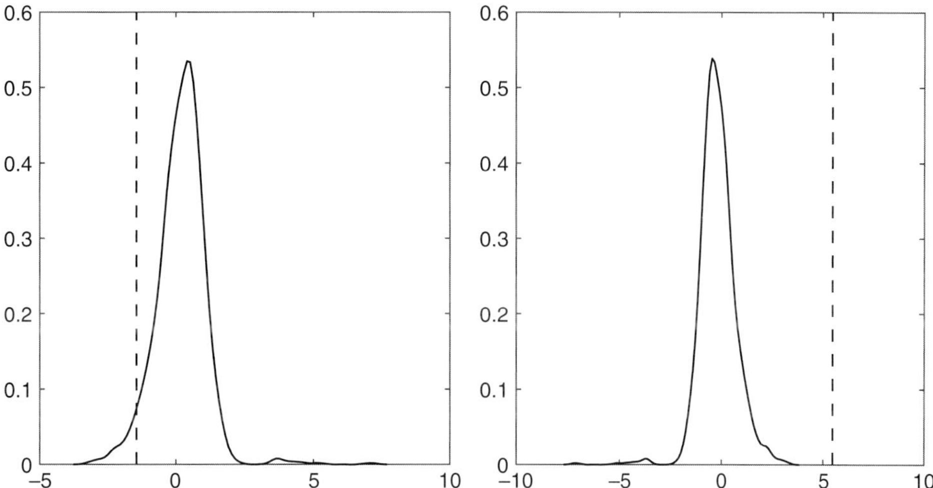

Figure 9.15 Bootstrap distributions for assessing the significance of the fit for the slope (left) and intercept (right). The vertical dashed line is the test statistic.

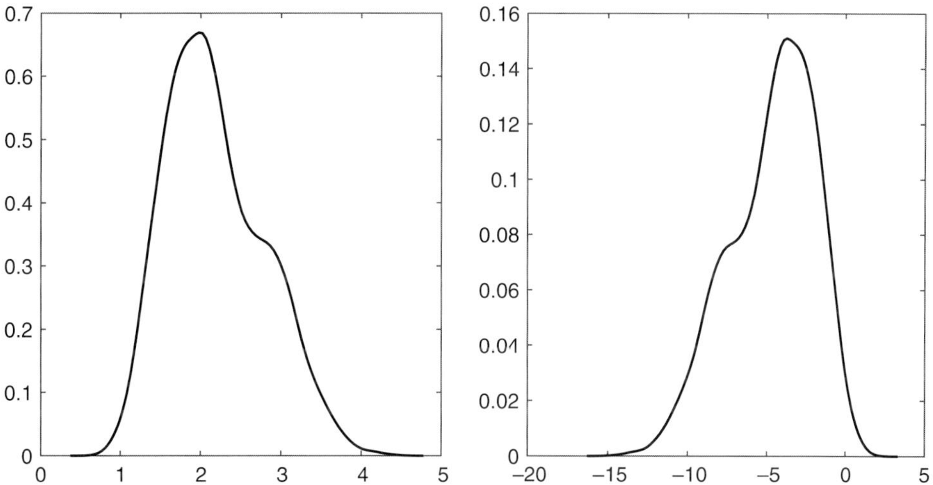

Figure 9.16 Bootstrap distributions for the slope (left) and intercept (right) regression coefficients for the star data after eliminating the four largest leverage points.

intercept distribution is substantial, and an eyeball 95% confidence interval ranges from ~ -3 to ~ -13.

Performing the bootstrap test of significance gives p-values of 0 and 0.0160 for the slope and intercept, so both parameters are required. Figure 9.19 shows the bootstrap test statistic distributions. For reference, the sample test statistics are 5.94 and -3.57, respectively. The distributions do not have the bimodality seen in Figure 9.17.

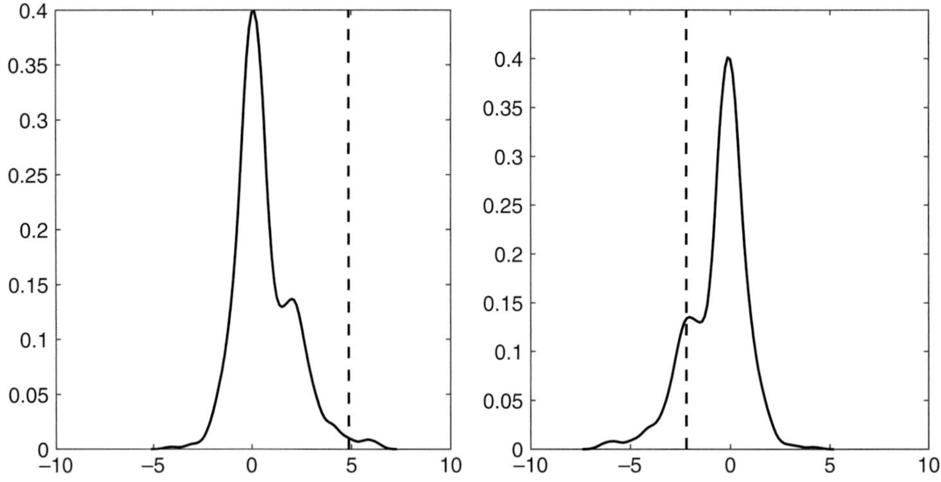

Figure 9.17 Bootstrap distributions of the hypothesis test for assessing the significance of the fit for the slope (left) and intercept (right) after removing the four largest leverage points. The vertical dashed line denotes the test statistic.

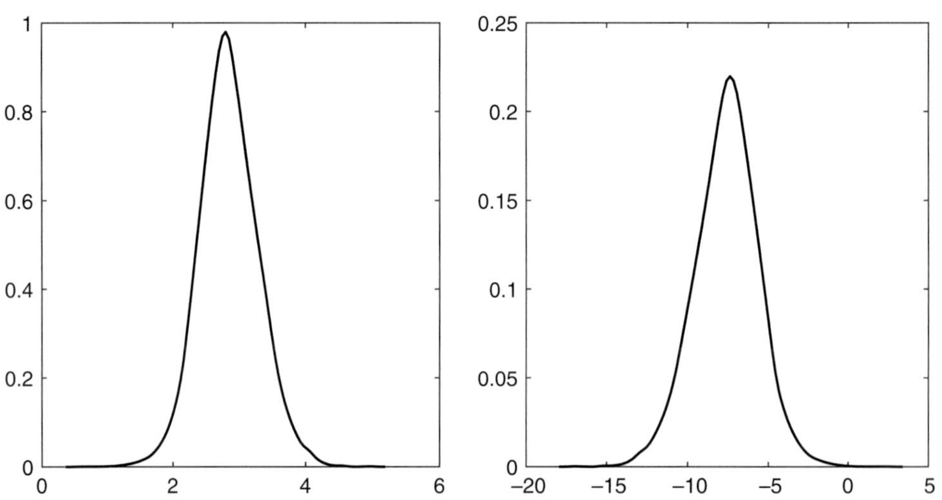

Figure 9.18 Bootstrap distributions for the slope (left) and intercept (right) regression coefficients for the star data after eliminating the six largest leverage points.

Note that the use of q-q plots to guide the process of detecting influential data should have been incorporated but was not because Example 9.2 revealed where the problems are.

Example 9.4 An experiment was performed to calibrate a near-infrared reflectance instrument to measure the protein content of ground wheat. The data consist of measurements of

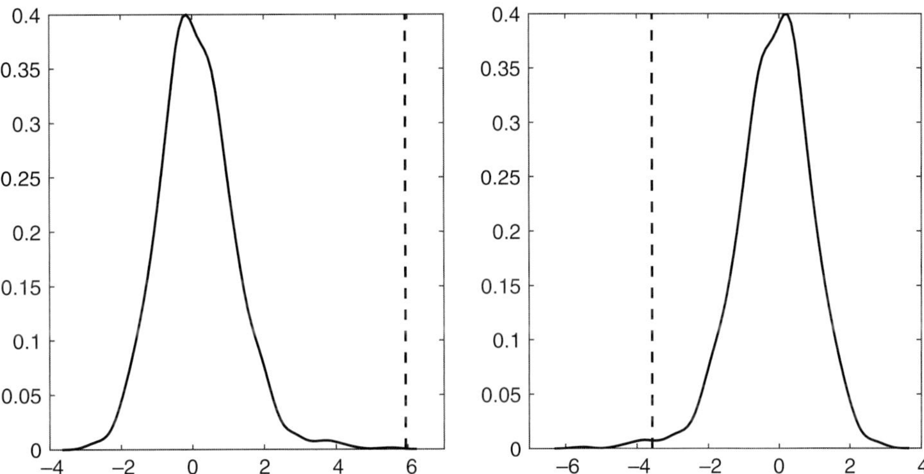

Figure 9.19 Bootstrap distributions for a hypothesis test statistic for the significance of the fit for the slope (left) and intercept (right) after removing the six largest leverage points.

the percent protein content of ground wheat against measurements of reflectance in six wavelength bands. These data were used by Ryan (1997) as an example of extreme multicollinearity, where the high correlation among the predictor variables results in numerical issues with conventional linear regression and hence requires the use of a penalized estimator called *ridge regression*. The data were taken from Ryan (1997, pp. 402–3).

It would be reasonable to expect (or at least be suspicious that) the predictors will be highly correlated because reflectance is not normally a strong function of wavelength. This can be easily confirmed by computing the correlation matrix of the six predictors.

```
wheat = importdata('wheat.dat');
y = wheat(:, 1);
x = wheat(:, 2:7);
corr(x)
ans =
     1.0000    0.9917    0.9951    0.9941    0.9247    0.9648
     0.9917    1.0000    0.9992    0.9769    0.9167    0.9677
     0.9951    0.9992    1.0000    0.9824    0.9241    0.9664
     0.9941    0.9769    0.9824    1.0000    0.9400    0.9689
     0.9247    0.9167    0.9241    0.9400    1.0000    0.9102
     0.9648    0.9677    0.9664    0.9689    0.9102    1.0000
```

There is indeed a lot of correlation in the predictor data.

Solving and evaluating the least squares solution including a column of ones gives

```
x = [x ones(size(y))];
w = ones(size(y));
```

```
a = Anovar(y, x, w);
[b, bse, t, pval] = Treg(y, x, w);
r2 = a(1,2)/a(3,2)
r2 =
   0.9834
[h, p] = runstest(y - x*b)
h =
   1
p =
   0.0617
dwtest(y - x*b, x)
ans =
   0.2417
```

The first, second, and fifth regression coefficients are not significant at the 0.05 level, and the correlation coefficient is nearly one. The runs and Durbin-Watson tests accept the null hypothesis of random and uncorrelated residuals. The residual and hat matrix q-q plots are shown in Figures 9.20 and 9.21.

The residual q-q plot (Figure 9.20) is unremarkable, but the hat matrix diagonal q-q plot (Figure 9.21) reveals some leverage effects, with the largest three values corresponding to data points 17, 25, and 34.

Eliminating points 17, 25, and 34 and repeating the exercise gives a correlation coefficient of 0.9834 (which is identical to that with all the data), a runs test that weakly rejects the null hypothesis of random residuals with a p-value of 0.0319, and a Durbin-Watson test

Table 9.13 ANOVA Table for Wheat Reflectance Data

Source	*df*	Sum of squares	Mean squares	*F*	*p*-Value
Due to regression	6	103.6679	17.2780	423.8930	0.0000
Due to error	43	1.7527	0.0408		
Total	49	105.4206			

Table 9.14 Regression t Table for Wheat Reflectance Data

Predictor	Coefficient	Standard error	t-Statistic	p-Value
X_1	−0.0314	0.0405	−0.7772	0.4413
X_2	−0.0620	0.0398	−1.5565	0.1269
X_3	0.3431	0.0466	7.3664	0
X_4	−0.2060	0.0275	−7.4908	0
X_5	0.0027	0.0031	0.8623	0.3933
X_6	−0.0445	0.0119	−3.7347	0.0005
X_7	22.6274	5.0498	4.4809	0.0001

Table 9.15 Regression t Table for Wheat Reflectance Data Less Three Rows

Predictor	Coefficient	Standard error	t-Statistic	p-Value
X_1	0.0035	0.0454	0.0770	0.9390
X_2	−0.0768	0.0475	−1.6419	0.1142
X_3	0.3398	0.0493	6.8976	0
X_4	−0.2304	0.0325	−7.0920	0
X_5	0.0054	0.0034	1.5770	0.1227
X_6	−0.0437	0.0188	−2.3226	0.0254
X_7	17.0288	5.9078	2.8824	0.0063

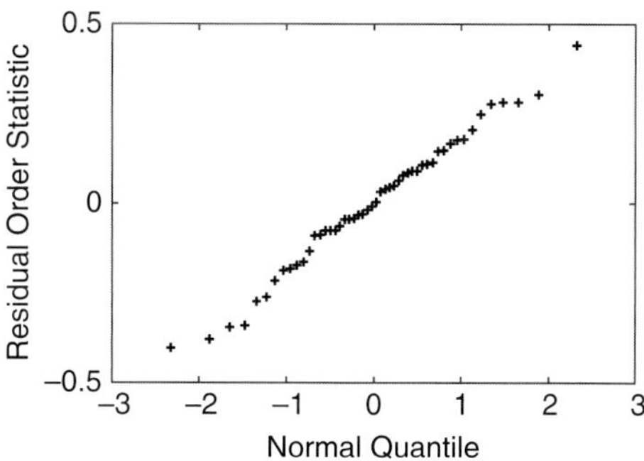

Figure 9.20 Gaussian q-q plot for the wheat reflectance data.

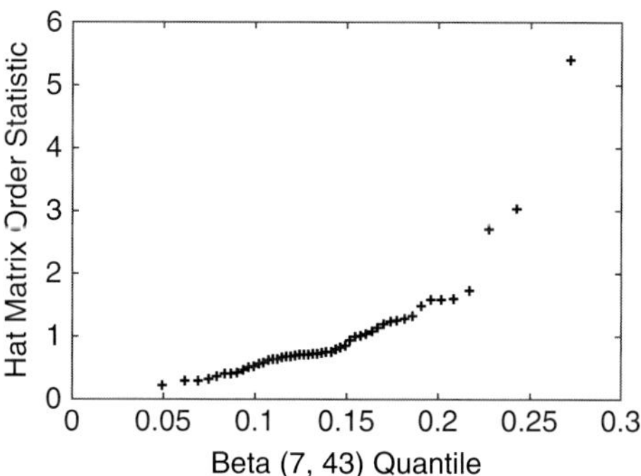

Figure 9.21 Hat matrix diagonal q-q plot for the wheat reflectance data.

that accepts uncorrelated residuals with a p-value of 0.3168. Note that these two tests must use the weighted residuals and predictor matrix. The t table is shown in Table 9.15.

The p-values for the first, second, and fifth predictors remain insignificant at the 0.05 level. Quantile-quantile plots for the residuals and hat matrix diagonal are free of influential data.

Table 9.16 Regression *t* Table for Reduced Wheat Reflectance Data

Predictor	Coefficient	Standard error	*t*-Statistic	*p*-Value
X_3	0.2620	0.0070	37.5984	0
X_4	−0.2107	0.0063	−33.5779	0
X_6	−0.0485	0.0121	−3.9947	0.0002
X_7	22.9873	2.3994	9.5805	0

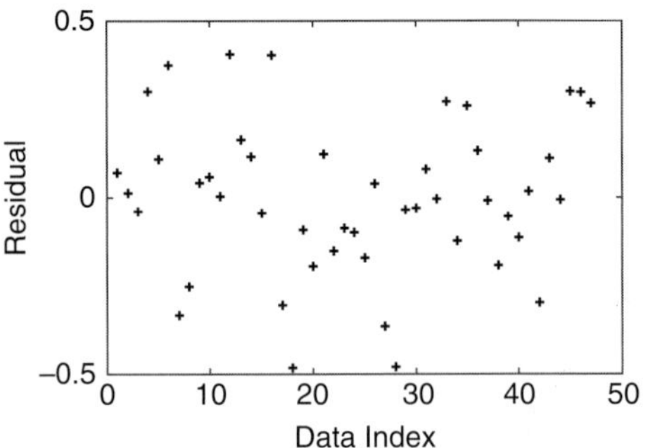

Figure 9.22 The regression residuals plotted against data index for the wheat reflectance data.

Eliminating the first, second, and fifth predictor variables with points 17, 25, and 34 suppressed and repeating the procedure give a correlation coefficient of 0.9772, a runs test that accepts the null hypothesis of random residuals with a *p*-value of 0.4998, and a Durbin-Watson test that weakly rejects uncorrelated residuals with a *p*-value of 0.0286. The *t* table is shown in Table 9.16.

The residual and hat matrix q-q plots are free of influential data. A plot of the residuals against their index reveals a random pattern (Figure 9.22), placing the Durbin-Watson test result in doubt. This example uses *backward stepwise selection* to reduce the number of predictors. The presence of multicollinearity in this case does not require the use of penalized least squares techniques. The difference between this result and that of Ryan (1997) is probably due to the pervasive use of 64-bit arithmetic in 2016 as opposed to 32-bit arithmetic in the late 1990s.

9.6 Robust and Bounded Influence Regression

Virtually all of the commonly used statistics (e.g., the sample mean and sample variance and the least squares estimator) are not robust to the presence of a small fraction of influential data. This problem can be obviated to a degree by using statistics based on the L_1 rather than the L_2 norm, but at a significant efficiency penalty. This is evident from Section 4.10.2, where it was shown that the variance of the median (an L_1 estimator) is $\pi/2$ times the variance of the mean (an L_2 estimator).

What are really needed are estimators that are robust but still efficient and preferably that approach the efficiency of standard estimators in the absence of influential data. Their operation should be automatic rather than requiring a lot of tedious or ad hoc data culling, as was carried out in Section 9.5.2. The definition of a robust estimator will be broadened to encompass ones that are relatively insensitive to a moderate amount of bad data or to model inadequacies and that react gradually rather than abruptly to perturbations of either.

The goal of a robust procedure is to eliminate the effect of influential data. As has already been shown, there are two main manifestations of influential data in a regression context: outliers (which are evident in the residuals) and leverage points (which are evident in the predictor data, for which the hat matrix diagonal is a useful detection statistic). In geophysical data, there are at least two classes of influential data: point defects and local nonstationarity. Point defects are isolated outliers or leverage points that exist independent of the statistics of the underlying process. These would include human error in counting, dropped or sticky bits in A/D converters and lightning strikes. Point defects are fairly easy to detect and remove using standard approaches. Local nonstationarity is the departure of the underlying statistics from a base state for a finite period of time. This would include (in a seismological context) noise on a buried sensor induced by tree motion driven by the wind, among many other things. The finite duration of local nonstationarity results in influential data that occur in patches, violating the randomness property of the residuals, and are much harder to detect and remove using standard robust estimators.

A key concept in robust estimation is that of the *finite sample breakdown point*, defined as the smallest fraction of anomalous data that can render the estimator useless. The smallest possible breakdown point is $1/N$ (i.e., a single anomalous point can destroy the estimate). The sample mean, the sample variance, and ordinary least squares all have breakdown points of $1/N$. However, if one believes that 1% to 10% (or more) of most data sets consist of anomalous values (and geophysical data are at the high end of this range or even beyond it), then better estimators are essential.

9.6.1 Robust Estimators

Starting around 1960, statisticians began to develop robust estimators, and this continues to be an active area of research. The most widely used robust estimators are maximum likelihood–like or M-estimators. Chave, Thomson, & Ander (1987) review the subject in a geophysical context. Rousseeuw & Leroy (2005) and Huber (2011) provide two recent texts on the subject.

Recall the maximum likelihood estimator (mle) for a location parameter λ given N independent samples $\{\mathbf{X}_i\}$ drawn from some joint pdf $f(x|\lambda)$, as described in Section 5.4.

Only the case where λ appears in $f(x|\lambda)$ in the form $x - \lambda$, as in the exponential family of distributions, will be considered. The log likelihood function is

$$\log \mathcal{L}(\lambda|x_i) = -\sum_{i=1}^{N} \log f(x_i - \lambda) \equiv -\sum_{i=1}^{N} \rho(x_i - \lambda) \tag{9.55}$$

where ρ is the *loss function*. For a Gaussian distribution, $\rho(x) = x^2/2 + c$, where c is a constant. If maximum likelihood principles are applied to this problem, the solution λ is the sample mean. This is the strict mle when $f(x|\lambda)$ is completely specified, but often a full description of $f(x|\lambda)$ is not available because it is very common for the bulk of a sample to be approximately Gaussian but contaminated with a small fraction of outliers whose distribution is unknown and typically longer tailed than Gaussian. The tail data are not well described by the random sample unless it is very large.

M-estimators turn this problem around: a loss function is chosen a priori that will ensure robustness and efficiency over a wide range of distributions. For example, it might be assumed that the real distribution is a hybrid, with normality in the center, but with longer tails than the Gaussian would exhibit. The actual choice of loss function rapidly becomes empirical, with validation by simulation.

Performing the minimization of $\log \mathcal{L}(\lambda)$ by taking the derivative with respect to λ yields the robust counterpart to the normal equations

$$\sum_{i=1}^{N} \psi(x_i - \lambda) = 0 \tag{9.56}$$

where $\psi(x) = \partial_x \rho(x)$ is the *influence function*. The influence function may be bounded or unbounded; Hampel et al. (1986, sec. 1.3) provide further details.

Common loss and influence function examples include that for the L_1 norm:

$$\begin{aligned} \rho(x) &= |x| + c \\ \psi(x) &= \operatorname{sgn}(x) \end{aligned} \tag{9.57}$$

where $\operatorname{sgn}(x)$ is -1, 0, or 1 as x is negative, zero, or positive, and that for the L_2 norm

$$\begin{aligned} \rho(x) &= x^2/2 + c \\ \psi(x) &= x \end{aligned} \tag{9.58}$$

Note that the L_1 norm influence function is a constant independent of the argument (except for a sign change at $x = 0$), whereas the L_2 norm influence function increases in an unbounded manner with the argument. In an implementation, that argument is $x - \lambda$ and is a measure of the distance between the parameter λ and any individual datum x_i. For the L_2 norm, the outcome is more strongly influenced by large values of $x - \lambda$ (or, in other words, by data that are farther away from the distribution center), whereas for the L_1 norm, the influence is independent of the relative size of each datum. Hence the name *influence function* and the greater robustness of the L_1 norm relative to the L_2 norm follow naturally.

The simplest robust estimator is due to Huber (1964) and is based on a conceptual model that is Gaussian in the center and double exponential (a natural distribution for the L_1 estimator, just as the Gaussian is the natural distribution for the L_2 norm) in the tails. The loss and influence functions for the Huber estimator are

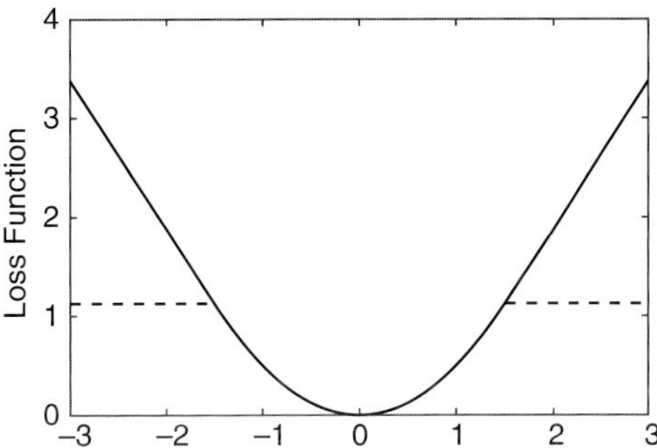

Figure 9.23 The Huber loss function (9.59) (solid line) compared with the trimmed influence function (9.61) (dashed line). The two loss functions are identical over the abscissa range of −1.5 to 1.5.

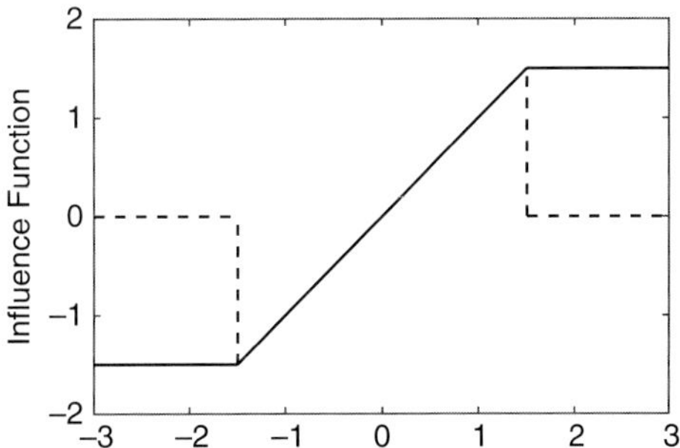

Figure 9.24 The Huber influence function (9.60) (solid line) compared with the trimmed influence function (9.62) (dashed line). The two influence functions are identical over the abscissa range of −1.5 to 1.5.

$$\begin{aligned} \rho(x) &= x^2/2 & |x| \le \alpha \\ &= \alpha|x| - \alpha^2/2 & |x| > \alpha \end{aligned} \tag{9.59}$$

$$\begin{aligned} \psi(x) &= x & |x| \le \alpha \\ &= \alpha \ \mathrm{sgn}(x) & |x| > \alpha \end{aligned} \tag{9.60}$$

The Huber loss and influence functions are shown in Figures 9.23 and 9.24, respectively, for $\alpha = 1.5$, which yields 95% efficiency with pure Gaussian data. Thus a 5% penalty is paid for using a robust procedure if all the data are good, as opposed to an ~60% variance penalty for the L_1 estimator. The Huber function bounds the influence of extreme data where the L_2 estimator does not.

More extreme loss/influence functions are possible. For example, trimming the data at $|x| = \alpha$ gives

$$\begin{aligned} \rho(x) &= x^2/2 & |x| \le \alpha \\ &= \alpha^2/2 & |x| > \alpha \end{aligned} \tag{9.61}$$

$$\begin{aligned} \psi(x) &= x & |x| \le \alpha \\ &= 0 & |x| > \alpha \end{aligned} \tag{9.62}$$

This is clearly less efficient than the Huber function but is proof against large outliers. The trimmed loss and influence functions are also shown in Figures 9.23 and 9.24.

Hybrid influence functions that descend to zero more gradually than this example also exist. A particularly useful influence function when extreme outliers are present [originally suggested by Thomson (1977) and motivated by the form of the Gumbel extreme value distribution] is

$$\psi(x) = x \exp\left(-e^{\beta(|x|-\beta)}\right) \tag{9.63}$$

This is somewhat like the trimmed mean but is continuous and continuously differentiable. It descends rapidly to zero as $|x|$ approaches β.

The implementation of a robust estimator requires several additional steps because the influence function analogue to the normal equations is nonlinear when the argument is the residual $r = x - \lambda$, and the residuals are not actually known until the equations are solved. A numerical solution requires two modifications to the equations. First, the argument in $\rho(x)$ or $\psi(x)$ must be made independent of the scale of the data. In the Huber example, the transition from an L_2 to an L_1 loss function occurs when the absolute value of the argument is 1.5 standard deviations, and hence the argument needs to be expressed in units of standard deviations for proper operation. This is achieved by replacing $x_i - \lambda$ with $(x_i - \lambda)/d$, where d is a sample scale estimate divided by a population scale estimate. It is important that d be computed in a manner that is not overly sensitive to bad data, and hence the use of (for example) the sample standard deviation divided by the population standard deviation is not advisable. A good choice is

$$d = \frac{s_{IQ}}{\sigma_{IQ}} \tag{9.64}$$

where the sample interquartile distance is given by (4.19), and σ_{IQ} is the theoretical interquartile distance for the target distribution (which is usually taken as the Gaussian, so $\sigma_{IQ} \approx 1.349$). The interquartile range has a breakdown point of $N/4$. An even better choice is the ratio of the sample and population median absolute deviation from the median (MAD)

$$d = \frac{s_{MAD}}{\sigma_{MAD}} \tag{9.65}$$

wheres s_{MAD} is given by (4.17) and σ_{MAD} satisfies (2.51). The population parameter $\sigma_{MAD} \approx 1.4826$ for a Gaussian distribution. The MAD is nearly bulletproof because it functions well when the fraction of bad data approaches $N/2$.

The second step in solving for a robust estimate of λ is linearization of the equations. For the influence function analogue to the normal equation, this is achieved by rewriting the equations using weighted least squares

$$\sum_{i=1}^{N} \psi(x_i - \lambda) \approx \sum_{i=1}^{N} w_i r_i = 0 \tag{9.66}$$

$$w_i = \frac{\psi(r_i/d)}{(r_i/d)} \tag{9.67}$$

and $r_i = x_i - \lambda$ is the ith residual. To linearize the equation, the weights w_i and scale d are computed using the residuals r_i from the prior iteration. The solution is initialized using standard least squares (i.e., with weights w_i of 1) or, for heavily contaminated data, using an L_1 estimator. At the second iteration, the residuals from the least squares solution are used to compute the weights and scale, and so on. This iterative process is terminated when the sum of squared residuals does not change above a threshold amount (typically, 1%).

For the Huber influence function, $w_i = 1$ for $|r_i/d| \le \alpha$, and $\alpha d \ \mathrm{sgn}(r_i/d)/r_i$, for $|r_i/d| > \alpha$. This leaves data within α "standard deviations" of the center alone and down-weights those farther away according to the inverse of their distance from the center. It has the advantage of having only a single minimum [i.e., $\psi(x)$ is never zero for large argument], hence guaranteeing that the linearized solution cannot converge to other than the true solution, but has the disadvantage that down-weighting of outliers is often not strong enough to remove their influence. A better approach with severely contaminated data is to use the Huber estimator for a few iterations to get an initial solution and final estimate of scale, and then to use the hard limiter (9.63) for the final iteration or two to kill bad outliers.

All this machinery extends to linear regression if the residual in the loss function is identified to be the regression residual $\hat{\mathbf{r}} = \mathbf{y} - \overleftrightarrow{\mathbf{X}} \cdot \hat{\boldsymbol{\beta}}$. Performing the minimization yields the regression normal equation analogue

$$\sum_{i=1}^{N} \psi(r_i/d)\ x_{ij} = 0 \tag{9.68}$$

whose solution using (9.66) is straightforward. The result should be assessed using a runs test, the Durbin-Watson test, ANOVA, and the t-statistic for consistency with the least squares model, just as for a conventional least squares solution. Care must be taken to use the weighted residuals to compute the sum of squared residuals and to incorporate the weights into the hat matrix. A flowchart for robust least squares is shown in Figure 9.25.

MATLAB supports robust regression through b = **robustfit**(x, y, 'wfun', tune, const), where x and y are the predictor matrix and response variable, "wfun" specifies the weight function, and *tune* is the tuning constant. The default for "wfun" is the "bisquare" weight, but "huber" is also supported, although with a default tuning constant of 1.345 instead of 1.5.

Caution: robustfit by default adds a column of ones to x. This behavior can be turned off by setting *const* to "off."

The following MATLAB function implements a robust least squares estimator based on the Huber weights. It returns three structures containing (1) the output from *Treg* by iteration; (2) the residuals, weights, and normalized residual sum of squares by iteration; and (3) $\hat{R}^2$, the ANOVA p-value, the runs test p-value, and the Durbin-Watson p-value by iteration.

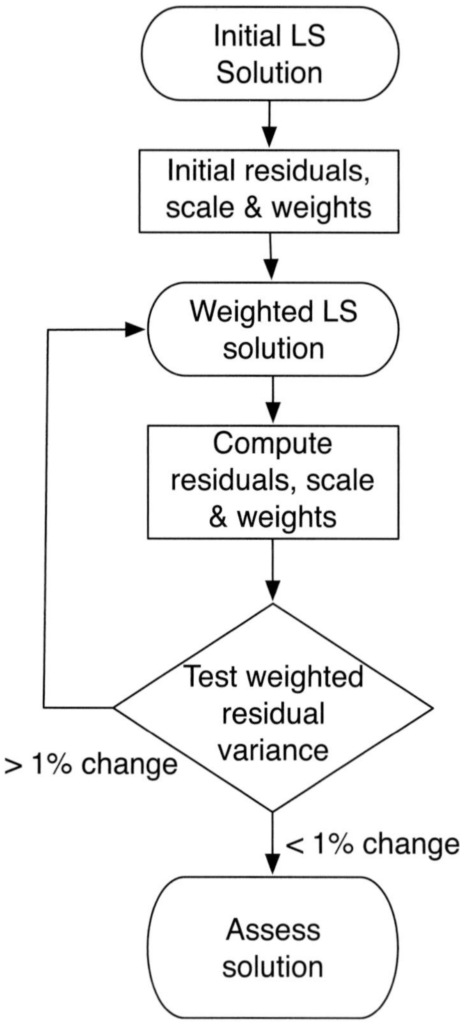

Figure 9.25 Flowchart for robust least squares estimation.

```
function [ tstruct, varargout] = Huber(y, x)
%Solves least squares problem using Huber robust algorithm
%input variables
% y - n by 1 response vector (required)
% x - n by p predictor matrix (required)
%output variables
%tstruct - structure containing number of iterations and
regression
%       parameters, se, t-stistic and p-value for each
%rstruct  -structure  containing  regression  residuals,
weights and
%       normalized sum of squared residuals for each iteration
```

```
%pstruct -  structure containing  r^2, and anova, runs test
and Durbin-Watson test
%      p-values for each iteration
n = length(y);
p = size(x, 2);
w = ones(size(y));
[b bse t tprob] = Treg(y, x, w);
r = y - x*b;
d = iqr(r)/1.34903;
ovar = r'*r/n;
rss = ovar;
bs = b;
bses = bse;
ts = t;
tprobs = tprob;
rs = [];
ws = [];
r2 = [];
apval = [];
rpval = [];
dwpval = [];
for k = 1:15 %allow 15 iterations max
    w = 1./max(1, abs(r)/(1.5*d));
    ws(:, k) = w;
    a = Anovar(y, x, w);
    r2 = [r2 a(1,2)/a(3,2)];
    apval = [apval a(1, 5)];
    [b bse t tprob] = Treg(y, x, w);
    bs = [bs b];
    bses = [bses bse];
    ts = [ts t];
    tprobs = [tprobs tprob];
    r = y - x*b;
    rs(:, k) = r;
    [h pval] = runstest(w.*r);
    rpval =[rpval pval];
    pval = dwtest(w.*r, repmat(w, 1, p).*x);
    dwpval - [dwpval pval];
    d = iqr(r)/1.34903;
    wsum = sum(w);
    nvar = sum((w.*r)^2)/wsum;
    rss = [rss nvar];
    if abs(nvar - ovar) <= 0.01*ovar
        break
```

```
    end
    ovar = nvar;
end
tstruct.b = bs;
tstruct.bse = bses;
tstruct.t = ts;
tstruct.pval = tprobs;
if nargout > 1
   rstruct.res = rs;
   rstruct.w = ws;
   rstruct.rss = rss;
   varargout{ 1} = rstruct;
   varargout{ 2} = [ ] ;
end
if nargout > 2
   pstruct.r2 = r2;
   pstruct.anova = apval;
   pstruct.runs = rpval;
   pstruct.dw = dwpval;
   varargout{ 2} = pstruct;
end
```

Example 9.5 Returning to the Hubble data from Example 9.1 (which possess neither outliers nor leverage points and do not require a column of ones in x), a MATLAB script implementing a robust M-estimator is

```
hubble = importdata('hubble.dat');
y = hubble(:, 2);
x = hubble(:, 1);
[tstruct rstruct pstruct] = Huber(y, x);
```

The normalized residual sum of squares decreases by only about 10% over two iterations.

```
rstruct.rss
ans =
  1.0e+04 *
   5.0267    4.5350    4.5426
```

The regression coefficient, standard error, and t test hardly budge.

```
tstruct
tstruct =
           b: [423.9373 418.7612 418.3056]
         bse: [42.1541 40.1465 40.1786]
           t: [10.0568 10.4308 10.4112]
        pval: [6.8686e-10 3.4288e-10 3.5551e-10]
```

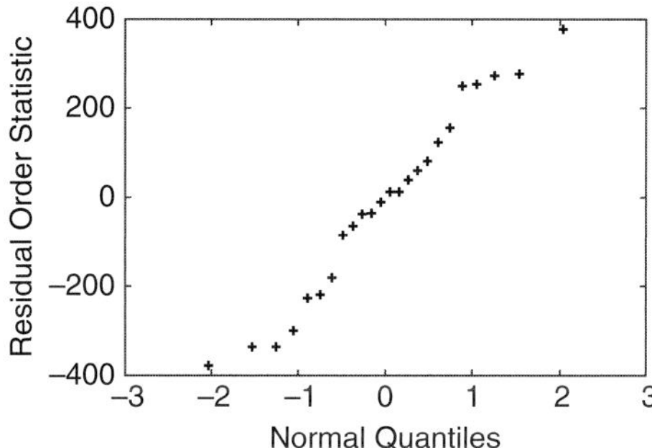

Figure 9.26 Normal q-q plot for the weighted residuals from a single-parameter fit to the Hubble data.

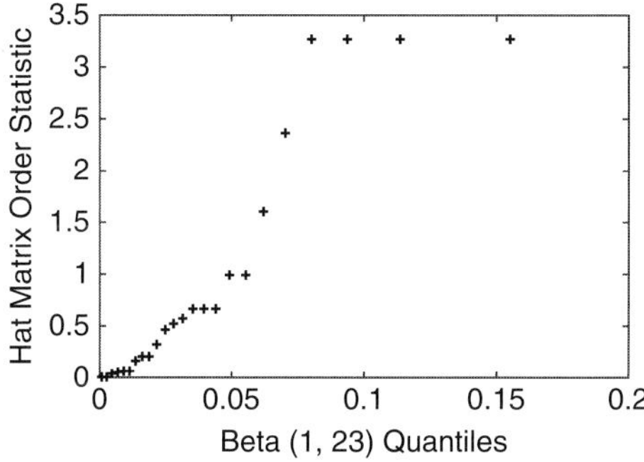

Figure 9.27 Weighted hat matrix diagonal q-q plot for a single-parameter robust fit to the Hubble data.

The weights are all 1 except for points 12 (0.9284) and 16 (0.7536). The runs and Durbin-Watson test p-values support the null hypotheses that the residuals are uncorrelated and random, with p-values of 1 and 0.9805, respectively. The robust procedure has not changed the solution significantly compared with ordinary least squares. Quantile-quantile plots of the weighted residuals and the hat matrix diagonal are outlier free (Figures 9.26 and 9.27) and should be compared with Figures 9.4 and 9.5. The hat matrix q-q plot is nearly unchanged, whereas the two most extreme residuals in the residual q-q plot have been moved closer to the distribution center in Figure 9.26.

This example demonstrates that a robust procedure does not carry a major penalty with uncontaminated data. However, to show that it does in fact remove outlying data, the largest response data variable will be replaced with a value 10 times larger, and the robust estimator will be applied to it. The estimator converges after four iterations, reducing the residual variance as follows:

```
rstruct.rss
ans =
   1.0e+06 *
   3.7147    0.0710    0.0491    0.0506    0.0505
```

Table 9.17 Regression t Table by Iteration for Hubble Data

Iteration	Coefficient	Standard error	t-Statistic	p-Value
OLS	1088.6	362.4	3.00	0.0063
1	463.5	71.48	6.48	1.3e−06
2	410.2	44.88	9.14	4.1e−09
3	398.3	44.71	8.91	6.5e−09
4	397.1	44.67	8.89	6.7e−09

Table 9.18 ANOVA Table for Star Data

Source	df	Sum of squares	Mean squares	F	p-Value
Due to regression	1	0.7430	0.7430	2.3335	0.1336
Due to error	45	14.3290	0.3184		
Total	46	15.0720			

Table 9.19 Regression t Table for Star Data

Predictor	Coefficient	Standard error	t-Statistic	p-Value
X_1	−0.4136	0.2861	−1.4458	0.1552
X_2	6.7952	1.2358	5.4987	0.0000

The final runs and Durbin-Watson test p-values are 1 and 0.8306, respectively. The final weights are nearly the same as for the original data, except that point 24 has a weight of 0.0375. The t table changes by iteration are contained in Table 9.17.

Example 9.6 The star data used in Example 9.2 will be revisited using the function *Huber*. Only one iteration is performed, yielding the results shown in Tables 9.18 and 9.19.

The correlation coefficient is $\hat{R}^2 = 0.0493$. The original sum of squared residuals is 0.3052, and the final sum of squared residuals is 0.3049. The runs and Durbin-Watson tests give p-values of 0.0982 and 0.1939, indicating that the residuals are random and uncorrelated. The regression is fitting nothing very well.

The robust result is statistically indistinguishable from the ordinary least squares result in Example 9.2. The robust weights are all one except for point 17, where it is 0.9929. The M-estimator has not detected any of the leverage points located previously. This should not come as a surprise because the weighted residual q-q plot shows no outliers, and an M-estimator detects influential points only based on the size of the residuals. Figures 9.28 and 9.29 show the residual and hat matrix q-q plots, which are nearly identical to Figures 9.6 and 9.7.

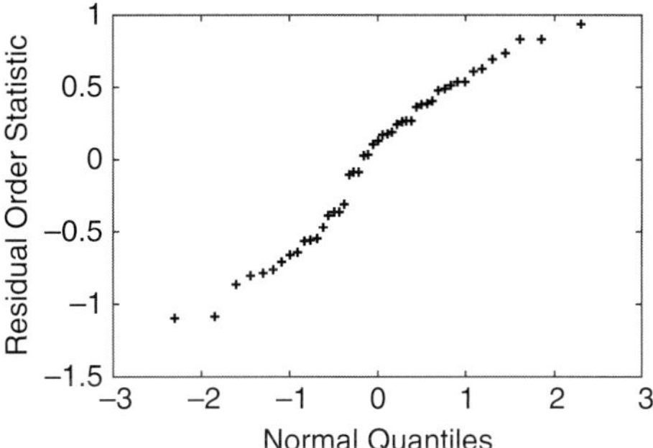

Figure 9.28 Weighted residual Gaussian q-q plot for a robust fit to the star data.

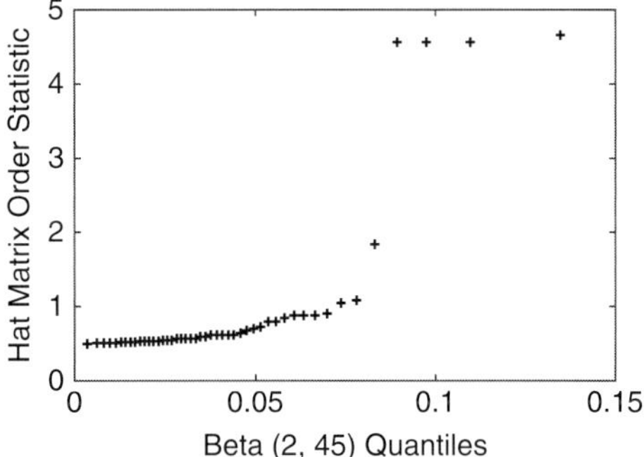

Figure 9.29 Weighted hat matrix diagonal q-q plot for a robust fit to the star data.

9.6.2 Bounded Influence Estimators

M-estimators can be quite effective in dealing with bad data that are manifest in the estimated residuals, but they may not be sensitive to outlying predictor variables that do not always result in outlying estimated residuals, as was evident in Example 9.6. In fact, M-estimators frequently make leverage worse or create leverage points that were not initially present. It can be shown that an M-estimate is dominated by a single leverage point; it has a breakdown point of $1/N$. This limitation has led to the development of generalized-M or GM or bounded influence estimators that guard against both leverage and outliers at the same time. Bounded influence estimators can have a breakdown point as large as $1/p$. This is an active area of research in applied statistics. However, most of the proposed solutions are unnecessarily complicated and/or computationally inefficient for large data sets. For example, Yohai (1987) describes a three-step algorithm that achieves a breakdown point approaching $N/2$, and Rousseeuw (1984) describes a least median of squares algorithm that has a high breakdown point. In the following, the approach advocated by Chave & Thomson (2003, 2004) will be outlined because it is computationally efficient, possesses a high breakdown point, and preserves the underlying statistical model for linear regression.

A generalization of the regression normal equation analog (9.68) that includes leverage weights is

$$\sum_{i=1}^{N} w_i v_i \psi\left(\frac{r_i}{u(x_i)d}\right) x_{ij} \tag{9.69}$$

where v_i is a leverage weight and $u(x_i)$ is a function of the i-th row of the predictor matrix. Two versions of (9.69) have been proposed

1. $u(x_i) = 1$ and $v_i = \sqrt{1 - h_{ii}}$ following on work by Mallows (1975);
2. $u(x_i) = v_i$ and $v_i = \sqrt{1 - h_{ii}}$ or $v_i = (1 - h_{ii})/\sqrt{h_{ii}}$ due to Handschin *et al.* (1975).

The first of these cleanly decouples weighting based on leverage (as detected from the hat matrix diagonal) and weighting based on the size of the residuals (i.e., M-estimation). The second is more efficient because large leverage points corresponding to small residuals are not heavily penalized but Carroll & Welsh (1988) showed that the resulting parameter estimates are not consistent. In the author's experience, neither works very well because the leverage weighting is very gentle. To see this, note that if $p = 2$ and $N = 100$, $v_i = \sqrt{1 - h_{ii}} = 0.99$ if h_{ii} has the expected value of p/N, and $v_i = 0.89$ if h_{ii} is 10 times the expected value (which is a very large leverage point). The reason this gentle approach was taken is presumably because the solution can quickly become unstable because of strong interaction between the robust and leverage weights where robust weighting frequently makes leverage worse.

Chave & Thomson (2003) solved this problem by focusing on the Mallows form of the bounded influence estimator defined earlier to avoid problems with consistency, rewriting it in iteratively reweighted form as

$$\sum_{i=1}^{N} w_i^{[k]} v_i^{[k]} r_i^{[k-1]} x_{ij} = 0 \tag{9.70}$$

for $j = 1, \ldots, p$, where the superscript in brackets denotes the iteration number from 1 to q, and the zeroth-order residuals are taken from the ordinary least squares solution. The first set of weights w_i are standard M-estimator types as defined in Section 9.6.1 (e.g., the Huber followed by the Thomson weight is a good choice) and using a robust estimate of the scale. The second set of weights v_i are based on a hat matrix diagonal measure of leverage. The form of v_i is iteratively multiplicative to solve the instability problem at the cost of reduced efficiency:

$$v_i^{[k]} = v_i^{[k-1]} \eta \exp\left(-e^{\chi(x_i - \chi)}\right) \tag{9.71}$$

where the zeroth-level values for the weights are all 1, and η is chosen so that the exponential term is 1 when $x_i = 0$. The statistic x_i is set to $Mh_{ii}^{[k]}/p$, where $h_{ii}^{[k]}$ is the diagonal element of the weighted hat matrix

$$\overleftrightarrow{\mathbf{H}}^{[k]} = \overleftrightarrow{\mathbf{U}}^{[k-1]} \cdot \overleftrightarrow{\mathbf{X}} \cdot \left(\overleftrightarrow{\mathbf{X}}^{H} \cdot \overleftrightarrow{\mathbf{U}}^{[k-1]} \cdot \overleftrightarrow{\mathbf{U}}^{[k-1]} \cdot \overleftrightarrow{\mathbf{X}}\right)^{-1} \cdot \overleftrightarrow{\mathbf{X}}^{H} \cdot \overleftrightarrow{\mathbf{U}}^{[k-1]} \tag{9.72}$$

and $\overleftrightarrow{\mathbf{U}}^{[k]} = \overleftrightarrow{\mathbf{W}}^{[k]} \cdot \overleftrightarrow{\mathbf{V}}^{[k]}$, where $\overleftrightarrow{\mathbf{W}}$ and $\overleftrightarrow{\mathbf{V}}$ are diagonal weight matrices, and M is the trace of $\overleftrightarrow{\mathbf{U}}^{[k]}$. The free parameter χ in (9.71) sets the point where down-weighting begins. A useful, albeit empirical choice for χ is the Nth quantile of the $\beta(p, N-p)$ distribution or the percentile of the distribution at some selected probability level. For example, 0.95 carries a 5% penalty for Gaussian data. As an additional step to make the estimator stable, down-weighting of leverage points is applied in half-decade steps beginning with χ set just below the largest normalized hat matrix diagonal element and ending with the final value rather than all at once. The breakdown point of the Chave-Thomson estimator approaches 50%. The following MATLAB function implements the Chave-Thomson bounded influence estimator

```
function [tstruct, varargout] = BI(y, x, ainu)
%Solves least squares problem using Chave-Thomson bounded
influence algorithm
%input variables
% y - n by 1 response vector (required)
% x - n by p predictor matrix (required)
%ainu - leverage point weighting scale factor
%output variables
%tstruct - structure containing number of iterations and
regression
%       parameters, se, t-stistic and p-value for each
%rstruct  -structure  containing  regression  residuals,
weights and
%       normalized sum of squared residuals for each iteration
%pstruct - structure containing  r^2, and anova, runs test
and Durbin- Watson test
%       p-values for each iteration
n = length(y);
p = size(x, 2);
w = ones(size(y));
[b bse t tprob] = Treg(y, x, w);
r = y - x*b;
d = iqr(r)/1.34903;
ovar = r'*r/n;
rss = ovar;
what = ones(size(y));
wsum = n;
swhat = exp(-0.3068528*exp(-ainu^2));
hat = diag(x*inv(x'*x)*x');
bs = b;
bses = bse;
ts = t;
tprobs = tprob;
```

```
rs = [];
ws = [];
whats = [];
r2 = [];
apval = [];
rpval = [];
dwpval = [];
for k = 1:15
    wh = 1./max(1, abs(r)/(1.5*d));
    ws(:, k) = wh;
    t1 = ainu*(wsum*hat/2 - ainu);
    what = what.*exp(-0.3068528*exp(t1))/swhat;
    whats(:, k) = what;
    w = wh.*what;
    a = Anovar(y, x, w);
    r2 = [r2 a(1,2)/a(3,2)];
    apval = [apval a(1, 5)];
    [b bse t tprob] = Treg(y, x, w);
    bs = [bs b];
    bses = [bses bse];
    ts = [ts t];
    tprobs = [tprobs tprob];
    [h pval] = runstest(w.*r);
    rpval = [rpval pval];
    wx = repmat(w, 1, p).*x;
    pval = dwtest(w.*r, wx);
    dwpval = [dwpval pval];
    hat = diag(wx*inv(wx'*wx)*wx');
    jj = find(w ~= 0);
    r = y - x*b;
    rs(:, k) = r;
    d = iqr(r(jj))/1.34903;
    wsum = sum(w);
    nvar = sum((w.*r).^2)/wsum;
    rss = [rss nvar];
    if abs(nvar - ovar) <= 0.01*ovar
       break
    end
    ovar = nvar;
end
tstruct.b = bs;
tstruct.bse = bses;
tstruct.t = ts;
```

```
tstruct.pval = tprobs;
if nargout > 1
   rstruct.res = rs;
   rstruct.w = ws;
   rstruct.what = whats;
   rstruct.rss = rss;
   varargout{ 1} = rstruct;
   varargout{ 2} = [ ] ;
end
if nargout > 2
   pstruct.r2 = r2;
   pstruct.anova = apval;
   pstruct.runs = rpval;
   pstruct.dw = dwpval;
   varargout{ 2} = pstruct;
end
end
```

Empirically, the leverage point weight scale factor *ainu* should usually be taken as the 0.995 quantile of the appropriate beta distribution given by

```
ainu = n*betainv(.995, p, n - p)/p;
```

where n is the number of data, and p is the number of predictors. However, if the data are systematically long tailed, a larger quantile may be needed and can be guided by the hat matrix diagonal q-q plot of the original data.

Example 9.7 Returning to the star data of Example 9.2, use the bounded influence estimator to compute a regression fit to the data.

```
star = importdata('star.dat');
y = star(:, 2);
x = [ star(:, 1) ones(size(y))] ;
n = length(y);
p = size(x, 2);
ainu = n*betainv(.995, p, n - p)/p;
[tstruct rstruct pstruct] = BI(y, x, ainu);
```

Four iterations occur, reducing the normalized residual sum of squares from 0.3052 to 0.1140, as seen in the returned variable *rstruct.rss*. The structure *pstruct* shows that the final runs and Durbin Watson p-values are 0.2485 and 3.0451e−04, respectively, which is in keeping with the results in Example 9.2. The final t table may be found in the last entries in *tstruct*.

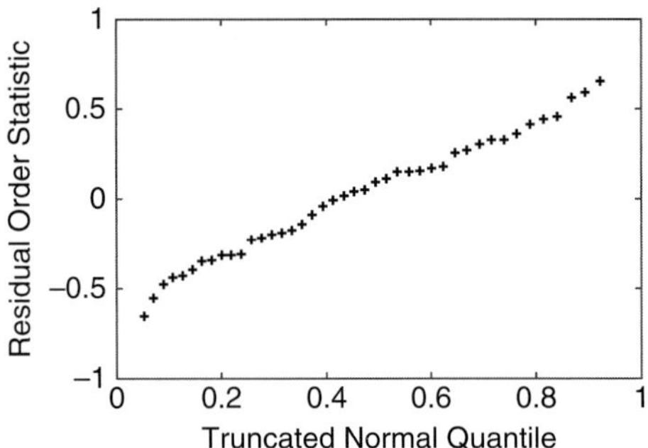

Figure 9.30 Weighted residual q-q plot from a bounded influence regression solution for the star data.

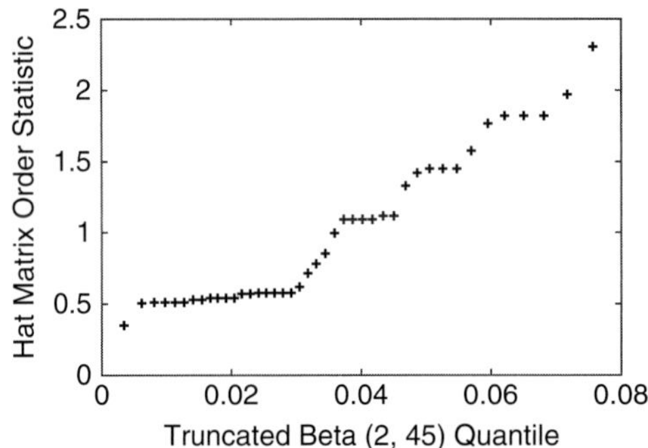

Figure 9.31 Weighted hat matrix diagonal q-q plot of a bounded influence solution for the star data.

```
[tstruct.b(:, 5) tstruct.bse(:, 5) tstruct.t(:, 5) tstruct.
pval(:, 5)]
ans =
   3.1360     0.4624     6.7823     0.0000
  -8.8864     2.0392    -4.3578     0.0001
```

The parameters compare very favorably to the final results in Example 9.2 once confidence limits are considered. The final leverage weights show that points 7, 11, 20, 30, and 34 have nearly been eliminated, and point 14 has been down-weighted by a factor of 2.

Quantile-quantile plots of the final residuals and hat matrix diagonal can be obtained from the following scripts and are shown in Figures 9.30 and 9.31:

```
w = rstruct.w(:, 4).*rstruct.what(:, 4);
ii = find(w > 0.00001); %this picks up data that are nearly
eliminated
res = rstruct.res(:, 4);
[res1 i1] = sort(w(ii).*res(ii));
%data weighted to zero are assumed evenly distributed
between the top and bottom of the %distribution
```

```
n = length(w);
m = length(ii);
mr1 = (n - m)/2;
mr2 = n - m - mr1;
i = 1:m;
p = (normcdf((n - mr2 - .5)/n) - normcdf((mr1 - .5)/n))*(i -
.5)/m + normcdf((mr1 - .5)/n);
plot(norminv(p), res1, '+')
```

The residual q-q plot is gratifyingly normal.

```
wx   = repmat(w, 1, 2).*x;
hat  = diag(wx*inv(wx'*wx)*wx');
mh1 = 0;
mh2 = n - m;
[hat1, i1]  = sort(hat(ii)*sum(w)/2);
p = betacdf(betainv((n - mh2 - .5)/n, 2, n - 2), 2, n - 2)*
(i - .5)/m;
plot(betainv(p, 2, n - 2), hat1, '+')
```

The hat matrix diagonal q-q plot is free of leverage points.

Example 9.8 Returning to the wheat data from Example 9.4 and applying the bounded influence estimator, it becomes apparent that the default tuning constant used for the star data is much too severe because it results in elimination of 90% of the wheat data, probably because the hat matrix diagonal distribution is systematically longer tailed than beta. By trial and error, it is increased to 7, although the exact value is not critical and anywhere above 5 works well. The result converges in two iterations, reducing the normalized sum of squares from 0.0351 to 0.0215. The final t table is

```
[ tstruct.b(:, 3) tstruct.bse(:, 3) tstruct.t(:, 3) tstruct.
pval(:, 3)]
ans =
  -0.0070    0.0354    -0.1979    0.8441
  -0.0598    0.0413    -1.4478    0.1555
   0.3338    0.0414     8.0598    0.0000
  -0.2219    0.0254    -8.7415    0.0000
   0.0050    0.0029     1.7176    0.0936
  -0.0548    0.0177    -3.0941    0.0036
  18.1344    4.7323     3.8320    0.0004
```

The leverage weights have eliminated points 17, 25, 34, and 50; the first three of these were removed manually in Example 9.4 based on their appearance in the hat matrix

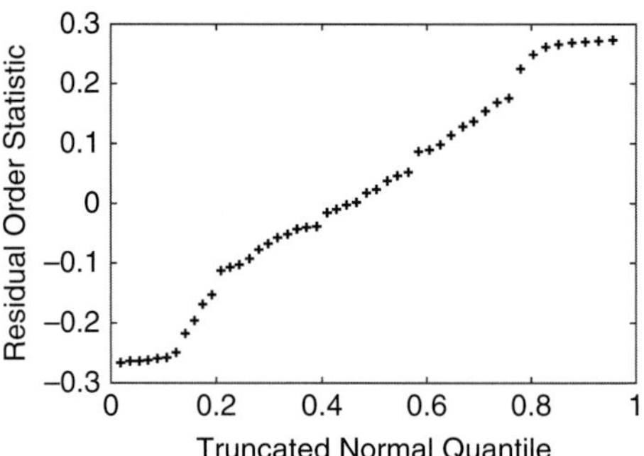

Figure 9.32 Weighted residual q-q plot from a bounded influence regression solution for the wheat data.

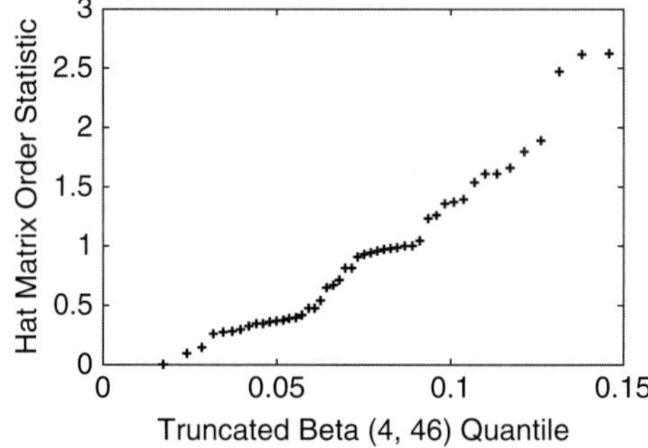

Figure 9.33 Weighted hat matrix diagonal q-q plot of a bounded influence solution for the wheat data.

diagonal q-q plot. As was the case there, the t tests on the first, second, and fifth predictors accept the null hypothesis that the coefficients are zero. Removing these three predictor variables and repeating the analysis, the bounded influence algorithm converges after three iterations, reducing the normalized sum of squares from 0.0454 to 0.0320. However, while it removes points 17 and 25, it does not affect point 34, and hence the hat matrix diagonal q-q plot still shows the effect of leverage. Reducing *ainu* from 7 to 6 solves that problem, and the normalized sum of squares decreases from 0.0454 to 0.0320 after three iterations. The final t table is

```
ans =
   0.2663    0.0057    46.4303         0
  -0.2147    0.0050   -42.8436         0
  -0.0492    0.0099    -4.9565    0.0000
  23.3819    1.9502    11.9896    0.0000
```

which is in reasonable agreement with the manually culled result in Example 9.4. The final runs and Durbin-Watson test p-values are 0.4736 and 0.0150, respectively, which is again consistent with the results in Example 9.4. Figures 9.32 and 9.33 show the final weighted residual and hat matrix diagonal q-q plots. The former is slightly short tailed, whereas the hat matrix diagonal q-q plot is quite clean.

9.7 Advanced Linear Regression

9.7.1 Errors in Variables

Throughout Sections 9.4 and 9.5 it was assumed that the predictor variables are free of random noise. In most cases, the true relationship is the linear model (9.12) with the predictors in $\overleftrightarrow{\mathbf{X}}$ given by

$$\overleftrightarrow{\mathbf{X}} = \overleftrightarrow{\mathbf{x}} + \overleftrightarrow{\boldsymbol{\delta}} \tag{9.73}$$

where it is assumed that $\overleftrightarrow{\boldsymbol{\delta}}$ is independent of $\overleftrightarrow{\mathbf{x}}$ and the random errors ε. If the least squares estimator (9.15) is applied to the observables $\left(\mathbf{y}, \overleftrightarrow{\mathbf{X}}\right)$, the resulting solution is downward biased by the predictor noise term $\overleftrightarrow{\boldsymbol{\delta}}$. To see this, note that $\overleftrightarrow{\mathbf{X}}^T \cdot \overleftrightarrow{\mathbf{X}} = \overleftrightarrow{\mathbf{x}}^T \cdot \overleftrightarrow{\mathbf{x}} + \overleftrightarrow{\boldsymbol{\delta}}^T \cdot \overleftrightarrow{\boldsymbol{\delta}}$, and hence $\left(\overleftrightarrow{\mathbf{X}}^T \cdot \overleftrightarrow{\mathbf{X}}\right)^{-1}$ in (9.15) will be systematically smaller than $\left(\overleftrightarrow{\mathbf{x}}^T \cdot \overleftrightarrow{\mathbf{x}}\right)^{-1}$ as the noise power from $\overleftrightarrow{\boldsymbol{\delta}}$ increases. In fact, it can be shown that the resulting estimator is asymptotically downward biased (i.e., it is downward biased from the true value even as $N \to \infty$). Further, the bias is not changed by substituting qr decomposition of the predictor matrix for solution of the normal equations.

There are two standard ways of dealing with this errors-in-variables problem. The first is called the *method of instrumental variables* and originated in the field of econometrics, being developed by Wald (1940), Reiersøl (1941, 1945) and Geary (1949). An auxiliary variable $\overleftrightarrow{\mathbf{z}}$ that is highly correlated with $\overleftrightarrow{\mathbf{x}}$ but not with $\overleftrightarrow{\boldsymbol{\delta}}$ is applied to yield the instrumental variables solution

$$\hat{\hat{\boldsymbol{\beta}}} = \left(\overleftrightarrow{\mathbf{z}}^T \cdot \overleftrightarrow{\mathbf{X}}\right)^{-1} \cdot \overleftrightarrow{\mathbf{z}}^T \cdot \mathbf{y} = \left(\overleftrightarrow{\mathbf{z}}^T \cdot \overleftrightarrow{\mathbf{x}}\right)^{-1} \cdot \overleftrightarrow{\mathbf{z}}^T \cdot \mathbf{y} \tag{9.74}$$

The instrumental variables approach was rediscovered by geophysicists working with magnetotelluric data in the 1970s, where it is called the *remote reference method* (Gamble, Goubau, & Clarke 1979). It works very effectively if the noise in $\overleftrightarrow{\mathbf{z}}$ and $\overleftrightarrow{\mathbf{X}}$ is uncorrelated, even when both data sets contain noise. However, it does require that $\overleftrightarrow{\mathbf{z}}$ and $\overleftrightarrow{\mathbf{X}}$ be highly correlated because weak correlation will introduce bias for two reasons: even a slight correlation between the instrumental variables in $\overleftrightarrow{\mathbf{z}}$ and the regression residuals $\hat{\mathbf{r}}$ can lead to large inconsistency, and for finite samples, the instrumental variables estimator (9.74) is always biased in the direction of ordinary least squares, with comparable bias as the correlation approaches zero (Bound, Jaeger, & Baker 1995).

The second approach to treating correlated noise is two-stage least squares. In the first stage, the local predictor variables are regressed on the instrumental variables by solving

$$\overleftrightarrow{\mathbf{X}} = \overleftrightarrow{\mathbf{z}} \cdot \overleftrightarrow{\chi} + \overleftrightarrow{\boldsymbol{\phi}} \tag{9.75}$$

where $\overleftrightarrow{\chi}$ is a matrix of regression parameters, and $\overleftrightarrow{\boldsymbol{\phi}}$ is a matrix of random errors. Equation (9.75) can accommodate a higher rank for $\overleftrightarrow{\mathbf{z}}$ than for $\overleftrightarrow{\mathbf{X}}$ and hence make use of more than one set of auxiliary variables. Using the solution to (9.75), the predicted value for the original predictor variables is computed

$$\hat{\overleftrightarrow{\mathbf{X}}} = \overleftrightarrow{\mathbf{Z}} \cdot \hat{\overleftrightarrow{\chi}} = \overleftrightarrow{\mathbf{Z}} \cdot \left(\overleftrightarrow{\mathbf{Z}}^T \cdot \overleftrightarrow{\mathbf{Z}} \right)^{-1} \cdot \overleftrightarrow{\mathbf{Z}}^T \cdot \overleftrightarrow{\mathbf{X}} = \overleftrightarrow{\mathbf{H}}_Z \cdot \overleftrightarrow{\mathbf{X}} \tag{9.76}$$

The predicted value $\hat{\overleftrightarrow{\mathbf{X}}}$ replaces $\overleftrightarrow{\mathbf{X}}$ in the original regression problem to yield the two-stage least squares solution

$$\hat{\hat{\hat{\boldsymbol{\beta}}}} = \left(\hat{\overleftrightarrow{\mathbf{X}}}^T \cdot \hat{\overleftrightarrow{\mathbf{X}}} \right)^{-1} \cdot \hat{\overleftrightarrow{\mathbf{X}}}^T \cdot \mathbf{y} \tag{9.77}$$

It can be easily shown that $\left(\hat{\overleftrightarrow{\mathbf{X}}}^T \cdot \hat{\overleftrightarrow{\mathbf{X}}} \right)^{-1} = \left(\hat{\overleftrightarrow{\mathbf{X}}}^T \cdot \overleftrightarrow{\mathbf{X}} \right)^{-1}$, and hence $\hat{\hat{\hat{\boldsymbol{\beta}}}}$ is not downward biased by noise in $\overleftrightarrow{\mathbf{X}}$.

Neither bias removal procedure comes without a cost: the variance of $\hat{\hat{\boldsymbol{\beta}}}$ and $\hat{\hat{\hat{\boldsymbol{\beta}}}}$ is always larger than that of the ordinary least squares solution $\hat{\boldsymbol{\beta}}$.

9.7.2 Shrinkage Estimators

The wheat reflectance data in Example 9.4 elaborated the process of predictor subset selection based on first removing influential data from the sample and then eliminating the predictors for which the p-values from a t test on the regression parameters are above a threshold amount. This is a variant of *backward stepwise selection* in which the least significant predictor is successively removed until all the remaining predictors are required. Alternative selection approaches are available, but backward selection typically works well unless the number of predictors is very large. However, it operates in a discrete fashion because predictors are retained (or not) according to a 0–1 selection rule. As a consequence, backward selection often exhibits high variance. By contrast, *shrinkage estimators* operate on a continuous basis and typically exhibit a smaller variance, at the cost of increased bias, following the usual goal of minimizing the mean squared error (mse). In the context of regression, the mse is given by

$$mse = \frac{1}{N} \sum_{i=1}^{N} (\hat{\mathbf{y}}_i - \mathbf{y}_i)^2 \tag{9.78}$$

where $\hat{\mathbf{y}}_i$ is the ith predicted value. Two approaches will be discussed: *ridge regression* and the *lasso*. A comprehensive treatment of both estimators is provided in Hastie, Tibshirani, & Friedman (2008, sec. 3.4).

Ridge regression shrinks the regression coefficients by penalizing their size in an L_2 norm sense

$$\hat{\beta}^{\text{ridge}} = \min \left(\left\| \mathbf{y} - \overleftrightarrow{\mathbf{X}} \cdot \boldsymbol{\beta} \right\|^2 + \lambda \|\boldsymbol{\beta}_1\|^2 \right) \tag{9.79}$$

where $\lambda \geq 0$ is an arbitrary constant that controls the amount of shrinkage, with a larger value producing a stronger effect, and $\boldsymbol{\beta}_1$ is the regression coefficients sans the intercept term. The imposition of the penalty term in (9.79) pushes the coefficients in $\boldsymbol{\beta}$ toward zero (and hence toward each other).

The normal equations solution to (9.79) is

$$\hat{\boldsymbol{\beta}}_1^{\text{ridge}} = \left(\overleftrightarrow{\mathbf{X}}_1^T \cdot \overleftrightarrow{\mathbf{X}}_1 + \lambda \overleftrightarrow{\mathbf{I}}_{p-1}\right)^{-1} \cdot \overleftrightarrow{\mathbf{X}}_1^T \cdot \mathbf{y} \tag{9.80}$$

where $\overleftrightarrow{\mathbf{X}}_1$ is the predictor matrix without a column of ones. The intercept term is obtained from ordinary least squares because the penalty term in (9.79) and (9.80) plays no role in defining it. Further, penalizing the intercept would make the solution dependent on the origin for $\mathbf{y}$. The ridge regression solution is not equivariant under affine transformation of the predictor variables; hence they are usually centered before solving (9.79) for $\hat{\boldsymbol{\beta}}_1$. For positive λ, the ridge regression solution exists even if the predictor matrix is not full rank and was its original motivation in the presence of multicollinearity.

MATLAB implements ridge regression as b = **ridge**(*y*, *x*, *lambda*, *scaled*),where y is the $n \times 1$ response vector, x is the $n \times p$ predictor matrix, *lambda* is a k vector of penalty terms, and b is a $p \times k$ matrix of solutions. By default, the predictor variables are standardized to have a mean of 0 and a standard deviation of 1, and x is assumed to not contain a column of ones. If the variable *scaled* is set to 0, then the solution will be restored to its original scale, and the first parameter in b is the intercept.

An alternative shrinkage estimator is the lasso, which stands for "least absolute shrinkage and selection operator" and was introduced by Tibshirani (1996); it differs from ridge regression in that the penalty term is based on the L_1 rather than L_2 norm. The lasso estimator is

$$\hat{\boldsymbol{\beta}}^{\text{lasso}} = \min\left[\left\|\mathbf{y} - \overleftrightarrow{\mathbf{X}} \cdot \boldsymbol{\beta}\right\|^2 + \lambda|\boldsymbol{\beta}_1|^T \cdot \mathbf{1}_{p-1}\right] \tag{9.81}$$

where $\mathbf{1}_{p-1}$ is a $p-1$ vector of ones. As for ridge regression, the parameter λ is free and controls the amount of shrinkage applied to the regression parameters. As it increases, the number of nonzero components in the regression parameters decreases. Equation (9.81) may be solved by quadratic programming with linear equality constraints, and no closed-form solution is possible. Obviously, as λ increases from zero, the coefficients $\hat{\boldsymbol{\beta}}^{\text{lasso}}$ will be biased, but this bias may be acceptable if the resulting mse decreases.

MATLAB supports a generalized version of the lasso called the *elastic net*, for which the lasso and ridge regression serve as end members. The function is [*b*, *FitInfo*] = **lasso** (*x*, *y*, 'alpha', *alpha*, 'Lambda', *lambda*, 'cv', *k*), which returns a set of fitted regression coefficients b using the regularization coefficients in *lambda*. The variable *alpha* controls the relative weights of lasso versus ridge regression, with a value approaching 0 corresponding to ridge regression and a value of 1 (default) corresponding to the lasso. The function **lasso** optionally computes the mse using a resampling technique called *cross-validation*, and the name-value pair "cv" 10 sets this to 10-fold cross-validation. If "cv" is set to "resubstitution" (which is the default), then the mse is estimated without cross-validation. There are many other options available. MATLAB also provides a plotting capability in *ax* = **lassoPlot**(*b*, *FitInfo*, 'PlotType', *x*), where x is set to "CV" to plot the mse against λ and "Lambda" to present a trace plot of the coefficients as a function of λ. However, this routine requires that **lasso** be operated in its default form. There are alternative ways to plot the results and other name-value pairs.

Example 9.9 Returning to the wheat reflectance data from Example 9.4, apply ridge regression and the lasso to estimate the required number of predictor variables after removing the leverage points identified earlier.

Apply ridge regression using the MATLAB function **ridge.**

```
lambda =[ 10:-.1:1.1 1:-.01:.11 .1:-.001:.011 .01:-.0001:.0011
.001:-.00001:.0001];
for i = 1:length(lambda)
    b = ridge(y, x, lambda(i), 0);
    yhat = y - x*b;
    mse(i) = sum((yhat - y).^2)/length(y);
end
b = ridge(y, x, lambda, 0);
```

Figure 9.34 presents the mse plotted against the regularization parameter λ. The mse rises slowly with λ from the OLS solution, peaking at $\lambda \approx 0.3$, and then descends to zero as the second part in the first term of (9.80) increasingly dominates the first part. Figure 9.35 shows a trace plot of the six coefficients (leaving out the constant term returned by **ridge**) against $\log \lambda$. The first, second, and fifth regression coefficients are all very small for $\lambda \approx 0.01$, yielding estimates for the remaining coefficients (and the intercept) that are in good agreement with the results of Example 9.4.

For the lasso, the MATLAB command is

```
[b fitinfo] = lasso(x, y, 'Lambda', lambda);
```

Figures 9.36 and 9.37 show the mse and the coefficients against the shrinkage parameter. The behavior of the lasso is quite distinct from that of ridge regression, producing abrupt changes to the coefficients and eliminating those that are not significant. A choice of λ slightly larger than 0.1 nearly eliminates the first, second, and fifth predictors at a minimum in the mse (Figure 9.36), resulting in good agreement with the backward selection result of Example 9.4.

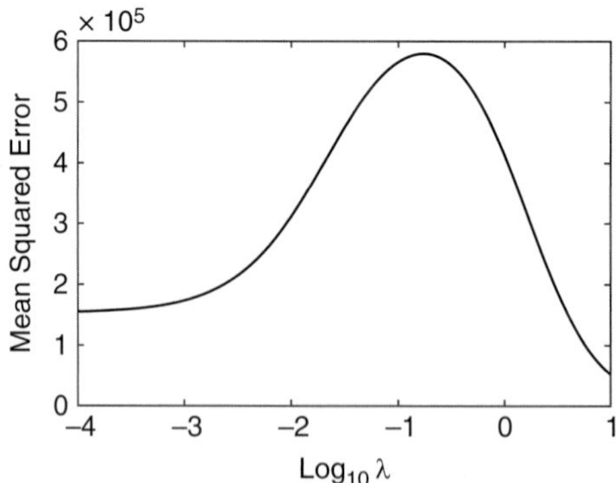

Figure 9.34 The mean square error as a function of the ridge regression shrinkage parameter for the wheat reflectance data.

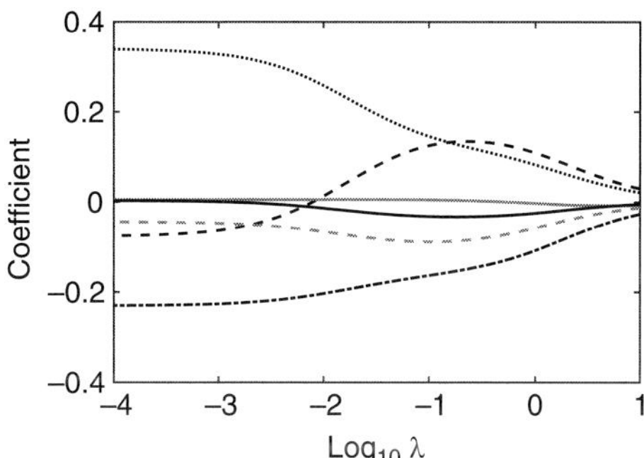

Figure 9.35 The regression parameters as a function of the ridge regression shrinkage parameter for the wheat reflectance data. The coefficients for the first (black solid line), second (black dashed line), third (black dotted line), fourth (black dash-dot line), fifth (gray solid line), and sixth (gray dashed line) wavelength bands are shown.

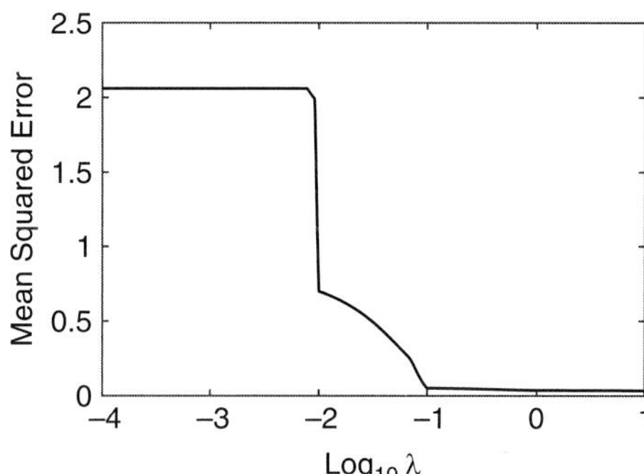

Figure 9.36 The mean square error as a function of the lasso shrinkage parameter for the wheat reflectance data.

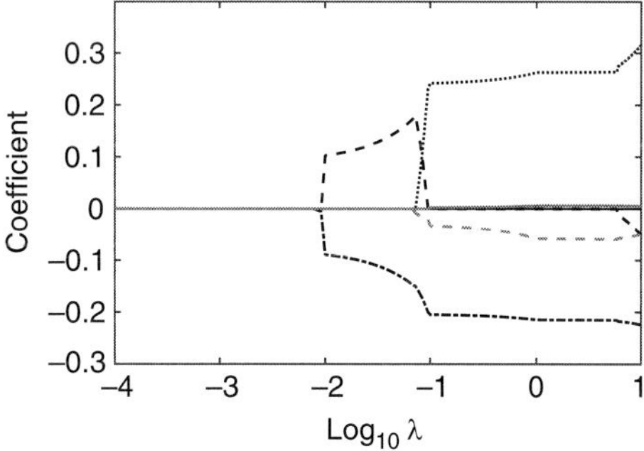

Figure 9.37 The regression parameters as a function of the lasso shrinkage parameter for the wheat reflectance data. The coefficients for the first (black solid line), second (black dashed line), third (black dotted line), fourth (black dash-dot line), fifth (gray solid line), and sixth (gray dashed line) wavelength bands are shown.

9.7.3 Logistic Regression

Suppose that the response variable **y** is dichotomous (i.e., it can take one of two possible values, such as 0 or 1) and the conditional probability $\Pr\left(\mathbf{y}=1|\overleftrightarrow{\mathbf{X}}=x\right)$ for a set of predictor variables is to be modeled using linear regression. *Logistic regression* relates the *logit link function f* to a linear combination of the predictors

$$f(p) = \log\frac{p}{1-p} = \overleftrightarrow{\mathbf{X}}\cdot\boldsymbol{\beta} \tag{9.82}$$

where $p/(1-p)$ is the *odds ratio*. The inverse of the logit is the *logistic function*

$$f^{-1}\left(\overleftrightarrow{\mathbf{X}}\cdot\boldsymbol{\beta}\right) = p = \frac{1}{1+e^{-\overleftrightarrow{\mathbf{X}}\cdot\boldsymbol{\beta}}} \tag{9.83}$$

Logically, if $p \geq 0.5$, then $\mathbf{y}=1$ should be predicted, and vice versa. Consequently, logistic regression is a linear classifier, with the decision boundary separating the two classes being the solution of $\overleftrightarrow{\mathbf{X}}\cdot\boldsymbol{\beta}=0$. Logistic regression defines the decision boundary, and (9.83) gives the class probabilities that depend on distance from the boundary, and that go to the limits of 0 or 1 more quickly as $\|\boldsymbol{\beta}\|$ increases.

The response variable $\mathbf{y}\sim\text{bin}\left[1,\ f^{-1}\left(\overleftrightarrow{\mathbf{X}}\cdot\boldsymbol{\beta}\right)\right]$ conditioned on a specific $\overleftrightarrow{\mathbf{X}}$, so the variance of **y** depends on the predictors. A weighted least squares approach is appropriate, with the weights being inversely proportional to the variance of specific measurements in **y**. This is similar to the robust estimator described in Section 9.6.1 except that the weights are derived directly from a model rather than being derived empirically from a loss or influence function. Since $\mathbf{y}=\{0,1\}, f(\mathbf{y})=\{-\infty,\infty\}$ and hence is unbounded. As a workaround, a first-order Taylor series expansion is applied to the link function

$$f(\mathbf{y}) \approx f(\boldsymbol{\mu}) + (\mathbf{y}-\boldsymbol{\mu})\ f'(\boldsymbol{\mu}) \equiv \boldsymbol{\zeta} \tag{9.84}$$

The estimated parameters $\hat{\boldsymbol{\beta}}$ are obtained by standard regression using $\boldsymbol{\zeta}$ in place of the response variables **y**. Further, the variance of $\boldsymbol{\zeta}$ conditioned on $\overleftrightarrow{\mathbf{X}}$ is given by $[f'(\boldsymbol{\mu})]^2$ times the conditional variance of **y**, which is $f^{-1}\left(\overleftrightarrow{\mathbf{X}}\cdot\boldsymbol{\beta}\right)\left[1-f^{-1}\left(\overleftrightarrow{\mathbf{X}}\cdot\boldsymbol{\beta}\right)\right]$ for a Bernoulli variable. Consequently, the variance of $\boldsymbol{\zeta}$ is $\left\{f^{-1}\left(\overleftrightarrow{\mathbf{X}}\cdot\boldsymbol{\beta}\right)\left[1-f^{-1}\left(\overleftrightarrow{\mathbf{X}}\cdot\boldsymbol{\beta}\right)\right]\right\}^{-1}$.

An iteratively reweighted estimation algorithm for logistic regression is:

1. Given a set of data $\left\{\mathbf{y}\ ,\ \overleftrightarrow{\mathbf{X}}\right\}$ and an initial estimate for the parameters $\hat{\boldsymbol{\beta}}$,
2. Continue until convergence:
   a. Compute $\hat{\boldsymbol{\mu}} = f^{-1}\left(\overleftrightarrow{\mathbf{X}}\cdot\boldsymbol{\beta}\right)$;
   b. Compute the transformed responses $\boldsymbol{\zeta}$ from (9.84);
   c. Compute the diagonal weight matrix $w_{ii} = 1/var(\zeta_i)$; and
   d. Compute the weighted least squares solution $\left(\overleftrightarrow{\mathbf{X}}^T\cdot\overleftrightarrow{\mathbf{w}}\cdot\overleftrightarrow{\mathbf{X}}\right)^{-1}\cdot\left(\overleftrightarrow{\mathbf{X}}^T\cdot\overleftrightarrow{\mathbf{w}}\cdot\boldsymbol{\zeta}\right)$ to derive new estimates for the parameters.

The weights defined in part 2c pertain to a solution using the normal equations, and if qr decomposition is used via the MATLAB **mldivide** function, the inverse of the standard deviation should be used instead.

Example 9.10 The iris flower data set is a classic one introduced by Fisher (1936) that is frequently used in discriminant analysis and data mining. The data consist of four taxonomic measurements (sepal length, sepal width, petal length, and petal width, all in centimeters) on three species of iris, with 50 samples per species. Logistic regression will be used to evaluate linear relationships among the second and third species of iris. These data are available from within MATLAB.

Compute a logistic regression solution on the second and third species, where y = 1 corresponds to species 2 and $y = 0$ corresponds to species 3

```
load fisheriris
x = [meas(51:150, :) ones(100, 1)];
y = [ones(1, 50) zeros(1, 50)]';
beta = x\y;
ovar = sum((y - x*beta).^2)/95;
for i = 1:15
    muhat = 1./(1 + exp(-x*beta));
    w = diag(muhat.*(1 - muhat));
    zeta = x*beta + (y - muhat)./(muhat.*(1 - muhat));
    beta = (x'*w*x)\(x'*w*zeta);
    nvar = sum((y - muhat).^2)/95;
    if abs(nvar - ovar) <= .01*ovar
       break
    end
    ovar = nvar;
end
```

The algorithm converges after eight iterations, reducing the residual variance from 0.1934 to 0.0198. The solution vector is

```
beta =
   2.4649
   6.6618
  -9.4034
 -18.2318
  42.4755
```

The standard error on the coefficients follows from the Fisher information matrix:

```
sevar = x'*w*x;
se = sqrt(diag(sevar))
```

```
se =
   8.4006
   3.8804
   6.7254
   2.3062
   1.3672
```

A Wald test (Section 6.5.4) may be used to assess the coefficients for significance:

```
wald = beta.^2./se.^2
wald =
  0.0861
  2.9474
  1.9549
  62.4959
  965.1345
1 - chi2cdf(wald, 1)
ans =
  0.7692
  0.0860
  0.1621
  0.0000
     0
```

The coefficients for the first three predictors accept the null hypothesis that the coefficients are zero (although the second one is weak acceptance), whereas the fourth coefficient (petal length) strongly rejects.

Logistic regression is one example of an approach called the *generalized linear model*, as proposed by Nelder & Wedderburn (1972). Other exemplars include Poisson regression, where the response variable is counts and the link function is the logarithm; gamma regression, where the response variable is gamma distributed and the link function is the inverse; and logistic regression using alternate link functions. MATLAB provides a straightforward function b = **glmfit**(x, y, distribution) that implements the general linear model, where distribution can be "binomial," "gamma," "inverse Gaussian," or "Poisson" in addition to the default "normal." By default, **glmfit** adds a leading column of ones to the predictor matrix. Additional name-value pairs are supported, of which the most useful is the name "link" followed by the name of a link function. Custom link functions can also be implemented. MATLAB also provides a function **fitglm** that is more general than **glmfit** but is at the same time bewilderingly difficult to comprehend.

Example 9.11 Apply logistic regression to the iris flower data of Example 9.10 using the MATLAB function **glmfit**.

```
beta = glmfit(x(:, 1:4), y, 'binomial', 'link', 'logit')
beta =
   42.6378
    2.4652
    6.6809
   -9.4294
  -18.2861
```

Recalling that the column of ones is in the last column in Example 9.10 but in the first column here, the results are nearly identical.

10 Multivariate Statistics

10.1 Concepts and Notation

Multivariate statistics began to be investigated in the 1920s through the work of Wishart, Hotelling, Wilks, and Mahalanobis, among others. The standard but somewhat advanced textbook on the subject is Anderson (2003), which is now in its third edition. Two more approachable general texts are Mardia, Kent, & Bibby (1979) and Rencher (1998). The scope of multivariate statistics is vast, and this chapter covers only a subset of the available techniques.

One of the major issues in multivariate statistics is notation because most statistical entities are vectors or matrices rather than the scalars typical of univariate statistics. Suppose that there are N random observation vectors of p variables. These may be written as the $N \times p$ *data matrix* $\overleftrightarrow{\mathbf{X}}$. The $1 \times p$ *sample mean vector* is $\overline{\mathbf{X}}_N = \mathbf{j}_N \cdot \overleftrightarrow{\mathbf{X}} / N$, where $\mathbf{j}_N$ is a $1 \times N$ vector of ones, and $\cdot$ denotes the inner product.

The *population mean vector* or expected value of the data matrix $\overleftrightarrow{\mathbf{X}}$ is defined to be the $1 \times p$ vector of expected values of the p variables

$$\mathcal{E}(\overleftrightarrow{\mathbf{X}}) = \mathcal{E}(\mathbf{x}_1, \ldots, \mathbf{x}_p) = (\mu_1, \ldots, \mu_p) = \boldsymbol{\mu} \tag{10.1}$$

where $\mathbf{x}_i$ is $N \times 1$. It directly follows that $\mathcal{E}(\overline{\mathbf{X}}_N) = \boldsymbol{\mu}$.

The symmetric matrix of sample variances and covariances is the *sample covariance matrix*

$$\overleftrightarrow{\mathbf{S}} = \begin{pmatrix} s_{11} & s_{12} & \cdot & s_{1p} \\ s_{21} & s_{22} & \cdot & s_{2p} \\ \cdot & \cdot & \cdot & \cdot \\ s_{p1} & s_{p2} & \cdot & s_{pp} \end{pmatrix} \tag{10.2}$$

where the biased version of the covariance obtained by normalizing by $1/N$ is typically used. The unbiased sample covariance is obtained by normalizing by $N - 1$ because a single linear constraint from estimating the mean applies to each column of $\overleftrightarrow{\mathbf{X}}$ and will be denoted as $\overleftrightarrow{\mathbf{S}}'$. The diagonal elements of $\overleftrightarrow{\mathbf{S}}$ or $\overleftrightarrow{\mathbf{S}}'$ are the sample variances of the p variables, whereas the off-diagonal element in the (j, k) position is the sample covariance between variable j and k.

A measure of overall variability of the data set is the *generalized sample variance* $\det(\overleftrightarrow{\mathbf{S}})$ or the *total sample variance* $\mathrm{tr}(\overleftrightarrow{\mathbf{S}})$. The generalized sample variance is a more complete representation of the sample variance because it incorporates the off-diagonal terms, whereas the total variance is simply the sum of all the variances.

The biased form of the sample covariance matrix may be written most succinctly in terms of the data matrix

$$\overleftrightarrow{\mathbf{S}} = \frac{1}{N}\overleftrightarrow{\mathbf{X}}^T \cdot \left(\overleftrightarrow{\mathbf{I}}_N - \frac{1}{N}\overleftrightarrow{\mathbf{J}}_N\right) \cdot \overleftrightarrow{\mathbf{X}} \tag{10.3}$$

where $\overleftrightarrow{\mathbf{I}}_N$ is the $N \times N$ identity matrix, and $\overleftrightarrow{\mathbf{J}}_N$ is an $N \times N$ matrix of ones. It can be easily shown that $\left(\overleftrightarrow{\mathbf{I}}_N - \overleftrightarrow{\mathbf{J}}_N/N\right) \cdot \left(\overleftrightarrow{\mathbf{I}}_N - \overleftrightarrow{\mathbf{J}}_N/N\right) = \left(\overleftrightarrow{\mathbf{I}}_N - \overleftrightarrow{\mathbf{J}}_N/N\right)$ [i.e., $\left(\overleftrightarrow{\mathbf{I}}_N - \overleftrightarrow{\mathbf{J}}_N/N\right)$ is idempotent], and so $\left(\overleftrightarrow{\mathbf{I}}_N - \overleftrightarrow{\mathbf{J}}_N/N\right)$ is the centering matrix that changes $\overleftrightarrow{\mathbf{X}}$ into $\overleftrightarrow{\mathbf{X}} - \overline{\mathbf{X}}_N$. It can also be shown that $\overleftrightarrow{\mathbf{S}}$ is at least positive semidefinite, meaning that all its eigenvalues are nonnegative.

The *population covariance matrix* is

$$\overleftrightarrow{\boldsymbol{\Sigma}} = cov\left(\overleftrightarrow{\mathbf{X}}\right) = \mathcal{E}\left[\left(\overleftrightarrow{\mathbf{X}} - \boldsymbol{\mu}\right)^T \cdot \left(\overleftrightarrow{\mathbf{X}} - \boldsymbol{\mu}\right)\right] = \begin{pmatrix} \sigma_{11} & \sigma_{12} & \cdot & \sigma_{1p} \\ \sigma_{21} & \sigma_{22} & \cdot & \sigma_{2p} \\ \cdot & \cdot & \cdot & \cdot \\ \sigma_{p1} & \sigma_{p2} & \cdot & \sigma_{pp} \end{pmatrix} \tag{10.4}$$

The population covariance matrix is symmetric and positive definite if there are no linear relationships between the p variables. It follows that $\mathcal{E}\left(\overleftrightarrow{\mathbf{S}}\right) = N\,\overleftrightarrow{\boldsymbol{\Sigma}}/(N-1)$ or $E\left(\overleftrightarrow{\mathbf{S}}'\right) = \overleftrightarrow{\boldsymbol{\Sigma}}$.

The *sample correlation matrix* is

$$\overleftrightarrow{\mathbf{R}} = \begin{pmatrix} 1 & r_{12} & \cdot & r_{1p} \\ r_{21} & 1 & \cdot & r_{2p} \\ \cdot & \cdot & \cdot & \cdot \\ r_{p1} & r_{p2} & \cdot & 1 \end{pmatrix} \tag{10.5}$$

where $r_{jk} = s_{jk}/\sqrt{s_{jj}s_{kk}}$. $\overleftrightarrow{\mathbf{R}}$ may be related to $\overleftrightarrow{\mathbf{S}}$ by defining the diagonal matrix $\overleftrightarrow{\mathbf{D}}_s = \left[\mathrm{diag}\left(\overleftrightarrow{\mathbf{S}}\right)\right]^{1/2}$ so that $\overleftrightarrow{\mathbf{S}} = \overleftrightarrow{\mathbf{D}}_s \cdot \overleftrightarrow{\mathbf{R}} \cdot \overleftrightarrow{\mathbf{D}}_s$ and $\overleftrightarrow{\mathbf{R}} = \overleftrightarrow{\mathbf{D}}_s^{-1} \cdot \overleftrightarrow{\mathbf{S}} \cdot \overleftrightarrow{\mathbf{D}}_s^{-1}$. If the data matrix $\overleftrightarrow{\mathbf{X}}$ has been standardized to $\overleftrightarrow{\mathbf{Z}}$ by removing the sample mean $\overline{\mathbf{X}}_N$ from each entry and dividing by the corresponding sample standard deviation, then $\overleftrightarrow{\mathbf{S}} = \overleftrightarrow{\mathbf{R}}$.

The *population correlation matrix* is

$$\overleftrightarrow{\mathbf{P}} = \begin{pmatrix} 1 & \rho_{12} & \cdot & \rho_{1p} \\ \rho_{21} & 1 & \cdot & \rho_{2p} \\ \cdot & \cdot & \cdot & \cdot \\ \rho_{p1} & \rho_{p2} & \cdot & 1 \end{pmatrix} \tag{10.6}$$

where $\rho_{jk} = \sigma_{jk}/\sqrt{\sigma_{jj}\sigma_{kk}}$. A diagonal matrix $\overleftrightarrow{\mathbf{D}}_\sigma$ analogous to $\overleftrightarrow{\mathbf{D}}_s$ may be defined such that $\overleftrightarrow{\boldsymbol{\Sigma}} = \overleftrightarrow{\mathbf{D}}_\sigma \cdot \overleftrightarrow{\mathbf{P}} \cdot \overleftrightarrow{\mathbf{D}}_\sigma$ and $\overleftrightarrow{\mathbf{P}} = \overleftrightarrow{\mathbf{D}}_\sigma^{-1} \cdot \overleftrightarrow{\boldsymbol{\Sigma}} \cdot \overleftrightarrow{\mathbf{D}}_\sigma^{-1}$. However, $\overleftrightarrow{\mathbf{R}}$ is a biased estimator for $\overleftrightarrow{\mathbf{P}}$. The sample correlation matrix based on the unbiased sample covariance matrix will be denoted $\overleftrightarrow{\mathbf{R}}'$.

A multivariate measure of the distance between two vectors $\mathbf{y}_1$ and $\mathbf{y}_2$ must account for the covariances as well as the variances of the variables, so the simple Euclidean distance $(\mathbf{y}_1 - \mathbf{y}_2)^T \cdot (\mathbf{y}_1 - \mathbf{y}_2)$ does not suffice. The *Mahalanobis distance* standardizes distance using the inverse of the unbiased covariance matrix

$$d^2 = (\mathbf{y}_1 - \mathbf{y}_2)^T \cdot \overleftrightarrow{\mathbf{S}}'^{-1} \cdot (\mathbf{y}_1 - \mathbf{y}_2) \tag{10.7}$$

The inverse covariance matrix transforms the variables so that they are uncorrelated and have the same variance, resulting in a rational distance measure.

10.2 The Multivariate Gaussian Distribution

10.2.1 Derivation of the Multivariate Gaussian Distribution

Most multivariate (mv) inferential procedures are based on the mv Gaussian distribution or distributions derived from it. The mv Gaussian distribution is a direct generalization of the univariate or bivariate Gaussian covered in Sections 3.4.1 and 3.4.10.

Suppose that there is a random vector $\mathbf{z} = (z_1, \ldots, z_p)$, where each $z_i \sim N(0, 1)$, and the $\{z_i\}$ are mutually independent. Their joint pdf is just the product of the marginal pdfs

$$N_p\left(\mathbf{0}, \overleftrightarrow{\mathbf{I}}_p\right) = \frac{1}{\left(\sqrt{2\pi}\right)^p} e^{-\mathbf{z}\cdot\mathbf{z}^T/2} \tag{10.8}$$

A more general mv normal pdf with mean vector $\boldsymbol{\mu}$ and covariance matrix $\overleftrightarrow{\boldsymbol{\Sigma}}$ may be obtained in an analogous manner to the derivation of the bivariate normal distribution in Section 3.4.10. Let $\mathbf{x} = \overleftrightarrow{\boldsymbol{\Sigma}}^{1/2} \cdot \mathbf{z}^T + \boldsymbol{\mu}$, where the square root of the population covariance matrix can be obtained by finding its p eigenvalues $\overleftrightarrow{\boldsymbol{\Lambda}} = \mathrm{diag}(\lambda_i)$ and eigenvector matrix $\overleftrightarrow{\mathbf{C}}$, so $\overleftrightarrow{\boldsymbol{\Sigma}}^{1/2} = \overleftrightarrow{\mathbf{C}} \cdot \overleftrightarrow{\boldsymbol{\Lambda}}^{1/2} \cdot \overleftrightarrow{\mathbf{C}}^T$, with $\overleftrightarrow{\boldsymbol{\Lambda}}^{1/2} = \mathrm{diag}\left(\sqrt{\lambda_i}\right)$. The expected value and variance for $\mathbf{x}$ are $\boldsymbol{\mu}$ and $\overleftrightarrow{\boldsymbol{\Sigma}}$, respectively.

Writing $\mathbf{z}$ in terms of $\mathbf{x}$, substituting into the joint pdf, and performing the transformation gives

$$N_p\left(\boldsymbol{\mu}, \overleftrightarrow{\boldsymbol{\Sigma}}\right) = \frac{1}{\sqrt{(2\pi)^p |\overleftrightarrow{\boldsymbol{\Sigma}}|}} e^{-(\mathbf{x}-\boldsymbol{\mu})\cdot\overleftrightarrow{\boldsymbol{\Sigma}}^{-1}\cdot(\mathbf{x}-\boldsymbol{\mu})^T/2} \tag{10.9}$$

where the vertical bars in the denominator denote the determinant, not the absolute value. Note that the exponent is the Mahalanobis distance between $\mathbf{x}$ and $\boldsymbol{\mu}$, where for the univariate Gaussian it is the simple Euclidean distance divided by the variance, and the generalized variance replaces the simple variance in the denominator. The p-variate Gaussian distribution (10.9) contains p means, p variances, and $\binom{p}{2}$ covariances, for a total of $p(p+3)/2$ parameters.

The central and noncentral moments of $\mathbf{x}$ are defined analogously to those for the univariate distribution. The first moment is the mean vector $\boldsymbol{\mu}$. The second central moment is $\mathcal{E}\left[(x_j - \mu_j) \cdot (x_k - \mu_k)\right] = \sigma_{jk}$, or the entries in the population covariance matrix. The third central moment is zero, as are all further odd-order moments. The fourth central moment is $\mathcal{E}\left[(x_i - \mu_i) \cdot (x_j - \mu_j) \cdot (x_k - \mu_k) \cdot (x_l - \mu_l)\right] = \sigma_{ij}\sigma_{kl} + \sigma_{ik}\sigma_{jl} + \sigma_{il}\sigma_{jk}$. For higher orders, even central moments become more complicated. Fortunately, they are rarely needed.

MATLAB implements the cdf, pdf, and random number generation from the mv Gaussian distribution as **mvnpdf**(x, *mu*, *sigma*) and so on, where *mu* is a vector and *sigma* is a matrix. Neither distribution objects nor the quantile function are supported for the mv Gaussian.

10.2.2 Properties of the MV Gaussian Distribution

Some key properties of the mv normal distribution will be stated and discussed without proof (which can be found in standard texts). Suppose a linear transformation $\overleftrightarrow{\mathbf{A}}$ is applied to a p-variate random Gaussian vector $\mathbf{x}$. Further, suppose that $\overleftrightarrow{\mathbf{A}}$is $q \times p$ and has rank $q \le p$. The distribution of the linear transformation applied to $\mathbf{x}$ is

$$\overleftrightarrow{\mathbf{A}} \cdot \mathbf{x} \sim N_q\left(\overleftrightarrow{\mathbf{A}} \cdot \boldsymbol{\mu}, \overleftrightarrow{\mathbf{A}} \cdot \overleftrightarrow{\boldsymbol{\Sigma}} \cdot \overleftrightarrow{\mathbf{A}}^T\right) \tag{10.10}$$

Equation (10.10) holds when $q = 1$ and $\overleftrightarrow{\mathbf{A}}$ is a p-vector $\mathbf{a}$.

The linear transformation (10.10) can be used to standardize mv Gaussian variables in a different way than the transformation $(\mathbf{x} - \boldsymbol{\mu}) \cdot \left(\overleftrightarrow{\boldsymbol{\Sigma}}^{1/2}\right)^{-1}$ that was applied to get (10.9). Any symmetric positive definite matrix can be factored into the product of a lower triangular matrix and its transpose using the *Cholesky decomposition*. Let $\overleftrightarrow{\boldsymbol{\Sigma}} = \overleftrightarrow{\mathbf{E}} \cdot \overleftrightarrow{\mathbf{E}}^T$ be the Cholesky decomposition of the population covariance matrix. Then a standardized version of a Gaussian random vector $\mathbf{x}$ is given by $\mathbf{z} = (\mathbf{x} - \boldsymbol{\mu}) \cdot \overleftrightarrow{\mathbf{E}}^{-1}$. In both cases, $\mathbf{z}$ is distributed as $N_p(\mathbf{0},\ \mathbf{I}_p)$, meaning that each z_i is independently distributed as $N(0,\ 1)$.

If $\mathbf{x} \sim N_p(\boldsymbol{\mu}, \boldsymbol{\Sigma})$, then any subvector of $\mathbf{x}$ is mv normal with the corresponding means, variances, and covariances. For example, if the first r elements of $\mathbf{x}$ are the subvector, then the result is $N_r\left(\boldsymbol{\mu}_r, \overleftrightarrow{\boldsymbol{\Sigma}}_r\right)$, where $\boldsymbol{\mu}_r$ contains the first r elements in $\boldsymbol{\mu}$, and $\overleftrightarrow{\boldsymbol{\Sigma}}_r$ is the upper left $r \times r$ partition of $\overleftrightarrow{\boldsymbol{\Sigma}}$. This establishes that the marginal distributions for the mv Gaussian are also mv Gaussian. It follows that each x_i is distributed as $N(\mu_i, \sigma_{ii})$. The reverse is not true; if each variable x_i in a set of variables is distributed as $N(\mu_i, \sigma_{ii})$, it does not follow that $\mathbf{x}$ is mv Gaussian.

If $\mathbf{x}$ and $\mathbf{y}$ are jointly mv Gaussian with $\overleftrightarrow{\boldsymbol{\Sigma}}_{xy} = \mathbf{0}$, then $\mathbf{x}$ and $\mathbf{y}$ are independent. In other words, zero covariance implies independence, and the converse also holds. The mv Gaussian is the only mv distribution with this property. If $\mathbf{x} \sim N_p\left(\boldsymbol{\mu},\ \overleftrightarrow{\boldsymbol{\Sigma}}\right)$, then any two variables $\mathbf{x}_i$ and $\mathbf{x}_j$ are independent if $\sigma_{ij} = 0$. Further, when two Gaussian vectors $\mathbf{x}$ and $\mathbf{y}$ are the same size with zero covariance, they can be added or subtracted according to

$$\mathbf{x} \pm \mathbf{y} \sim N_p\left(\boldsymbol{\mu}_x \pm \boldsymbol{\mu}_y, \overleftrightarrow{\boldsymbol{\Sigma}}_{xx} + \overleftrightarrow{\boldsymbol{\Sigma}}_{yy}\right) \tag{10.11}$$

If $\mathbf{x}$ and $\mathbf{y}$ are jointly mv normal with $\overleftrightarrow{\boldsymbol{\Sigma}}_{xy} \ne \mathbf{0}$, then the conditional distribution of $\mathbf{x}$ given $\mathbf{y}$ is mv normal with conditional mean vector and covariance matrix given by

$$
\begin{aligned}
\mathcal{E}(\mathbf{x}|\mathbf{y}) &= \boldsymbol{\mu}_x + \overleftrightarrow{\boldsymbol{\Sigma}}_{xy} \cdot \overleftrightarrow{\boldsymbol{\Sigma}}_{yy}^{-1} \cdot (\mathbf{y} - \boldsymbol{\mu}_y) \\
cov(\mathbf{x}|\mathbf{y}) &= \overleftrightarrow{\boldsymbol{\Sigma}}_{xx} - \overleftrightarrow{\boldsymbol{\Sigma}}_{xy} \cdot \overleftrightarrow{\boldsymbol{\Sigma}}_{yy}^{-1} \cdot \overleftrightarrow{\boldsymbol{\Sigma}}_{yx}
\end{aligned} \tag{10.12}
$$

This defines the relationships of subvectors that are not independent.

The sum of the squares of p independent standardized Gaussian variables is chi square with p degrees-of-freedom. For mv standardized Gaussian variables, if $\mathbf{x} \sim N_p(\boldsymbol{\mu}, \overleftrightarrow{\boldsymbol{\Sigma}})$, then the quadratic form $(\mathbf{x} - \boldsymbol{\mu}) \cdot \overleftrightarrow{\boldsymbol{\Sigma}}^{-1} \cdot (\mathbf{x} - \boldsymbol{\mu})^T$ is χ_p^2.

10.2.3 The Sample Mean Vector and Sample Covariance Matrix

The maximum likelihood estimators (mles) for the mean vector and population covariance matrix may be derived in the same way as for the univariate normal distribution. If $\overleftrightarrow{\mathbf{X}}$ contains N samples from $N_p(\boldsymbol{\mu}, \overleftrightarrow{\boldsymbol{\Sigma}})$, then the mles for $\boldsymbol{\mu}$ and $\overleftrightarrow{\boldsymbol{\Sigma}}$ are

$$
\begin{aligned}
\hat{\boldsymbol{\mu}} &= \overline{\mathbf{X}}_N \\
\hat{\overleftrightarrow{\boldsymbol{\Sigma}}} &= \overleftrightarrow{\mathbf{W}}/N = \overleftrightarrow{\mathbf{S}}
\end{aligned} \tag{10.13}
$$

The total sum of squares matrix is

$$
\overleftrightarrow{\mathbf{W}} = \left(\overleftrightarrow{\mathbf{X}} - \overline{\mathbf{X}}_N\right)^T \cdot \left(\overleftrightarrow{\mathbf{X}} - \overline{\mathbf{X}}_N\right) \tag{10.14}
$$

Equations (10.13) and (10.14) can be proved in the same way as for the univariate Gaussian, with the slight added complexity of dealing with matrices.

As for the univariate case, the mle is equivariant, so the mle of a function is given by the function of the mle. For example, consider the mle for the population correlation matrix $\overleftrightarrow{\mathbf{P}} = \overleftrightarrow{\mathbf{D}}_\sigma^{-1} \cdot \overleftrightarrow{\boldsymbol{\Sigma}} \cdot \overleftrightarrow{\mathbf{D}}_\sigma^{-1}$, where $\overleftrightarrow{\mathbf{D}}_\sigma^{-1} = \mathrm{diag}(1/\sigma_{ii})$. The mle for $1/\sigma_{ii}$ is $1/\hat{\sigma}_{ii}$, so $\hat{\overleftrightarrow{\mathbf{P}}} = \hat{\overleftrightarrow{\mathbf{D}}}_\sigma^{-1} \cdot \hat{\overleftrightarrow{\boldsymbol{\Sigma}}} \cdot \hat{\overleftrightarrow{\mathbf{D}}}_\sigma^{-1}$, which is the same as the mle for the sample correlation matrix $\overleftrightarrow{\mathbf{R}}$.

The properties of the sample mean vector and sample covariance matrix also resemble those for the univariate case. For example, for an $N \times p$ data matrix $\overleftrightarrow{\mathbf{X}}$, the sample mean vector $\overline{\mathbf{X}}_N \sim N_p(\boldsymbol{\mu}, \overleftrightarrow{\boldsymbol{\Sigma}}/N)$. Further, $\overline{\mathbf{X}}_N$ and $\overleftrightarrow{\mathbf{S}}$ are independent and jointly sufficient for $\boldsymbol{\mu}$ and $\overleftrightarrow{\boldsymbol{\Sigma}}$. This can be shown by factoring the mv normal pdf and means that all the information necessary to describe a random mv Gaussian sample $\overleftrightarrow{\mathbf{X}}$ is contained in $\overline{\mathbf{X}}_N$ and $\overleftrightarrow{\mathbf{S}}$.

The mv central limit theorem states that if $\overline{\mathbf{X}}_N$ is derived from a random sample $\overleftrightarrow{\mathbf{X}}$ from a population with mean vector $\boldsymbol{\mu}$ and covariance matrix $\overleftrightarrow{\boldsymbol{\Sigma}}$, then $\overline{\mathbf{X}}_N \xrightarrow{p} N_p(\boldsymbol{\mu}, \overleftrightarrow{\boldsymbol{\Sigma}}/N)$. The distribution of $\overline{\mathbf{X}}_N$ is asymptotically $N_p(\boldsymbol{\mu}, \overleftrightarrow{\boldsymbol{\Sigma}}/N)$ regardless of the distribution of $\overleftrightarrow{\mathbf{X}}$, provided that the covariance matrix $\overleftrightarrow{\boldsymbol{\Sigma}}$ is finite.

The distribution of the sample covariance matrix $\overleftrightarrow{\mathbf{S}}$ is a multivariate analogue of the chi square distribution called the *Wishart distribution*, as originally derived by Wishart (1928). If the variables in $\overleftrightarrow{\mathbf{X}}$ are independent and distributed as $N_p(\boldsymbol{\mu}, \overleftrightarrow{\boldsymbol{\Sigma}})$, then the $p(p+1)/2$ distinct variables in $\left(\overleftrightarrow{\mathbf{X}} - \boldsymbol{\mu}\right)^T \cdot \left(\overleftrightarrow{\mathbf{X}} - \boldsymbol{\mu}\right)$ are jointly distributed as the Wishart distribution $W_p(N, \overleftrightarrow{\boldsymbol{\Sigma}})$, where N is the degrees-of-freedom. Consequently, the $p(p+1)/2$ elements of $\overleftrightarrow{\mathbf{W}} = \left(\overleftrightarrow{\mathbf{X}} - \overline{\mathbf{X}}_N\right)^T \cdot \left(\overleftrightarrow{\mathbf{X}} - \overline{\mathbf{X}}_N\right) = N\,\overleftrightarrow{\mathbf{S}}$ are distributed as $W_p(N-1, \overleftrightarrow{\boldsymbol{\Sigma}})$, which reflects the loss of one degree-of-freedom from estimating the sample mean of each column of $\overleftrightarrow{\mathbf{X}}$.

The Wishart distribution has a reproductive property like that for the chi square distribution. If $\overleftrightarrow{\mathbf{W}}_1 \sim W_p\left(N_1, \overleftrightarrow{\Sigma}\right)$ and $\overleftrightarrow{\mathbf{W}}_2 \sim W_p\left(N_2, \overleftrightarrow{\Sigma}\right)$ and $\overleftrightarrow{\mathbf{W}}_1$ and $\overleftrightarrow{\mathbf{W}}_2$ are independent, then $\overleftrightarrow{\mathbf{W}}_1 + \overleftrightarrow{\mathbf{W}}_2 \sim W_p\left(N_1 + N_2, \overleftrightarrow{\Sigma}\right)$. Further, if $\overleftrightarrow{\mathbf{W}} \sim W_p\left(N, \overleftrightarrow{\Sigma}\right)$ and $\overleftrightarrow{\mathbf{C}}$ is a $q \times p$ constant matrix of rank $q \le p$, then $\overleftrightarrow{\mathbf{C}} \cdot \overleftrightarrow{\mathbf{W}} \cdot \overleftrightarrow{\mathbf{C}}^T \sim W_q\left(N, \overleftrightarrow{\mathbf{C}} \cdot \overleftrightarrow{\Sigma} \cdot \overleftrightarrow{\mathbf{C}}^T\right)$. The degrees-of-freedom for the Wishart distribution have decreased, as evidenced by the fact that $\overleftrightarrow{\mathbf{C}} \cdot \overleftrightarrow{\mathbf{W}} \cdot \overleftrightarrow{\mathbf{C}}^T$ is $q \times q$.

MATLAB supports the generation of random numbers from a Wishart distribution using **wishrnd**(*sigma*, *df*), where *sigma* is the covariance matrix and *df* is the degrees-of-freedom.

10.2.4 The Complex Multivariate Gaussian Distribution

The complex mv Gaussian distribution is important in spectral analysis and signal processing. However, some of the concepts in Section 10.1 must be extended to handle complex data before introducing the complex form of the Gaussian distribution.

A complex random vector $\mathbf{x} = \mathbf{x}_r + i\,\mathbf{x}_i$ has p dimensions, where $\mathbf{x}_r$ and $\mathbf{x}_i$ are, respectively, its real and imaginary parts. The probability distribution of a complex random vector is the joint distribution of the real and imaginary parts and hence is at least bivariate. Its second-order statistics are described by the covariance and *pseudocovariance* (sometimes called the *complementary covariance*) matrices (Van Den Bos 1995; Picinbono 1996):

$$
\begin{aligned}
\overleftrightarrow{\Gamma} &= \mathcal{E}\left[(\mathbf{x} - \boldsymbol{\mu}_x)^H (\mathbf{x} - \boldsymbol{\mu}_x)\right] = \overleftrightarrow{\Sigma}_{x_r x_r} + \overleftrightarrow{\Sigma}_{x_i x_i} + i\left(\overleftrightarrow{\Sigma}^T_{x_r x_i} - \overleftrightarrow{\Sigma}_{x_r x_i}\right) \\
\breve{\overleftrightarrow{\Gamma}} &= \mathcal{E}\left[(\mathbf{x} - \boldsymbol{\mu}_x)^T (\mathbf{x} - \boldsymbol{\mu}_x)\right] = \overleftrightarrow{\Sigma}_{x_r x_r} - \overleftrightarrow{\Sigma}_{x_i x_i} + i\left(\overleftrightarrow{\Sigma}^T_{x_r x_i} + \overleftrightarrow{\Sigma}_{x_r x_i}\right)
\end{aligned} \tag{10.15}
$$

where the superscript H denotes the complex conjugate transpose, and $\overleftrightarrow{\Sigma}$ denotes the real covariance matrix of the elements given by its subscript. The covariance matrix $\overleftrightarrow{\Gamma}$ is complex, Hermitian, and positive semidefinite, whereas the pseudocovariance matrix $\breve{\overleftrightarrow{\Gamma}}$ is complex and symmetric.

A complex random vector $\mathbf{x}$ is proper if $\breve{\overleftrightarrow{\Gamma}}$ is identically zero and otherwise is improper. The conditions on the covariance matrices of the real and imaginary parts for propriety reduce to

$$
\begin{aligned}
\overleftrightarrow{\Sigma}_{x_r x_r} &= \overleftrightarrow{\Sigma}_{x_i x_i} \\
\overleftrightarrow{\Sigma}_{x_r x_i} &= -\overleftrightarrow{\Sigma}^T_{x_r x_i}
\end{aligned} \tag{10.16}
$$

where the second equation in (10.16) requires that the diagonal elements of $\overleftrightarrow{\Sigma}_{x_r x_i}$ vanish. The complex covariance matrix for proper data is then given equivalently by the sum or difference of the two equations in (10.15). However, in the improper case, both $\overleftrightarrow{\Gamma}$ and $\breve{\overleftrightarrow{\Gamma}}$ are required for a complete description of the second-order statistics of a complex random vector.

Extending statistical definitions from the real line to the complex plane is a topic that is not without controversy. In the signal processing literature, where there has been extensive recent work, the guiding approach has been that definitions and principles should be the same in the real and complex domains. Schreier & Scharf (2010) provide a comprehensive recent survey. This has led to the concept of so-called augmented variables and covariance matrices. The augmented complex random variable for $\mathbf{x}$ is given by $\underset{\frown}{\mathbf{x}} = (\mathbf{x}\ \mathbf{x}^*)$, where the superscript * denotes the complex conjugate and is obtained by adding the p variables of $\mathbf{x}^*$ to those of $\mathbf{x}$. Its elements $\mathbf{x}$ and $\mathbf{x}^*$ are clearly not independent, but augmented variables simplify statistical algebra. The augmented covariance matrix is given by

$$\underset{\frown}{\overleftrightarrow{\boldsymbol{\Gamma}}} = \begin{bmatrix} \overleftrightarrow{\boldsymbol{\Gamma}} & \breve{\overleftrightarrow{\boldsymbol{\Gamma}}} \\ \breve{\overleftrightarrow{\boldsymbol{\Gamma}}}^* & \overleftrightarrow{\boldsymbol{\Gamma}}^* \end{bmatrix} \tag{10.17}$$

where $\underset{\frown}{\overleftrightarrow{\boldsymbol{\Gamma}}}$ is block structured, Hermitian, and positive semidefinite.

Using the augmented variable notation, the pdf of a complex Gaussian random vector is

$$\underset{\frown}{C}_p\left(\underset{\frown}{\boldsymbol{\mu}}, \underset{\frown}{\overleftrightarrow{\Gamma}}\right) = \frac{e^{-(1/2)\left(\underset{\frown}{\mathbf{x}}-\underset{\frown}{\boldsymbol{\mu}}\right)\cdot \underset{\frown}{\overleftrightarrow{\Gamma}}^{-1}\cdot\left(\underset{\frown}{\mathbf{x}}-\underset{\frown}{\boldsymbol{\mu}}\right)^H}}{\pi^p\sqrt{\left|\underset{\frown}{\overleftrightarrow{\Gamma}}\right|}} \tag{10.18}$$

and is viewed as the joint pdf of $\mathbf{x}_r$ and $\mathbf{x}_i$ pertaining to both proper and improper variables. By the block matrix inversion lemma, the inverse of the augmented covariance matrix is

$$\underset{\frown}{\overleftrightarrow{\Gamma}}^{-1} = \begin{Bmatrix} \left[\overleftrightarrow{\Gamma} - \breve{\overleftrightarrow{\Gamma}}\cdot\left(\overleftrightarrow{\Gamma}^*\right)^{-1}\cdot\breve{\overleftrightarrow{\Gamma}}^*\right]^{-1} & -\left[\overleftrightarrow{\Gamma} - \breve{\overleftrightarrow{\Gamma}}\cdot\left(\overleftrightarrow{\Gamma}^*\right)^{-1}\cdot\breve{\overleftrightarrow{\Gamma}}^*\right]^{-1}\cdot\breve{\overleftrightarrow{\Gamma}}\cdot\left(\overleftrightarrow{\Gamma}^*\right)^{-1} \\ -\left[\overleftrightarrow{\Gamma}^* - \breve{\overleftrightarrow{\Gamma}}^*\cdot\overleftrightarrow{\Gamma}^{-1}\cdot\breve{\overleftrightarrow{\Gamma}}\right]^{-1}\cdot\breve{\overleftrightarrow{\Gamma}}^*\cdot\overleftrightarrow{\Gamma}^{-1} & \left[\overleftrightarrow{\Gamma}^* - \breve{\overleftrightarrow{\Gamma}}^*\cdot\overleftrightarrow{\Gamma}^{-1}\cdot\breve{\overleftrightarrow{\Gamma}}\right]^{-1} \end{Bmatrix} \tag{10.19}$$

In older works, the complex Gaussian distribution is usually implicitly taken to be the proper form, in which case $\breve{\overleftrightarrow{\Gamma}}=0$ and $\underset{\frown}{\overleftrightarrow{\Gamma}}$ is block diagonal. The proper complex Gaussian distribution is given by

$$C_p\left(\boldsymbol{\mu}, \overleftrightarrow{\Gamma}\right) = \frac{e^{-(1/2)(\mathbf{x}-\boldsymbol{\mu})\cdot\overleftrightarrow{\Gamma}^{-1}\cdot(\mathbf{x}-\boldsymbol{\mu})^H}}{\pi^p\left|\overleftrightarrow{\Gamma}\right|} \tag{10.20}$$

10.3 Hotelling's T^2 Tests

Multivariate hypothesis testing is more complicated than its univariate counterpart because the number of hypotheses that can be posed is much larger and because it is generally preferable to test p variables with a single test rather than performing p univariate tests. The single test preserves the significance level, generally has a higher power because multivariate tests include the effect of correlation between variables, and enables testing of the contribution of each variable to the result.

Consider the test H_0: $\boldsymbol{\mu} = \boldsymbol{\mu}_0$ versus H_1: $\boldsymbol{\mu} \neq \boldsymbol{\mu}_0$ when the covariance matrix $\overleftrightarrow{\boldsymbol{\Sigma}}$ is known. The inequality in the alternate hypothesis implies that at least one of the parameters $\mu_j \neq \mu_{0j}$. A random sample of N observations $\overleftrightarrow{\mathbf{X}}$ is distributed as $N_p\left(\boldsymbol{\mu}, \overleftrightarrow{\boldsymbol{\Sigma}}\right)$, from which the sample mean vector $\overline{\mathbf{X}}_N$ is calculated. The test statistic is

$$\hat{Z}^2 = N\left(\overline{\mathbf{X}}_N - \boldsymbol{\mu}_0\right) \cdot \overleftrightarrow{\boldsymbol{\Sigma}}^{-1} \cdot \left(\overline{\mathbf{X}}_N - \boldsymbol{\mu}_0\right)^T \tag{10.21}$$

or the Mahalanobis distance between $\overline{\mathbf{X}}_N$ and $\boldsymbol{\mu}_0$ scaled by N and is distributed as χ_p^2 if the null hypothesis is true. The test rejects H_0 if $\hat{Z}^2 \geq \chi_p^2(1 - \alpha/2) \cap \hat{Z}^2 < \chi_p^2(\alpha/2)$, and p-values can be obtained in the usual way. For the univariate case using (10.21), the test statistic is the square of the z statistic of Section 6.3.1 that is tested against the chi square distribution with one degree-of-freedom. This test is based on a likelihood ratio and hence has concomitant optimality properties.

When the covariance matrix $\overleftrightarrow{\boldsymbol{\Sigma}}$ is unknown, the multivariate counterpart to the one-sample t test described in Section 6.3.2 pertains. Given a random sample of N observations $\overleftrightarrow{\mathbf{X}}$ distributed as $N_p\left(\boldsymbol{\mu}, \overleftrightarrow{\boldsymbol{\Sigma}}\right)$, where $\overleftrightarrow{\boldsymbol{\Sigma}}$ is unknown, the test is again H_0: $\boldsymbol{\mu} = \boldsymbol{\mu}_0$ against H_1: $\boldsymbol{\mu} \neq \boldsymbol{\mu}_0$. The test statistic is

$$\hat{T}^2 = N\left(\overline{\mathbf{X}}_N - \boldsymbol{\mu}_0\right) \cdot \overleftrightarrow{\mathbf{S}}'^{-1} \cdot \left(\overline{\mathbf{X}}_N - \boldsymbol{\mu}_0\right)^T \tag{10.22}$$

or, equivalently, may be obtained from (10.21) by replacing the population covariance matrix with the unbiased sample covariance matrix $\overleftrightarrow{\mathbf{S}}'$. The test statistic is simply the Mahalanobis distance between the observed and postulated means scaled by N.

The statistic in (10.22) has Hotelling's T^2 distribution with dimension p and degrees-of-freedom $N - 1$ if the null hypothesis is true and was first derived by Hotelling (1931). The test rejects H_0 if $\hat{T}^2 \geq T_{p,N-1}^2(1 - \alpha/2) \cap \hat{T}^2 < T_{p,N-1}^2(\alpha/2)$, and the two-sided p-value follows in the usual way. When $p = 1$, the T^2 statistic reduces to the square of the univariate t statistic t_{N-1}^2, which, in turn, is equivalent to $F_{1,N-1}$. As $N \to \infty$, Hotelling's T^2 becomes χ_p^2. Just as the univariate t statistic is derived from a likelihood ratio, Hotelling's T^2 is a likelihood ratio test with concomitant optimality properties. The T^2 statistic is invariant under affine transformations, and the test is UMP.

Let $\overleftrightarrow{\mathbf{X}}$ be distributed as $N_p\left(\boldsymbol{\mu}, \overleftrightarrow{\boldsymbol{\Sigma}}\right)$, and let $\hat{T}^2$ be given by (10.22). The distribution of T^2 can be derived in an analogous manner to the univariate t distribution through the method introduced in Section 2.9. A more useful approach in practice is to convert the distribution of T^2 to the F distribution using

$$\frac{N - p}{(N - 1)p} T_{p,N-1}^2 \sim F_{p,N-p} \tag{10.23}$$

This allows tests to be made using the F distribution with argument given by (10.22). It also provides the null and alternate distribution for hypothesis tests and power calculations. Matlab has no built-in functions to compute Hotelling's T^2 statistic, presumably for this reason.

Since $N(\overline{\mathbf{X}}_N - \boldsymbol{\mu}_0) \cdot \overleftrightarrow{\mathbf{S}}'^{-1} \cdot (\overline{\mathbf{X}}_N - \boldsymbol{\mu}_0)^T \sim T^2_{p,\ N-1}$, the probability statement

$$\Pr\left[N(\overline{\mathbf{X}}_N - \boldsymbol{\mu}) \cdot \overleftrightarrow{\mathbf{S}}'^{-1} \cdot (\overline{\mathbf{X}}_N - \boldsymbol{\mu})^T \le T^2_{p,N-1}(\alpha)\right] = 1 - \alpha \tag{10.24}$$

can be used to derive a $1 - \alpha$ confidence region on $\boldsymbol{\mu}$ given by all vectors $\boldsymbol{\mu}$ that satisfy $N(\overline{\mathbf{X}}_N - \boldsymbol{\mu}) \cdot \overleftrightarrow{\mathbf{S}}'^{-1} \cdot (\overline{\mathbf{X}}_N - \boldsymbol{\mu})^T \le T^2_{p,N-1}(\alpha)$, which describes a hyperellipsoid centered at $\boldsymbol{\mu} = \overline{\mathbf{X}}_N$ and is equivalent to mapping out those $\boldsymbol{\mu}_0$ that are not rejected by the upper tail T^2 test of H_0: $\boldsymbol{\mu} = \boldsymbol{\mu}_0$. However, this process is difficult to visualize for $p > 2$.

The confidence interval on a single linear combination $\mathbf{a} \cdot \boldsymbol{\mu}^T$ for a fixed vector $\mathbf{a}$ is given by

$$\mathbf{a} \cdot \overline{\mathbf{X}}_N^T - t_{N-1}(\alpha/2)\sqrt{\frac{\mathbf{a} \cdot \overleftrightarrow{\mathbf{S}}' \cdot \mathbf{a}^T}{N}} \le \mathbf{a} \cdot \boldsymbol{\mu}^T \le \mathbf{a} \cdot \overline{\mathbf{X}}_N^T + t_{N-1}(\alpha/2)\sqrt{\frac{\mathbf{a} \cdot \overleftrightarrow{\mathbf{S}}' \cdot \mathbf{a}^T}{N}} \tag{10.25}$$

This encompasses a single mean value μ_j by making all the elements in $\mathbf{a}$ zero except for the jth and also allows for linear combinations of the parameters. The outcome is identical to the univariate case and is appropriate when the variables are independent, meaning that $\overleftrightarrow{\mathbf{S}}'$ is diagonal or at least diagonally dominant.

Simultaneous confidence intervals for all possible values of $\mathbf{a} \cdot \boldsymbol{\mu}^T$ obtained by varying $\mathbf{a}$ are given by

$$\mathbf{a} \cdot \overline{\mathbf{X}}_N^T - T_{p,N-1}(\alpha)\sqrt{\frac{\mathbf{a} \cdot \overleftrightarrow{\mathbf{S}}' \cdot \mathbf{a}^T}{N}} \le \mathbf{a} \cdot \boldsymbol{\mu}^T \le \mathbf{a} \cdot \overline{\mathbf{X}}_N^T + T_{p,N-1}(\alpha)\sqrt{\frac{\mathbf{a} \cdot \overleftrightarrow{\mathbf{S}}' \cdot \mathbf{a}^T}{N}} \tag{10.26}$$

The probability that all such intervals generated by all choices of $\mathbf{a}$ will simultaneously contain $\mathbf{a} \cdot \boldsymbol{\mu}^T$ is $1 - \alpha$. For example, simultaneous confidence intervals on a set of p sample mean estimates could be obtained by setting $\mathbf{a}$ in turn to $(1, 0, \ldots, 0), (0, 1, \ldots, 0)$, ..., $(0, \ldots, 0, 1)$. These intervals are much wider than would be obtained under univariate reasoning [i.e., $T_{p,N-1}(\alpha) >> t_{N-1}(\alpha/2)$ except when $p = 1$]. For example, with $N = 25$ and $p = 10$, $T_{p,N-1}(\alpha) = 6.38$ and $t_{N-1}(\alpha/2) = 1.71$ when $\alpha = 0.05$. Using a Bonferroni approach where α is replaced by $\alpha/10$ yields $t_{N-1}(\alpha/20) = 3.09$, which is still much smaller than the simultaneous confidence interval. The Hotelling's T^2 confidence interval is more conservative than the Bonferroni one because it accounts for correlations among the variables.

Example 10.1 For the medical tablet data in Example 6.18, suppose that it is postulated that the test value should be 4.1. This is very nearly the grand mean of all the data, or 4.0917. The mean for each laboratory can be tested individually against this value with a univariate t test to get the p-values, and the hypothesis that all the laboratory means simultaneously have this value can be evaluated using Hotelling's T^2 test.

```
tablet = importdata('tablet.dat');
[n, m] = size(tablet);
```

```
ybar = mean(tablet)
ybar =
   4.0590   4.0180   4.0570   4.1210   4.1700   4.1150   4.1020
ystd = std(tablet);
t = sqrt(n)*(ybar - 4.1)./ystd
t =
   -2.5487  -4.4646  -5.0952   1.2287   5.1542   1.6270   0.1206
2*(1 - tcdf(abs(t), n - 1))
ans =
   0.0313   0.0016   0.0006   0.2503   0.0006   0.1382   0.9067
```

The t tests reject at the 0.05 level for the first, third, and fifth laboratories. Using the Bonferroni method, the rejection threshold is 0.005, and the test rejects for the first, second, third, and fifth laboratories. Using the Benjamini-Hochberg method from Section 6.6, the rejection threshold is 0.0016, and the test rejects for the first, second, third, and fifth laboratories, just as for the Bonferroni approach.

For the multivariate test using Hotelling's T^2,

```
s = cov(tablet);
t2 = n*(ybar - 4.1*ones(size(ybar)))*inv(s)*(ybar - 4.1*ones
(size(ybar)))'
t2 =
  422.2370
2*min(fcdf((n - m)*t2/(m*(n - 1)), m, n - m), 1 - fcdf((n -
m)*t2/(m*(n - 1)), m, n - m))
ans =
  0.0317
```

The p-value is below the 0.05 threshold, and hence the null hypothesis that all the laboratories simultaneously return a value of 4.1 is rejected, although not strongly. Simultaneous confidence intervals can be placed on the sample means of the data from each of the seven laboratories.

```
t2 = (n - 1)*m/(n - m)*finv(.95, 7, 3)
t2 =
  186.6216
```

For this example, $\sqrt{T^2_{7,3}(0.95)} = 13.661$, whereas $t_9(0.975) = 2.2622$, which is about a factor of 6 difference, so the simultaneous confidence intervals will be over an order of magnitude wider than Bonferroni confidence intervals.

```
for i = 1:7
     a = zeros(size(ybar));
     a(i) = 1;
     [a*ybar' - sqrt(t2)*sqrt(a*s*a'/n), a*ybar '+ sqrt(t2)
     *sqrt(a*s*a'/n)]
end
```

```
ans =
   3.8392    4.2788
ans =
   3.7671    4.2689
ans =
   3.9417    4.1723
ans =
   3.8875    4.3545
ans =
   3.9845    4.3555
ans =
   3.9891    4.2409
ans =
   3.8754    4.3286
```

Two-sample T^2 tests can also be derived that are analogous to the two-sample t test. In this case, the mean vectors from two populations are being compared. Assume that N_1 samples in $\overleftrightarrow{\mathbf{X}}$ are distributed as $N_p(\boldsymbol{\mu}_1, \overleftrightarrow{\boldsymbol{\Sigma}})$ and that N_2 samples in $\overleftrightarrow{\mathbf{Y}}$ are distributed as $N_p(\boldsymbol{\mu}_2, \overleftrightarrow{\boldsymbol{\Sigma}})$, where $\overleftrightarrow{\boldsymbol{\Sigma}}$ is unknown but the same for both samples. The sample means $\overline{\mathbf{X}}_{N_1}$ and $\overline{\mathbf{Y}}_{N_2}$ are computed along with the sample sum of squared residuals $\overleftrightarrow{\mathbf{W}}_1 = N_1\overleftrightarrow{\mathbf{S}}_1$ and $\overleftrightarrow{\mathbf{W}}_2 = N_2\overleftrightarrow{\mathbf{S}}_2$. A pooled estimator for the covariance matrix is $\overleftrightarrow{\mathbf{S}}'_p = (\overleftrightarrow{\mathbf{W}}_1 + \overleftrightarrow{\mathbf{W}}_2)/(N_1 + N_2 - 2)$, for which $\mathcal{E}(\overleftrightarrow{\mathbf{S}}'_p) = \overleftrightarrow{\boldsymbol{\Sigma}}$. Then a test of H_0: $\boldsymbol{\mu}_1 = \boldsymbol{\mu}_2$ versus H_1: $\boldsymbol{\mu}_1 \neq \boldsymbol{\mu}_2$ is based on the test statistic

$$\hat{T}^2 = \frac{N_1 N_2}{N_1 + N_2 - 2}\left(\overline{\mathbf{X}}_{N_1} - \overline{\mathbf{Y}}_{N_2}\right) \cdot \overleftrightarrow{\mathbf{S}}_p^{\prime -1} \cdot \left(\overline{\mathbf{X}}_{N_1} - \overline{\mathbf{Y}}_{N_2}\right)^T \tag{10.27}$$

that is distributed as T^2_{p,N_1+N_2-2} when the null hypothesis is true. It rejects if the test statistic exceeds the critical value of $T^2_{p,N_1+N_2-2}(\alpha)$.

The univariate two-sample t test is remarkably robust to departures from the model assumptions of equal variance of the two populations and normality. The multivariate T^2test is also robust, although not to the same degree.

10.4 Multivariate Analysis of Variance

A data set contains N random samples of p-variate observations from each of M Gaussian populations with equal covariance matrices. The one-way multivariate analysis of variance (MANOVA) hypothesis is H_0: $\boldsymbol{\mu}_1 = \cdots = \boldsymbol{\mu}_M$ versus H_1: at least two $\boldsymbol{\mu}_i$ are unequal. Note that the test is against the populations rather than the variables, in contrast to Section 10.3, and that there are MN observations in total. The distinction between ANOVA, as described in Section 6.3.7, and MANOVA is that the samples are p-variate instead of univariate. The test procedure is analogous to that for ANOVA, with additional complexity because the between and within estimators are matrices rather than scalars, and the null distribution is a multivariate generalization of the F distribution.

Let $\overleftrightarrow{\mathbf{X}}$ denote the $N \times M \times p$ data matrix, which can be represented as a parallelepiped, and denote the $N \times M$ face at the ith variable index by $\overleftrightarrow{\mathbf{X}}_i$. The i,j element of the between or hypothesis matrix is

$$H_{ij} = \frac{1}{N}\mathbf{j}_N \cdot \overleftrightarrow{\mathbf{X}}_i \cdot \left(\overleftrightarrow{\mathbf{I}}_M - \frac{1}{M}\overleftrightarrow{\mathbf{J}}_M\right) \cdot \overleftrightarrow{\mathbf{X}}_j^T \cdot \mathbf{j}_N^T \tag{10.28}$$

whereas the corresponding element of the within or error matrix is

$$E_{ij} = \mathrm{tr}\left(\overleftrightarrow{\mathbf{X}}_i \cdot \overleftrightarrow{\mathbf{X}}_j^T\right) - \frac{1}{N}\mathbf{j}_N \cdot \overleftrightarrow{\mathbf{X}}_i \cdot \overleftrightarrow{\mathbf{X}}_j^T \cdot \mathbf{j}_N^T \tag{10.29}$$

$\overleftrightarrow{\mathbf{H}}$ is $p \times p$ and has the sum of squares explained for each variable on its diagonal. The off-diagonal components are analogous sums for all possible pairs of variables. $\overleftrightarrow{\mathbf{E}}$ is $p \times p$ and has the sum of squared errors on its main diagonal with analogous sums for all possible pairs of variables in the off-diagonal entries. There are $n_H = M - 1$ degrees-of-freedom for the between and $n_E = M(N-1)$degrees-of-freedom for the within estimator. The rank of $\overleftrightarrow{\mathbf{H}}$ is the smaller of $M - 1$ and p, whereas the rank of $\overleftrightarrow{\mathbf{E}}$ must be p.

Four standard test statistics have been proposed to test the pair of MANOVA hypotheses. Only the likelihood ratio test originally presented by Wilks (1932) will be considered here, and it has test statistic

$$\hat{\Lambda} = \frac{\left|\overleftrightarrow{\mathbf{E}}\right|}{\left|\overleftrightarrow{\mathbf{E}} + \overleftrightarrow{\mathbf{H}}\right|} \tag{10.30}$$

Equation (10.30) is known as *Wilks' Λ* and has a range of (0, 1). Wilks' Λ is distributed as Wilks' Λ distribution if (1) $\overleftrightarrow{\mathbf{E}}$ is distributed as the Wishart distribution $W_p\left(n_E, \overleftrightarrow{\boldsymbol{\Sigma}}\right)$, (2) $\overleftrightarrow{\mathbf{H}}$ is distributed as $W_p\left(n_H, \overleftrightarrow{\boldsymbol{\Sigma}}\right)$ under the null hypothesis, and (3) $\overleftrightarrow{\mathbf{E}}$ and $\overleftrightarrow{\mathbf{H}}$ are independent, all of which hold if the data are multivariate normal with the same covariance matrix. The test rejects if the test statistic $\hat{\Lambda} \le \Lambda_{p,n_H,n_E}(\alpha)$. Rejection occurs when the test statistic is smaller than the critical value, in contrast with Hotelling's T^2.

The Wilks' Λ distribution is analogous to the F distribution in univariate statistics and is indexed by three parameters: the dimensionality p and the degrees-of-freedom for the between and within estimators. The degrees-of-freedom parameters are the same as for one-way ANOVA described in Section 6.3.7, so the addition of the dimensionality parameter distinguishes MANOVA from it. The Wilks' Λ distribution has symmetries such that Λ_{p,n_H,n_E} and $\Lambda_{n_H,p,n_E+n_H-p}$ are the same. Further, for fixed values of n_E and n_H, the critical values of Wilks' Λ decrease as the dimensionality p rises. This means that the addition of variables that do not improve the ability to reject the null hypothesis will lead to decreased power.

The distribution can be written as a product of beta distributions, which is difficult to work with. However, exact relationships between Wilks' Λ distribution and the F distribution exist when $p = 1$ or 2 or when $n_H = 1$ or 2, as given in Table 10.1.

Table 10.1 Relation between Λ and F Statistics

Parameters	F	Degrees-of-freedom
Any p, $n_H = 1$	$\frac{(n_E + n_H - p)(1 - \Lambda)}{p\Lambda}$	$p, n_E + n_H - p$
Any p, $n_H = 2$	$\frac{(n_E + n_H - p - 1)(1 - \sqrt{\Lambda})}{p\sqrt{\Lambda}}$	$2p, 2(n_E + n_H - p - 1)$
Any n_H, $p = 1$	$\frac{n_E(1 - \Lambda)}{n_H\Lambda}$	n_H, n_E
Any n_H, $p = 2$	$\frac{(n_E - 1)(1 - \sqrt{\Lambda})}{n_H\sqrt{\Lambda}}$	$2n_H, 2(n_E - 1)$

For other values of p and M, an approximate relationship due to Rao (1951) is

$$\hat{F} = \frac{(1 - \hat{\Lambda}^{1/t})n_2}{\Lambda^{1/t}n_1} \tag{10.31}$$

$$w = n_H + n_E - (p + n_H + 1)/2 \tag{10.32}$$

$$t = \sqrt{(p^2 n_H{}^2 - 4)/(p^2 + n_H^2 - 5)} \tag{10.33}$$

$$n_1 = pn_H \tag{10.34}$$

$$n_2 = wt - (pn_H - 2)/2 \tag{10.35}$$

$\hat{F}$ is approximately distributed as F_{n_1,n_2} and may be assessed against the F distribution in the usual way. This approach is more accurate than using the χ^2 approximation for a likelihood ratio test. MATLAB does not provide support for Wilks' Λ distribution, so an approximate method must be applied directly. The following function provides that support.

```
function [pval, lambda] = WilksLambda(h, e, p, n, m)
%Computes the Wilks lambda statistic and double sided p-value
%from the error and hypothesis matrices using Rao (1951) F
approximation
%Input arguments
% e error matrix
% h hypothesis matrix
% p dimensionality
% n hypothesis degrees of freedom
% m error degrees of freedom
lambda = det(e)/det(e + h);
if p == 1
   f = m* (1 - lambda)/(n* lambda)
   n1 = n;
   n2 = m;
```

```
elseif p == 2
   f = (m - 1)*(1 - sqrt(lambda))/(n*sqrt(lambda));
   n1 = 2*n;
   n2 = 2*(m - 1);
elseif n == 1
   f = (n + m - p)*(1 - lambda)/(p*lambda);
   n1 = p;
   n2 = n + m - p;
elseif n == 2
   f = (n + m - p - 1)*(1 - sqrt(lambda))/(p*sqrt(lambda));
   n1 = 2*p;
   n2 = 2*(n + m - p - 1);
else
   n1 = p*n;
   w = n + m - (p + n + 1)/2;
   t = sqrt((p^2*n^2 - 4)/(p^2 + n^2 -5));
   n2 = w*t - (p*n -2)/2;
   f = n2*(1 - lambda^(1/t))/(n1*lambda^(1/t));
end
pval = 2*min(fcdf(f, n1, n2), 1 - fcdf(f, n1, n2));
end
```

MATLAB does provide support for one-way MANOVA through the function [*d*, *p*, *stats*] = **manova1** (*x*, *groups*). The matrix *x* is $n \times p$, with each row corresponding to a single observation of *p* variables. The matrix *groups* sorts the variables in *x* into groups in a variety of ways. The returned variable *d* is the dimensionality of the space containing the group means. If $d = 0$, there is no evidence to reject the null hypothesis that the means of the groups are identical. If $d = 1$, the null hypothesis is rejected at the 0.05 level, but the multivariate means may lie on a line. If $d = 2$, the null hypothesis is rejected at the 0.05 level, and the means do not lie on a line, but they may lie on a plane, and so on. The returned variables *p* are the corresponding *p*-values. The returned variable *stats* is a structure that contains 14 additional variables. However, it appears that **manova1** tests the result against a less accurate chi square approximation to the Wilks' $\wedge$ distribution rather than the *F* approximation (10.31)–(10.35).

Example 10.2 Zonally averaged temperature data are available at http://data.giss.nasa.gov/gistemp/ for various ranges of latitudes starting in 1880 and extending to date. The data through 2015 will be used for this example. The data are available with and without the inclusion of sea surface temperature (SST) and are presented as a temperature anomaly relative to a base period of 1951–80. Such data have figured strongly in the current debate about global warming. The data that include SST will be evaluated for significance of change over time by binning them into age groups and performing a MANOVA test. The data are plotted in Figure 10.1.

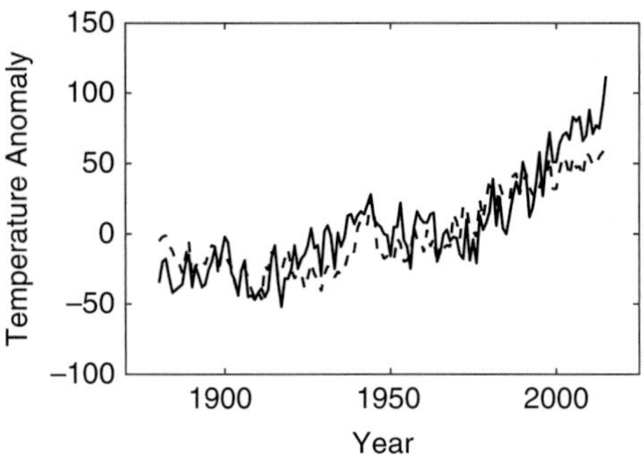

Figure 10.1 Zonally averaged temperature anomaly data for the northern (solid) and southern (dashed) hemispheres.

The northern and southern hemisphere data are the most strongly averaged results. Dividing the 136 years of data into 10 year groups, Wilks' Λ with $p = 2$ measurements, $M = 13$ groups, and $N = 10$ data per group is evaluated.

```
x = importdata('ZonAnnTs+dSST.dat'); %N and S hemisphere
data in columns 3 and 4
p = 2;
m = 13;
n = 10;
xh = [];
i = 1:n;
for j = 1:m
    xh(i, j, 1:p)= x((j - 1)*n + i, 3:4);
end
e = [];
h = [];
for i = 1:p
   for j = 1:p
        h(i, j) = sum(sum(xh(:, :, i)*(eye(m) - ones(m,m)/m)*
                  xh(:, :, j)'))/n;
        e(i, j)=trace(xh(:, :, i)*xh(:, :, j)') - sum(sum
                  (xh(:, :, i)*xh(:, :, j)'))/n;
   end
end
[pval, lambda] = WilksLambda(h, e, p, m - 1, n*(m - 1))
pval =
   0
lambda =
   0.0368
```

The p-value is floating-point zero; hence the null hypothesis that the means of the northern and southern hemisphere data are the same over time is rejected. This ought not to be a surprise because there is a long-term trend in the data, especially since 1980. However, this result omits the data from 2010 to 2015. Repeating the result after skipping the first six years of data gives a p-value of 0 and a Wilks' Λ statistic of 0.0396, yielding the same conclusion.

The same analysis can be carried out using the MATLAB function, although the result is not as easy to interpret. For the northern and southern hemisphere data,

```
groups = [];
for i = 1:13;
     groups = [groups i*ones(1,10)];
end
x1 = x(7:136, 3:4);
[d, p, stats] = manova1(x1, groups)
d =
    2
p =
   1.0e-17 *
   0.0000
   0.7690
```

A value $d = 0$ would support the null hypothesis that the means are the same, and hence $d = 2$ rejects the null hypothesis.

The eight variable data set consisting of zonal averages extending from 90–64, 64–44, 44–24, and 24–0 latitude on both sides of the equator will be considered next. Figures 10.2 and 10.3 show the data. The variability is highest at polar latitudes, especially for the southern hemisphere.

For the northern hemisphere data, $p = 4$, $m = 13$, and $n = 10$ for 10 year groups, yielding $\hat{\Lambda} = 0.0525$. The p-value is 0, and hence the null hypothesis that the 10 year averages are the same is rejected.

For both the northern and southern hemisphere data, the result is $\hat{\Lambda} = 0.0063$. The p-value is again 0, and hence the null hypothesis that the 10-year averages are the same is rejected.

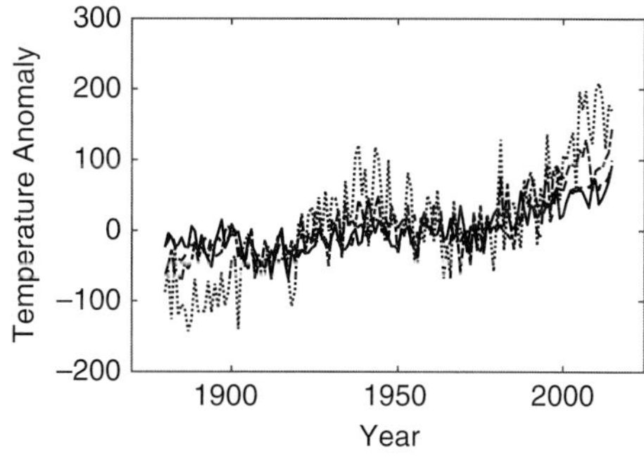

Figure 10.2 Northern hemisphere temperature anomaly estimates with a solid line indicating the 0–24, dashed line indicating 24–44, dot-dash line indicating 44–64, and dotted line indicating 64–90 latitude bands.

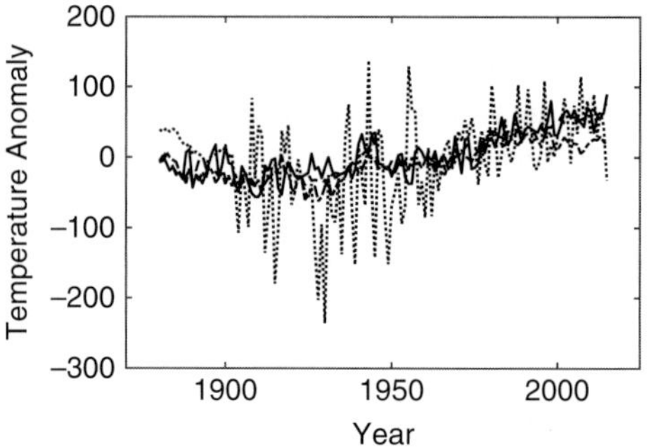

Figure 10.3 Southern hemisphere temperature anomaly estimates with a solid line indicating the 0—24, a dashed line indicating 24—44, a dash-dot line indicating 44—64, and a dotted line indicating 64—90 latitude bands.

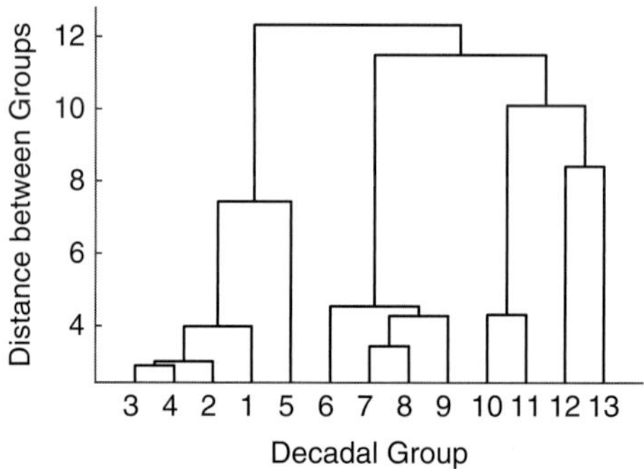

Figure 10.4 Dendrogram for the MANOVA analysis of the NOAA temperature data.

The variability is much larger at polar latitudes, and in addition, the actual measurements are sparsest at those locales. This is especially apparent for the most poleward southern hemisphere data that are very smooth prior to 1904 because they are based only on SST data and exhibit a lot of variability after that time as sparse land data became available. If the most poleward data series for each hemisphere are excluded, for 10-year intervals the result is $\hat{\Lambda} = 0.0140$. The *p*-value is still 0, so the null hypothesis is rejected. The mean temperature has changed over time based on this data set. However, the change is far from just an increase and has spatial patterns that cannot be discerned from a simple comparison of the means.

Applying the MATLAB **manova1** function to the eight zonally averaged temperature data from 1887 to 2015 gives a dimensionality of 4. MATLAB also provides the function **manovacluster**(stats) to plot a dendrogram for the results. A *dendrogram* is a hierarchical binary cluster tree. Figure 10.4 shows the result. The bottom of the diagram corresponds to the individual decadal groups (1 is 1886–95, 2 is 1896–1905, and so on), and these are joined together at a height that is proportional to the distance between the clusters. The dendrogram clearly pulls out the four dimensions corresponding to the nearly constant temperature from 1890 to 1930 (groups 1–5), the weak warming and cooling trend from

1940 to 70, and the marked warming trend since that time in the northern hemisphere. The latter has been broken into two clusters. This result is not changed much by omitting the two poleward-most temperature anomalies.

Example 10.3 Four types of measurements were made on male Egyptian skulls recovered from tombs from five different time periods ranging from 4000 BC to AD 150. The measurements are maximal breadth of skull, basibregmatic (or maximum vertical) height of skull, basialveolar (fore and aft) length of skull, and nasal height of skull. There are 30 measurements for each of the five time periods. The goal was to determine whether there is any difference in skull size between the time periods that might indicate interbreeding of the Egyptians with immigrant groups. This is a classic data set from Thomson & Randall-Maciver (1905) that was made familiar to statisticians by Sir Ronald Fisher in the 1930s.

```
skulls = importdata('skulls.dat');
p = 4;
m = 5;
n = 30;
x = zeros(n, m, p);
for i = 1:n
    for j = 1:m
        x(i, j, 1:p) = skulls((j - 1)*n + i, 1:p);
    end
end
e = [];
h = [];
for i = 1:p
    for j = 1:p
        h(i, j) = sum(sum(x(:, :, i)*(eye(m) - ones(m, m)/m)*x
              (:, :, j)'))/n;
        e(i, j)=trace(x(:, :, i)*x(:, :, j)') - sum(sum(x(:,
              :, i)*x(:, :, j)'))/n;
   end
end
[pval, lambda] = WilksLambda(h, e, p, m - 1, n*(m - 1))
pval =
  1.4020e-06
lambda =
   0.6636
```

The p-value is very small; hence the null hypothesis that there is no difference in skull size between the time periods is rejected.

ANOVA tables for each variable over time will be examined to assess the univariate relations among the data.

```
for j = 1:p
    [pp, anovatab] = anova1(x(:, :, j))
end
pp =
  1.8263e-04
anovatab =
    'Source'   'SS'           'df'  'MS'        'F'       'Prob>F'
    'Columns'  [ 502.8267]    [ 4]  [125.7067]  [5.9546]  [1.8263e-04]
    'Error'    [3.0611e+03]   [145] [21.1108]   []        []
    'Total'    [3.5639e+03]   [149] []          []        []
pp =
  0.0490
anovatab =
    'Source'   'SS'           'df'  'MS'        'F'       'Prob>F'
    'Columns'  [ 229.9067]    [ 4]  [57.4767]   [2.4474]  [0.0490]
    'Error'    [3.4053e+03]   [145] [23.4846]   []        []
    'Total'    [3.6352e+03]   [149] []          []        []
pp =
   4.6364e-06
anovatab =
    'Source'   'SS'           'df'  'MS'        'F'       'Prob>F'
    'Columns'  [803.2933]     [4]   [200.8233]  [8.3057]  [4.6364e-06]
    'Error'    [ 3.5060e+03]  [ 145] [ 24.1791] []        []
    'Total'    [ 4.3093e+03]  [ 149] []         []        []
pp =
  0.2032
anovatab =
    'Source'   'SS'            'df'    'MS'         'F'         'Prob>F'
    'Columns'  [   61.2000]    [   4]  [ 15.3000]   [ 1.5070]   [ 0.2032]
    'Error'    [ 1.4721e+03]   [ 145]  [ 10.1526]   []          []
    'Total'    [ 1.5333e+03]   [ 149]  []           []          []
```

Only the last variable (nasal height of skull) accepts the univariate null hypothesis, although the second one (fore and aft length of skull) only weakly rejects it. If a Bonferroni approach is used, the second and last variables accept the null hypothesis. This remains true using the Benjamini-Hochberg method. The remaining variables all demonstrate significant change through time.

10.5 Hypothesis Tests on the Covariance Matrix

Statistical testing of the structure of the covariance matrix is often used to verify assumptions for other tests, just as tests for the homogeneity of variance (either the F test or

Bartlett's M test) were used to verify the assumptions behind a t test in Chapter 6. This includes tests that the covariance matrix has a particular structure, tests for the equality of two or more covariance matrices, and tests of independence.

10.5.1 Sphericity Test

The hypothesis that the p variables in a multivariate data set are independent and share a common variance can be expressed as H_0: $\overleftrightarrow{\mathbf{\Sigma}} = \sigma^2 \overleftrightarrow{\mathbf{I}}_p$ versus H_1: $\mathbf{\Sigma} \neq \sigma^2 \overleftrightarrow{\mathbf{I}}_p$, where σ^2 is the unknown common variance. This is called a *sphericity test* because the Mahalanobis distance between the variables becomes independent of direction in p-space when the null hypothesis is true. The test was introduced by Mauchly (1940).

The likelihood ratio for the sphericity test assuming multivariate Gaussian variables is

$$\hat{\lambda} = \left\{ \frac{\left|\overleftrightarrow{\mathbf{S}}'\right|}{\left[\mathrm{tr}\left(\overleftrightarrow{\mathbf{S}}'\right)/p\right]^p} \right\}^{N/2} \quad (10.36)$$

The distribution of $\hat{\lambda}$ under the null hypothesis is unknown, and hence the Wilks' chi square approximation must be used. Consequently, the log likelihood

$$-2\log\hat{\lambda} = -N\log\left\{ \frac{\left|\overleftrightarrow{\mathbf{S}}'\right|}{\left[\mathrm{tr}\left(\overleftrightarrow{\mathbf{S}}'\right)/p\right]^p} \right\} = -N\log\hat{U} \quad (10.37)$$

is asymptotically chi square with $p(p+1)/2 - 1$ degrees-of-freedom under the null hypothesis. The degrees-of-freedom follow from the number of parameters in $\overleftrightarrow{\mathbf{\Sigma}}$, or $p(p+1)/2$, minus the number due to estimation of σ^2.

One application of the sphericity test arises in determining whether a univariate or multivariate ANOVA is appropriate. In general, one would expect p variables to be correlated, and hence a multivariate test of the null hypothesis would be required. However, if the sphericity test accepts, then the variables are statistically uncorrelated, and univariate ANOVA would suffice.

Example 10.4 Returning to the Egyptian skull data in Example 10.3, apply the sphericity test to determine if the variables are correlated.

```
p = 4;
m = 5;
n = 30;
s = cov(skulls);
u = p^p*det(s)/trace(s)^p;
test = -n*m*log(u)
test =
  4.1000e+03
```

```
2*min(chi2cdf(test, p*(p+1)/2-1), 1-chi2cdf(test, p*(p+1)/2-1))
ans =
   0
```

The p-value is zero, and hence the sphericity test rejects. A MANOVA approach has to be used, as was done in Example 10.3.

10.5.2 Comparing Covariance Matrices

An assumption behind the Hotelling's T^2 or Wilks' Λ tests is that the population covariance matrices for the samples being compared are equal. This allows the sample covariance matrices to be pooled to get an estimate of the population covariance matrix. If the population covariance matrices are not the same, the Hotelling's T^2 and Wilks' Λ tests are still reliable if the samples are both large and the same size. When these conditions do not hold, it is useful to be able to test H_0: $\overleftrightarrow{\mathbf{\Sigma}}_1 = \cdots = \overleftrightarrow{\mathbf{\Sigma}}_k$ versus H_1: not H_0.

In the univariate case, the equality of two variances can be tested using the F test of Section 6.3.4, and the equality of many variances can be tested using Bartlett's M test, as described in Section 6.3.5. The latter can be generalized to the multivariate case, as shown in Rencher (1998, sec. 4.3). Assume that there are M independent samples of size n_i from multivariate normal distributions, where $N = \sum_{i=1}^{M} n_i$. The likelihood ratio test statistic is

$$\hat{\lambda} = \frac{\left|\overleftrightarrow{\mathbf{S}}_1'\right|^{(n_1-1)/2} \cdots \left|\overleftrightarrow{\mathbf{S}}_M'\right|^{(n_M-1)/2}}{\left|\overleftrightarrow{\mathbf{S}}_p'\right|^{(N-M)/2}} \tag{10.38}$$

Where $\overleftrightarrow{\mathbf{S}}_p'$ is the pooled covariance matrix

$$\overleftrightarrow{\mathbf{S}}_p' = \frac{\sum_{i=1}^{M} (n_i - 1)\overleftrightarrow{\mathbf{S}}_i'}{N - M} \tag{10.39}$$

The test statistic lies between 0 and 1, with a larger value favoring the null hypothesis. Further, each $n_i > p$, or else (10.38) would be zero.

The statistic

$$\begin{aligned} \hat{U} &= -2(1-\delta)\log\hat{\lambda} \\ &= (1-\delta)\left[(N-M)\log\left|\hat{\overleftrightarrow{S}}_p'\right| - \sum_{i=1}^{M}(n_i-1)\log\left|\hat{\overleftrightarrow{S}}_i'\right|\right] \end{aligned} \tag{10.40}$$

where

$$\delta = \frac{2p^2+3p-1}{6(p+1)(M-1)}\left[\sum_{i=1}^{M}\frac{1}{n_i-1} - \frac{1}{\sum_{i=1}^{M}(n_i-1)}\right] \tag{10.41}$$

has a chi square distribution with $(M - 1)p(p + 1)/2$ degrees-of-freedom. This result is due to Box (1949). There is also an F approximation for the null distribution.

This test should be used with caution because it has been demonstrated that the Box M test with equal sample sizes may reject when the heterogeneity of covariance is small, but this does not affect the Hotelling's or Wilks' tests. The Box M test is also sensitive to departures from normality that the T^2 and Λ tests are more robust against, notably kurtosis of the distribution.

Example 10.5 Evaluate the equivalence of the covariance matrices for the global temperature data in Example 10.2.

For the four northern hemisphere data sets,

```
p = 4;
m = 13;
n = 10;
i = 1:n;
for j = 1:m
    xh(i, j, 1:p)= x((j - 1)*n + i, 8:8 + p - 1);
end
sp = 0;
s1 = 0;
for i = 1:m
    s = cov(squeeze(xh(:, i, :)));
    sp = sp + (n - 1)*s;
    s1 = s1 + (n - 1)*log(det(s));
end
sp = sp/(m*(n  - 1));
delta= (2*p^2 + 3*p-1)/(6* (p+1)* (m-1))* (m/(n-1) -1/(m* (n-1)));
u = (1 - delta)* (m* (n - 1)*log(det(sp)) - s1)
u =
   129.4492
```

The test accepts the null hypothesis that the covariance matrices for the 13 groups are the same. Consequently, it appears that the 10-year means of the data are different, but their covariance structures are similar, and hence the MANOVA analysis in Example 10.2 is solid.

10.5.3 Test of Independence

Suppose that an $N \times p$ matrix of observations $\overleftrightarrow{\mathbf{X}}$ is partitioned into an $N \times p_1$ submatrix $\overleftrightarrow{\mathbf{X}}_1$ and an $N \times p_2$ submatrix $\overleftrightarrow{\mathbf{X}}_2$. The population covariance matrix then partitions into

$$\overleftrightarrow{\mathbf{\Sigma}} = \begin{bmatrix} \overleftrightarrow{\mathbf{\Sigma}}_{11} & \overleftrightarrow{\mathbf{\Sigma}}_{12} \\ \overleftrightarrow{\mathbf{\Sigma}}_{21} & \overleftrightarrow{\mathbf{\Sigma}}_{22} \end{bmatrix} \tag{10.42}$$

and the sample covariance matrix has a similar form. The null hypothesis that $\overleftrightarrow{\mathbf{X}}_1$ and $\overleftrightarrow{\mathbf{X}}_2$ are independent is

$$H_0 : \quad \overleftrightarrow{\mathbf{\Sigma}} = \begin{bmatrix} \overleftrightarrow{\mathbf{\Sigma}}_{11} & \\ & \overleftrightarrow{\mathbf{\Sigma}}_{22} \end{bmatrix} \tag{10.43}$$

Equivalently, the test is H_0: $\overleftrightarrow{\mathbf{\Sigma}}_{12} = \mathbf{0}$ versus H_1: not H_0. The null hypothesis holds that every variable in $\overleftrightarrow{\mathbf{X}}_1$ is independent of every variable in $\overleftrightarrow{\mathbf{X}}_2$.

The likelihood ratio test statistic is

$$\hat{\lambda} = \frac{|\overleftrightarrow{\mathbf{S}}'|}{|\overleftrightarrow{\mathbf{S}}'_{11}|\,|\overleftrightarrow{\mathbf{S}}'_{22}|} \tag{10.44}$$

and is distributed as $\Lambda_{p_1,p_2,N-p_2-1}$ if the null hypothesis is true. The test can be easily extended to more than two partitions.

The test for independence of variables can also be adapted as a test for propriety of complex Gaussian variables by applying it to (10.17). A generalized likelihood ratio test for impropriety was proposed by Schreier, Scharf, & Hanssen (2006) and elaborated by Walden & Rubin-Delanchy (2009). However, their impropriety hypothesis test is identical to the present test of independence for multivariate data applied to the augmented covariance matrix. The null hypothesis holds that the sample version of (10.17) is block diagonal versus the alternate hypothesis that it is not or, equivalently, the null hypothesis holds that $\breve{\overleftrightarrow{\mathbf{\Gamma}}} = \mathbf{0}$ versus the alternate hypothesis that it is not. The likelihood ratio test statistic for impropriety is

$$\hat{\Lambda} = \frac{\left|\hat{\breve{\overleftrightarrow{\mathbf{\Gamma}}}}\right|}{\left|\hat{\overleftrightarrow{\mathbf{\Gamma}}}\right|^2} \tag{10.45}$$

Equation (10.45) is distributed as Wilks' Λ distribution $\Lambda_{p,p,N-p-1}$ if the null hypothesis is true. This test was used by Chave (2014) in evaluating the propriety of magnetotelluric response function estimates.

10.6 Multivariate Regression

The linear relationship between a single response variable $\mathbf{y}$ and one or more predictor variables $\overleftrightarrow{\mathbf{X}}$ was considered in Chapter 9. This can be generalized to the case of more than one response variable $\overleftrightarrow{\mathbf{Y}}$ to give multivariate multiple linear regression. The extension is straightforward, although statistical inference quickly gets messy. The regression problem is

$$\overleftrightarrow{\mathbf{Y}} = \overleftrightarrow{\mathbf{X}} \cdot \overleftrightarrow{\boldsymbol{\beta}} + \overleftrightarrow{\boldsymbol{\Xi}} \tag{10.46}$$

Where $\overleftrightarrow{\mathbf{Y}}$ is $N \times q$, $\overleftrightarrow{\mathbf{X}}$ is $N \times p$, $\overleftrightarrow{\boldsymbol{\beta}}$ is $p \times q$, and $\overleftrightarrow{\boldsymbol{\Xi}}$ is $N \times q$. Each of the q columns of $\overleftrightarrow{\mathbf{Y}}$ depends on the predictor variables $\overleftrightarrow{\mathbf{X}}$ in its own way, resulting in the parameters in $\overleftrightarrow{\boldsymbol{\beta}}$ becoming a matrix rather than a vector. Under the same assumptions as for univariate regression, the solution that minimizes $\overleftrightarrow{\boldsymbol{\Xi}}^T \cdot \overleftrightarrow{\boldsymbol{\Xi}}$ is

$$\hat{\overleftrightarrow{\boldsymbol{\beta}}} = \left(\overleftrightarrow{\mathbf{X}}^T \cdot \overleftrightarrow{\mathbf{X}}\right)^{-1} \cdot \overleftrightarrow{\mathbf{X}}^T \cdot \overleftrightarrow{\mathbf{Y}} \tag{10.47}$$

This is identical to solving q separate univariate regressions with the same predictor variables but different response variables. The least squares solution minimizes both the trace and the determinant of $\overleftrightarrow{\boldsymbol{\Xi}}^T \cdot \overleftrightarrow{\boldsymbol{\Xi}}$. The parameters $\hat{\overleftrightarrow{\boldsymbol{\beta}}}$ are the minimum variance unbiased linear estimator if the conditions on the least squares model in Section 9.2 are met. However, the column elements of $\hat{\overleftrightarrow{\boldsymbol{\beta}}}$ will be correlated unless the columns of the predictor matrix $\overleftrightarrow{\mathbf{X}}$ are orthogonal, as is also true of univariate linear regression. The row elements of $\hat{\overleftrightarrow{\boldsymbol{\beta}}}$ will be correlated if the response data are correlated. As a result, hypothesis tests and confidence limit estimates typically must be multivariate.

The estimator $\overleftrightarrow{\boldsymbol{\beta}}$ will be partitioned so that the constant term (if present) is excluded. This is necessary because otherwise multivariate tests would require that the columns of $\overleftrightarrow{\mathbf{Y}}$ have zero mean. Let the first $p-1$ rows of $\hat{\overleftrightarrow{\boldsymbol{\beta}}}$ be $\hat{\overleftrightarrow{\boldsymbol{\beta}}}_1$. The hypotheses H_0: $\hat{\overleftrightarrow{\boldsymbol{\beta}}}_1 = \mathbf{0}$ versus H_1: $\hat{\overleftrightarrow{\boldsymbol{\beta}}}_1 \neq \mathbf{0}$ can be tested using Wilks' Λ. The alternate hypothesis holds that at least one of the elements in $\boldsymbol{\beta}_1$ is nonzero. The within $\overleftrightarrow{\mathbf{E}}$ and between $\overleftrightarrow{\mathbf{H}}$ matrices in the Wilks' Λ test become

$$\overleftrightarrow{\mathbf{E}} = \left(\overleftrightarrow{\mathbf{Y}} - \overleftrightarrow{\mathbf{X}} \cdot \hat{\overleftrightarrow{\boldsymbol{\beta}}}\right)^T \cdot \overleftrightarrow{\mathbf{Y}} \tag{10.48}$$

$$\overleftrightarrow{\mathbf{H}} = \hat{\overleftrightarrow{\boldsymbol{\beta}}}^T \cdot \overleftrightarrow{\mathbf{X}}^T \cdot \overleftrightarrow{\mathbf{Y}} - N \overline{\mathbf{Y}}_N^T \cdot \overline{\mathbf{Y}}_N \tag{10.49}$$

so that

$$\overleftrightarrow{\mathbf{E}} + \overleftrightarrow{\mathbf{H}} = \overleftrightarrow{\mathbf{Y}}^T \cdot \overleftrightarrow{\mathbf{Y}} - N \overline{\mathbf{Y}}_N^T \cdot \overline{\mathbf{Y}}_N \tag{10.50}$$

The usual test statistic (10.30) is assessed against $\Lambda_{q,p,N-p-1}$ when the null hypothesis is true. The test rejects if $\hat{\Lambda} \leq \Lambda_{q,p,N-p-1}$ at the α level. This test is analogous to the F test for regression that is described in Section 9.3.1.

Example 10.6 In a classic data set, Siotani et al. (1963) present data containing two response variables (y_1 = taste, y_2 = odor) against eight predictor variables (x_1 = pH, x_2 = acidity 1, x_3 = acidity 2, x_4 = sake meter, x_5 = direct reducing sugar, x_6 = total sugar, x_7 = alcohol, x_8 = formyl nitrogen) for 30 brands of Japanese sake. The predictive power of the data will be evaluated.

```
sake = importdata('sake.dat');
y = sake(:, 1:2);
x = [ sake(:, 3:10) ones(30, 1)] ;
[n, m] = size(x);
beta = x\y;
e = (y - x*beta)'*y;
h = beta'*x'*y - n*mean(y)'*mean(y);
[pval, lambda] = WilksLambda(h, e, 2, m - 1, n - m - 2)
pval =
  0.8620
lambda =
  0.4642
```

The p-value is large; hence the null hypothesis that all the regression coefficients are zero is accepted. There is no point in undertaking further analysis.

The significance of a subset of the rows of $\hat{\overleftrightarrow{\boldsymbol{\beta}}}$ can also be tested. Let $\hat{\overleftrightarrow{\boldsymbol{\beta}}}$ be partitioned so that the first r rows are $\hat{\overleftrightarrow{\boldsymbol{\beta}}}_r$ and the last s rows are $\hat{\overleftrightarrow{\boldsymbol{\beta}}}_s$, where $r+s=p$. The hypothesis to be tested is H_0: $\hat{\overleftrightarrow{\boldsymbol{\beta}}}_s = \mathbf{0}$ versus H_1: $\hat{\overleftrightarrow{\boldsymbol{\beta}}}_s \neq \mathbf{0}$. Let $\overleftrightarrow{\mathbf{X}}_r$ denote the columns of $\overleftrightarrow{\mathbf{X}}$ corresponding to $\hat{\overleftrightarrow{\boldsymbol{\beta}}}_r$ so that $\overleftrightarrow{\mathbf{Y}} = \overleftrightarrow{\mathbf{X}}_r \cdot \overleftrightarrow{\boldsymbol{\beta}}_r + \overleftrightarrow{\boldsymbol{\Xi}}'$. The $\overleftrightarrow{\mathbf{E}}$ matrix is given by (10.48), whereas the $\overleftrightarrow{\mathbf{H}}$ matrix becomes

$$\overleftrightarrow{\mathbf{H}} = \hat{\overleftrightarrow{\boldsymbol{\beta}}}^T \cdot \overleftrightarrow{\mathbf{X}}^T \cdot \overleftrightarrow{\mathbf{Y}} - \hat{\overleftrightarrow{\boldsymbol{\beta}}}_r^T \cdot \overleftrightarrow{\mathbf{X}}_r^T \cdot \overleftrightarrow{\mathbf{Y}} \tag{10.51}$$

The test statistic is the usual estimator for $\hat{\Lambda}$ (10.30) and is assessed against $\Lambda_{q,s,N-p-1}$.

When the number of predictor variables p or the number of response variables q is large, it is often desirable to find an optimal subset, often called the *adequate set* of variates. While there is no bulletproof way to achieve this, a standard approach is backward elimination beginning with all p of the predictor variables. The variables are deleted one at a time with replacement, and each instance is assessed against a partial Wilks' Λ. For the first deletion, the partial Λ when the jth predictor is omitted is given by the conditional distribution

$$\Lambda\left(x_j \middle| x_1 \ldots, x_{j-1}, x_{j+1}, \ldots, x_p\right) = \frac{\Lambda\left(\overleftrightarrow{\mathbf{Y}}, x_1, \ldots, x_p\right)}{\Lambda\left(\overleftrightarrow{\mathbf{Y}}, x_1, \ldots, x_{j-1}, x_{j+1}, \ldots, x_p\right)} \tag{10.52}$$

and has $\Lambda_{q,1,N-p-1}$ as its distribution. The variable with the largest Λ is omitted, and the process continues with the $p-1$ remaining variables. It terminates when the variable corresponding to the largest Λ is significant at a chosen probability level. A similar procedure can be used to determine whether the reduced set of predictor variables predicts all or only some of the response variables.

Example 10.7 A data set from Rencher (1995, p. 295) relates temperature, humidity, and evaporation, and contains the following variables:

- Maximum daily temperature
- Minimum daily temperature
- Integrated daily temperature
- Maximum daily soil temperature
- Minimum daily soil temperature
- Integrated daily soil temperature
- Maximum daily relative humidity
- Minimum daily relative humidity
- Integrated daily relative humidity
- Total wind in miles per day
- Evaporation

The last two variables will be regressed on the first nine, and backward stepwise selection will be applied to cull the set of predictors.

```
the = importdata('the.dat');
x = [the(:, 1:9) ones(size(the(:, 1)))];
y = the(:, 10:11);
[n, m] = size(x);
m = m - 1;
beta = x\y;
e = (y - x*beta)'*y;
h = beta'*x'*y - length(y)*mean(y)'*mean(y);
[pval, lambda] = WilksLambda(h, e, 2, m, n - m - 1)
pval =
  5.7607e-11
lambda =
  0.1011
```

The p-value is very small; hence the null hypothesis that all the regression coefficients are zero is rejected.

The backward elimination procedure will be applied to determine if all the predictor variables are required.

```
q = 2;
nh = 1;
p = 9;
ne = length(y) - p - 1;
lambda2 = [];
for i=1:p
    if i == 1
        x1 = x(:, 2:p + 1);
```

```
        elseif i < p
            x1 = [x(:, 1:i) x(:, i + 2:p + 1)];
        else
            x1 = [x(:, 1:p - 1) x(:, p + 1)];
        end
        beta1 = x1\y;
        e = (y - x1*beta1)'*y;
        h = beta1'*x1'*y - length(y)*mean(y)'*mean(y);
        lambda1 = lambda*det(e + h)/det(e);
        lambda2 = [ lambda2 lambda1];
    end
    f = (ne - p + 1)*(1 - lambda2)./(p*lambda2);
    pval = 2*min(fcdf(f, p, ne - p + 1), 1 - fcdf(f, p, ne - p + 1));
    lambda2
    lambda2 =
      0.9877  0.9197  0.9497  0.8988  0.8772  0.9738  0.9888  0.8498  0.8498
```

Eliminate the seventh variable and repeat.

```
lambda2 =
  0.9651  0.8943  0.9286  0.8424  0.8474  0.9692          0.7955  0.7955
```

Eliminate the sixth variable and repeat.

```
lambda2 =
  0.8405  0.7959  0.7436  0.6571  0.8405                  0.6578  0.6578
```

Eliminate the first variable (noting that there is a tie and hence the choice between the first and fifth is arbitrary) and repeat.

```
lambda2 =
          0.7874  0.7382  0.6559  0.8352                  0.6303  0.6303
```

Eliminate the fifth variable and repeat.

```
lambda2 =
          0.5940  0.5639  0.5570                          0.4442  0.4442
```

At this point, the p-values are all below 0.05, and the process terminates. The maximum daily temperature, minimum daily soil temperature, integrated daily soil temperature, and maximum daily relative humidity were eliminated from the predictors.

MATLAB supports multivariation regression through the function [*b*, *sigma*, *r*, *covb*] = **mvregress**(*y*, *x*), where *y* is the response matrix and *x* is the predictor matrix. The returned parameters are the regression coefficients in *b*, the covariance matrix in *sigma*, the residuals in *r*, and the regression coefficient covariance matrix in *covb*.

10.7 Canonical Correlation

The correlation coefficient $\hat{R}^2$ introduced in Section 9.3.1 is a measure of the extent of a linear relationship between a single response variable and one or more predictor variables. Canonical correlation is a generalization to the case of more than one response and more than one predictor variable. It differs from principal component analysis (Section 10.8) in that the focus is on the correlation between two sets of random variables rather than on the correlation within a single data set. Canonical correlation was first proposed by Hotelling (1936).

Consider a random sample of N observation vectors $\overleftrightarrow{\mathbf{Y}}$ and $\overleftrightarrow{\mathbf{X}}$, where the former is $N \times q$ and the latter is $N \times p$. The sample covariance matrix can be partitioned as

$$\overleftrightarrow{\mathbf{S}} = \begin{pmatrix} \overleftrightarrow{\mathbf{S}}_{yy} & \overleftrightarrow{\mathbf{S}}_{yx} \\ \overleftrightarrow{\mathbf{S}}_{xy} & \overleftrightarrow{\mathbf{S}}_{xx} \end{pmatrix} \tag{10.53}$$

Just as the correlation coefficient is the maximum squared correlation between a single response and multiple predictor variables, the squared canonical correlation $\hat{\mathrm{P}}^2$ is the maximum squared correlation between a linear combination of response and a linear combination of predictor variables. Let $\boldsymbol{\alpha}$ be $1 \times q$ and $\boldsymbol{\beta}$ be $1 \times p$ vectors of coefficients chosen to maximize the sample squared correlation given by

$$\hat{\rho}_{\alpha\beta} = \frac{\boldsymbol{\alpha} \cdot \overleftrightarrow{\mathbf{S}}_{yx} \cdot \boldsymbol{\beta}^T}{\left(\boldsymbol{\alpha} \cdot \overleftrightarrow{\mathbf{S}}_{yy} \cdot \boldsymbol{\alpha}^T\right)^{1/2} \left(\boldsymbol{\beta} \cdot \overleftrightarrow{\mathbf{S}}_{xx} \cdot \boldsymbol{\beta}^T\right)^{1/2}} \tag{10.54}$$

such that

$$\hat{\mathrm{P}}^2 = \max_{\boldsymbol{\alpha},\boldsymbol{\beta}} \hat{\rho}^2_{\boldsymbol{\alpha}\boldsymbol{\beta}} \tag{10.55}$$

It can be shown that (10.55) is the largest eigenvalue of either $\overleftrightarrow{\mathbf{S}}_{yy}^{-1} \cdot \overleftrightarrow{\mathbf{S}}_{yx} \cdot \overleftrightarrow{\mathbf{S}}_{xx}^{-1} \cdot \overleftrightarrow{\mathbf{S}}_{xy}$ or $\overleftrightarrow{\mathbf{S}}_{xx}^{-1} \cdot \overleftrightarrow{\mathbf{S}}_{xy} \cdot \overleftrightarrow{\mathbf{S}}_{yy}^{-1} \cdot \overleftrightarrow{\mathbf{S}}_{yx}$, and the canonical coefficients $\boldsymbol{\alpha}$ and $\boldsymbol{\beta}$ are the corresponding eigenvectors. When $q = 1$, $\overleftrightarrow{\mathbf{S}}_{yy}^{-1} \cdot \overleftrightarrow{\mathbf{S}}_{yx} \cdot \overleftrightarrow{\mathbf{S}}_{xx}^{-1} \cdot \overleftrightarrow{\mathbf{S}}_{xy}$ reduces to $\mathbf{s}_{yx} \cdot \overleftrightarrow{\mathbf{S}}_{xx}^{-1} \cdot \mathbf{s}_{yx}^T / \mathbf{s}_{yy}$, which is the multiple correlation coefficient $\hat{R}^2$. Note also that there are $\min(p, q)$ distinct eigenvalues for either $\overleftrightarrow{\mathbf{S}}_{yy}^{-1} \cdot \overleftrightarrow{\mathbf{S}}_{yx} \cdot \overleftrightarrow{\mathbf{S}}_{xx}^{-1} \cdot \overleftrightarrow{\mathbf{S}}_{xy}$ or $\overleftrightarrow{\mathbf{S}}_{xx}^{-1} \cdot \overleftrightarrow{\mathbf{S}}_{xy} \cdot \overleftrightarrow{\mathbf{S}}_{yy}^{-1} \cdot \overleftrightarrow{\mathbf{S}}_{yx}$ that can be used to provide additional information about the relationships between the response and predictor variables. The result is a simplification of the covariance matrix structure: the qp covariances in $\overleftrightarrow{\mathbf{S}}_{yx}$ have been reduced to $\min(p, q)$ canonical correlations. The $\min(p, q)$ canonical variates are given by $\overleftrightarrow{\mathbf{Y}} \cdot \boldsymbol{\alpha}$ and $\overleftrightarrow{\mathbf{X}} \cdot \boldsymbol{\beta}$, respectively. Because $\boldsymbol{\alpha}$ and $\boldsymbol{\beta}$ are eigenvectors, they are unique only up to a scale factor that can be chosen such that $\overleftrightarrow{\mathbf{Y}} \cdot \boldsymbol{\alpha}$ and $\overleftrightarrow{\mathbf{X}} \cdot \boldsymbol{\beta}$ are orthonormal. The canonical variable scores are defined as $\overleftrightarrow{\mathbf{u}} = \left(\overleftrightarrow{\mathbf{X}} - \overline{\mathbf{X}}_N\right) \cdot \boldsymbol{\alpha}$ and $\overleftrightarrow{\mathbf{v}} = \left(\overleftrightarrow{\mathbf{Y}} - \overline{\mathbf{Y}}_N\right) \cdot \boldsymbol{\beta}$ and are akin to the canonical variates after centering the variables.

MATLAB supports canonical correlation through the function [*a*, *b*, *r*] = **canoncorr** (*x*, *y*), where x is $n \times p$ and y is $n \times q$. The returned variables a and b are the canonical coefficients $\boldsymbol{\alpha}$ and $\boldsymbol{\beta}$, whereas r contains the canonical correlations. [*a*, *b*, *r*, *u*, *v*, *stats*] = **canoncorr**(*x*, *y*) also returns the canonical variates u and v and a structure containing test information about the canonical correlations.

Example 10.8 Returning to the sake data of Example 10.6, compute the canonical correlations among the variables.

```
sake = importdata('sake.dat');
y = sake(:, 1:2);
x = sake(:, 3:10);
[a, b, r, u, v, stats] = canoncorr(x, y);
r
r =
   0.6208    0.4946
```

The canonical correlations are moderate, but they are not significant, as can be shown by examining the structure stats.

```
stats
stats =
      Wilks: [0.4642 0.7553]
      df1: [16 7]
      df2: [40 21]
      F: [1.1692 0.9718]
      pF: [0.3321 0.4766]
      chisq: [18.0330 7.0419]
      pChisq: [0.3220 0.4245]
      dfe: [16 7]
      p: [0.3220 0.4245]
```

The fifth entry is the p-value for the F approximation to Wilks' Λ, whereas the seventh entry is the p-value for a chi square approximation. In both cases, the p-values are large; hence the null hypothesis that the data are uncorrelated is accepted. There is no discernible correlation in these data.

Example 10.9 Returning to the temperature-humidity-evaporation data from Example 10.7, compute and assess the canonical correlations among the data.

```
the = importdata('the.dat');
x = the(:,1:9);
y = the(:,10:11);
[a, b, r, u, v, stats] = canoncorr(x, y);
r
r =
   0.9095    0.6442
```

```
stats
stats =
    Wilks: [0.1011 0.5849]
    df1: [18 8]
    df2: [70 36]
    F: [8.3438 3.1930]
    pF: [2.8804e-11 0.0077]
    chisq: [89.3869 21.0253]
    pChisq: [1.8602e-11 0.0071]
    dfe: [18 8]
    p: [1.8602e-11 0.0071]
```

The canonical correlations are fairly high, and the p-values are all very small, rejecting the null hypothesis that the correlations are zero. Figure 10.5 shows the canonical variable scores. The higher canonical correlation of the first canonical variable compared with the second one is readily apparent.

10.8 Empirical Orthogonal Functions

A significant weakness of linear regression in a multivariate context is that the distinction between response and predictor variables is frequently arbitrary, yet their choice plays a major role in the testing of hypotheses about the parameters. A second weakness is that the sensitivity of least squares estimators to influential data in the response and predictor variables is different, and hence the distinction between them can influence the effect of outlying data. The lack of consistency when the predictors contain random noise is also an issue. Linear regression makes sense when the underlying physics demands a linear relationship between specific variables but is less tenable when the relationship between the variables is unknown and hence a statistical parameter to be determined.

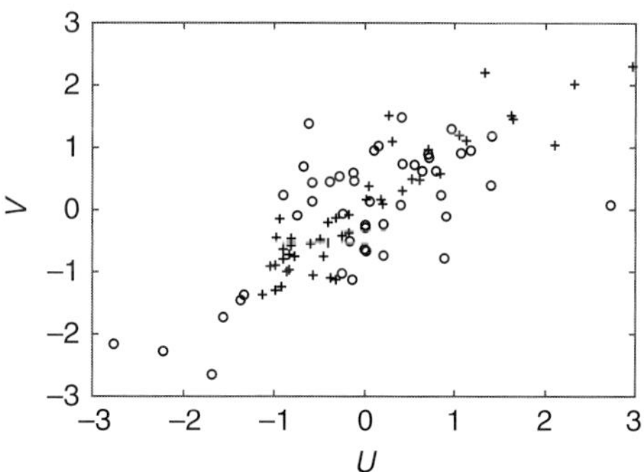

Figure 10.5 The first (crosses) and second (circles) canonical variable scores for the temperature-humidity-wind-speed data.

There are several multivariate techniques available that elucidate the relationship between different data sets. *Discriminant analysis* is used to describe the differences between two or more groups of linear functions of the variables and is an extension of MANOVA that is typically nonparametric. *Classification analysis* uses discriminant functions to reveal the dimensionality of separation and the contribution of the variables to each group. Neither of these is completely data driven, and the grouping remains to some degree arbitrary. *Canonical correlation* (Section 10.7) is a measure of the linear relationship between several response and several predictor variables and hence is an extension of the multiple correlation coefficient. It is valuable in assessing the results from multivariate multiple regression and encompasses much of MANOVA and discriminant/classification analysis, but the distinction between response and predictor variables remains subjective. What are needed are approaches that treat all the variables identically and yield their linear relationships using all possible correlations between all the variables. Two such tools are *empirical orthogonal function (eof)/principal components analysis* and *factor analysis*. The eof terminology has its origin in meteorology, where the technique is commonly applied to spatial data sets or to frequency domain analysis of array data, whereas principal components originated in the social sciences. The former uses terminology that is easily parsed and will be emphasized in what follows. However, eofs and principal components inherently are identical in concept. Principal component analysis was introduced by Hotelling (1933). A key reference about principal components is Jolliffe (2002), and Preisendorfer (1988) provides a thorough description of their use in meteorology and physical oceanography.

10.8.1 Theory

In eof analysis, the original p variables are linearly transformed into combinations whose variances are maximized while being orthogonal and hence mutually uncorrelated. Eof analysis is a dimension reduction technique that simplifies the p variables into a few linear combinations that explain most of the data variance and that constitute the core structure of the data set. In general, the more highly correlated that data are, the smaller is the number of eofs required to describe them or, conversely, if the data are only weakly correlated, the resulting eofs will largely reflect the original data, and limited parsimony will be achieved. Eof analysis is a one-sample technique with no subjective groupings (unlike discriminant/classification analysis) and no partitioning of the data into response and predictor variables.

Consider N samples of p observation vectors. Each sample is a swarm in a p-dimensional space. For example, there may be N time points (not necessarily equally spaced) and p locations at which some variable is measured, although there is not a requirement that the variables all be the same or that they be scalar quantities. For simplicity, scalar data will be considered for the present. The N samples of p observation vectors constitute an $N \times p$ data matrix $\overleftrightarrow{\mathbf{X}}$ that is assumed to be centered (i.e., have the sample mean removed from each column by multiplying by the centering matrix $\overleftrightarrow{\mathbf{I}}_N - \overleftrightarrow{\mathbf{J}}_N/N$).

For a given sample (row in $\overleftrightarrow{\mathbf{X}}$), the p variables (columns of $\overleftrightarrow{\mathbf{X}}$) may be (and typically are) correlated, and the swarm of points may not be oriented parallel to any of the p axes of the data. The goal of eof analysis is to find the natural axes of the data. This is accomplished by rotating the axes in a p-dimensional space, after which the new axes become the natural ones for the data, and the new variables (which are linear combinations of the original data) will be mutually uncorrelated. In some cases, the eofs are directly interpretable as physical phenomena, but there is no guarantee that the output will be physically meaningful entities.

The first eof is the linear combination of the variables with maximum sample variance among all possible linear combinations of the variables. An N-vector $\mathbf{z} = \overleftrightarrow{\mathbf{X}} \cdot \mathbf{a}^T$ is sought whose sample variance $s_z^2 = \mathbf{a} \cdot \overleftrightarrow{\mathbf{S}}' \cdot \mathbf{a}^T$, where $\mathbf{a}$ is $1 \times p$ and $\overleftrightarrow{\mathbf{S}}'$ is the $p \times p$ unbiased sample covariance matrix, is a maximum. The vector $\mathbf{a}$ cannot be arbitrary because s_z^2 has no maximum if the elements in $\mathbf{a}$ increase without limit, and hence s_z^2 must be maximized as the coefficients in $\mathbf{a}$ vary relative to each other, or equivalently, the constraint $\mathbf{a} \cdot \mathbf{a}^T = 1$ must be imposed. The quantity to be maximized is the Rayleigh quotient

$$t^2 = \frac{\mathbf{a} \cdot \overleftrightarrow{\mathbf{S}}' \cdot \mathbf{a}^T}{\mathbf{a} \cdot \mathbf{a}^T} \tag{10.56}$$

According to Rayleigh's principle, the vector $\mathbf{a}$ that maximizes (10.56) is given by $\mathbf{a}_1$, the eigenvector of $\overleftrightarrow{\mathbf{S}}'$ that corresponds to its largest eigenvalue (usually called the *first eigenvector*). The maximum value of (10.56) is given by the corresponding eigenvalue of $\overleftrightarrow{\mathbf{S}}'$. The resulting $\mathbf{z}_1 = \overleftrightarrow{\mathbf{X}} \cdot \mathbf{a}_1^T$ is the first *empirical orthogonal function amplitude* of $\overleftrightarrow{\mathbf{X}}$.

$\overleftrightarrow{\mathbf{S}}'$ has p eigenvectors or eofs (originally called *loadings* in principal components terminology), and hence there are p eof amplitudes (originally called *scores*) that can be written as the $N \times p$ matrix

$$\overleftrightarrow{\mathbf{Z}} = \overleftrightarrow{\mathbf{X}} \cdot \overleftrightarrow{\mathbf{A}}^T \tag{10.57}$$

$\overleftrightarrow{\mathbf{S}}'$ is symmetric; hence the eigenvectors are mutually orthogonal and unitary $\left(\overleftrightarrow{\mathbf{A}}^T \cdot \overleftrightarrow{\mathbf{A}} = \overleftrightarrow{\mathbf{A}} \cdot \overleftrightarrow{\mathbf{A}}^T = \overleftrightarrow{\mathbf{I}}_p\right)$, and if $\overleftrightarrow{\mathbf{S}}'$ is positive definite, the eigenvalues are all positive. If $\overleftrightarrow{\mathbf{S}}'$ is only positive semidefinite, then the number of positive eigenvalues corresponds to its rank.

By the spectral decomposition theorem,

$$\overleftrightarrow{\mathbf{S}}' = \overleftrightarrow{\mathbf{A}}^T \cdot \overleftrightarrow{\mathbf{D}} \cdot \overleftrightarrow{\mathbf{A}} \tag{10.58}$$

where $\overleftrightarrow{\mathbf{D}} = \mathrm{diag}(\lambda_i)$ is the diagonal matrix of eigenvalues that give the sample variances of the empirical orthogonal function amplitudes. The sample covariance matrix is $\overleftrightarrow{\mathbf{S}}' = \overleftrightarrow{\mathbf{X}}^T \cdot \overleftrightarrow{\mathbf{X}} / (N-1)$ because the variables are centered, and hence

$$\overleftrightarrow{\mathbf{S}}_Z = \overleftrightarrow{\mathbf{Z}}^T \cdot \overleftrightarrow{\mathbf{Z}} / N = \overleftrightarrow{\mathbf{A}} \cdot \overleftrightarrow{\mathbf{X}}^T \cdot \overleftrightarrow{\mathbf{X}} \cdot \overleftrightarrow{\mathbf{A}}^T / (N-1) = \overleftrightarrow{\mathbf{A}} \cdot \overleftrightarrow{\mathbf{S}}' \cdot \overleftrightarrow{\mathbf{A}}^T = \overleftrightarrow{\mathbf{D}} \tag{10.59}$$

The empirical orthogonal function amplitudes are mutually orthogonal.

The eigenvalues and eigenvectors of the sample covariance matrix may be found by solving the eigenvalue problem

$$\overleftrightarrow{\mathbf{S}}' \cdot \overleftrightarrow{\mathbf{A}} = \overleftrightarrow{\lambda} \cdot \overleftrightarrow{\mathbf{A}} \tag{10.60}$$

where $\overleftrightarrow{\lambda}$ is diagonal, and the columns of $\overleftrightarrow{\mathbf{A}}$ are the eigenvectors of $\overleftrightarrow{\mathbf{S}}'$. The eigenvectors are uncertain by a sign because the result is unchanged if both sides of (10.60) are multiplied by -1. The columns of $\overleftrightarrow{\mathbf{A}}$ are the eofs that constitute an orthogonal basis set empirically determined from the data. The eof amplitudes are the projections of the data set onto the eof basis functions.

The eigenvectors of $\overleftrightarrow{\mathbf{X}}^T \cdot \overleftrightarrow{\mathbf{X}}$ and $\overleftrightarrow{\mathbf{S}}'$ are identical, whereas the eigenvalues of $\overleftrightarrow{\mathbf{X}}^T \cdot \overleftrightarrow{\mathbf{X}}$ are $N-1$ times the eigenvalues of $\overleftrightarrow{\mathbf{S}}'$. Consequently, it is often more convenient to work directly with $\overleftrightarrow{\mathbf{X}}^T \cdot \overleftrightarrow{\mathbf{X}}$ and correct for the scale factor $N-1$ at the conclusion.

Let $\overleftrightarrow{\mathbf{D}}_k$ be the diagonal matrix of eigenvalues with all entries beyond $i = k$ set to zero. Then $\overleftrightarrow{\mathbf{S}}'_k = \overleftrightarrow{\mathbf{A}}^T \cdot \overleftrightarrow{\mathbf{D}}_k \cdot \overleftrightarrow{\mathbf{A}}$ is the best approximation of rank k to $\overleftrightarrow{\mathbf{S}}'$, in the sense that the sum of squares $\| \overleftrightarrow{\mathbf{S}}' - \overleftrightarrow{\mathbf{S}}'_k \|^2$ is minimized. Consequently, the vector subspace spanned by the first k eofs has a smaller mean square deviation from the sample variables than any other k-dimensional subspace. Further, if $\overleftrightarrow{\mathbf{S}}'$ has rank $k < p$, then the total variation in the data may be explained by the first k eofs.

The sum of the first k eigenvalues divided by the sum of all the eigenvalues is the proportion of the variance represented by the first k eof amplitudes. Consequently, if one variable has a much larger variance than the remaining data, then this variable will dominate the first eof amplitude, which will, in turn, dominate the variance. Conversely, if one variable is uncorrelated with the remaining variables, then its variance will be one of the eigenvalues, and the corresponding eof will be all zeroes except for a one at the variable index. In this case, the variable is itself an eof amplitude. Finally, because no inverse or determinant is involved in extracting the eofs, the covariance matrix can be singular, in which case some of the eigenvalues will be zero.

The eof amplitudes are not scale invariant. As a result, it is desirable for the data to be commensurable (i.e., have similar measurement scales and variances). Eof analysis can be carried out on the sample covariance matrix $\overleftrightarrow{\mathbf{S}}'$ or the corresponding sample correlation matrix $\overleftrightarrow{\mathbf{R}}'$, and the latter may be preferred if the data are incommensurable because the eof amplitudes extracted from $\overleftrightarrow{\mathbf{R}}'$ are scale invariant. The eof amplitudes from $\overleftrightarrow{\mathbf{R}}'$ will not be the same as those from $\overleftrightarrow{\mathbf{S}}'$ because of the lack of scale invariance of the latter. In general, the percent variance explained by a given eof amplitude will be different for $\overleftrightarrow{\mathbf{R}}'$ as compared to $\overleftrightarrow{\mathbf{S}}'$.

As an example illustrating these issues, suppose that bivariate data have

$$\overleftrightarrow{\mathbf{S}}' = \begin{bmatrix} 1 & 4 \\ 4 & 25 \end{bmatrix} \quad \text{and} \quad \overleftrightarrow{\mathbf{R}}' = \begin{bmatrix} 1 & 0.8 \\ 0.8 & 1 \end{bmatrix}$$

The eigenvalues and eigenvectors for $\overleftrightarrow{\mathbf{S}}'$ are $\boldsymbol{\lambda} = (25.6,\ 0.35)$, $\mathbf{a}_1 = (0.160,\ 0.987)$, and $\mathbf{a}_2 = (0.987,\ -0.160)$. The large variance of the second variable ensures that the first eof amplitude $\mathbf{z}_1 = 0.160\mathbf{x}_1 + 0.987\mathbf{x}_2$ nearly duplicates $\mathbf{x}_2$ and does not reflect the correlation of the two variables. Note also that $\lambda_1/(\lambda_1 + \lambda_2) = 0.99$, and hence the second eof amplitude $\mathbf{z}_2$ is not very important. By contrast, the eigenvalues and eigenvectors for $\overleftrightarrow{\mathbf{R}}'$ are $\boldsymbol{\lambda} = (1.8,\ 0.2)$, $\mathbf{a}_1 = (0.707,\ 0.707)$, and $\mathbf{a}_2 = (-0.707,\ 0.707)$. The variables are equally weighted by the eofs. However, $\lambda_1/(\lambda_1 + \lambda_2) = 0.9$, so the first eof amplitude is still dominant.

While eof analysis can be carried out from the sample covariance or correlation matrices, there is an easier way using the singular value decomposition described in Section 9.3 to directly estimate them from the data

$$\overleftrightarrow{\mathbf{X}} \cdot \overleftrightarrow{\mathbf{B}} = \overleftrightarrow{\mathbf{U}} \cdot \overleftrightarrow{\Upsilon} \cdot \overleftrightarrow{\mathbf{V}}^T \tag{10.61}$$

Where $\overleftrightarrow{\mathbf{B}}$ is a $p \times p$ diagonal matrix that preweights the columns of $\overleftrightarrow{\mathbf{X}}$. If $\overleftrightarrow{\mathbf{B}} = \overleftrightarrow{\mathbf{I}}_p$, then the result is equivalent to working with the sample covariance matrix, whereas if $\overleftrightarrow{\mathbf{B}} = \text{diag}(1/s_{ii})$, then the result is equivalent to working with the sample correlation matrix. Note that $\overleftrightarrow{\mathbf{X}} \cdot \overleftrightarrow{\mathbf{B}}$ contains the z-scores of the data in this instance because $\overleftrightarrow{\mathbf{X}}$ is centered. A third way to preweight the columns of $\overleftrightarrow{\mathbf{X}}$ is to first regress each column in turn on the remaining $p - 1$ columns and place the inverse of the residual standard deviation in the diagonal elements of $\overleftrightarrow{\mathbf{B}}$. This equalizes the noise level, where *noise* is defined to be the component of a given sample that is uncorrelated with all the remaining samples across the data under the assumption that the noise covariance matrix is diagonal, meaning that there is no correlated noise present. The $p \times p$ matrix of *right singular vectors* $\overleftrightarrow{\mathbf{V}}$ and the $N \times N$ matrix of *left singular vectors* $\overleftrightarrow{\mathbf{U}}$ are both orthogonal and unitary and are ambiguous in sign because multiplying both by -1 leaves (10.61) unchanged. The $N \times p$ matrix $\overleftrightarrow{\Upsilon}$ of *singular values* is zero except for the p diagonal locations starting at the upper left corner and by convention ranked from largest to smallest down the diagonal. The relationships between the matrices in (10.61) are

$$\overleftrightarrow{\mathbf{X}} \cdot \overleftrightarrow{\mathbf{B}} \cdot \overleftrightarrow{\mathbf{V}} = \overleftrightarrow{\mathbf{U}} \cdot \overleftrightarrow{\Upsilon} \tag{10.62}$$

$$\overleftrightarrow{\mathbf{B}}^T \cdot \overleftrightarrow{\mathbf{X}}^T \cdot \overleftrightarrow{\mathbf{U}} = \overleftrightarrow{\mathbf{V}} \cdot \overleftrightarrow{\Upsilon}^T \tag{10.63}$$

$$\overleftrightarrow{\mathbf{B}}^T \cdot \overleftrightarrow{\mathbf{X}}^T \cdot \overleftrightarrow{\mathbf{X}} \cdot \overleftrightarrow{\mathbf{B}} \cdot \overleftrightarrow{\mathbf{V}} = \overleftrightarrow{\mathbf{V}} \cdot \overleftrightarrow{\Upsilon}^T \cdot \Upsilon \tag{10.64}$$

$$\overleftrightarrow{\mathbf{X}} \cdot \overleftrightarrow{\mathbf{B}} \cdot \overleftrightarrow{\mathbf{B}}^T \cdot \overleftrightarrow{\mathbf{X}}^T \cdot \overleftrightarrow{\mathbf{U}} = \overleftrightarrow{\mathbf{U}} \cdot \overleftrightarrow{\Upsilon} \cdot \overleftrightarrow{\Upsilon}^T \tag{10.65}$$

hence the columns of $\overleftrightarrow{\mathbf{V}}$ are the variable eofs, and the squared singular values are the corresponding eigenvalues [i.e., $\overleftrightarrow{\mathbf{D}} = \overleftrightarrow{\Upsilon}^T \cdot \overleftrightarrow{\Upsilon} / (N-1)$in (10.59)]. The eof amplitudes are given by (10.62). A second type of eof amplitude in observation rather than variable space is obtained in the first p columns of (10.63). Equations (10.64) and (10.65) define eof analysis in terms of the covariance or correlation structure rather than the data structure, recognizing that the squared singular values must be scaled by $1/(N-1)$.

MATLAB directly supports eof analysis through the functions **pca** and **pcacov** that operate, respectively, on the raw data and the covariance matrix. The call [*coeff*, *score*, *latent*] = **pca**(x) uses the svd by default on the $n \times p$ data matrix x and returns the eigenvectors ($\overleftrightarrow{\mathbf{V}}$ from the svd) in *coeff*, the eof amplitudes in *score*, and the eigenvalues or squared singular values divided by the degrees-of-freedom $n - 1$ in *latent*. A variety of name-value arguments can be added to control the algorithm, whether the data are centered (default) or not, weighting, and how missing data are handled. The call [*coeff*, *latent*] = **pcacov**(*cov*) provides fewer capabilities based on the sample covariance matrix *cov* and is not recommended by comparison to **pca**.

10.8.2 Choosing the Number of Eofs

A key decision that the analyst has to make is how many eofs to retain in representing a given data set. There are three main approaches that often produce contradictory results in part because they are largely qualitative rather than quantitative.

The simplest but most arbitrary is retaining enough eofs to account for a fixed percentage of the total variance (e.g., 90%, with typical values lying between 70% and 90%). The choice of threshold is subjective.

A more objective approach is to retain those eofs whose eigenvalues are larger than the average $\mathrm{tr}\left(\overleftrightarrow{\mathbf{S}}\right)/p$ of all the eigenvalues. This works well in general and is likely to err on the side of retaining too many rather than too few eofs, which is a conservative approach. It is the default in many software packages.

A third method is to plot the eigenvalues against their index (called a *scree plot*) and look for the natural break between large and small eigenvalues. An alternative that is often used in the atmospheric sciences is to plot the log of the eigenvalues rather than the eigenvalues themselves. These methods work well with data that are highly correlated so that a few eofs is sufficient.

Once the number of eofs is determined, and presuming that this is two or three, the pertinent empirical orthogonal function amplitudes and the variables can be plotted as a *biplot*. MATLAB supports this through the function **biplot**(*v*(:, 1:*p*), 'scores', *eofa*(:, 1:*p*)), where p is either 2 or 3. Additional name-value pairs can be used to label the eigenvector or eof amplitude axes. Biplots feature prominently in compositional data analysis in Chapter 11.

Once a subset of the eofs has been selected as representative of the data, the subset eofs can be rotated using one of many criteria to simplify the interpretation of the entire subset. This is an integral part of factor analysis but is less commonly used in eof analysis.

Let $\overleftrightarrow{\mathbf{V}}_k$ be a $p \times k$ submatrix of right singular vectors defined by eliminating the last $p - k$ columns corresponding to the smallest eigenvalues in (10.62). Applying an orthogonal rotation matrix $\overleftrightarrow{\mathbf{T}}$ yields a rotated subset of eigenvectors $\overleftrightarrow{\mathbf{V}}_k \cdot \overleftrightarrow{\mathbf{T}}$ that remain orthogonal because $\overleftrightarrow{\mathbf{T}}^T \cdot \overleftrightarrow{\mathbf{V}}_k^T \cdot \overleftrightarrow{\mathbf{V}}_k \cdot \overleftrightarrow{\mathbf{T}} = \overleftrightarrow{\mathbf{T}}^T \cdot \overleftrightarrow{\mathbf{T}} = \overleftrightarrow{\mathbf{I}}_k$. However, the rotated eigenvalue matrix $\overleftrightarrow{\mathbf{T}}^T \cdot \overleftrightarrow{\mathbf{\Upsilon}}_k \cdot \overleftrightarrow{\mathbf{T}}$ is not diagonal, and hence the rotated eof amplitudes are not uncorrelated.

The rotation matrix is chosen based on one of many (there are at least twenty) criteria. The most widely used is the *varimax* rotation that maximizes the sum of the variances of the eigenvectors (i.e., it simplifies the columns of $\overleftrightarrow{\mathbf{V}}_k$). The goal is to minimize the complexity of the eofs by making the large eigenvectors larger and the small ones smaller. The result is a set of principal components that are as independent of each other as possible. An alternative is the *quartimax* rotation, which makes large eigenvectors larger within each variable (i.e., it simplifies the rows of $\overleftrightarrow{\mathbf{V}}_k$). The *equimax* rotation is a compromise between varimax and quartimax. It is also possible to apply oblique (i.e., nonorthogonal) rotations, although this is done less commonly.

MATLAB supports rotation through *v*1 = **rotatefactors**(*v*(:, 1:*p*)), where *v* is the eigenvector matrix from **svd** or the coefficient matrix from **pca**. By default, the varimax rotation is applied. The keyword value pair "Method," "quartimax," or "equimax" can be used to specify the quartimax or equimax methods.

10.8.3 Example

Example 10.10 Returning to the zonally averaged temperature data of Example 10.2, an eof analysis will be applied to the eight zonal bands extending from the north to the south pole.

```
z = importdata('ZonAnnTs+dSST.dat');
x = z(:, 8:15);
var(x)
ans =
   1.0e+03 *
   6.4492 2.3541 1.0484 0.9533 1.0833 0.9371 0.6632 4.4149
```

The variance range covers an order of magnitude, with the largest values occurring toward the poles, so the covariance eofs will probably be dominated by those variables.

```
x1 =  x - repmat(mean(x), length(x), 1);
[u, s, v]  = svd(x1);
diag(s.^2)'/sum(diag(s.^2))
ans =
     0.6318 0.2398 0.0696 0.0272 0.0158 0.0097 0.0042 0.0019
```

The first two eofs account for 87.2% of the variance. By criteria 1 and 2 earlier, only three and two eofs would be retained, respectively. A scree plot suggests that three or four eofs are required (Figure 10.6). Consequently, four will be retained for further analysis. A plot of the logarithm of the eigenvalues against the index is nearly a straight line and hence uninformative.

Examine the eofs.

```
v
v =
   0.697  0.345  0.568 -0.265  0.040 -0.013  0.006  0.020
   0.416  0.089 -0.188  0.765 -0.108  0.428 -0.02  -0.052
   0.269  0.010 -0.242  0.245 -0.061 -0.817 -0.297  0.221
   0.239 -0.026 -0.385 -0.206  0.529  0.012 -0.255 -0.638
   0.254 -0.042 -0.450 -0.251  0.366  0.202  0.249  0.657
   0.247 -0.088 -0.289 -0.146 -0.429 -0.191  0.708 -0.325
   0.165 -0.113 -0.264 -0.393 -0.615  0.269 -0.532  0.061
   0.242 -0.922  0.283  0.057  0.088  0.004 -0.010  0.017
```

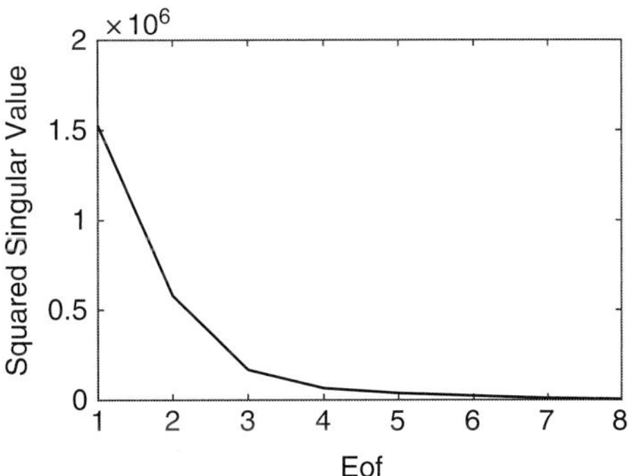

Figure 10.6 Scree plot for the zonally averaged temperature data.

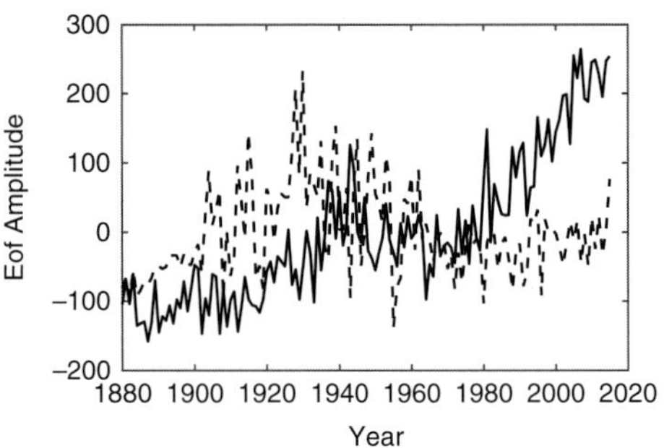

Figure 10.7 First two eof amplitudes for the unscaled zonal temperature data. The solid line is the first and the dashed line is the second eof amplitude.

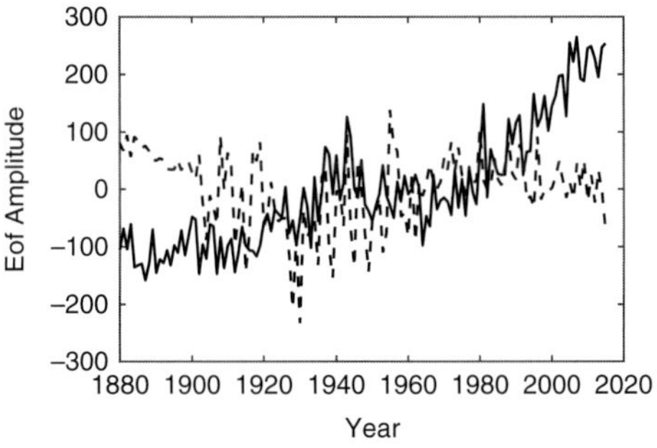

Figure 10.8 First two eof amplitudes for the unscaled zonal temperature data after reversing the sign of the second eof. The solid line is the first and the dashed line is the second eof amplitude.

The first eof is dominated by the northern hemisphere polar and subpolar zonal bands. The second eof is dominated by anticorrelation of the northern and southern polar zonal bands. The third eof represents anticorrelation of the northern polar and two equatorial zonal bands. The fourth eof represents anticorrelation of the northern and southern subpolar bands. The first two eof amplitudes are plotted in Figure 10.7.

```
eofa = u*s;
t = z(:, 1);
plot(t, eofa(:, 1), t, eofa(:, 2))
```

The first eof amplitude reproduces the generally increasing temperature at northern polar latitudes seen in Figure 10.2. However, the second eof amplitude is largest at the south pole yet has the opposite sense to the south pole data (see Figure 10.3) until about 1960. Reversing the sign of the second eof amplitude gives Figure 10.8. The first eof amplitude shows a generally increasing temperature anomaly, with a slight decrease from 1940 to 1970, and explains 64.5% of the data variance. The second eof amplitude reveals a weak decrease until approximately 1930, followed by nearly constant temperature, and explains 23.5% of the variance.

Apply a varimax rotation to the first four eigenvectors.

```
v1 = rotatefactors(v(:,1:4))
v1 =
   0.999    0.002   -0.005   -0.003
   0.014   -0.006    0.115    0.888
  -0.011    0.001   -0.153    0.410
  -0.009    0.042   -0.491    0.077
  -0.030    0.046   -0.569    0.068
   0.015   -0.048   -0.402    0.094
   0.027   -0.044   -0.487   -0.155
  -0.002   -0.996   -0.007    0.004
```

The result maximizes the differences between the eigenvector columns, resulting in dominance at the north and south poles in the first two eigenvectors. A broad average over the northern equatorial to southern subpolar bands is seen in the third eigenvector. The fourth eigenvector is dominated by the northern hemisphere midlatitude and subpolar bands. A very similar result obtains using the quartimax or equimax rotation.

Repeat the exercise using the correlation eofs rather than the covariance eofs.

```
b = diag(1./std(x));
[u, s, v] = svd(x1*b);
diag(s.^2)/sum(diag(s.^2))
ans =
   0.7025  0.1305  0.0666  0.0429  0.0276  0.0154  0.0105  0.0041
```

By criterion 1, three eofs would be retained. By criterion 2, only two eofs would be retained. A scree plot (Figure 10.10) suggests that only two eofs should be retained.

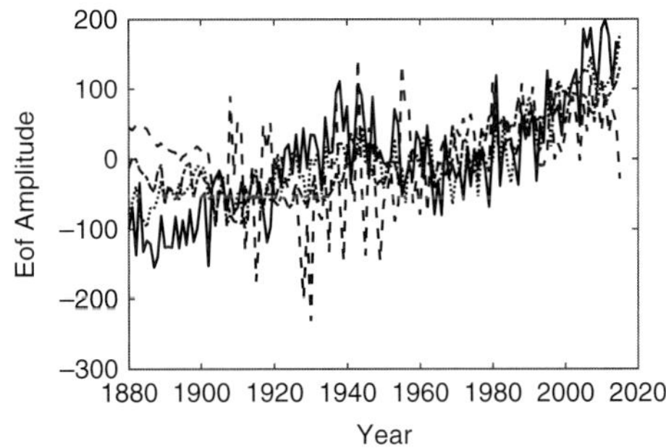

Figure 10.9 First four eof amplitudes obtained after varimax rotation of the first four eigenvectors. The first (solid line), second (dashed line), third (dash-dot line), and fourth (dotted line) eof amplitudes are shown.

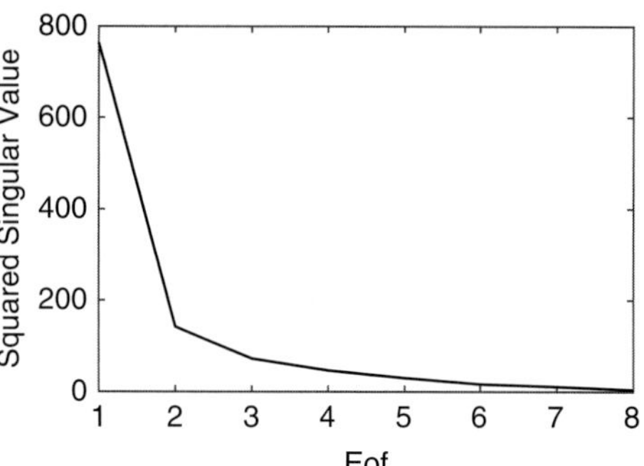

Figure 10.10 Scree plot for the zonally averaged temperature data with the columns normalized by their standard deviation.

Examine the eofs.

```
v
v =
  0.332   0.375   0.455  -0.307  -0.616  -0.253   0.022  -0.060
  0.371   0.275   0.331   0.016   0.278   0.768  -0.018   0.098
  0.386   0.163   0.155   0.114   0.579  -0.503  -0.387  -0.223
  0.385   0.042  -0.296   0.496  -0.272  -0.070  -0.243   0.616
  0.390  -0.005  -0.360   0.355  -0.215   0.132   0.249  -0.684
  0.397  -0.130  -0.010  -0.245   0.269  -0.215   0.741   0.297
  0.338  -0.330  -0.393  -0.644  -0.063   0.155  -0.422  -0.036
  0.177  -0.794   0.527   0.216  -0.107   0.009  -0.035  -0.037
```

The first eof represents a nearly global average of the temperature anomaly, with some reduction of the weighting at south polar latitudes. The second eof is dominated by anticorrelation of the northern and southern subpolar and polar bands. The third eof represents a complicated global pattern. Plotting the eof amplitudes gives a first one that very closely resembles the global average temperature anomaly. The second eof amplitude is similar in shape to the second covariance eof amplitude and hence has the wrong sign. The correct sign of the third eof amplitude is not obvious. However, comparing the average of the two northernmost and equatorial zonal averages suggests that the sign is correct. The corrected eof amplitudes are shown in Figure 10.11. Note the much smaller ordinate by comparison with Figure 10.8.

Apply a varimax rotation to the first three eigenvectors.

```
v1 =
  -0.107   0.016   0.668
   0.020  -0.012   0.568
   0.162  -0.017   0.415
   0.465   0.116   0.086
   0.519   0.109   0.026
   0.390  -0.131   0.122
   0.568  -0.135  -0.191
  -0.023  -0.969   0.020
```

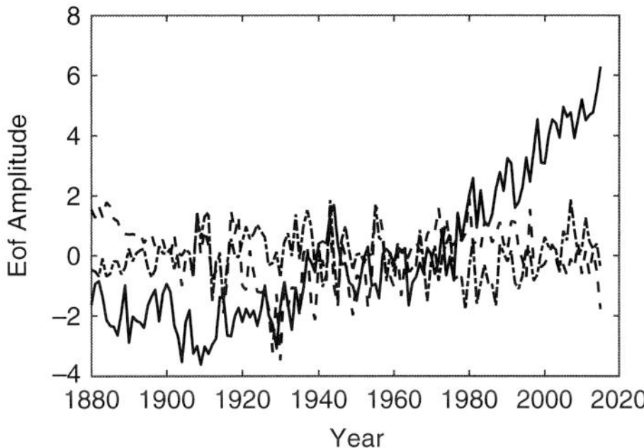

Figure 10.11 First three eof amplitudes for the zonally averaged temperature anomaly data with the data matrix columns normalized by their standard deviations. The first (solid line), second (dashed line), and third (dash-dot line) eof amplitudes are shown.

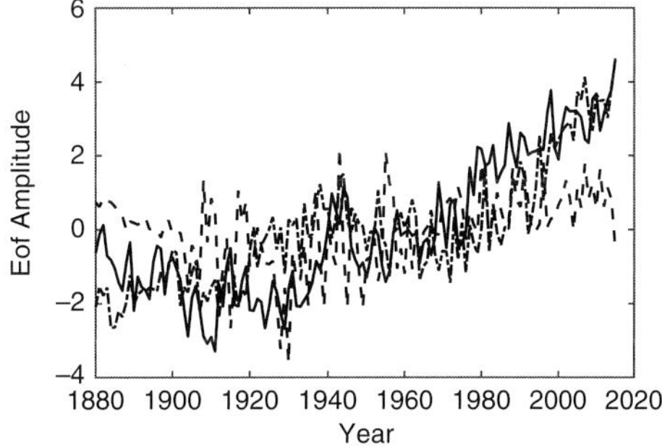

Figure 10.12 First three varimax rotated eof amplitudes for the zonally averaged temperature anomaly data with the data matrix columns normalized by their standard deviations. The first (solid line), second (dashed line), and third (dash-dot line) eof amplitudes are shown.

The first rotated eof is dominated by the equatorial northern hemisphere and equatorial through subpolar southern hemisphere data and is similar in shape to the third rotated eof for the covariance eofs. The second rotated eof is dominated by the southern hemisphere polar temperature anomaly and hence is similar in shape to the second rotated eof for the covariance eofs. The third rotated eof represents the average of the northern hemisphere temperature anomaly poleward of equatorial latitudes. The eof amplitudes after reversing the sign of the second one are shown in Figure 10.12.

Apply the third version of preweighting based on regressing a given column of $\overleftrightarrow{\mathbf{X}}$ against the remaining ones.

```
res = [];
for i = 1:p
    y = x1(:, i);
    if i == 1
        x2 = x1(:, 2:p);
   elseif i < p
        x2 = [x1(:, 1:i - 1) x1(:, i + 1:p)];
```

```
    else
        x2 = x1(:, 1:p - 1);
    end
    b1 = x2\y;
    res = [res std(y - x2*b1)];
end
b = diag(1./res);
var(x1*b)
ans =
    3.1738 5.7292 7.0055 13.5203 15.9300 10.1663 4.0031 1.3753
```

This form of weighting emphasizes the equatorward zones relative to the polar ones and hence has the opposite sense to the covariance approach.

```
[u, s, v] = svd(x1*b);
eofa = u*s;
diag(s.^2/sum(diag(s.^2)))
ans =
   0.8204  0.0636  0.0528  0.0188  0.0154  0.0125  0.0101  0.0065
```

By the first criterion, three eofs would be used. By the second criterion, only a single eof would be used. A scree plot (Figure 10.13) suggests that two or three eofs should be retained.

Examine the eofs.

```
v
v =
   0.186   0.429   0.150  -0.541   0.312  -0.421  -0.385  -0.211
   0.287   0.533   0.170  -0.138   0.208   0.532   0.438   0.259
   0.336   0.442   0.052   0.577  -0.318  -0.266   0.122  -0.412
   0.496  -0.337   0.3959  0.1537  0.1621 -0.421   0.1245  0.491
   0.545  -0.422   0.179  -0.171  -0.077   0.433  -0.163  -0.494
   0.418   0.130  -0.583   0.038  -0.227   0.116  -0.475   0.420
   0.221  -0.152  -0.563  -0.354  -0.020  -0.294   0.610  -0.163
   0.061  -0.067  -0.316   0.420   0.821   0.070  -0.074  -0.176
```

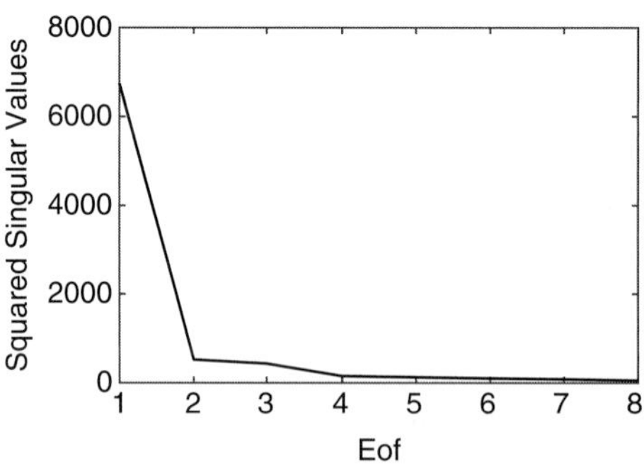

Figure 10.13 Scree plot for the zonally averaged temperature data with the columns normalized by the noise variance.

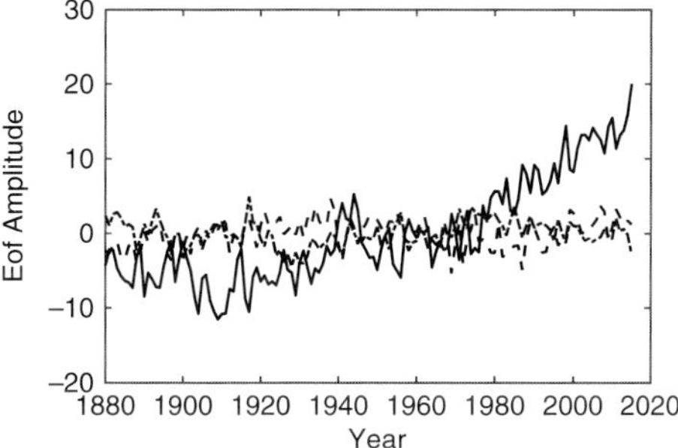

Figure 10.14 First three eof amplitudes for the globally averaged temperature anomaly data normalized by their noise variance. The first (solid line), second (dashed line), and third (dash-dot line) eof amplitudes are shown.

The first eof is an average over equatorial and subtropical latitudes. The second eof represents anticorrelation of the northern hemisphere poleward of subtropical latitudes and the equatorial data. The third eof is dominantly southern hemisphere poleward of subtropical latitudes. The first three eof amplitudes are plotted in Figure 10.14 after comparing them with suitable averages of the data, suggesting that the first two have the right sign and the last one does not.

Apply a varimax rotation to the first three eofs.

```
v1 =
 -0.033   0.482   0.085
 -0.005   0.625   0.067
  0.035   0.552  -0.066
  0.706   0.040   0.127
  0.704  -0.046  -0.096
  0.018   0.219  -0.695
  0.029  -0.115  -0.613
 -0.041  -0.082  -0.316
```

The first rotated eof is dominated by the equatorial zonal bands. The second rotated eof is dominated by midlatitude to polar northern hemisphere zonal bands. The third rotated eof is dominated by the poleward southern hemisphere. After reversing the sign of the third eof, the results are shown in Figure 10.15. An increasing temperature anomaly at equatorial latitudes is quite apparent in the first eof amplitude.

10.8.4 Empirical Orthogonal Function Regression

Empirical orthogonal function analysis is frequently used in conjunction with other statistical techniques, most notably linear regression when the predictor variables are highly correlated. This problem is called *multicollinearity* and occurs when two or more of the

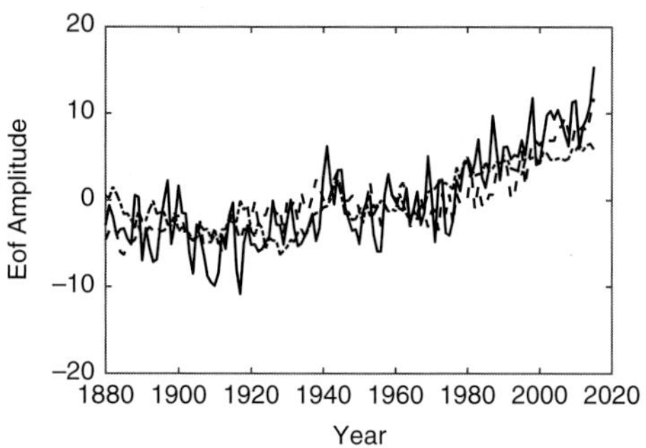

Figure 10.15 First three varimax rotated eof amplitudes for the zonally averaged temperature anomaly data normalized by their noise standard deviations. The first (solid line), second (dashed line), and third (dash-dot line) eof amplitudes are shown.

predictor variables are nearly constant linear functions of the others. The outcome can be regression parameters with very large variances.

Consider the linear regression

$$\mathbf{y} = \overleftrightarrow{\mathbf{X}} \cdot \boldsymbol{\beta} + \boldsymbol{\varepsilon} \tag{10.66}$$

where both the response variable and the predictor matrix are assumed to be centered. The eof amplitudes for the observations are

$$\overleftrightarrow{\mathbf{Z}} = \overleftrightarrow{\mathbf{X}} \cdot \overleftrightarrow{\mathbf{V}} \tag{10.67}$$

Because $\overleftrightarrow{\mathbf{V}}$ is orthogonal, $\overleftrightarrow{\mathbf{X}} \cdot \boldsymbol{\beta} = \overleftrightarrow{\mathbf{X}} \cdot \overleftrightarrow{\mathbf{V}} \cdot \overleftrightarrow{\mathbf{V}}^T \cdot \boldsymbol{\beta} = \overleftrightarrow{\mathbf{Z}} \cdot \boldsymbol{\gamma}$, where $\boldsymbol{\gamma} = \overleftrightarrow{\mathbf{V}}^T \cdot \boldsymbol{\beta}$. Equation (10.66) becomes

$$\mathbf{y} = \overleftrightarrow{\mathbf{Z}} \cdot \boldsymbol{\gamma} + \boldsymbol{\varepsilon} \tag{10.68}$$

where the predictors are replaced by the eof amplitudes or principal component scores. Eof regression is the solution of (10.68) or, more likely, a reduced-rank version obtained by retaining only the first k eofs in $\overleftrightarrow{\mathbf{V}}$. The least squares principles developed in Chapter 9 can be used to solve (10.68) for $\hat{\boldsymbol{\gamma}}$ from which an estimate for $\boldsymbol{\beta}$ is

$$\hat{\boldsymbol{\beta}} = \overleftrightarrow{\mathbf{V}} \cdot \hat{\boldsymbol{\gamma}} \tag{10.69}$$

The advantage of the eof regression approach is that the columns of $\overleftrightarrow{\mathbf{Z}}$ in (10.68) are orthogonal, simplifying the behavior in the presence of multicollinearity. The solution to (10.68) is

$$\hat{\boldsymbol{\gamma}} = \left(\overleftrightarrow{\mathbf{Z}}^T \cdot \overleftrightarrow{\mathbf{Z}}\right)^{-1} \cdot \overleftrightarrow{\mathbf{Z}}^T \cdot \mathbf{y} = \left(\overleftrightarrow{\boldsymbol{\Upsilon}}^T \cdot \overleftrightarrow{\boldsymbol{\Upsilon}}\right)^{-1} \cdot \overleftrightarrow{\mathbf{Z}}^T \cdot \mathbf{y} \tag{10.70}$$

using (10.64). $\left(\overleftrightarrow{\boldsymbol{\Upsilon}}^T \cdot \overleftrightarrow{\boldsymbol{\Upsilon}}\right)^{-1}$ is a diagonal matrix of inverse squared singular values, eliminating any problems from matrix inversion, assuming that the original data are full rank. The regression coefficient estimates follow from combining (10.69) and (10.70):

$$\hat{\beta} = \overleftrightarrow{\mathbf{V}} \cdot \left(\overleftrightarrow{\mathbf{Y}}^T \cdot \overleftrightarrow{\mathbf{Y}}\right)^{-1} \cdot \overleftrightarrow{\mathbf{V}}^T \cdot \overleftrightarrow{\mathbf{X}}^T \cdot \mathbf{y} \tag{10.71}$$

The product of the first three terms on the right side of (10.71) replaces $\left(\overleftrightarrow{\mathbf{X}}^T \cdot \overleftrightarrow{\mathbf{X}}\right)^{-1}$ and does not involve inversion of the predictor matrix. Using the svd, (10.71) can be written entirely in terms of the eigenvalue and eigenvector matrices as

$$\hat{\beta} = \overleftrightarrow{\mathbf{V}} \cdot \left(\overleftrightarrow{\mathbf{Y}}^T \cdot \overleftrightarrow{\mathbf{Y}}\right)^{-1} \cdot \overleftrightarrow{\mathbf{S}}^T \cdot \overleftrightarrow{\mathbf{U}}^T \cdot \mathbf{y} \tag{10.72}$$

When multicollinearity is substantial, then a reduced-rank solution of (10.71) obtained by deleting the eofs having small variances can yield more stable estimates of $\boldsymbol{\beta}$. The improved estimate is obtained by removing the eofs in $\overleftrightarrow{\mathbf{V}}$ corresponding to small singular values in $\overleftrightarrow{\mathbf{S}}$, yielding

$$\hat{\boldsymbol{\beta}}' = \overleftrightarrow{\mathbf{V}}_{1:k} \cdot \left(\overleftrightarrow{\mathbf{Y}}_{1:k}^T \cdot \overleftrightarrow{\mathbf{Y}}_{1:k}\right)^{-1} \cdot \overleftrightarrow{\mathbf{S}}_{1:k}^T \cdot \overleftrightarrow{\mathbf{U}}^T \cdot \mathbf{y} \tag{10.73}$$

However, (10.73) is typically biased, as is evident from

$$\hat{\boldsymbol{\beta}}' - \hat{\boldsymbol{\beta}} = \overleftrightarrow{\mathbf{V}}_{k+1:p} \cdot \left(\overleftrightarrow{\mathbf{Y}}_{k+1:p}^T \cdot \overleftrightarrow{\mathbf{Y}}_{k+1:p}\right)^{-1} \cdot \overleftrightarrow{\mathbf{S}}_{k+1:p}^T \cdot \overleftrightarrow{\mathbf{U}}^T \cdot \mathbf{y} \tag{10.74}$$

Since $\mathcal{E}\left(\hat{\boldsymbol{\beta}}\right) = \boldsymbol{\beta}$, the bias is given by the right-hand side of (10.74) and is nonzero unless the singular values at indices of $k + 1$ to p are all zero, in which case the predictors are rank deficient.

By analogy to (9.19), the variance of the eof regression estimator is

$$var\left(\hat{\boldsymbol{\beta}}\right) = \sigma^2 \overleftrightarrow{\mathbf{V}} \cdot \left(\overleftrightarrow{\mathbf{Y}}^T \cdot \overleftrightarrow{\mathbf{Y}}\right)^{-1} \cdot \overleftrightarrow{\mathbf{V}}^T \tag{10.75}$$

whereas the variance of the reduced rank version is

$$var\left(\hat{\boldsymbol{\beta}}'\right) = \sigma^2 \overleftrightarrow{\mathbf{V}}_{1:k} \cdot \left(\overleftrightarrow{\mathbf{Y}}_{1:k}^T \cdot \overleftrightarrow{\mathbf{Y}}_{1:k}\right)^{-1} \cdot \overleftrightarrow{\mathbf{V}}_{1:k}^T \tag{10.76}$$

Since the entries in $\left(\overleftrightarrow{\mathbf{Y}}_{k+1:p}^T \cdot \overleftrightarrow{\mathbf{Y}}_{k+1:p}\right)^{-1}$ are all larger (and typically much more so) than those in $\left(\overleftrightarrow{\mathbf{Y}}_{1:k}^T \cdot \overleftrightarrow{\mathbf{Y}}_{1:k}\right)^{-1}$, it follows that $var\left(\hat{\boldsymbol{\beta}}'\right) < var\left(\hat{\boldsymbol{\beta}}\right)$. Consequently, the reduced-rank eof regression estimator has a lower variance than the ordinary one at the penalty of increased bias. This may be an acceptable tradeoff if the bias is small while the variance reduction is large.

Example 10.11 Returning to the sake data of Example 10.6, use eof regression to evaluate the predictors for multicollinearity by applying (10.73) and using the Wilks' Λ test for multivariate regression to test the result.

```
sake = importdata('sake.dat');
y = sake(:, 1:2);
y = y - repmat(mean(y), length(y), 1);
x = sake(:, 3:10);
```

```
x = x - repmat(mean(x), length(x), 1);
[n, m] = size(x);
[u, s, v] = svd(x);
for i = m:-1:2
    beta=v(:, 1:i)*inv(s(:, 1:i)'*s(:, 1:i))*s(:, 1:i)'*u'*y;
    e = (y - u*s(:, 1:i)*v(:, 1:i)'*beta)'*y;
    h = beta'*v(:, 1:i)*s(:, 1:i)'*u'*y;
    p = WilksLambda(h, e, 2, i, n - i - 1)
end
p =
   0.6642
p =
   0.7940
p =
   0.3109
p =
   0.1352
p =
   0.2453
p =
   0.3027
p =
   0.2173
```

All the p-values are well above 0.05, and hence there is no need for regression even when only two eigenvalues are retained. The problem with these data is not multicollinearity. Further, the result is not changed if the variables in x and y are standardized so that the principal components are those from the correlation matrix.

Example 10.12 A standard data set to illustrate the importance of multicollinearity is due to Longley (1967) and is available at www.itl.nist.gov/div898/strd/lls/data/LINKS/DATA/Longley.dat. There are one response variable, six predictor variables, and 16 observations in the seven columns of the data file. The predictors are highly incommensurable, and hence the data will be standardized for analysis.

```
data = importdata('longley.dat');
y = data(:, 1);
y = y - repmat(mean(y), length(y), 1);
y = y./repmat(std(y), length(y), 1);
x = data(:, 2:7);
x = x - repmat(mean(x), length(x), 1);
x = x./repmat(std(x), length(x), 1);
[n p] = size(x);
```

```
w = ones(size(y));
[b bse t pval] = Treg(y, x, w);
b'
ans =
  0.0463   -1.0137   -0.5375   -0.2047   -0.1012    2.4797
bse'
ans =
  0.2475    0.8992    0.1233    0.0403    0.4248    0.5858
t'
ans =
  0.1870   -1.1274   -4.3602   -5.0828   -0.2383    4.2331
pval'
ans =
  0.8554    0.2859    0.0014    0.0005    0.8165    0.0017
```

The standard errors for the first and fifth predictors are substantially larger than the coeffients, and the corresponding t statistics are very small. A test that the coefficients are zero rejects only for the third, fourth, and sixth predictors. The problem is that the predictors are quite highly correlated.

```
corr(x)
ans =
  1.0000    0.9916    0.6206    0.4647    0.9792    0.9911
  0.9916    1.0000    0.6043    0.4464    0.9911    0.9953
  0.6206    0.6043    1.0000   -0.1774    0.6866    0.6683
  0.4647    0.4464   -0.1774    1.0000    0.3644    0.4172
  0.9792    0.9911    0.6866    0.3644    1.0000    0.9940
  0.9911    0.9953    0.6683    0.4172    0.9940    1.0000
[u s v] = svd(x);
diag(s.^2)/sum(diag(s.^2))
ans =
  0.7672
  0.1959
  0.0339
  0.0025
  0.0004
  0.0001
```

The first three eigenvalues account for 99.7% of the variance and will be retained for further analysis. The first three eigenvectors are

```
v(:, 1:3)
ans =
 0.4618  -0.0578   0.1491
 0.4615  -0.0532   0.2777
 0.3213   0.5955  -0.7283
```

```
 0.2015  -0.7982  -0.5616
 0.4623   0.0455   0.1960
 0.4649  -0.0006   0.1281
```

The solution from eof regression using the first three eigenvectors is

```
bp = v(:, 1:3)*inv(s(:, 1:3)'*s(:, 1:3))*s(:, 1:3)'*u'*y;
yhat = u*s(:, 1:3)*v(:, 1:3)'*bp;
sigma2 = (y - yhat)'*(y - yhat)/(n - p);
bpse = sqrt(sigma2*diag(v(:, 1:3)*inv(s(:, 1:3)'*s(:, 1:3))
*v(:, 1:3)'));
tp = bp./bpse;
pvalp = 2*(1 - tcdf(abs(tp), n - p));
bp'
ans =
  0.2913    0.3587   -0.3090   -0.1186    0.3048    0.2751
bpse'
ans =
  0.0149    0.0245    0.0641    0.0543    0.0182    0.0134
tp'
ans =
  19.5302   14.6412   -4.8183   -2.1848   16.7107   20.5575
pvalp'
ans =
  0.0000    0.0000    0.0007    0.0538    0.0000    0.0000
```

The p-values indicate that the regression coefficients for all but the fourth predictor are required, and acceptance of the null hypothesis is weak for that variable. The regression coefficients have changed dramatically, indicating the influence of multicollinearity on these data.

11 Compositional Data

11.1 Introduction

Compositional data are vectors whose elements are mutually exclusive and exhaustive percentages or proportions of some quantity. They are statistically unusual because the constraint that the sum of the components of an observation be a constant is explicitly imposed; for example, percentages must sum to 100 and proportions to 1. On first look, this appears not to be a major issue, and compositional data give the appearance of being ordinary statistical data. As a consequence, statistical tools that do not take account of the constrained form of compositional data have frequently been applied to them over the past century despite warnings from statisticians and domain scientists that this could lead to incorrect inferences.

Compositional data are ubiquitous in the earth and ocean sciences. Some common examples include:

- Major or trace element analyses of rocks (expressed as wt%, ppm, or ppb) or fluids (usually in molar concentration);
- Grain size analyses of sediments;
- Micropaleontological assemblages (e.g., taxa abundance);
- Proportions of oil, gas, brine, and solids in petroleum-bearing formations; and
- Surface composition of planetary bodies from remote spectral sensing.

In each case, a data set may be obtained from a suite of samples, either as a way to ensure that real variability is adequately represented (e.g., in rock composition or trace element data) or to measure the variability as a function of time (e.g., sediment or foraminifera composition).

The most obvious problem in treating compositions using statistical tools meant for unconstrained data occurs for the Pearson correlation coefficient. Let the elements of a D-part composition be $\{x_i\}$, where $\sum_{i=1}^{D} x_i = \kappa$. It follows that

$$cov\left(x_i, \sum_{i=1}^{D} x_i\right) = 0 \quad (11.1)$$

because κ is invariant, so

$$cov(x_i,x_1) + \cdots + cov(x_i,x_{i-1}) + cov(x_i,x_{i+1}) + \cdots + cov(x_i,x_D) = -var(x_i) \quad (11.2)$$

The right side of (11.2) is negative unless x_i is constant; hence at least one of the covariances on the left side must be negative. Consequently, there must be at least one negative entry in each row or column of the $D \times D$ covariance matrix. The result is a negative bias on the covariance entries because the variables are not free to wander over the standard range constrained only by nonnegative definiteness of the covariance matrix. This occurs because a D-part composition has only $D-1$ degrees-of-freedom. Inevitable problems in interpretation ensue. Compositional data analysis has moved through four phases. In the first phase, lasting until about 1960, many of the tools of multivariate data analysis were under development and were often applied to compositional data without regard to whether or not they were appropriate and, in addition, largely ignoring the prescient warning of Karl Pearson (1896) just before the turn of the twentieth century:

> "Beware of attempts to interpret correlations between ratios whose numerators and denominators contain common parts."

In the second phase, there at least was recognition of (primarily) the correlation problem, particularly by the petrologist Felix Chayes, but ad hoc workarounds using multivariate statistics continued to be the focus. The third phase was almost entirely the work of the Scottish statistician John Aitchison in the 1980s and 1990s. Aitchison realized that compositions provide only information about the relative rather than absolute values of its components and that the constrained compositional sample space precluded the direct application of multivariate statistical tools to compositions. Aitchison advocated the use of the additive and centered log ratio transformations (alr and clr) to carry compositions into an unconstrained real space where multivariate statistical tools could be applied. This required care because the alr does not preserve distance and the clr yields singular covariance matrices. Opposition to this approach was and is rampant, especially among petrologists. The fourth phase is based on working within the constrained sample space of compositional data, an evolution that has largely occurred in Europe, particularly the University of Girona in Spain.

Given the recent rapid evolution of the statistics of compositional data, texts on the subject are limited, and developments are being published on an ongoing basis. Aitchison (1986) wrote the first text on compositional data, and it remains relevant. Pawlowsky-Glahn & Buccianti (2011) put together a collection of original contributions covering theory and applications in a large number of domains. Pawlowsky-Glahn, Egozcue, & Tolosana-Delgado (2015) produced a text covering the theory and implementation of compositional data analysis written from a mathematical rather than application-oriented perspective. This chapter draws on these sources.

11.2 Statistical Concepts for Compositions

11.2.1 Definitions and Principles

A *composition* is a sample that is broken down into D mutually exclusive and exhaustive *parts*. Each part has a corresponding *label* stating what physical attribute it represents (e.g., sand, silt, or clay for sediment). The numerical elements, or *components*, of the composition are strictly positive and have a constant sum κ that depends on the measurement units. For example, x_i may be the weight percent of a particular oxide, so $\sum_{i=1}^{D} x_i = 100$ for each sample. Samples with a known and constant sum are called *closed data*. Two compositional vectors $\mathbf{x}, \mathbf{y}$ are *compositionally equivalent* if $\mathbf{x} = \alpha \mathbf{y}$ for a positive scalar α. As a consequence, any change of units for the closure constant κ is compositionally equivalent.

The *closure* operation on a D-part compositional vector is

$$\mathcal{C}(\mathbf{x}) = \left[\frac{\kappa x_1}{\sum_{i=1}^{D} x_i}, \ldots, \frac{\kappa x_D}{\sum_{i=1}^{D} x_i} \right] \tag{11.3}$$

Closure rescales a compositional vector so that its sum is a positive constant κ that depends on the units of measurement (e.g., proportions, percent, ppm, etc). An equivalent statement to $\mathbf{x}, \mathbf{y}$ being compositionally equivalent is that $\mathcal{C}(\mathbf{x}) = \mathcal{C}(\mathbf{y})$. Under closure, the vectors (1, 3, 5) and (2, 6, 10) are equivalent. A MATLAB function to implement closure is

```
function [Result] = Close(x, varargin)
% closes a composition x to the constant kappa that
defaults to one if not provided
if nargin == 1
   kappa = 1;
else
   kappa = varargin{1} ;
end
[n D] = size(x);
Result = kappa* x./repmat(sum(x')', 1, D)
end
```

The sample space of compositional data is the *simplex* defined as

$$\mathcal{S}^D = \left\{ \mathbf{x} = (x_1, \ldots, x_D) | x_i > 0, \sum_{i=1}^{D} x_i = \kappa \right\} \tag{11.4}$$

This definition explicitly excludes elements of zero size, although those can be handled with some difficulty. The most basic example of a simplex is ternary because two-component compositions are trivial.

An s-part *subcomposition* $\mathbf{x}_s$ of a D-part composition $\mathbf{x}$ is $\mathcal{C}(\mathbf{x}_s)$. This operation is a projection from one subsimplex to another. Conversely, compositions can be built by combining subcompositions.

While compositions frequently contain many parts (e.g., the approximately 16 major oxides for rock major element compositions), the focus in the earth sciences is often on groups of three. These can be represented on a *ternary diagram* comprising an equilateral triangle. For a ternary diagram with unit edges, let $B = (0,\ 0)$ denote the location of the left lower vertex. Then the remaining vertices are at $C = (1,\ 0)$ and $A = \left(0.5, \sqrt{3}/2\right)$, respectively. A three-part closed composition (α, β, γ) maps onto Cartesian coordinates on the unit ternary through the isometric transformation

$$x = \frac{\alpha + 2\beta}{2(\alpha + \beta + \gamma)} \quad y = \frac{\sqrt{3}\,\alpha}{2(\alpha + \beta + \gamma)} \tag{11.5}$$

so the composition $(1, 0, 0)$ corresponds to the top apex, the composition $(0, 1, 0)$ corresponds to the lower right apex, and the composition $(0, 0, 1)$ corresponds to the lower left apex. The α axis increases from the lower right apex to the top one, the β axis increases from the lower left apex to the lower right one, and the γ axis increases from the top apex to the lower left one. The *barycenter* of a ternary diagram is the point $(1/3,\ 1/3,\ 1/3)$ and comprises its natural center. A set of MATLAB functions to support ternary diagrams is contained in Appendix 11A.

A less useful way to reduce the dimensionality of a composition is *amalgamation*, which is the summation of a subset of the composition components into a single new part. Let a D-component composition be separated into two groups comprising r and s parts, respectively. The amalgamation of the parts contained in the s group is

$$x_A = \sum_{i \in s} x_i \tag{11.6}$$

so that the new composition is $\{\mathbf{x}_r, x_A\}$ possessing a sample space on the simplex S^{D-s+1}. Information about the original composition is lost through the process of amalgamation.

Following Aitchison (1986), there are three conditions that must be fulfilled by any statistical method used on compositions. The first is *scale invariance*, and will be illustrated by example. Suppose that marine sediment consists of sand, silt, and clay at weight percentages (in g/100 g) of 50, 40, and 10, respectively. The sediment is known to have been moved and sorted via water current activity, and the redeposited sediment contains sand, silt, and clay in weight percentages of 40, 53, and 7, respectively. This could have happened by one of three end-member changes or by a mixture of them. In the first end member, sand is unchanged, but silt increases by 26 g/100 g and clay decreases by 1 g/100 g. In the second end member, silt is unchanged, but sand decreases by 20 g/100 g and clay decreases by 4.5 g/100 g. In the third end member, clay is unchanged, but sand increases by 7 g/100 g and silt increases by 36 g/100 g. There is no way to distinguish these three end members (or mixtures of them) based purely on compositional data because the final composition is only available after closure, and hence compositional data contain only relative information. As a corollary, a function is scale invariant if it yields the same result for all compositionally equivalent vectors [i.e., $f(\alpha\mathbf{x}) = f(\mathbf{x})$]. This property is only possible if $f(\mathbf{x})$ is a function of ratios of the components in $\mathbf{x}$.

To illustrate the last point, there is a one-to-one relationship between the components $x_i \in S^D$ and the set of independent and exhaustive ratios

$$\begin{aligned} y_i &= \frac{x_i}{x_1 + \cdots + x_D} \qquad (i = 1, \ldots, D-1) \\ y_D &= \frac{1}{x_1 + \cdots + x_D} \end{aligned} \tag{11.7}$$

and the inverse relationship

$$x_i = \frac{y_i}{y_1 + \cdots + y_{D-1} + 1} \qquad (i = 1, \ldots, D-1)$$
$$x_D = \frac{1}{y_1 + \cdots + y_{D-1} + 1} \tag{11.8}$$

Consequently, working with the compositions or the ratios $\{y_i/y_D\}$ is equivalent.

The second condition on compositional statistical procedures is *permutation invariance*. A function is permutation invariant if it yields equivalent results when the ordering of the parts of a composition is changed. This property is closely tied to the requirement in most of statistics that entities be exchangeable.

The final condition is *subcompositional coherence*, where the ratio of two components remains unchanged as a full composition is decomposed into subcompositions. As a corollary, the distance between two compositions must be greater than or equal to the distance between two subcompositions derived from them. This is called *subcompositional dominance*. Ordinary Euclidean distance does not satisfy this condition, so it is not an appropriate measure of the distance between compositions. A second consequence is that if a noninformative part of a composition is removed (e.g., an oxide in a rock composition that does not vary between samples), the results of compositional analysis will not change.

11.2.2 Compositional Geometry

In an unrestricted real vector space such as is used in classical statistics for either nonnegative or unrestricted real random variables, operations such as addition, scaling, orthogonality, and distance are intuitive due to the familiar unconstrained Euclidean geometry. The restriction to the simplex for compositional data imposes constraints that are not compatible with such a geometry. Consequently, it is necessary to define a set of operations on the simplex that give it a linear vector structure analogous to the Euclidean geometry of real space. The compositional geometry, or geometry on the simplex, has been given the name *Aitchison geometry on the simplex*.

There are two operations that give a simplex the characteristics of a vector space. *Perturbation* of a composition $\mathbf{x} \in S^D$ by a second composition $\mathbf{y} \in S^D$ is defined by

$$\mathbf{x} \oplus \mathbf{y} = \mathcal{C}(x_1 y_1, \ldots, x_D y_D) \tag{11.9}$$

Perturbation is commutative

$$\mathbf{x} \oplus \mathbf{y} = \mathbf{y} \oplus \mathbf{x} \tag{11.10}$$

and associative

$$(\mathbf{x} \oplus \mathbf{y}) \oplus \mathbf{z} = \mathbf{x} \oplus (\mathbf{y} \oplus \mathbf{z}) \tag{11.11}$$

The unique *neutral element* defines the *barycenter* of the simplex

$$\mathbf{n} = \mathcal{C}(1, \ldots, 1) = \left(\frac{1}{D}, \ldots, \frac{1}{D}\right) \tag{11.12}$$

so the *inverse* of $\mathbf{x}$ becomes

$$\mathbf{x}^{-1} = \mathcal{C}\left(x_1^{-1}, \ldots, x_D^{-1}\right) \tag{11.13}$$

and $\mathbf{x} \oplus \mathbf{x}^{-1} = \mathbf{n}$. A MATLAB function that implements perturbation is

```
function [Result] = Perturbation(x, y, varargin)
%Performs the compositional operation perturbation on two
compositions
if nargin == 2
   kappa = 1;
else
   kappa = varargin{1} ;
end
[nx Dx] = size(x);
[ny Dy] = size(y);
x1 = x;
y1 = y;
if Dy == 1
   y1 = repmat(y, 1, Dx);
elseif nx == 1
   x1 = repmat(x, ny, 1);
elseif ny == 1
   y1 = repmat(y, nx, 1);
if Dx ~= Dy
   warning('Perturbation: compositions must have the same length')
   return
elseif nx ~= ny
   warning('Perturbation: number of observations must be the same')
   return
end
Result = Close(x1.* y1, kappa)
end
```

Power transformation or *powering* of a composition $\mathbf{x} \in S^D$ by a real scalar α is defined by

$$\alpha \odot \mathbf{x} = \mathcal{C}\left(x_1^{\alpha}, \ldots, x_D^{\alpha}\right) \tag{11.14}$$

Powering has the unique neutral element $1 \odot \mathbf{x} = \mathbf{x}$. Powering is associative

$$[\alpha \odot (\beta \odot \mathbf{x}) = (\alpha\beta) \odot \mathbf{x}] \tag{11.15}$$

and has two distributive properties

$$\alpha \odot (\mathbf{x} \oplus \mathbf{y}) = (\alpha \odot \mathbf{x}) \oplus (\alpha \odot \mathbf{y}) \tag{11.16}$$

$$(\alpha + \beta) \odot \mathbf{x} = (\alpha \odot \mathbf{x}) \oplus (\beta \odot \mathbf{x}) \tag{11.17}$$

In addition, power transformation has the inverse relationship

$$(-1) \odot \mathbf{x} = \mathbf{x}^{-1} \tag{11.18}$$

A MATLAB function that implements powering is

```
function [ Result] = Powering(alpha, x, vargin)
%Computes the power transformation of a compositional vector x
if nargin == 2
   kappa = 1;
else
   kappa = varagin{1} ;
end
Result = Close(x.^alpha, kappa);
end
```

The combination of the simplex with perturbation and power transformation defines a vector space, and the two operations are analogous to the roles played by addition and multiplication in an unconstrained space. Perturbation and power transformation are, respectively, internal and external operations, and the standard rules of precedence in vector spaces would place power transformation first and perturbation second.

To complete the specification of the Aitchison geometry, the *Aitchison inner product* of $\mathbf{x}, \mathbf{y} \in \mathcal{S}^D$ is defined by

$$\begin{aligned} \langle \mathbf{x}, \mathbf{y} \rangle_a &= \frac{1}{2D} \sum_{i=1}^{D} \sum_{j=1}^{D} \log\left(\frac{x_i}{x_j}\right) \log\left(\frac{y_i}{y_j}\right) \\ &= \sum_{i=1}^{D} \log\frac{x_i}{g(\mathbf{x})} \log\frac{y_i}{g(\mathbf{y})} \end{aligned} \tag{11.19}$$

where $g(\mathbf{x})$ is the geometric mean of the parts of the composition $\mathbf{x}$. The log ratios in the second line of (11.19) are the centered log ratio, which will be described later, and the subscript a differentiates $\langle \mathbf{x}, \mathbf{y} \rangle_a$ from the ordinary inner product $\langle \mathbf{x}, \mathbf{y} \rangle$. A MATLAB function implementing the Aitchison inner product is

```
function [ Result] = AInnerProduct(x, y)
%Computes the Aitchison inner product of the compositions x and y
[nx Dx] = size(x);
[ny Dy] = size(y);
if Dx ~= Dy
   warning('Ainnerprod: compositions must have the same length')
   return
elseif nx ~= ny
   warning('AInnerProduct: number of observations must be the same')
   return
end
Result = zeros(nx, 1);
```

```
x1 = log(x./repmat(geomean(x')', 1, Dx));
y1 = log(y./repmat(geomean(y')', 1, Dy));
for i = 1:nx
   Result(i) = x1(i, :)*y1(i, :)';
end
end
```

It follows from (11.19) that the *Aitchison norm* of **x** is

$$\|\mathbf{x}\|_a = \sqrt{\frac{1}{2D}\sum_{i=1}^{D}\sum_{j=1}^{D}\left[\log\left(\frac{x_i}{x_j}\right)\right]^2} = \sqrt{\sum_{i=1}^{D}\left[\log\frac{x_i}{g(\mathbf{x})}\right]^2} \tag{11.20}$$

A MATLAB function *ANorm*(*x*) follows directly from *AInnerProd*(*x*, *y*).

The *Aitchison distance* between **x** and **y** is given by

$$d_a(\mathbf{x},\mathbf{y}) = \left\|\mathbf{x}\oplus\mathbf{y}^{-1}\right\|_a = \sqrt{\frac{1}{2D}\sum_{i=1}^{D}\sum_{j=1}^{D}\left[\log\left(\frac{x_i}{x_j}\right) - \log\left(\frac{y_i}{y_j}\right)\right]^2} \tag{11.21}$$

Note that the Aitchison distance is compatible with perturbation and power transformation because

$$d_a(\mathbf{s}\oplus\mathbf{x},\mathbf{s}\oplus\mathbf{y}) = d_a(\mathbf{x},\mathbf{y}) \tag{11.22}$$

$$d_a(\alpha\odot\mathbf{x},\alpha\odot\mathbf{y}) = |\alpha| d_a(\mathbf{x},\mathbf{y}) \tag{11.23}$$

In addition, the Aitchison distance is subcompositionally coherent and subcompositionally dominant. A MATLAB function that implements the Aitchison distance is

```
function [Result] = ADistance(x, y)
%Computes the Aitchison distance between compositions x and y
Result = ANorm(Perturbation(x, 1./y));
end
```

Within the Aitchison geometry on the simplex, a compositional line is defined by

$$\mathbf{y} = \mathbf{x}_0 \oplus (\alpha\odot\mathbf{x}) \tag{11.24}$$

where $\mathbf{x}_0$ is the starting point, and **x** is the leading vector. A set of parallel lines comprises (11.24) with different starting points but identical leading vectors. A set of orthogonal lines occurs when the Aitchison inner product of the leading vectors vanishes and intersects at the starting point. Figure 11.1 shows two sets of parallel lines on $\mathcal{S}^3$ produced by the following script

```
x = [ (0:.01:1)' (0:.01:1)' .5*ones(101, 1)];
Ternary;
x0 = [ 1/3 1/3 1/3];
alpha = 10;
Result = Perturbation(x0, Powering(alpha, x));
```

```
TernaryLine(Result)
x0 = [ 1/6 1/2 1/3];
Result = Perturbation(x0, Powering(alpha, x));
TernaryLine(Result)
x0 = [ 1/2 1/6 1/3];
Result = Perturbation(x0, Powering(alpha, x));
TernaryLine(Result)
x = [ (0:.01:1)' .5*ones(101,1) (1:-.01:0)'];
x0 = [ 1/3 1/3 1/3];
Result = Perturbation(x0, Powering(alpha, x));
TernaryLine(Result)
x0 = [ 1/6 1/2 1/3];
Result = Perturbation(x0, Powering(alpha, x));
TernaryLine(Result)
x0 = [ 1/2 1/6 1/3];
Result = Perturbation(x0, Powering(alpha, x));
TernaryLine(Result)
```

Perturbation of a straight line in a ternary diagram results in a straight line with different beginning and ending points. Figure 11.2 compares ternary diagrams before and after perturbation of the center by the composition [2 3 6].

```
subplot(1, 2, 1)
Ternary;
subplot(1, 2, 2)
Ternary(7, [ 2 3 6]);
```

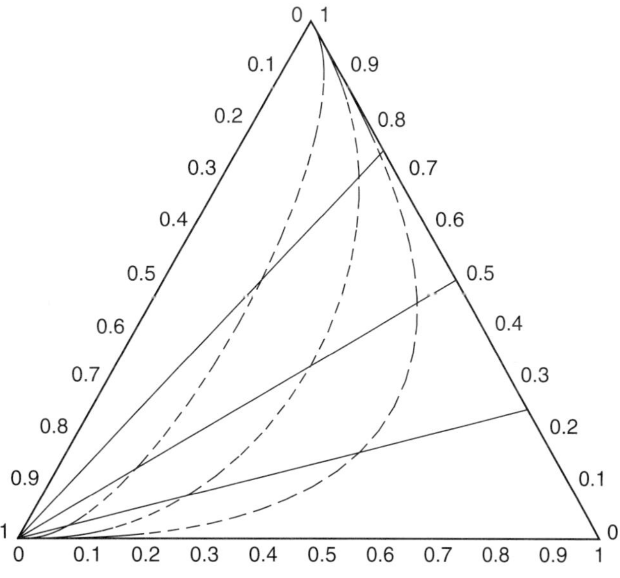

Figure 11.1 Two sets of parallel lines on $\mathcal{S}^3$.

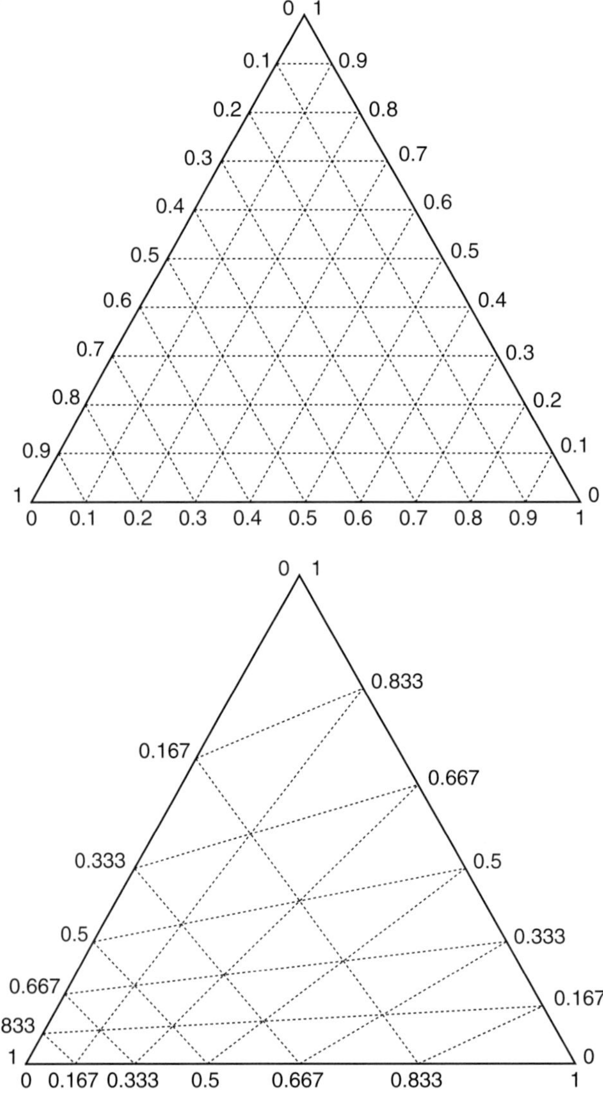

Figure 11.2 Ternary diagram showing grid lines before (left) and after perturbation by the composition [2 3 6].

11.2.3 Compositional Transformations

Representation of compositional data using scale invariant log ratios or *log contrasts* was introduced by Aitchison in the early 1980s and is defined by

$$\log\left(\prod_{i=1}^{D} x_i^{\alpha_i}\right) = \sum_{i=1}^{D} \alpha_i \log x_i \tag{11.25}$$

$$\sum_{i=1}^{D} \alpha_i = 0 \tag{11.26}$$

The first log contrast that was implemented is the *additive log ratio transformation* (alr). If $\mathbf{x} \in \mathcal{S}^D$, the alr is given by

$$\mathrm{alr}(\mathbf{x}) = \left(\log \frac{x_1}{x_D}, \ldots, \log \frac{x_{D-1}}{x_D} \right) = \Theta \tag{11.27}$$

and is easily inverted to yield the composition

$$\mathbf{x} = \mathrm{alr}^{-1}(\Theta) = \mathcal{C}\left(e^{\Theta_1}, \ldots, e^{\Theta_D}\right) \tag{11.28}$$

The alr changes perturbation and power transformation into standard operations in a $(D - 1)$-dimensional real space (hence matching the dimensionality of $\mathcal{S}^D$) because

$$\mathrm{alr}[(\alpha \odot \mathbf{x}) \oplus (\beta \odot \mathbf{y})] = \alpha \, \mathrm{alr} \, \mathbf{x} + \beta \, \mathrm{alr} \, \mathbf{y} \tag{11.29}$$

The chief limitation to the alr is that it is not permutation invariant because the Dth part (which can be chosen to be any part of the composition) plays a special role, and hence many standard statistical procedures based on it fail due to the absence of exchangeability. In addition, standard distance measures cannot be easily used on the alr. The alr is now primarily of historical interest, but it serves to introduce the concept of compositional transformations that move compositions from the simplex to a real vector space where standard operations and statistical tools can be applied, following which the result can be reverse transformed back to the simplex.

The problems with the alr led Aitchison to also introduce the *centered log ratio transformation* (clr) from $\mathcal{S}^D$ to $\mathcal{R}^D$ given by

$$\mathrm{clr} \, \mathbf{x} = \left(\log \frac{x_1}{g(\mathbf{x})}, \ldots, \log \frac{x_D}{g(\mathbf{x})} \right) \equiv \xi \tag{11.30}$$

where $\sum_{i=1}^{D} \xi_i = 0$. The D elements of the clr are log contrasts as in (11.25) but with a natural rather than arbitrary reference. The inverse clr is

$$\mathrm{clr}^{-1}(\xi) = \mathcal{C}\left[e^{\xi_1}, \ldots, e^{\xi_D}\right] = \mathbf{x} \tag{11.31}$$

The clr is symmetric in the components (unlike the alr) but imposes the constraint that the sum of the transformed sample is zero. This leads to a singular covariance matrix for ξ. In addition, clr coefficients are not subcompositionally coherent because the geometric mean of a subcomposition will not in general be the same as that of the entire composition, and hence the clr coefficients of a subcomposition will not be consistent with that of the whole. The following functions implement the clr and its inverse in MATLAB:

```
function [ Result] = Clr(x)
%Applies the centered log ratio transformation to the composition x
[n D] = size(x);
Result = log(x./repmat(geomean(x')', 1, D));
end
```

```
function [ Result] = ClrI(x, varargin)
%Computes the inverse clr of x
if nargin == 1
   kappa = 1;
else
   kappa = varargin{ 1} ;
end
Result = Close(exp(x), kappa);
end
```

The clr has several key properties that make it useful for many purposes:

$$\begin{aligned} &\operatorname{clr}(\alpha \odot \mathbf{x}_1 \oplus \beta \odot \mathbf{x}_2) = \alpha \operatorname{clr} \mathbf{x}_1 + \beta \operatorname{clr} \mathbf{x}_2 \\ &\langle \mathbf{x}_1, \mathbf{x}_2 \rangle_a = \langle \operatorname{clr} \mathbf{x}_1, \operatorname{clr} \mathbf{x}_2 \rangle \\ &\|\mathbf{x}_1\|_a = \|\operatorname{clr} \mathbf{x}_1\| \\ &d_a(\mathbf{x}_1, \mathbf{x}_2) = d(\operatorname{clr} \mathbf{x}_1, \operatorname{clr} \mathbf{x}_2) \end{aligned} \tag{11.32}$$

The first equation in (11.32) shows that perturbation and power transformation on the simplex correspond to addition and multiplication by a scalar in a D-dimensional real space, whereas the remaining three equations show that the Aitchison inner product, norm, and distance measure on the simplex correspond to the standard ones on the a real vector space. However, the real vector space is not fully unconstrained because the sum of the clr elements must be zero and hence in reality exist on a particular $(D - 1)$-dimensional subspace of $\mathcal{R}^D$. In addition, the components of (11.32) apply to the entire composition and not to pairs from it because the clr does not yield an orthonormal basis set.

The Aitchison geometry and standard practice in a real vector space suggest that the usual approach of defining an orthonormal basis and coordinates would be useful. Let $\{\mathbf{e}_1, \ldots, \mathbf{e}_{D-1}\}$ be an orthonormal basis on $\mathcal{S}^D$ that must satisfy [using (11.32)]

$$\langle \mathbf{e}_i, \mathbf{e}_j \rangle_a = \langle \operatorname{clr} \mathbf{e}_i, \operatorname{clr} \mathbf{e}_j \rangle = \delta_{ij} \tag{11.33}$$

where $\|\mathbf{e}_i\|_a = 1$. Once an orthornormal basis has been constructed, a composition $\mathbf{x} \in \mathcal{S}^D$ may be written

$$\mathbf{x} = \operatorname{ilr}^{-1} \mathbf{x}^* = (x_1^* \odot \mathbf{e}_1) \oplus \cdots \oplus (x_{D-1}^* \odot \mathbf{e}_{D-1}) = \bigoplus_{j=1}^{D-1} \left(x_j^* \odot \mathbf{e}_j\right) \tag{11.34}$$

$$\mathbf{x}^* = \operatorname{ilr} \mathbf{x} = \left(\langle \mathbf{x}, \mathbf{e}_1 \rangle_a, \ldots, \langle \mathbf{x}, \mathbf{e}_{D-1} \rangle_a\right) \tag{11.35}$$

Equation (11.35) defines the coordinates of $\mathbf{x}$ relative to the orthonormal basis using the Aitchison inner product, and (11.34) is an expansion of $\mathbf{x}$ in terms of the basis.

Equation (11.35) is the *isometric log ratio transformation* (ilr; where isometric means transformation without change of shape or size) that assigns the coordinates $\mathbf{x}^*$ to $\mathbf{x}$ and is a mapping from $\mathcal{S}^D \to \mathcal{R}^{D-1}$, where clr is a constrained mapping from $\mathcal{S}^D \to \mathcal{R}^D$. The ilr has the following properties that should be compared with (11.32)

$$\begin{aligned}
&\operatorname{ilr}(\alpha \odot \mathbf{x}_1 \oplus \beta \odot \mathbf{x}_2) = \alpha \operatorname{ilr}\mathbf{x}_1 + \beta \operatorname{ilr}\mathbf{x}_2 = \alpha \mathbf{x}_1^* + \beta \mathbf{x}_2^* \\
&\langle \mathbf{x}_1, \mathbf{x}_2 \rangle_a = \langle \operatorname{ilr}\mathbf{x}_1, \operatorname{ilr}\mathbf{x}_2 \rangle = \langle \mathbf{x}_1^*, \mathbf{x}_2^* \rangle \\
&\| \mathbf{x} \|_a = \| \operatorname{ilr}\mathbf{x} \| = \| \mathbf{x}^* \| \\
&d_a(\mathbf{x}_1, \mathbf{x}_2) = d(\operatorname{ilr}\mathbf{x}_1, \operatorname{ilr}\mathbf{x}_2) = d(\mathbf{x}_1^*, \mathbf{x}_2^*)
\end{aligned} \tag{11.36}$$

Let $\boldsymbol{\Psi}$ be the $(D-1) \times D$ *contrast matrix* whose rows are $\operatorname{clr}\mathbf{e}_i$ so that

$$\begin{aligned}
&\boldsymbol{\Psi} \cdot \boldsymbol{\Psi}^T = \mathbf{I}_{D-1} \\
&\boldsymbol{\Psi}^T \cdot \boldsymbol{\Psi} = \mathbf{I}_D - \mathbf{J}_D / D
\end{aligned} \tag{11.37}$$

In a compact form, the ilr becomes

$$\mathbf{x}^* = \operatorname{ilr}\mathbf{x} = \operatorname{clr}\mathbf{x} \cdot \boldsymbol{\Psi}^T \tag{11.38}$$

so that the inverse is, using (11.31),

$$\mathbf{x} = \operatorname{ilr}^{-1}\mathbf{x}^* = \mathcal{C}[\exp(\mathbf{x}^* \cdot \boldsymbol{\Psi})] \tag{11.39}$$

where the last term is just $\operatorname{clr}^{-1}(\mathbf{x}^* \cdot \boldsymbol{\Psi})$ from (11.31).

The remaining issue is how one computes the orthonormal basis $\{\mathbf{e}_i\}$ and contrast matrix $\boldsymbol{\Psi}$ because there are an infinite number of orthornormal bases in S^D. This can be done using empirical orthogonal functions (see Section 10.8), although the result may not have a clear physical interpretation. The alternative is construction of a basis using a Gram-Schmidt procedure starting from a partition of the data that is driven by domain knowledge. A starting point is a *sequential binary partition* (SBP) of a composition vector, which is a hierarchy obtained by splitting the parts into two groups and then dividing each group into two groups and continuing until all groups have a single part. For a D-part composition, there are $D-1$ divisions to complete a SBP. Obviously, there are many ways to apply a SBP, and it should be done in a manner that makes interpretation of the data easier, requiring some degree of insight into the underlying physics or chemistry. For example, chemical species may be divided into cations and anions, or oxides may be divided into silicates and nonsilicates.

Once a sequential binary partition is defined, an orthonormal basis can be constructed using the Gram-Schmidt procedure. The Cartesian coordinates of a composition using such a basis is called a *balance*, and the composition comprises *balancing elements*. A balance differs from the clr only in that the composition is not centered and is given by

$$\boldsymbol{\Upsilon} = \log(\mathbf{x}) \cdot \boldsymbol{\Psi}^T \tag{11.40}$$

Let a sequential binary sequence be encoded for each part using 1 or -1, with 0 indicating parts that are omitted. For a D-part composition, let there be r elements encoded with 1, s elements encoded with -1, and t elements encoded with 0, where $r + s + t = D$. Table 11.1 is an exemplar SBP for a five-part composition.

The contrast matrix $\boldsymbol{\Psi}$ is obtained by counting the number of +1 entries r and -1 entries s in a given row and computing

Table 11.1 Exemplar Sequential Binary Partition

Order	x_1	x_2	x_3	x_4	x_5	r	s
1	+1	+1	+1	−1	−1	3	2
2	+1	+1	−1	0	0	2	1
3	+1	−1	0	0	0	1	1
4	0	0	0	+1	−1	1	1

Table 11.2 Exemplar Contrast Matrix

Order	x_1	x_2	x_3	x_4	x_5
1	$\sqrt{2/15}$	$\sqrt{2/15}$	$\sqrt{2/15}$	$-\sqrt{3/10}$	$-\sqrt{3/10}$
2	$\sqrt{1/6}$	$\sqrt{1/6}$	$-\sqrt{2/3}$	0	0
3	$\sqrt{1/2}$	$-\sqrt{1/2}$	0	0	0
4	0	0	0	$\sqrt{1/2}$	$-\sqrt{1/2}$

$$\begin{aligned} \psi^+ &= \sqrt{\frac{s}{r(r+s)}} \\ \psi^- &= -\sqrt{\frac{r}{s(r+s)}} \end{aligned} \tag{11.41}$$

The column entries are given by ψ^+ or ψ^-, where the SBP has +1 or −1 entries, respectively, and 0 where the SBP has a corresponding 0 entry. The contrast matrix is shown in Table 11.2 for the SBP of Table 11.1.

Caution: It is very easy to make a mistake when constructing an SBP. It is prudent to check the contrast matrix for orthonormality by evaluating (11.37) before using it for data analysis.

Once the contrast matrix has been constructed, the ilr follows from (11.38). The contrast matrix can be constructed from an SBP using the following MATLAB function:

```
function [Result] = Contrast(sbp)
%Computes the contrast matrix from the sequential binary
partition sbp
[Dm1 D] = size(sbp);
if D ~= Dm1 + 1
   warning('Contrast: sbp has wrong number of rows')
   return
end
Result = zeros(Dm1, D);
for i = 1:Dm1
   r = find(sbp(i, :) == 1);
   rs = sum(sbp(i, r));
```

```
    s = find(sbp(i, :) == -1);
    ss = -sum(sbp(i, s));
    psip = sqrt(ss/(rs*(rs + ss)));
    psim = -sqrt(rs/(ss*(rs + ss)));
    Result(i, r) = psip;
    Result(i, s) = psim;
  end
  end
```

The ilr and its inverse can be obtained using the following MATLAB functions:

```
function [Result] = Ilr(x, Psi)
%Computes the ilr of the composition x using the contrast
matrix Psi
Result = Clr(x)*Psi';
end

function [Result] = IlrI(x, Psi, varargin)
%Computes the inverse ilr using the contrast matrix Psi
if nargin == 2
    kappa = 1
else
    kappa = varargin{1};
end
Result = ClrI(x*Psi, kappa);
end
```

11.3 Exploratory Compositional Data Analysis

The initial steps in compositional data analysis comprise the exploratory ones: computing descriptive statistics such as the center and variation matrix, centering the data so that subcompositions can be viewed in ternary diagrams, identifying patterns of variability, constructing an orthonormal coordinate system, and projecting the data into it and computing summary statistics. The remainder of this chapter covers these topics in some detail. At a more advanced stage, distributions on the simplex need to be introduced so that statistical tests (which are inevitably multivariate) can be devised. This leads to hypothesis testing and confidence interval estimation as well as linear regression. These topics are covered in the last few chapters of Pawlowsky-Glahn, Egozcue, & Tolosana-Delgado (2015).

For the purposes of illustration, a data set consisting of the major oxides from basalts from 23 to 35°N on the Mid-Atlantic Ridge (MAR) was downloaded from the PetDB data archive (www.earthchem.org/petdb). This section of the ridge was selected because it is known to display significant compositional variability proximal to the Azores hotspot. About 41 samples were chosen based on the absence of zeros in the 11 major element oxides SiO_2, TiO_2, Al_2O_3, Fe_2O_3, FeO, MnO, MgO, CaO, Na_2O, K_2O, and P_2O_5. The parts have that order. An exploratory compositional data analysis will be carried out on them.

Example 11.1 Using the MAR data, demonstrate their inconsistency with standard estimators for correlation and the requirement to use compositional concepts.

Input and close the data.

```
mar = xlsread('MAR.xls');
mar = mar(2:length(mar), :);
%keep the major oxides
mar = mar(:, 3:13);
%keep only data that have all of the major oxides
mar = [ mar(1:4, :)' mar(22, :)' mar(47:88, :)']';
%remove redundant outliers at 37-38 and 39-40
%remove redundant data at 24-26 and 30-36
mar = [ mar(1:25, :)' mar(27:35, :)' mar(41:47, :)']';
mar = Close(mar, 100);
```

In classical statistics, the standard descriptive statistics are the sample mean and variance (or covariance). However, these are not suitable as measures of central tendency and dispersion for compositions, in particular, for the reason elaborated in (11.2), as can be readily shown. The standard correlation coefficients between (CaO, K_2O), (Fe_2O_3, FeO), (FeO, MgO), and (Fe_2O_3, MgO) are

```
[corr(mar(:, 8), mar(:, 10)) corr(mar(:, 4), mar(:, 5))
corr(mar(:, 5), mar(:, 7)) corr(mar (:, 4), mar(:, 7))]
ans =
   0.2850   -0.5243    0.4712   -0.3587
```

Delete SiO_2 and Al_2O_3 from the data set, reclose and recompute the correlation coefficients.

```
mar1=[ mar(:, 2) mar(:, 4:11)];
mar1 = Close(mar1, 100);
[corr(mar1(:, 6), mar1(:, 8)) corr(mar1(:, 2), mar1(:, 3))
corr(mar1(:, 3), mar1(:, 5)) corr(mar1 (:, 2), mar1(:, 5))]
ans =
   0.1889   -0.6507    0.2554   -0.4698
```

The correlation matrix between the mafic constituents (MgO, FeO, Fe_2O_3) before closing the subcomposition is

```
corr([ mar(:, 4:5) mar(:, 7)])
ans =
   1.0000   -0.5243   -0.3587
  -0.5243    1.0000    0.4712
  -0.3587    0.4712    1.0000
```

and after closure

```
mar1 = Close([ mar(:, 4:5) mar(:, 7)] , 100);
corr(mar1)
ans =
   1.0000   -0.6311   -0.4919
  -0.6311    1.0000   -0.3649
  -0.4919   -0.3649    1.0000
```

The differences are dramatic and illustrate the incompatibility of classical summary statistics with the Aitchison geometry. Classical summary statistics for the entire data set would be the sample mean (by columns) and the covariance matrix.

```
mean(mar)
ans =
   Columns 1 through 7
   50.1381   1.3846  16.1387   1.4690  7.9954  0.1662  8.1798
   Columns 8 through 11
   11.4451   2.7109  0.2264    0.1459
cov(mar)
ans =
   Columns 1 through 7
    0.9874    0.1222   -0.2330    0.1257   -0.3626   -0.0008   -0.8500
    0.1222    0.0947   -0.1605    0.0953   -0.0272    0.0003   -0.1292
   -0.2330   -0.1605    0.8230   -0.1259   -0.2958   -0.0040   -0.0273
    0.1257    0.0953   -0.1259    0.3992   -0.3045   -0.0019   -0.2406
   -0.3626   -0.0272   -0.2958   -0.3045    0.8447    0.0071    0.4599
   -0.0008    0.0003   -0.0040   -0.0019    0.0071    0.0004    0.0036
   -0.8500   -0.1292   -0.0273   -0.2406    0.4599    0.0036    1.1273
    0.0388   -0.0950    0.0496   -0.0898   -0.1361   -0.0043   -0.0879
    0.1311    0.0564    0.0666    0.0880   -0.1225    0.0010   -0.1815
    0.0239    0.0307   -0.0609    0.0452   -0.0586   -0.0013   -0.0576
    0.0173    0.0122   -0.0318    0.0092   -0.0044   -0.0001   -0.0168

   Columns 8 through 11
    0.0388    0.1311    0.0239    0.0173
   -0.0950    0.0564    0.0307    0.0122
    0.0496    0.0666   -0.0609    0.0318
   -0.0898    0.0880    0.0452    0.0092
   -0.1361   -0.1225   -0.0586   -0.0044
   -0.0043    0.0010   -0.0013   -0.0001
   -0.0879   -0.1815   -0.0576   -0.0168
    0.4385   -0.1540    0.0372    0.0031
   -0.1540    0.1216   -0.0061   -0.0006
    0.0372   -0.0061    0.0388    0.0088
    0.0031   -0.0006    0.0088    0.0031
```

Note that each row or column of the covariance matrix has one or more negative entries, illustrating (11.2). The data are quite multivariate, and the compositional constraint is clearly playing a significant statistical role.

The replacement for the sample mean in the Aitchison geometry is the *center* of the data given by the closed geometric mean

$$\mathbf{g} = \mathcal{C}(g_1, \dots, g_D)$$
$$g_j = \left(\prod_{i=1}^{N} x_{ij}\right)^{1/N} \tag{11.42}$$

The product in (11.42) is taken by rows (i.e., over all samples for a given component), where in the definition of clr (11.30) it is taken by columns (i.e., over all components for a given sample). The center (11.42) is an unbiased minimum variance estimator for the expected value of a random composition.

Example 11.2 For the MAR data, evaluate the geometric mean.

```
Close(geomean(mar), 100)
ans =
 Columns 1 through 7
 50.3376    1.3533    16.1814   1.3658   7.9745   0.1655   8.1521
 Columns 8 through 11
 11.4750    2.6985    0.1617    0.1345
```

Dispersion of a compositional data set is obtained through the $D \times D$ *variation matrix*

$$\mathbf{T} = \{t_{ij}\}$$
$$t_{ij} = var\left[\log\left(\frac{x_i}{x_j}\right)\right] \tag{11.43}$$

This is sometimes given as the *normalized variation matrix* $\mathbf{T}^*$, where

$$t_{ij}^* = var\left[\log(x_i/x_j)/\sqrt{2}\right] \tag{11.44}$$

in which case the log ratio is a balance. In either instance, the entries are the variances of the log ratio between the ith and jth parts. Note that $\mathbf{T}^* = \mathbf{T}/2$, where the normalized entries have units of squared Aitchison distance. $\mathbf{T}$ and $\mathbf{T}^*$ are symmetric and have zeros on the main diagonal. Further, the entries in either matrix are independent of the units of the composition because it involves only their ratios, so rescaling has no influence.

A summary statistic for overall dispersion is given by the total variance

$$\begin{aligned} \operatorname{totvar}(\mathbf{x}) &= \frac{1}{2D}\sum_{i=1}^{D}\sum_{j=1}^{D} var\left[\log\left(\frac{x_i}{x_j}\right)\right] \\ &= \frac{1}{2D}\sum_{i=1}^{D}\sum_{j=1}^{D} t_{ij} = \frac{1}{D}\sum_{i=1}^{D}\sum_{j=1}^{D} t_{ij}^{*} \end{aligned} \tag{11.45}$$

Example 11.3 For the MAR data, compute and analyze the normalized variation matrix and total variance.

The following MATLAB function produces this matrix

```
function [ T, totvar] = NVariation(x)
%computes the normalized variation matrix for the composition x.
[n D] = size(x);
k = 1:D;
ratio = zeros(D, D, n);
for i = 1:n
  for j = 1:D
      ratio(j, k, i) = log(x(i, j)./x(i, k));
  end
end
T = var(ratio/sqrt(2), 1, 3);
totvar = sum(sum(T))/D;
end
```

Only the upper triangle needs presentation and is as follows:

```
[T, totvar] = NVariation(mar);
T
T =
```

SiO_2	TiO_2	Al_2O_3	Fe_2O_3	FeO	MnO	MgO	CaO	Na_2O	K_2O	P_2O_5	
0	0.027	0.002	0.073	0.008	0.010	0.009	0.002	0.008	0.419	0.096	SiO_2
	0	0.038	0.054	0.040	0.037	0.049	0.036	0.020	0.291	0.040	TiO_2
		0	0.082	0.011	0.012	0.009	0.003	0.009	0.458	0.119	Al_2O_3
			0	0.105	0.085	0.101	0.082	0.062	0.320	0.102	Fe_2O_3
				0	0.010	0.007	0.010	0.021	0.483	0.118	FeO
					0	0.013	0.013	0.016	0.456	0.113	MnO
						0	0.010	0.024	0.498	0.133	MgO
							0	0.016	0.428	0.103	CaO
								0	0.416	0.101	Na_2O
									0	0.147	K_2O
										0	P_2O_5

The values are coded as <0.05 (dark gray), 0.05 to 0.3 (light gray), and >0.3 (white). The highest values (>0.3) are K_2O with the other oxides, and the intermediate values are

P_2O_5 with the other oxides save TiO_2. Both of these may reflect the low concentrations of K_2O and P_2O_5. The other element with intermediate variability is Fe_2O_3.

```
totvar
totvar =
  1.0823
```

A common way to plot data in a ternary diagram is to shift and rescale the data so that they are approximately in the middle, and their range over the three coordinates is approximately the same. This is a perturbation, although applied in an ad hoc manner. Recall from (11.12) and (11.13) that $\mathbf{x} \oplus \mathbf{x}^{-1} = \mathbf{n}$, where $\mathbf{n}$ is the barycenter of the simplex. A composition can be moved to the barycenter by computing the center of the data from (11.42) and perturbing the data by $\mathbf{g}^{-1}$. Such a transformation changes gridlines on ternary diagrams by the same perturbation (see Figure 11.2). By analogy to studentization of a classical statistical variable, a composition can be scaled by power transformation with $\mathrm{totvar}(\mathbf{x})^{-1/2}$. This results in data with the same relative contribution from each log ratio.

Example 11.4 Plot ternary diagrams of the oxides (Fe_2O_3, K_2O, P_2O_5) that exhibit the highest covariation in Example 11.3 before and after centering.

```
mar1 = Close([ mar(:, 4), mar(:, 10), mar(:, 11)]);
subplot(1, 2, 1)
Ternary;
TernaryPlot(mar1, 'd', 'k')
TernaryLabel('Fe2O3', 'K2O', 'P2O5')
subplot(1, 2, 2)
g = Close(geomean(mar1));
Ternary(0, 1./g, [ 0 .1 .5 .9 1]);
cmar1 = Perturbation(mar1, 1./g);
TernaryPlot(cmar1, 'd', 'k')
TernaryLabel('Fe2O3', 'K2O', 'P2O5')
```

The center and variation matrix provide summary information about compositions, and perturbation and power transformation using the data center and total variance allow their plotting in subcompositions of three, although without guidance from domain understanding or statistical insight, this would be a time-consuming task because there are $\binom{D}{3}$ unique ternary diagrams for a D-part composition. The next step in exploring compositional data is use of a dendrogram to evaluate the relationships among the clrs of the data. The dendrogram groups the clrs into a hierarchical binary cluster tree with the heights of the U-shaped sections proportional to the distance between the data being connected.

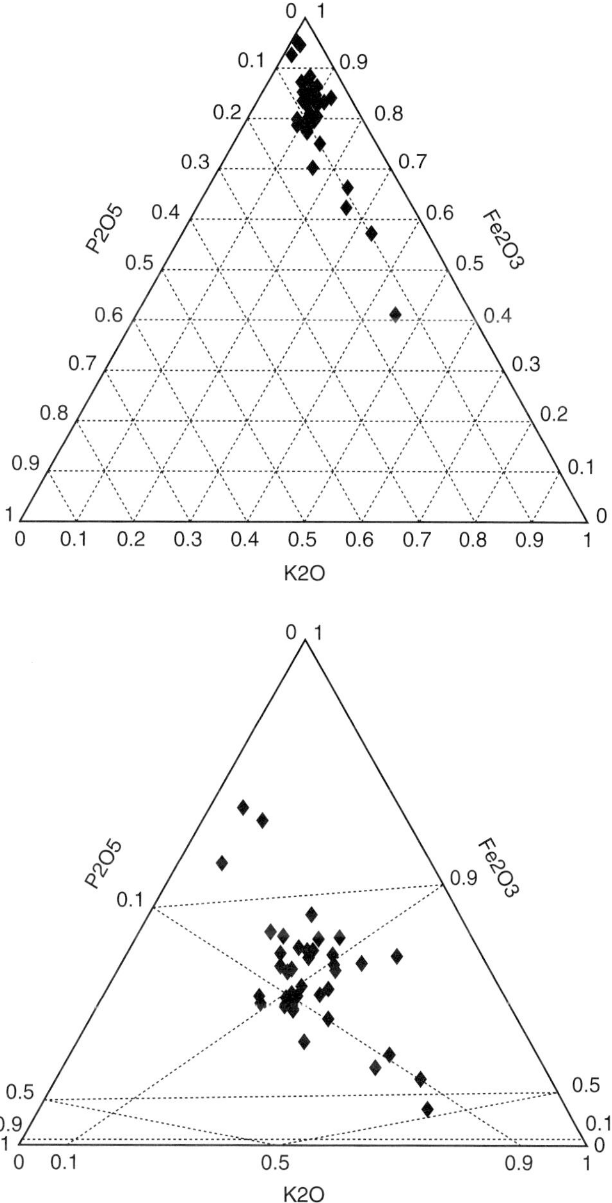

Figure 11.3 A ternary diagram of Fe_2O_3-K_2O-P_2O_5 before (left) and after (right) centering the data.

Example 11.5 Present and describe a dendrogram for the clr of the centered MAR data.

```
g = Close(geomean(mar));
tree = linkage(Clr(Perturbation(mar, 1./g))');
dendrogram(tree)
```

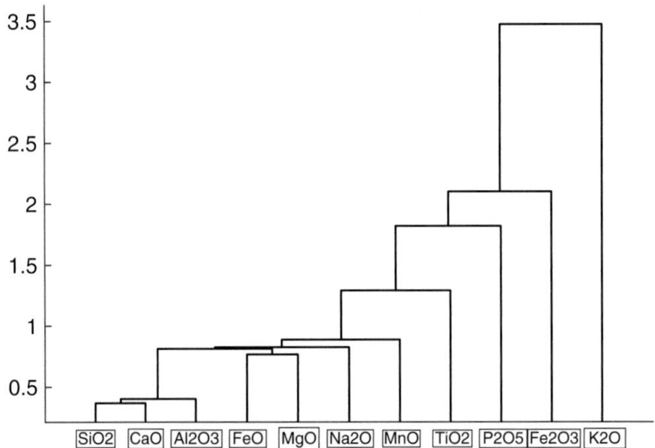

Figure 11.4 Dendrogram of the clr of the centered MAR data.

Figure 11.4 shows the result. It is interesting that the incompatible elements Fe^{3+}, K, and P cluster at the right and are together a large distance from the remaining ones but with Ti the closest of them, followed by Mn. The remaining elements are close together in the clusters (Fe^{+2}, Mg, Na) and (Si, Al, Ca).

Empirical orthogonal functions summarize the interrelation of all the clrs of the data taken together. These are obtained by using the singular value decomposition on the clr (11.30) after centering the data about their geometric mean. First compute

$$\overleftrightarrow{\xi} = \text{clr}\left(\mathbf{x} \oplus \mathbf{g}^{-1}\right) \tag{11.46}$$

This quantity is decomposed as

$$\overleftrightarrow{\xi} = \overleftrightarrow{\mathbf{U}} \cdot \overleftrightarrow{\mathbf{S}} \cdot \overleftrightarrow{\mathbf{V}}^T \tag{11.47}$$

Example 11.6 Evaluate the svd of the MAR data.

```
g = Close(geomean(mar));
[u, s, v] = svd(Clr(Perturbation(mar, 1./g)));
diag(s).^2./sum(diag(s).^2)'
ans =
  0.8158
  0.1018
  0.0321
  0.0218
  0.0134
  0.0081
  0.0047
```

```
0.0018
0.0004
0.0002
0.0000
```

The first two eigenvalues explain 91.8% of the variance. The first three eigenvalues explain 95% of the variance. The last singular value is precisely zero, reflecting the fact that the clr results in a singular covariance matrix. The first 10 right singular vectors are

```
-0.143  0.066 -0.169  0.084  0.153 -0.130  0.099 -0.082  0.499 -0.737
 0.042 -0.016  0.561  0.291  0.109  0.374  0.093 -0.568 -0.150 -0.050
-0.187  0.043 -0.295  0.240  0.209 -0.170 -0.123  0.106 -0.764 -0.214
 0.064 -0.901  0.004 -0.289  0.051 -0.042  0.059  0.042 -0.021  0.003
-0.210  0.225  0.117 -0.247 -0.030  0.161  0.751  0.364 -0.083  0.100
-0.179  0.069  0.053 -0.103 -0.862 -0.246 -0.128 -0.185 -0.063 -0.034
-0.227  0.141 -0.170 -0.354  0.056  0.636 -0.502  0.110  0.090  0.059
-0.152  0.117 -0.371 -0.056  0.267 -0.297  0.096 -0.477  0.193  0.546
-0.128 -0.117  0.123  0.665 -0.052 -0.025 -0.151  0.460  0.298  0.309
 0.830  0.115 -0.353  0.106 -0.166  0.184  0.103  0.031 -0.008  0.004
 0.291  0.258  0.497 -0.338  0.264 -0.446 -0.297  0.199  0.008  0.014
```

The three largest elements in each are highlighted in gray. The eigenvectors in rank order show (1) the largest value at K_2O but correlation with P_2O_5 and anticorrelation with MgO, (2) the largest value at Fe_2O_3 but anticorrelation with FeO and P_2O_5, (3) the largest value at TiO_2 but correlation with P_2O_5 and anticorrelation with CaO, (4) the largest value at Na_2O but anticorrelation with MgO and P_2O_5, (5) the largest value at MnO but anticorrelation with TiO_2 and P_2O_5, (6) the largest value at MgO but correlation with TiO_2 and anticorrelation with P_2O_5, (7) the largest value at FeO but anticorrelation with MgO and P_2O_5, (8) the largest value at TiO_2 but correlation with Na_2O and anticorrelation with CaO, (9) the largest value at Al_2O_3 but anitcorrelation with SiO_2 and Na_2O, and (10) the largest value at SiO_2 but anticorrelation with CaO and Na_2O.

A biplot can be used to evaluate the variability in the data and can be easily obtained with MATLAB. A biplot of compositional data (Aitchison & Greenacre 2002) can take on two end-member forms depending on whether the eigenvectors in $\overleftrightarrow{\mathbf{V}}$ are scaled by the corresponding singular values or not. The former are called *principal coordinates*, and the latter are called *standard coordinates*. The biplot in these two types of coordinates are called *covariance* and *form biplots*, respectively. The length of the rays in a covariance biplot is proportional to the standard deviation of the clr components of the centered data, whereas the form biplot should have rays that are identical in length if all the data are represented. There are corresponding observation biplots that will not be considered here.

In each case, the *origin* is the center of the compositional data set. The join of the origin to each *vertex* is called a *ray*, and the join of two vertices is called a *link* (not shown in the figures). The lengths of links and rays are a measure of relative variability of the composition, whereas their orientation provides information about their correlation. The square of the length of the i, j link is approximately $var\left[\log\left(x_i/x_j\right)\right]$, whereas the square of the ith ray is approximately $var[\mathrm{clr}(x_i)]$. If two links are at right angles, then the corresponding pair of log ratios is uncorrelated. If two or more vertices coincide, then the ratio of those two parts of the composition is constant, and hence the two parts are redundant. Finally, if two or more rays are collinear, this suggests that their subcomposition constitutes a compositional line.

Example 11.7 Compute covariance and form biplots for the first two eigenvectors of the MAR data, and present the key insights they provide as ternary diagrams before and after centering.

The covariance biplot is given by the MATLAB script

```
labels = {'SiO2', 'TiO2', 'Al2O3', 'Fe2O3', 'FeO', 'MnO',
'MgO', 'CaO', 'Na2O', 'K2O', 'P2O5'};
biplot([ s(1, 1)*v(:, 1) s(2, 2)*v(:, 2)], 'VarLabels', labels)
axis square
```

and is shown in Figure 11.5. The form biplot is shown in Figure 11.6.

The covariance biplot in Figure 11.5 is somewhat one-dimensional, reflecting the size of the first singular value. It is dominated by rays from K_2O, Fe_2O_3, and P_2O_5, which are the same pattern seen in the dendrogram in Figure 11.4. The rays for the oxides SiO_2, Al_2O_3, MnO, and CaO are nearly coincident and are not labeled in Figure 11.5. As a consequence, they are redundant. They are coincident in direction but not magnitude with FeO and MgO. The links for FeO-P_2O_5 and Fe_2O_3-P_2O_5 are approximately orthogonal, so their ratios are independent. The same holds for FeO-Na_2O and Na_2O-K_2O.

In Figure 11.6, the rays for the predominant oxides Fe_2O_3, K_2O, and P_2O_5 are more nearly the same size compared with Figure 11.5, but the entire data set is not evenly represented, suggesting that more eigenvectors might be needed. It is possible to represent a three-eigenvector biplot in MATLAB. The three-eigenvector covariance and form biplots are shown in Figures 11.7 and 11.8, respectively. The covariance biplot is difficult to

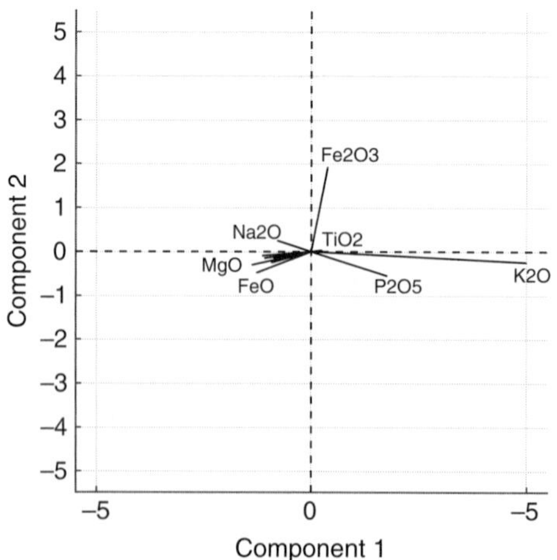

Figure 11.5 Covariance biplot of the MAR oxide data.

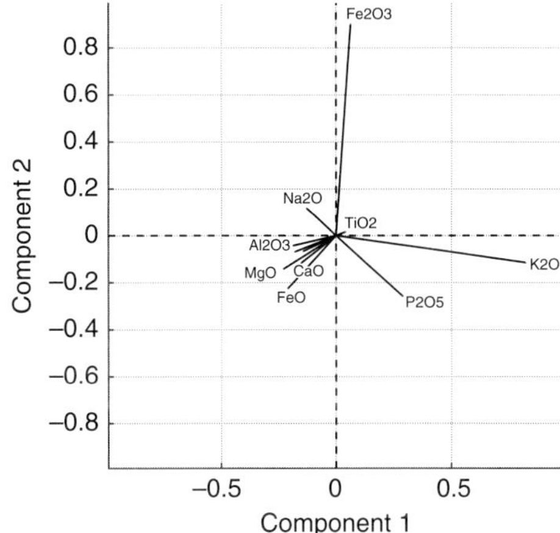

Figure 11.6 Form biplot of the MAR oxide data.

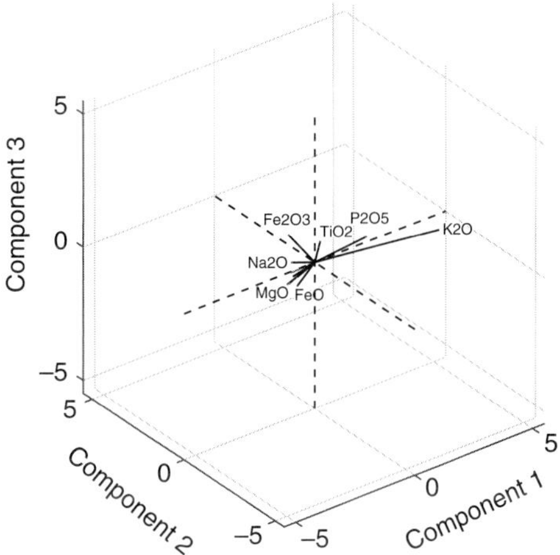

Figure 11.7 Covariance biplot for the MAR data using three eigenvectors.

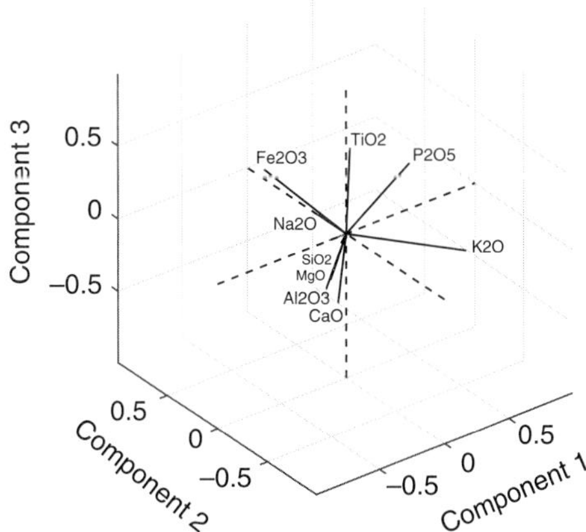

Figure 11.8 Form biplot for the MAR data using three eigenvectors.

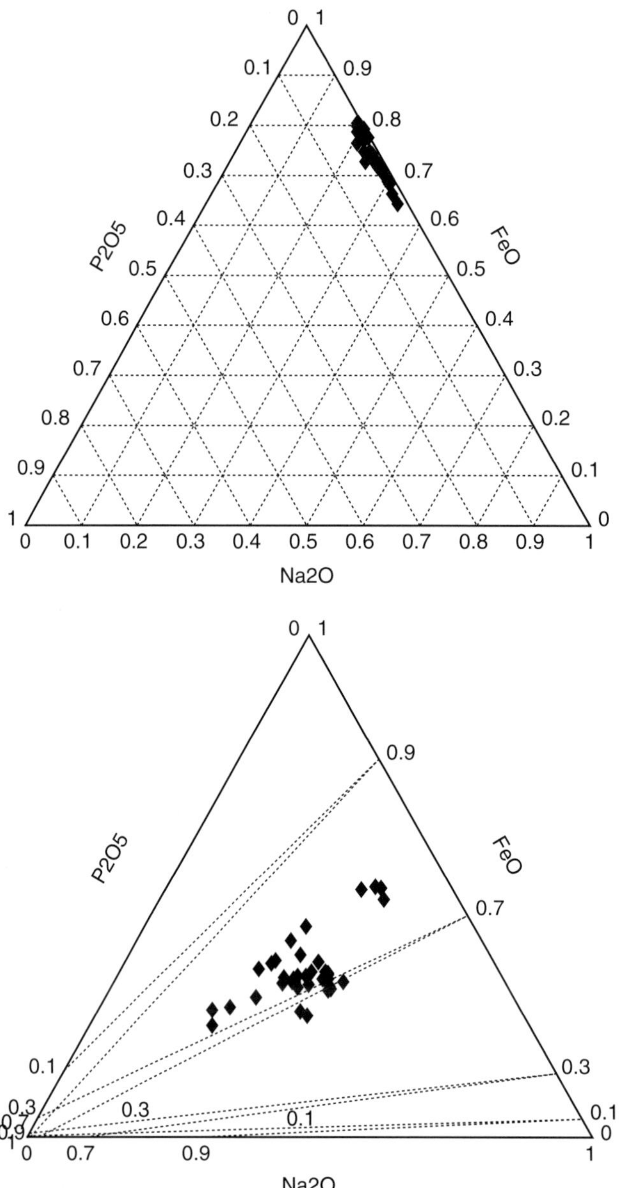

Figure 11.9 Ternary diagrams for FeO-Na_2O-P_2O_5 before (left) and after (right) centering.

evaluate because so many of the components are nearly coincident, but the form biplot is more evenly distributed. These figures can be rotated in 3D within MATLAB, and that is necessary to understand the relationships in them.

Figure 11.9 shows the FeO-Na_2O-P_2O_5 ternary that shows little correlation of the data either before or after centering, consistent with what is seen in Figures 11.5 through 11.8.

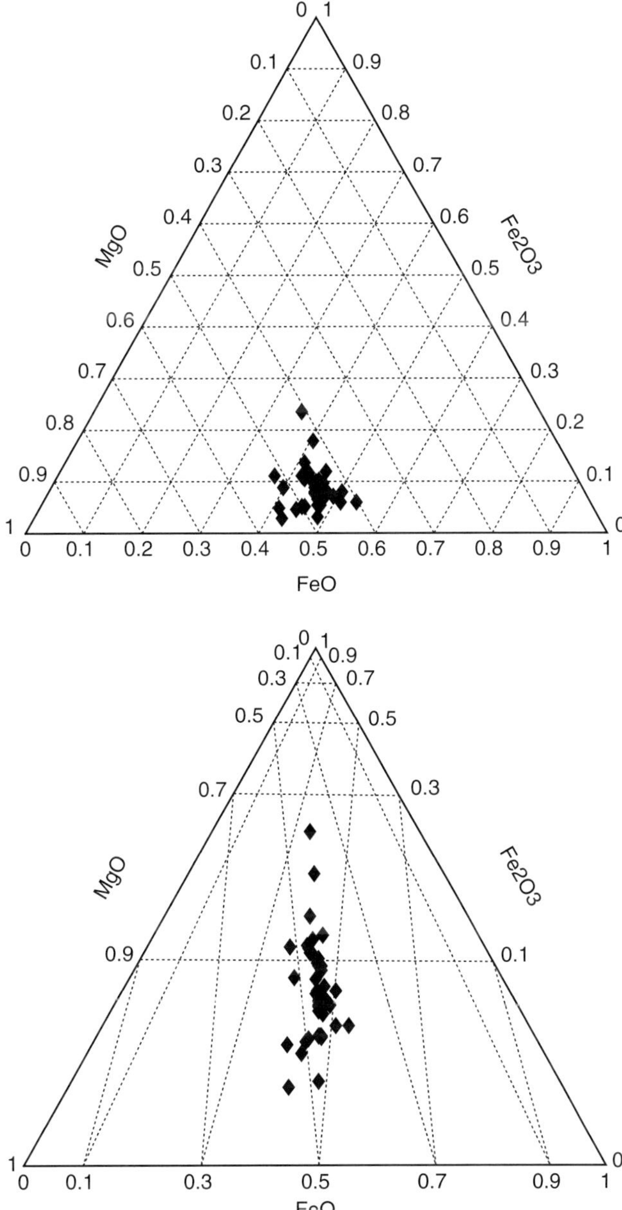

Figure 11.10 Ternary diagram for Fe_2O_3-FeO-MgO before (left) and after (right) centering.

Figure 11.10 shows a ternary diagram of the ferric minerals Fe_2O_3-FeO-MgO for which the biplots suggest the existence of a compositional line that is readily apparent in the centered data plot.

Figure 11.11 shows a ternary diagram of the feldspar minerals CaO-Na_2O-K_2O on which the data plot as nearly constant proportions of 0.8 CaO to 0.2 Na_2O and K_2O.

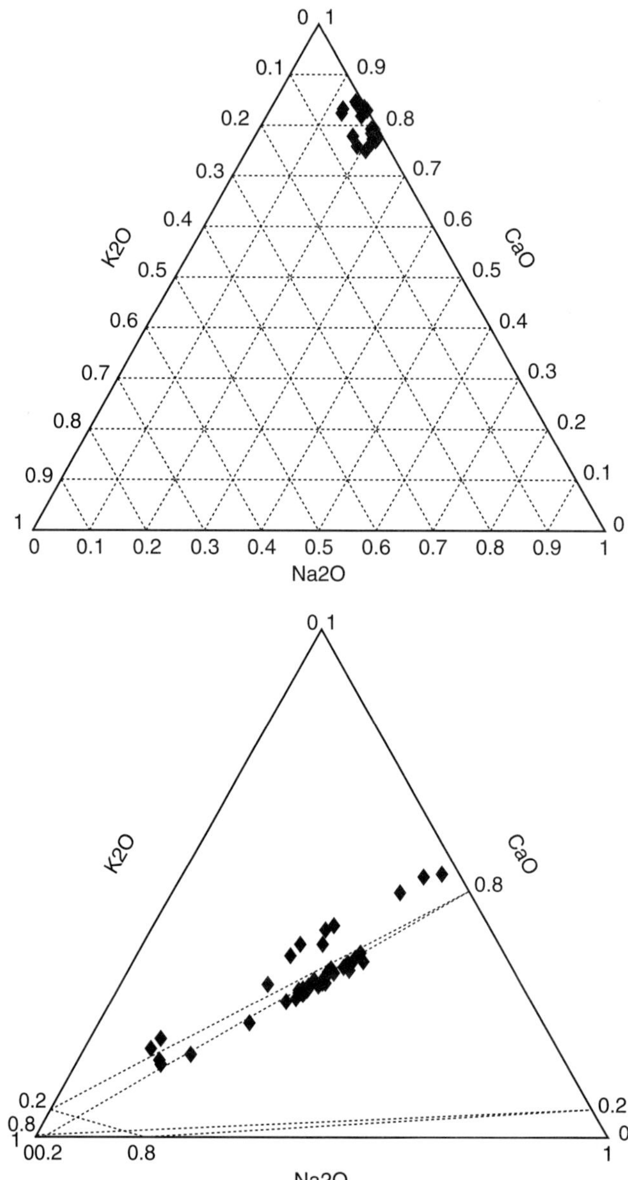

Figure 11.11 Ternary diagram for CaO-Na_2O-K_2O before (left) and after (right) centering.

Example 11.8 Define a sequential binary partition, and apply it to the MAR data.

Insight into the underlying scientific problem is required to suggest a sequential binary partition of the composition. This should be driven by an understanding of the chemistry and mineralogy of basalt. A candidate set of criteria taken from Pawlosky-Glahn et al. (2015) consists of

1. Fe_2O_3 versus FeO (Fe^{3+} versus Fe^{2+} oxidation state proxy);
2. SiO_2 versus Al_2O_3 (silica saturation proxy; when Si is lacking, Al takes its place);
3. Distribution within heavy minerals (rutile versus apatite);
4. Silicates versus heavy minerals;
5. Distribution within plagioclase (anorthite versus albite);
6. Plagioclase versus potassium feldspar;
7. Distribution within mafic nonferric minerals (MnO versus MgO);
8. Ferric (Fe_2O_3 and FeO) versus nonferrous (MnO and MgO) mafic minerals;
9. Mafic versus felsic minerals; and
10. Structure filling or cation oxides (those filling the crystalline structure of minerals) versus frame-forming oxides (i.e., those forming the structure).

Such criteria will account for changes in composition related to variations in partial melting of the mid-ocean ridge basalt (MORB) mantle source and for any effects of shallow level fractionation. A MATLAB script to compute the ilr and present the results initially as a boxplot is

```
sbp = [ 0 0 0 1 -1 0 0 0 0 0 0;
1 0 -1 0 0 0 0 0 0 0 0;
0 1 0 0 0 0 0 0 0 0 -1;
1 -1 1 0 0 0 0 0 0 0 -1;
0 0 0 0 0 0 0 1 -1 0 0;
0 0 0 0 0 0 0 1 1 -1 0;
0 0 0 0 0 1 -1 0 0 0 0;
0 0 0 1 1 -1 -1 0 0 0 0;
0 0 0 1 1 1 1 -1 -1 -1 0;
1 1 1 -1 -1 -1 -1 -1 -1 -1 1];
Psi = Contrast(sbp);
xstar = Ilr(mar, Psi);
boxplot(xstar)
```

The result is shown in Figure 11.12. The center of each boxplot element is very nearly the ilr of the geometric mean of the observations in the composition. The larger the distance a given boxplot center is from zero on the y-axis, the more one part of the balance is larger than the other. Moving through the balances (1) Fe_2O_3 versus FeO is strongly toward the ferrous lower oxidation state with some variability, (2) silica saturation is toward silica with little variability, (3) distribution within heavy minerals is strongly toward rutile with little variability, (4) silicates versus heavy minerals is very strongly toward silicates with some variability, (5) anorthite versus albite is strongly toward the Ca end member with little variability, (6) plagioclase versus K feldspar is strongly toward plagioclase with considerable variability, (7) MnO versus MgO is very strongly toward MgO with little variability, (8) ferric versus nonferric mafics is toward MnO and MgO with little variability, (9) mafic versus nonmafic minerals is neutral with considerable variability, and (10) structure-filling versus frame-filling oxides is toward structure-filling with little variability.

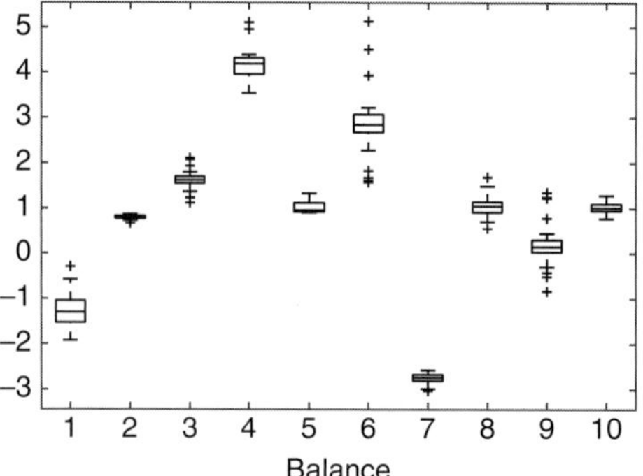

Figure 11.12 Boxplot of the ilr computed from the MAR data.

The boxplot defines the marginal (i.e., within balance) behavior. Understanding the relationships between them requires the conventional covariance matrix and the corresponding correlation matrix. These are obtained by transforming the normalized form of the total variation matrix (11.45) in the standard way. The following list shows the covariance matrix in the upper triangle including diagonal, whereas the correlation matrix is below the main diagonal. Both are obtained from the MATLAB script.

```
Cov = -Psi* NVariation(mar)* Psi';
ds = sqrt(diag(diag(Cov)));
Corr = inv(ds)* Cov* inv(ds);
```

1	2	3	4	5	6	7	8	9	10
0.105	0.003	-0.015	-0.049	-0.016	-0.127	0.010	0.059	-0.055	-0.019
0.226	0.002	-0.006	-0.012	0.000	-0.022	0.001	0.003	-0.009	0.001
-0.229	-0.630	0.040	0.053	-0.007	0.125	-0.004	-0.012	0.065	-0.004
-0.439	-0.780	0.767	0.119	0.011	0.228	-0.012	-0.038	0.114	-0.011
-0.388	0.032	-0.285	0.249	0.016	0.007	-0.006	-0.009	0.006	-0.001
-0.525	-0.655	0.835	0.887	0.075	0.557	-0.022	-0.085	0.293	0.002
0.257	0.258	-0.162	-0.314	-0.425	-0.263	0.013	0.005	-0.011	-0.002
0.89	0.37	-0.28	-0.53	-0.351	-0.550	0.198	0.043	-0.039	-0.007
-0.407	-0.494	0.776	0.797	0.108	0.942	-0.239	-0.449	0.174	-0.007
-0.485	0.133	-0.182	-0.266	-0.040	0.019	-0.113	-0.269	-0.140	0.015

Some observations about the correlation matrix in rank order:

1. Balance 1 (Fe_2O_3 versus FeO) and balance 8 (ferric versus nonferric oxides) are strongly correlated.
2. Balance 4 (silicate versus heavy minerals) and balance 6 (plagioclase versus K feldspar) are strongly correlated.
3. Balance 3 (distribution within heavy minerals) and balance 6 (plagioclase versus K feldspar) are strongly correlated.

4. Balance 4 (silicate versus heavy minerals) and balance 9 (mafic versus felsic minerals) are strongly correlated.
5. Balance 2 (silica saturation proxy) and balance 4 (silicates versus heavy minerals) are strongly anticorrelated.
6. Balance 3 (distribution within heavy minerals) and balance 9 (mafic versus felsic minerals) are strongly correlated.
7. Balance 3 (distribution within heavy minerals) and balance 4 (silicates versus heavy minerals) are strongly correlated.
8. Balance 2 (silica saturation proxy) and balance 6 (plagioclase versus K feldspar) are correlated.
9. Balance 2 (silica saturation proxy) and balance 3 (distribution within heavy minerals) are correlated.

Overall, the correlation structure is quite complex.

As a final step, a biplot is computed from the ilr.

```
g = geomean(mar);
xstar = Ilr(Perturbation(mar, 1./g), Psi);
[u s v] = svd(xstar);
labels = {'1', '2', '3', '4', '5', '6', '7', '8', '9', '10'};
biplot([s(1, 1)*v(:, 1) s(2, 2)*v(:, 2)], 'VarLabels', labels)
axis square
```

The eigenvalues from the ilr data are identical to those for the clr svd, but the eigenvectors are different due to the use of $\overleftrightarrow{\boldsymbol{\Psi}}$ to weight the clr. The covariance biplot is shown in Figure 11.13. Figure 11.14 shows the form biplot.

The covariance biplot is dominated by balances 1 (Fe oxidation state), 6 (plagioclase versus K feldspar), and 9 (mafic minerals versus feldspar). Conversely, balances 2 (silica saturation proxy), 5 (anorthite versus albite), 7 (MnO versus MgO), and 10 (structural

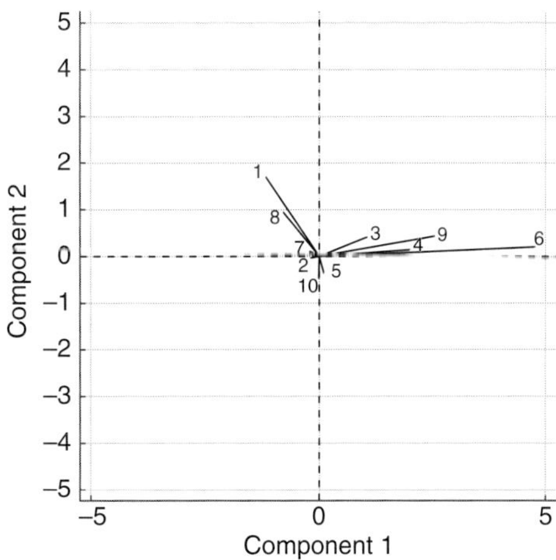

Figure 11.13 Covariance biplot for the ilr computed from the centered MAR data.

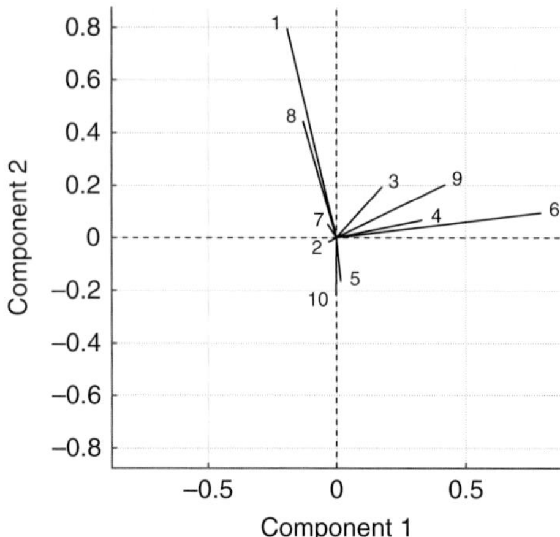

Figure 11.14 Form biplot for the ilr computed from the centered MAR data.

versus frame oxides) are very small. A compositional line relation is present between balances 3 (distribution within heavy minerals), 4 (heavy minerals versus silicates), 9 (mafic minerals versus feldspar), and 10 (structural versus frame oxides), on the one hand, and balances 1 (Fe oxidation state) and 8 (ferric versus nonferric mafic minerals), on the other. None of the large links are close to orthogonal; hence there is dependence among the balances.

The form biplot in Figure 11.14 is far from evenly distributed among the balances, suggesting that more than two eigenvalues are needed to explain the data. This can be addressed by producing 3D biplots, as in Figures 11.7 and 11.8. Finally, the ilr of the data can be transformed back to the simplex using ilr^{-1} and then plotted on ternary diagrams.

Example 11.9 The file komatiite.dat contains the major oxides from komatiites taken from the Petdb database that were selected for Archaean age, the presence of NiO, the presence of separate measurements for ferric and ferrous iron, and the absence of zeros in the 13 major oxides SiO_2, TiO_2, Al_2O_3, Cr_2O_3, NiO, Fe_2O_3, FeO, CaO, MgO, MnO, K_2O, Na_2O, and P_2O_5. Komatiites are ultramafic mantle-derived rocks of predominantly Archaean age that exhibit compositions that are consistent with melting at temperatures of over 1600°C, or at least 200°C higher than anything known today, reflecting the much higher internal heat production in Earth at the time. Komatiites are characterized by low Si, K, and Al and high Mg compared with modern day ultramafics and were formed as a result of extensive (~50%) partial melting. Exploratory compositional data analysis will be carried out on the data.

```
x = importdata('komatiite.dat');
komatiite = [x(:, 1:4) x(:, 10) x(:, 5:6) x(:, 9) x(:, 8) x
(:, 7) x(:, 12) x(:, 11) x(:, 13)];
komatiite = Close(komatiite);
```

```
center = Close(geomean(komatiite), 100)
center =
 Columns 1 through 9
 47.4147    0.3235    3.9727    0.4038    0.2039    2.8546
 8.6417     0.1984   30.7611
 Columns 10 through 13
 5.0783     0.0997    0.0192    0.0283
```

The center of these data is decidedly different from that of MORB in Example 11.2.

```
[T totvar] = NVariation(komatiite);
```

SiO_2	TiO_2	Al_2O_3	Cr_2O_3	NiO	Fe_2O_3	FeO	MnO	MgO	CaO
0	0.034	0.061	0.065	0.077	0.254	0.027	0.017	0.021	0.170
	0	0.049	0.107	0.151	0.352	0.032	0.051	0.087	0.110
		0	0.154	0.242	0.374	0.064	0.090	0.132	0.110
			0	0.094	0.234	0.087	0.082	0.053	0.352
				0	0.307	0.123	0.099	0.030	0.369
					0	0.397	0.204	0.206	0.718
						0	0.051	0.068	0.129
							0	0.035	0.216
								0	0.286
									0

Na_2O	K_2O	P_2O_5
0.386	0.312	0.250
0.335	0.327	0.262
0.287	0.260	0.279
0.595	0.417	0.242
0.658	0.445	0.337
0.897	0.558	0.165
0.371	0.355	0.342
0.430	0.340	0.212
0.535	0.364	0.246
0.217	0.421	0.557
0	0.415	0.752
	0	0.501
		0

The color code is <0.2 (dark gray), 0.2 to 0.4 (light gray), and >0.4 (white). The highest value occurs for Na_2O and Fe_2O_3, followed by Na_2O and P_2O_5, CaO and Fe_2O_3, Na_2O and NiO, Na_2O and Cr_2O_3, CaO and P_2O_5, and Na_2O and MgO. This is heavily tilted toward the felsic elements (Na, Ca, K) paired with mafic ones (Ni, Cr, ferric iron, M), although felsic concentration in komatiites is very low. Conversely, SiO_2, TiO_2, and Al_2O_3 are not highly correlated with any other oxide. The total variation is 3.08.

The clr of the centered data is computed and decomposed using the svd. The eigenvalues are

```
0.4849
0.2104
0.1285
0.0692
0.0401
0.0338
0.0123
0.0107
0.0059
0.0026
0.0011
0.0004
0.0000
```

The first two eigenvalues explain 69.5%, the first three eigenvalues explain 82.3%, and the first four eigenvalues explain 89.3% of the variance. A scree plot is shown in Figure 11.15 and suggests that at least five eigenvalues are needed to explain the data. This is substantially different from that for the MORB data in Example 11.6.

The first nine columns of the variable eigenvectors are

SiO_2	0.025	-0.143	0.041	-0.065	-0.097	0.062	0.095	-0.158	0.016
TiO_2	-0.070	-0.144	-0.110	0.198	-0.010	-0.013	-0.429	0.374	0.708
Al_2O_3	-0.133	0.023	-0.124	0.290	-0.322	-0.215	0.550	-0.331	0.329
Cr_2O_3	0.189	-0.178	0.110	-0.008	0.117	-0.747	0.031	0.360	-0.296
NiO	0.190	-0.304	0.369	-0.175	0.448	0.340	0.201	0.087	0.139
Fe_2O_3	0.475	0.312	-0.179	-0.394	-0.493	0.103	0.044	0.268	-0.003
FeO	-0.069	-0.275	0.052	0.111	-0.073	-0.214	-0.304	-0.385	-0.143
MnO	0.080	-0.091	-0.046	-0.107	-0.179	0.176	-0.515	-0.372	-0.205
MgO	0.156	-0.174	0.174	-0.154	0.030	0.136	0.266	-0.164	0.035
CaO	-0.389	-0.233	-0.194	0.296	-0.188	0.383	0.138	0.431	-0.449
Na_2O	-0.589	0.224	-0.264	-0.598	0.277	-0.126	0.024	-0.034	0.042
K_2O	-0.194	0.625	0.649	0.230	-0.038	0.032	-0.108	0.047	-0.060
P_2O_5	0.328	0.358	-0.477	0.375	0.528	0.082	0.007	-0.123	-0.114

It is likely that most of the variability described by the first few eigenvectors is due to alteration or metamorphism over the two billion years since the komatiites erupted, and in fact, the first mantle signal is the signature of Cr_2O_3 and NiO that does not appear until the fifth and sixth eigenvectors, representing 4% and 3% of the variance, respectively. The clr eigenvectors do not appear to be a source of great insight into these data.

Rather than examining biplots of the clr eofs, it makes more sense to go straight to a sequential binary partition. A candidate set is

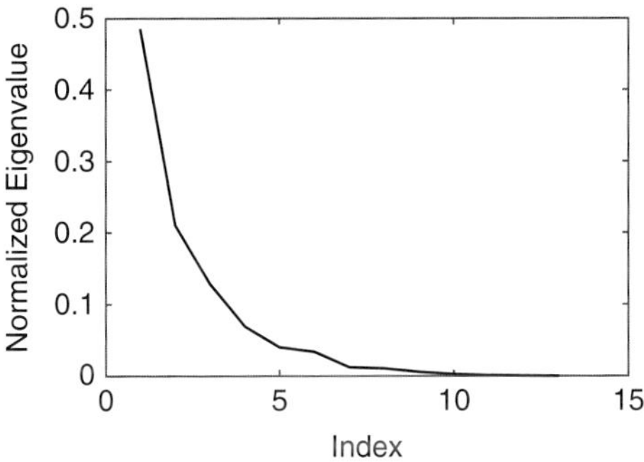

Figure 11.15 Scree plot for the svd of the clr of the centered komatiite data.

1. SiO_2 versus Al_2O_3 (silica saturation proxy);
2. TiO_2 versus NiO;
3. TiO_2 and NiO versus P_2O_5 (specific types of heavy mineral);
4. Silicate versus heavy minerals;
5. Albite versus anorthite;
6. Plagioclase versus potassium feldspar;
7. Fe_2O_3 versus FeO (oxidation state proxy);
8. MnO versus MgO (distribution within mafic nonferric minerals);
9. Ferric versus nonferric mafic minerals;
10. Cr_2O_3 versus mafic minerals;
11. Mafic versus felsic minerals; and
12. Structure-filling (cation) oxides versus frame-forming oxides.

```
sbp = [1 0 -1 0 0 0 0 0 0 0 0 0 0;
0 1 0 0 -1 0 0 0 0 0 0 0 0;
0 1 0 0 1 0 0 0 0 0 0 0 -1;
1 -1 1 0 -1 0 0 0 0 0 0 0 -1;
0 0 0 0 0 0 0 0 0 1 -1 0 0;
0 0 0 0 0 0 0 0 0 1 1 -1 0;
0 0 0 0 0 1 -1 0 0 0 0 0 0;
0 0 0 0 0 0 0 1 -1 0 0 0 0;
0 0 0 0 0 1 1 -1 -1 0 0 0 0;
0 0 0 1 0 -1 -1 -1 -1 0 0 0 0;
0 0 0 1 0 1 1 1 1 -1 -1 -1 0;
1 1 1 -1 1 -1 -1 -1 -1 -1 -1 -1 1];
Psi = Contrast(sbp);
xstar = Ilr(komatiite, Psi);
boxplot(xstar)
```

Figure 11.16 shows a boxplot of the ilr of the centered komatiite data. Moving through the results, balance 1 (SiO_2 versus Al_2O_3) is toward silica with limited variability, balance 2 (TiO_2 versus NiO) is approximately neutral, balance 3 (TiO_2 and NiO versus P_2O_5) is toward the metals, balance 4 (silicate versus heavy minerals) is very strongly toward

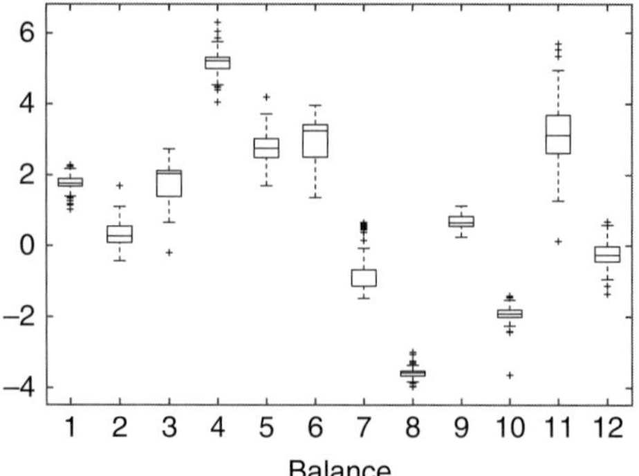

Figure 11.16 Boxplot of the ilr of the komatiite data.

silicates, balance 5 (albite versus anorthite) is strongly toward albite, balance 6 (plagioclase versus K feldspar) is strongly toward plagioclase, balance 7 (iron oxidation state) is weakly ferrous, balance 8 (MnO versus MgO) is very strongly toward MgO with little variability, balance 9 (ferric versus nonferric mafic minerals) is weakly toward the ferric side, balance 10 (Cr_2O_3 versus mafic minerals) is strongly toward chromium, balance 11 (mafic versus felsic minerals) is strongly toward mafic minerals with a large amount of variability, and balance 12 (structure-filling versus frame-forming oxides) is neutral.

```
Cov = -Psi*NVariation(komatiite)*Psi';
ds = sqrt(diag(diag(Cov)));
Corr = inv(ds)*Cov*inv(ds);
```

The correlation matrix is

1	2	3	4	5	6	7	8	9	10	11	12
1.000	-0.783	0.241	-0.593	0.168	-0.092	0.263	-0.418	-0.179	0.029	0.618	0.203
	1.000	-0.187	0.503	-0.174	0.369	-0.277	0.731	0.227	-0.126	-0.573	-0.256
		1.000	0.467	0.130	0.279	-0.666	-0.260	-0.709	0.152	-0.221	-0.417
			1.000	-0.176	0.302	-0.532	0.253	-0.207	-0.093	-0.613	-0.663
				1.000	0.010	-0.106	-0.199	-0.147	0.106	0.475	0.597
					1.000	-0.465	0.287	-0.308	-0.045	-0.182	-0.006
						1.000	-0.065	0.745	-0.255	0.543	0.257
							1.000	0.169	-0.395	-0.356	-0.267
								1.000	-0.032	0.325	0.110
									1.000	0.206	0.144
										1.000	0.660
											1.000

Some observations about the result in rank order are

1. Balance 1 (silica versus alumina) and balance 2 (TiO_2 versus NiO) are strongly inversely correlated.
2. Balance 7 (iron oxidation state) and balance 9 (ferric versus nonferrous mafic minerals) are strongly correlated.

3. Balance 2 (TiO_2 versus NiO) and balance 8 (MnO versus MgO) are strongly correlated. Because MgO is relatively constant, this suggest a tradeoff between TiO_2 and MnO.
4. Balance 3 (TiO_2 and NiO versus P_2O_5) and balance 9 (ferric versus nonferric mafic minerals) are strongly inversely correlated.
5. Balance 3 (TiO_2 and NiO versus P_2O_5) and balance 7 (oxidation state proxy) are negatively correlated.
6. Balance 4 (silicate versus heavy minerals) and balance 12 (structure- versus frame-filling oxides) are negatively correlated.
7. Balance 11 (mafic versus felsic minerals) and balance 12 (structure- versus frame-filling oxides) are correlated.
8. Balance 1 (silica saturation proxy) and balance 11 (mafic versus felsic minerals) are correlated.
9. Balance 4 (silicate versus heavy minerals) and balance 11 (mafic versus felsic minerals) are negatively correlated.

```
g = geomean(komatiite);
xstar = Ilr(Perturbation(komatiite, 1./g), Psi);
[u s v] = svd(xstar);
labels={'1','2','3','4','5','6','7','8','9','10','11','12'};
biplot([s(1, 1)*v(:, 1) s(2, 2)*v(:, 2) s(3,3)*v(:,3)],
'VarLabels', labels)
axis square
```

Figures 11.17 and 11.18 show covariance biplots corresponding to the first and second and second and third eigenvectors of the ilr of the centered komatiite data, respectively. Given the scree plot in Figure 11.15, this is not an adequate description of this data set. Figure 11.17 shows that the longest ray is for the eleventh balance (mafic versus felsic minerals), followed by the sixth (plagioclase versus K feldspar) and seventh (iron oxidation state). The rays for balances 3 (TiO_2 and NiO versus P_2O_5) and 6 (plagioclase versus

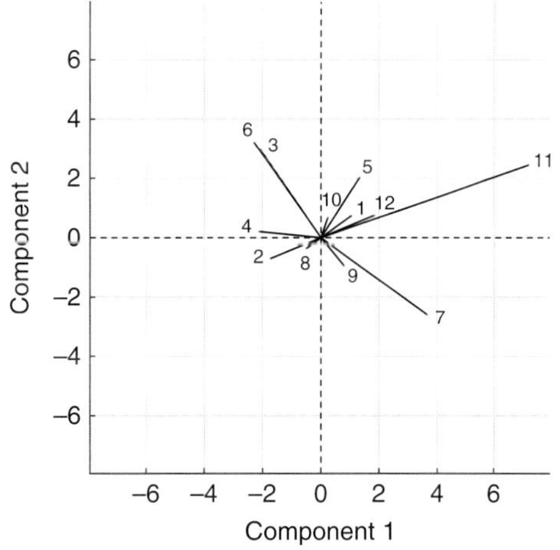

Figure 11.17 Covariance biplot corresponding to the first two eigenvectors of the ilr of the centered komatiite data.

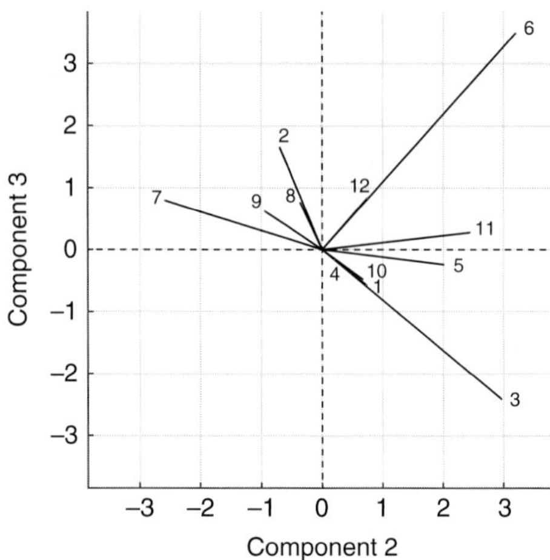

Figure 11.18 Covariance biplot corresponding to the second and third eigenvectors of the ilr of the centered komatiite data.

K feldspar) are nearly coincident and hence redundant. There are compositional lines between balance 2 (TiO_2 versus NiO) and balance 11 (structure versus frame filling oxides) and balances 3/6 and 7 (oxidation state proxy). It is likely that most of what is observed is due to weathering.

Figure 11.18 shows the covariance biplot corresponding to the second and third eigenvectors of the ilr of the centered komatiite data. The longest rays occur for balance 6 (plagioclase versus K feldspar), followed by balances 3 (TiO_2 and NiO versus P_2O_5) and 7 (oxidation state proxy). Balances 1 (silica saturation proxy) and 10 (Cr_2O_3 versus mafic minerals) are nearly coincident, suggesting redundancy. Compositional lines exist between balances 3 (TiO_2 and NiO versus P_2O_5) and 9 (ferric versus nonferric mafic minerals) and balances 5 (albite versus anorthite) and 7 (oxidation state proxy).

11A Appendix 11A MATLAB Functions to Produce Ternary Diagrams

The function *Ternary* is called initially to set up a ternary diagram, plot its frame, and place gridlines through it. It optionally allows perturbation to facilitate centering. Section 11.2.2 and Example 7.4 illustrate its use.

```
function [h] = Ternary(varargin)
%sets up axes for ternary diagram
%arguments
%varargin{1} = ntick - number of tick marks on the axes
(default is 11); if set to zero,
%then tick values are contained in varargin{ 3}
%varargin{2} = composition - composition by which the grid-
lines are perturbed
if nargin == 0
   ntick = 11;
else
   ntick = varargin{ 1} ;
end
hold off
h = fill([ 0 1 0.5 0] , [ 0 0 sqrt(3)/2 0] , 'w', 'linewidth', 2);
axis image
axis off
hold on
label = linspace(0, 1, ntick);
if nargin < 2
   label = linspace(0, 1, ntick);
   for i = 1:ntick
      plot([ label(i) 0.5* (1 + label(i))] , [ 0 0.866* (1 - label
      (i))] , ':k', 'linewidth', 0.75)
      plot([ 0.5* (1 - label(i)) 1 - label(i)] , [ 0.866* (1 - label
      (i)) 0] , ':k', 'linewidth', 0.75)
      plot([ 0.5* label(i) 1 - 0.5* label(i)] , [ 0.866* label(i)
      0.866* label(i)] , ...
           ':k', 'linewidth', 0.75)
      text(1 - 0.5* label(i) + .025, sqrt(3)* label(i)/2 +
      0.01, ...
          num2str(label(i),3), 'FontSize', 14, 'FontWeight',
          'bold');
```

```
            str = num2str(label(i), 3);
            text(label(i) - 0.02*length(str)/2, -.025, str, 'Font-
            Size', 14, 'FontWeight', 'bold');
            str = num2str(label(i), 3);
            text(0.5*(1 - label(i)) - 0.015*(length(str) - 1) -
            .035, ...
                sqrt(3)/2*(1 - label(i)) + .01, str, 'FontSize',
                14, 'FontWeight', 'bold');
        end
    elseif nargin == 2
        label = linspace(0, 1, ntick);
        y = varargin{2} ;
        for i = 1:ntick
            x1 = Comp_to_Cart(Perturbation(Cart_to_Comp([ label(i)
            0] ), y));
            x2 = Comp_to_Cart(Perturbation( ...
                Cart_to_Comp([ 0.5*(1 + label(i)) 0.866*(1 - label
                (i))] ), y));
            plot([x1(1) x2(1)], [x1(2) x2(2)], ':k', 'linewidth',
            0.75)
            x1 = Comp_to_Cart(Perturbation( ...
                Cart_to_Comp([ 0.5*(1 - label(i)) 0.866*(1 - label
                (i))] ), y));
            x2 = Comp_to_Cart(Perturbation( ...
                Cart_to_Comp([ 1 - label(i) 0] ), y));
            plot([x1(1) x2(1)], [x1(2) x2(2)], ':k', 'linewidth',
            0.75)
            x1 = Comp_to_Cart(Perturbation(Cart_to_Comp([ 0.5*label
            (i) ...
                0.866*label(i)] ), y));
            x2 = Comp_to_Cart(Perturbation( ...
                Cart_to_Comp([ 1-0.5*label(i) 0.866*label(i)] ), y));
            plot([ x1(1) x2(1)], [ x1(2) x2(2)], ':k', 'linewidth',
            0.75)
            x1 = Comp_to_Cart(Perturbation( ...
                Cart_to_Comp([ 1-0.5*label(i) 0.866*label(i)] ), y));
            text(x1(1) + .025, x1(2) + 0.01, num2str(label(i),
            3), ...
                'FontSize', 14, 'FontWeight', 'bold');
            x1 = Comp_to_Cart(Perturbation(Cart_to_Comp([ label(i)
            0] ), y));
            str = num2str(label(i), 3);
            text(x1(1) - 0.02*length(str)/2, x1(2) - .025, str,
            'FontSize', ...
```

```
            14, 'FontWeight', 'bold');
        x1 = Comp_to_Cart(Perturbation( ...
            Cart_to_Comp([ 0.5*(1 - label(i)) 0.866*(1 - label
            (i))]), y));
        str = num2str(label(i), 3);
        text(x1(1) - .015*(length(str) - 1) - .035, x1(2) +
        .01, str, ...
            'FontSize', 14, 'FontWeight', 'bold');
    end
else
    label = varargin{3};
    ntick = length(label);
    for i = 1:ntick
        y = varargin{2};
        x1 = Comp_to_Cart(Perturbation(Cart_to_Comp([ label(i)
        0]), y));
        x2 = Comp_to_Cart(Perturbation( ...
            Cart_to_Comp([ 0.5*(1 + label(i)) 0.866*(1 - label
            (i))]), y));
        plot([x1(1) x2(1)],[x1(2) x2(2)], ':k', 'linewidth', 0.75)
        x1 = Comp_to_Cart(Perturbation( ...
            Cart_to_Comp([ 0.5*(1 - label(i)) 0.866*(1 - label
            (i))]), y));
        x2 = Comp_to_Cart(Perturbation( ...
            Cart_to_Comp([ 1 - label(i) 0]), y));
        plot([ x1(1) x2(1)], [ x1(2) x2(2)], ':k', 'linewidth',
        0.75)
        x1 = Comp_to_Cart(Perturbation(Cart_to_Comp([ 0.5*label
        (i) ...
            0.866*label(i)]), y));
        x2 = Comp_to_Cart(Perturbation( ...
            Cart_to_Comp([ 1 - 0.5*label(i) 0.866*label(i)]), y));
        plot([ x1(1) x2(1)],[ x1(2) x2(2)], ':k', 'linewidth', 0.75)
        x1 = Comp_to_Cart(Perturbation( ...
            Cart_to_Comp([1 - 0.5*label(i) 0.866*label(i)]), y));
        text(x1(1) + .025, x1(2) + 0.01, num2str(label(i),
        3), ...
            'FontSize', 14, 'FontWeight', 'bold');
        x1 = Comp_to_Cart(Perturbation(Cart_to_Comp([ label(i)
        0]), y));
        str = num2str(label(i), 3);
        text(x1(1) - 0.02*length(str)/2, x1(2) - .025, str,
        'FontSize', ...
            14, 'FontWeight', 'bold');
```

```
            x1 = Comp_to_Cart(Perturbation( ...
               Cart_to_Comp([ 0.5*(1 - label(i)) 0.866*(1 - label
               (i))]), y));
            str = num2str(label(i), 3);
            text(x1(1) - .015*(length(str) - 1) - .035, x1(2) +
            .01, str, ...
               'FontSize', 14, 'FontWeight', 'bold');
        end
    end
end
function [ Result] = Comp_to_Cart(comp)
%Converts a closed composition comp to Cartesian components
Result on a
%ternary with unit edges
[ n D] = size(comp);
if D ~= 3
   warning('Comp_to_Cart: composition must have 3 parts')
   return
end
Result = zeros(n, 2);
for i = 1:n
    x = (comp(i, 1) + 2*comp(i, 2))/(2*(comp(i, 1) + comp(i,
    2) + ...
       comp(i, 3)));
    y = sqrt(3)*comp(i, 1)/(2*(comp(i, 1) + comp(i, 2) + comp
    (i, 3)));
    Result(i, :) = [ x y] ;
end
end
function [ Result] = Cart_to_Comp(xx)
%Converts Cartesian components xx on the ternary with unit
edges to a
%composition Result
n = size(xx, 1);
Result = zeros(n, 3);
for i = 1:n
    Result(i, 1:3) =[ 2*xx(i, 2)/sqrt(3) xx(i, 1) - xx(i, 2)/sqrt
    (3) ...
                1 - xx(i, 1) - xx(i, 2)/sqrt(3)] ;
end
end
```

The function *TernaryLabel* places labels on the *x*, *y*, and *z* axes of a ternary diagram created using *Ternary.*

```
function [ h] = TernaryLabel(x, y, z)
%Places labels on the x, y and z axes of a ternary diagram
if nargout == 0
   text(0.85, sqrt(3)/4 + 0.05, x, 'horizontalalignment', ...
       'center', 'rotation', -60, 'FontSize', 16, 'FontWeight',
       'bold')
   text(0.5, -0.075, y, 'horizontalalignment', 'center',
   'FontSize', 16, ...
       'FontWeight', 'bold')
   text(0.15, sqrt(3)/4 + 0.05, z, 'horizontalalignment', ...
       'center', 'rotation', 60, 'FontSize', 16, 'FontWeight',
       'bold')
else
   h = zeros(1, 3);
   h(1) = text(0.85, sqrt(3)/4 + 0.05, x, 'horizontalalignment', ...
       'center', 'rotation', -60, 'FontSize', 16, 'Font-
       Weight', 'bold');
   h(2) = text(0.5, -0.075, y, 'horizontalalignment',
   'center', ...
       'FontSize', 16, 'FontWeight', 'bold');
   h(3) = text(0.15, sqrt(3)/4 + 0.05, z, 'horizontalalignment', ...
       'center', 'rotation', 60, 'FontSize', 16, 'FontWeight',
       'bold');
end
end
```

The function *TernaryLine* plots a compositional line given the three-part composition in the argument *comp*.

```
function TernaryLine(comp)
%Plots a compositional line given the 3 part composition in
comp
xo = (comp(1, 1) + 2*comp(1, 2))/(2*sum(comp(1, :)));
yo = sqrt(3)*comp(1, 1)/(2*sum(comp(1, :)));
for i = 2:length(comp)
    x = (comp(i, 1) + 2*comp(i, 2))/(2*sum(comp(i, :)));
    y = sqrt(3)*comp(i, 1)/(2*sum(comp(i, :)));
    plot([ xo x] , [ yo y] , 'k', 'linewidth', 1.0)
    xo = x;
    yo = y;
end
end
```

The function *TernaryPlot* plots compositional data in the three-vector *comp* using blue plusses by default. Both the symbol type and color can be changed through optional input arguments.

```
function TernaryPlot(comp, varargin)
%Plots the compositional data in comp using blue + by
default
%varargin{ 1} specifies alternate symbols (e.g.,'d') and var-
argin{ 2}
%specifies alternate colors (e.g.,'k')
xx = Comp_to_Cart(comp);
if nargin == 1
   mrk = '+';
   mcol = 'b';
elseif nargin == 2
   mrk = varargin{ 1} ;
   mcol = 'b';
else
   mrk = varargin{ 1} ;
   mcol = varargin{ 2} ;
end
for i = 1:length(xx)
   plot(xx(i, 1), xx(i, 2), 'Marker', mrk, 'MarkerSize', 8, ...
      'MarkerEdgeColor', mcol, 'MarkerFaceColor', mcol)
end
end
```

References

Aitchison, J. (1986). *The Statistical Analysis of Compositional Data*. London: Chapman & Hall.

Aitchison, J., & M. Greenacre (2002). Biplots of compositional data. *Appl. Stat.*, **51**, 375–92.

Anderson, T. W. (2003). *An Introduction to Multivariate Statistical Analysis*, 3rd edn. New York: Wiley.

Anderson, T. W., & D. Darling (1952). Asymptotic theory of certain goodness of fit criteria based on stochastic processes. *Ann. Math. Stat.*, **23**, 193–212.

Anderson, T. W., & D. Darling (1954). A test of goodness of fit. *J. Am. Stat. Assoc.*, **49**, 765–9.

Andrews, D. F., & A. M. Herzberg (1985). *Data: A Collection of Problems from Many Fields for the Student and Research Worker*. New York: Springer.

Ansari, A. R., & R. A. Bradley (1960). Rank-sum tests for dispersions. *Ann. Math. Stat.*, **31**, 1174–89.

Arbuthnott, J. (1710). An argument for divine providence, taken from the constant regularity in the births of both sexes. *Philos. Trans. R. Soc. Lond.*, **27**, 186–90.

Azzalini, A., & A. W. Bowman (1990). A look at some data on the Old Faithful geyser. *J. R. Stat. Soc.*, **C39**, 357–65.

Bartlett, M. S. (1937). Properties of sufficiency and statistical tests. *Proc. R. Soc. Lond.*, **A160**, 268–82.

Bayes, T. (1763). An essay towards solving a problem in the doctrine of chances. *Philos. Trans. R. Soc. Lond.*, **53**, 370–418.

Benjamini, Y., & Y. Hochberg (1995). Controlling the false discovery rate: a practical and powerful approach to multiple testing. *J. R. Stat. Soc.*, **B57**, 289–300.

Benjamini, Y., A. M. Krieger, & D. Yekutieli (2006). Adaptive linear step-up procedures that control the false discovery rate. *Biometrika*, **93**, 491–507.

Berry, K. J., P. W. Mielke, Jr., & J. E. Johnston (2016). *Permutation Statistical Methods: An Integrated Approach*. New York: Springer.

Birnbaum, A. (1954). Combining independent tests of significance. *J. Am. Stat. Assoc.*, **49**, 559–74.

Blackwell, D. (1947). Conditional expectation and unbiased sequential estimation. *Ann. Math. Stat.*, **18**, 105–10.

Bound, J., D. A. Jaeger, & R. M. Baker (1995). Problems with instrumental variables estimation when the correlation between the instruments and the endogenous explanatory variable is weak. *J. Am. Stat. Assoc.*, **90**, 443–50.

Box, G. E. P. (1949). A general distribution theory for a class of likelihood criteria. *Biometrika*, **36**, 317–46.

Brofitt, J. D., & R. H. Randles (1977). A power approximation for the chi-square goodness-of-fit test: simple hypothesis case. *J. Am. Stat. Assoc.*, **72**, 604–7.

Carroll, R. J., & A. H. Welsh (1988). A note on asymmetry and robustness in linear regression. *Am. Stat.*, **42**, 285–7.

Carter, M., & B. van Brunt (2000). *The Lebesgue-Stieltjes Integral: A Practical Introduction*. New York: Springer.

Chave, A. D. (2014). Magnetotelluric data, stable distributions and impropriety: an existential combination. *Geophys. J. Int.*, **198**, 622–36.

Chave, A. D. (2015). A note about Gaussian statistics on a sphere. *Geophys. J. Int.*, **203**, 893–5.

Chave, A. D., & D. J. Thomson (2003). A bounded influence regression estimator based on the statistics of the hat matrix. *J. R. Stat. Soc.*, **C52**, 307–22.

Chave, A. D., & D. J. Thomson (2004). Bounded influence estimation of magnetotelluric response functions. *Geophys. J. Int.*, **157**, 988–1006.

Chave, A. D., D. J. Thomson, & M. E. Ander (1987). On the robust estimation of power spectra, coherences and transfer functions. *J. Geophys. Res.*, **92**, 633–48.

Clarke, R. D. (1946). An application of the Poisson distribution. *J. Inst. Actuaries*, **22**, 32.

Cochran, W. G. (1934). The distribution of quadratic forms in a normal system, with applications to the analysis of covariance. *Math. Proc. Camb. Philos. Soc.*, **30**, 178–91.

Cochran, W. G. (1952). The χ^2 test of goodness of fit. *Ann. Math. Stat.*, **23**, 35–45.

Cramér, H. (1945). *Mathematical Methods of Statistics*. Uppsala: Almqvist & Wiksell.

Csörgő, S., & J. J. Faraway (1996). The exact and asymptotic distributions of Cramér–von Mises statistics. *J. R. Stat. Soc.*, **B58**, 221–34.

David, H. A. (1981). *Order Statistics*, 2nd edn. New York: Wiley.

David, H. A. (2009). A historical note on zero correlation and independence. *Am. Stat.*, **63**, 185–6.

David, H. A., & H. N. Nagaraja (2003). *Order Statistics*, 3rd edn. New York: Wiley.

Davison, A. C., & D. V. Hinkley (1997). *Bootstrap Methods and Their Application*. Cambridge University Press.

De Groot, M. H., & M. J. Schervish (2011). *Probability and Statistics*, 4th edn. London: Pearson.

De Moivre, A. (1711). De mensura sortis. *Philos. Trans. R. Soc. Lond.*, **27**, 213–64.

Doob, J. L. (1993). *Measure Theory*. New York: Springer.

DuMouchel, W. H. (1975). Stable distributions in statistical inference. 2. Information from stably distributed samples. *J. Am. Stat. Assoc.*, **70**, 386–93.

Durbin, J., & G. S. Watson (1950). Testing for serial correlation in least squares regression, part I. *Biometrika*, **37**, 409–28.

Durbin, J. & G. S. Watson (1951). Testing for serial correlation in least squares regression, part II. *Biometrika*, **38**, 159–77.

Durbin, J., & G. S. Watson (1971). Testing for serial correlation in least squares regression, part III. *Biometrika*, **58**, 1–19.

Efron, B. (1979). Bootstrap methods: another look at the jackknife. *Ann. Stat.*, **7**, 1–26.

Efron, B., & C. Stein (1981). The jackknife estimate of variance. *Ann. Stat.*, **9**, 586–96.

Efron, B., & R. Tibshirani (1998). *An Introduction to the Bootstrap*. London: Chapman & Hall.

Ernst, M. D. (2004). Permutation methods: a basis for exact inference. *Stat. Sci.*, **19**, 676–85.

Feller, W. (1948). On the Kolmogorov-Smirnov limit theorems for empirical distributions. *Ann. Math. Stat.*, **19**, 177–89.

Feller, W. (1950). Errata: On the Kolmogorov-Smirnov limit theorems for empirical distributions. *Ann. Math. Stat.*, **21**, 301–2.

Feller, W. (1971). *An Introduction to Probability Theory and Its Applications*, vol. 2. New York: Wiley.

Fieller, E. C. (1940). The biological standardization of insulin. *J. R. Stat. Soc.*, **7**(Suppl.), 1–64.

Fieller, E. C. (1944). A fundamental formula in the statistics of biological assays and some applications. *Q. J. Pharm. Pharmacol.*, **17**, 117–23.

Fieller, E. C. (1954). Some problems in interval estimation. *J. R. Stat. Soc.*, **16**, 175–85.

Fisher, N. I. (1995). *Statistical Analysis of Circular Data*. Cambridge University Press.

Fisher, R. A. (1915). Frequency distribution of the values of the correlation coefficient in samples from an indefinitely large population. *Biometrika*, **10**, 507–21.

Fisher, R. A. (1922). On the mathematical foundations of theoretical statistics. *Philos. Trans. R. Soc. Lond.*, **A222**, 309–68.

Fisher, R. A. (1928). The general sampling distribution of the multiple correlation coefficient. *Proc. R. Stat. Soc.*, **A121**, 654–73.

Fisher, R. A. (1932). *Statistical Methods for Research Workers*, 4th edn. Edinburgh: Oliver & Boyd.

Fisher, R. A. (1936). The use of multiple measurements in taxonomic problems. *Ann. Eugenics*, **7**, 179–88.

Fisher, R. A. (1953). Dispersion on a sphere. *Proc. R. Soc. Lond.*, **A217**, 295–305.

Gamble, T. D., W. M. Goubau, & J. Clarke (1979). Magnetotellurics with a remote reference. *Geophysics*, **44**, 53–68.

Gauss, K. F. (1823). *Theoria Combinationis Observationum Erroribus Minimis Oboxiae*. Göttingen: Dieterich.

Geary, R. C. (1949). Determination of linear relationships between systematic parts of variables with errors of observation the variances of which are unknown. *Econometrica*, **17**, 30–58.

Gelman, A., J. B. Carlin, H. S. Stern, D. B. Dunson, A. Vehtari, et al. (2013). *Bayesian Data Analysis*, 3rd edn. Boca Raton, FL: Taylor & Francis.

Genovese, C., & L. Wasserman (2002). Operating characteristics and extension of the false discovery rate procedure. *J. R. Stat. Soc.*, **B64**, 499–517.

Gleser, L. J., & J. T. Hwang (1987). The nonexistence of $100(1-\alpha)\%$ confidence sets of finite expected diameter in errors-in-variables and related models. *Ann. Stat.*, **15**, 1351–62.

Good, P. I. (2000). *Permutation Methods*, 2nd edn. New York: Springer.

Good, P. I. (2005). *Permutation, Parametric, and Bootstrap Tests of Hypotheses*, 3rd edn. New York: Springer.

Gossett, W. S. (1908). The probable error of a mean. *Biometrika*, **6**, 1–25.

Gradshteyn, I. S., & I. M. Ryzhik (1980). *Table of Integrals, Series and Products*. San Diego: Academic Press.

Guenther, W. C. (1977). Power and sample size for approximate chi-square tests. *Am. Stat.*, **31**, 83–5.

Hampel, F. R., E. M. Ronchetti, P. J. Rousseeuw, & W. A. Stahel (1986). *Robust Statistics*. New York: Wiley.

Hänggi, P., F. Roesel, & P. Trautmann (1978). Continued fraction expansion in scattering theory and statistical non-equilibrium mechanics. *Z. Naturforsch.*, **33a**, 402–17.

Hand, D. J., F. Daly, K. McConway, D. Lunn, & E. Ostrowski (1994). *A Handbook of Small Data Sets*. London: Chapman & Hall.

Handschin, E., F. C. Schweppe, J. Kohlas, & A. Fiechter (1975). Bad data analysis of power system state analysis. *IEEE Trans. Power Appar. Syst.*, **PAS-94**, 329–37.

Hanley, J. A., M. Julien, & E. E. M. Moodie (2008). Student's *z*, *t*, and *s*: what if Gossett had *R*?. *Am. Stat.*, **62**, 64–9.

Hastie, T., R. Tibshirani, & J. Friedman (2008). *The Elements of Statistical Learning*, 2nd edn. New York: Springer.

Herschel, J. F. W. (1850). Quetelet on probabilities. *Edinb. Rev.*, **92**, 1–57.

Hettmansperger, T. P., & J. W. McKean (1998). *Robust Nonparametric Statistical Methods*. London: Edward Arnold.

Hirschberg, J., & J. Lye (2010). A geometric comparison of the delta and Fieller confidence intervals. *Am. Stat.*, **64**, 234–41.

Hoaglin, D. C., F. Mosteller, & J. W. Tukey (1991). *Fundamentals of Exploratory Analysis of Variance*. New York: Wiley.

Hodges, J. L., & E. L. Lehmann (1956). The efficiency of some nonparametric competitors of the *t*-test. *Ann. Math. Stat.*, **27**, 324–35.

Hoeffding, W. (1952). The large-sample power of tests based on permutations of observations. *Ann. Math. Stat.*, **23**, 169–92.

Hoeffding, W. (1963). Probability inequalities for sums of bounded random variables. *J. Am. Stat. Assoc.*, **58**, 13–30.

Hogg, R. V., & A. T. Craig (1995). *Introduction to Mathematical Statistics*, 5th edn. Saddle River, NJ: Prentice-Hall.

Hotelling, H. (1931). The generalization of Student's ratio. *Ann. Math. Stat.*, **2**, 360–78.

Hotelling, H. (1933). Analysis of a complex of statistical variables into principal components. *J. Educ. Psychol.*, **24**, 417–41.

Hotelling, H. (1936). Relations between two sets of variates. *Biometrika*, **28**, 321–77.

Hotelling, H. (1953). New light on the correlation coefficient and its transforms. *Proc. R. Stat. Soc.*, **B15**, 193–232.

Hubble, E. (1929). A relation among distance and radial velocity among extra-galactic nebulae. *Proc. Nat. Acad. Sci. USA*, **15**, 168–73.

Huber, P. (1964). Robust estimation of a location parameter. *Ann. Math. Stat.*, **35**, 73–101.

Huber, P. (2011). *Data Analysis: What Can Be Learned from the Past 50 Years*. New York: Wiley.

Jarque, C. M., & A. K. Bera (1987). A test for normality of observations and regression residuals. *Int. Stat. Rev.*, **55**, 163–72.

Johnson, N. L., S. Kotz, & N. Balikrishnan (1994). *Continuous Univariate Distributions*, vol. 1, 2nd edn. New York: Wiley.

Johnson, N. L., S. Kotz, & N. Balikrishnan (1995). *Continuous Univariate Distributions*, vol. 2, 2nd edn. New York: Wiley.

Johnson, N. L., S. Kotz, & N. Balakrishnan (1997). *Discrete Multivariate Distributions*. New York: Wiley.

Johnson, N. L., S. Kotz, & A. W. Kemp (1993). *Univariate Discrete Distributions*, 2nd edn. New York: Wiley.

Jolliffe, I. T. (2002). *Principal Component Analysis*, 2nd edn. New York: Springer.

Kolmogorov, A. (1933). Sulla determinazione empirica di una legge di distribuzione. *Inst. Ital. Attuarai, Giorn.*, **4**, 1–11.

Kotz, S., & N. L. Johnson (eds.) (1992). *Breakthroughs in Statistics,* vol. II: *Methodology and Distributions*. New York: Springer.

Kotz, S., N. Balakrishnan, & N. L. Johnson (2000). *Continuous Multivariate Distributions,* vol 1: *Models and Applications*, 2nd edn. New York: Wiley.

Kvam, P. H., & B. Vidakovic (2007). *Nonparametric Statistics with Applications to Science and Engineering*. Hoboken, NJ: Wiley.

Lagarias, J. C., J. A. Reeds, M. H. Wright, & P. E. Wright (1998). Convergence properties of the Neider-Mead simplex method in low dimensions. *SIAM J. Optim.*, **9**, 112–47.

Langevin, P. (1905). Magnétisme et théorie des électrons. *Ann. Chim. Phys.*, **5**, 71–127.

Lanzante, J. R. (2005). A cautionary note on the use of error bars. *J. Climate*, **18**, 3699–703.

Lehmann, E. L. (1953). The power of rank tests. *Ann. Math. Stat.*, **24**, 23–43.

Lehmann, E. L., & J. P. Schaffer (1988). Inverted distributions. *Am. Stat.*, **42**, 191–4.

Lévy, P. P. (1925). *Calcul des Probabilités*. Paris: Gauthier-Villars.

Lilliefors, H. W. (1967). On the Kolmogorov-Smirnov test for normality with mean and variance unknown. *J. Am. Stat. Assoc.*, **62**, 399–402.

Longley, J. W. (1967). An appraisal of least squares programs for electronic computers from the viewpoint of the user. *J. Am. Stat. Assoc.*, **62**, 819–41.

Love, J. J., & C. G. Constable (2003). Gaussian statistics for paleomagnetic vectors. *Geophys. J. Int.*, **152**, 515–65.

Mallows, C. L. (1975). On some topics in robustness. Tech. Mem., Bell Telephone Laboratories.

Mann, H. B., & D. R. Whitney (1947). On a test of whether one of two random variables is stochastically larger than the other. *Ann. Math. Stat.*, **18**, 50–60.

Mardia, K. V., & P. E. Jupp (2000). *Directional Statistics*. New York: Wiley.

Mardia, K. V., J. T. Kent, & J. Bibby (1979). *Multivariate Analysis*. London: Academic Press.

Mauchly, J. W. (1940). Significance test for sphericity of a normal n-variate distribution. *Ann. Math. Stat.*, **11**, 204–9.

Meerschaert, M. M. (2012). Fractional calculus, anomalous diffusion and probability. In *Fractional Dynamics*, ed. S. C. Lim, J. Klafter, & R. Metler. Singapore: World Science Press.

Michael, J. R. (1983). The stabilized probability plot. *Biometrika*, **70**, 11–17.

Mitra, S. K. (1958). On the limiting power function of the frequency chi-square test. *Ann. Math. Stat.*, **29**, 1221–33.

Moore, D. S., & G. P. McCabe (1989). *Introduction to the Practice of Statistics*. New York: W.H. Freeman.

Mosteller, F. (1946). On some useful "inefficient" statistics. *Ann. Math. Stat.*, **17**, 377–408.

Murphy, K. R., B. Myors, & A. Wolach (2014). *Statistical Power Analysis*. New York: Routledge.

Nelder, J. A., & R. W. M. Wedderburn (1972). Generalized linear models. *J. R. Stat. Soc.*, **A135**, 370–84.

Neyman, J., & E. S. Pearson (1933). On the problem of the most efficient tests of statistical hypotheses. *Philos. Trans. R. Soc. Lond.*, **A231**, 289–337.

Nolan, J. P. (1997). Numerical calculation of stable densities and distribution functions. *Comm. Stat. Stoch. Mod.*, **13**, 759–74.

Nolan, J. P. (1998). Parameterizations and modes of stable distributions. *Stat. Prob. Lett.*, **38**, 187–95.

Nolan, J. P. (2001). Maximum likelihood estimation and diagnostics for stable distributions. In *Lévy Processes: Theory and Applications*, ed. O. E. Barnsdorff-Nielsen, T. Mikosch, & S. I. Resnick. Basel, Birkhäuser.

Oldham, K. B., & J. Spanier (1974). *The Fractional Calculus*. San Diego: Academic Press.

Patnaik, P. B. (1949). The non-central χ^2- and F-distributions and their applications. *Biometrika*, **36**, 202–32.

Pawlowsky-Glahn, V., & A. Buccianti (eds.) (2011). *Compositional Data Analysis: Theory and Applications*. New York: Wiley.

Pawlowsky-Glahn, V., J. J. Egozcue, & R. Tolosana-Delgado (2015). *Modeling and Analysis of Compositional Data*. New York: Wiley.

Pearson, K. (1896). On a form of spurious correlation which may arise when indices are used in the measurement of organs. *Philos. Trans. R. Soc. Lond.*, **60**, 489–98.

Pearson, K. (1900). On the criterion that a given system of deviations from the probable in the case of a correlated system of variables is such that it can be reasonably supposed to have arisen from random sampling. *Philos. Mag.*, **50**, 339–57.

Pesarin, F., & L. Salmaso (2010). *Permutation Tests for Complex Data: Theory, Applications and Software*. New York: Wiley.

Phipson, B., & G. K. Smyth (2010). Permutation p-values should never be zero: calculating exact p-values when permutations are randomly drawn. *Stat. Appl. Genet. Mol. Biol.*, **9**(1), art. 39.

Picinbono, B. (1996). Second order complex random vectors and normal distributions. *IEEE Trans. Sig. Proc.*, **44**, 2637–40.

Pitman, E. J. G. (1937a). Significance tests which may be applied to samples from any population. *J. R. Stat. Soc. Suppl.*, **4**, 119–30.

Pitman, E. J. G. (1937b). Significance tests which may be applied to samples from any population. II. The correlation coefficient test. *J. R. Stat. Soc. Suppl.*, **4**, 225–32.

Pitman, E. J. G. (1938). Significance tests which may be applied to samples from any population. III. The analysis of variance test. *Biometrika*, **29**, 322–35.

Poisson, S. D. (1837). *Recherches sur la Probabilité des Jugements en Matiére Criminelle et en Matiére Civile, Précédées des Regles Générales du Calcul des Probabilités*. Paris: Bachelier.

Preisendorfer, R. W. (1988). *Principal Component Analysis in Meteorology and Oceanography*. Amsterdam: Elsevier.

Rao, C. R. (1945). Information and the accuracy attainable in the estimation of statistical parameters. *Bull. Calcutta Math. Soc.*, **37**, 81–91.

Rao, C. R. (1947). Large sample tests of statistical hypotheses concerning several parameters with applications to problems of estimation. *Proc. Camb. Philos. Soc.*, **44**, 50–7.

Rao, C. R. (1951). An asymptotic expansion of the distribution of Wilks' criterion, *Bull. Inter. Stat. Inst.*, **33**, 177–80.

Reiersøl, O. (1941). Confluence analysis by means of lag moments and other methods of confluence analysis. *Econometrica*, **9**, 1–24.

Reiersøl, O. (1945). Confluence analysis by means of instrumental sets of variables. *Ark. Mat. Astron. Fys.*, **32**, 1–119.

Rencher, A. C. (1995). *Methods of Multivariate Analysis*. New York: Wiley.

Rencher, A. C. (1998). *Multivariate Statistical Inference and Applications*. New York: Wiley.

Rice, J. A. (2006). *Mathematical Statistics and Data Analysis*, 3rd edn. Independence, KY: Cengage Learning.

Romano, J. P. (1990). On the behavior of randomization tests without a group invariance assumption. *J. Am. Stat. Assoc.*, **85**, 686–92.

Rousseeuw, P. J. W. (1984). Least median of squares regression. *J. Am. Stat. Assoc.*, **79**, 871–80.

Rousseeuw, P. J. W., & A. M. Leroy (1987). *Robust Regression and Outlier Detection*. New York: Wiley.

Rousseeuw, P. J. W., & A. M. Leroy (2005). *Robust Regression and Outlier Detection*, 2nd edn. New York: Wiley.

Rutherford, E., H. Geiger, & H. Bateman (1910). The probability variations in the distribution of α particles. *Philos. Mag.*, **20**, 698–707.

Ryan, T. P. (1997). *Modern Regression Methods*. New York: Wiley.

Samorodnitsky, G., & M. Taqqu (1994). *Stable Non-Gaussian Random Processes*. London: Chapman & Hall.

Satterthwaite, F. E. (1946). An approximate distribution of estimates of variance components. *Biomed. Bull.*, **2**, 110–14.

Schenker, J., & J. F. Gentleman (2001). On judging the significance of differences by examining the overlap between confidence intervals. *Am. Stat.*, **55**, 182–6.

Scheffé, H. (1959). *Analysis of Variance*. New York: Wiley.

Schreier, P. J., & L. L. Scharf (2010). *Statistical Signal Processing of Complex-Valued Data*. Cambridge University Press.

Schreier, P. J., L. L. Scharf, & A. Hanssen (2006). A generalized likelihood ratio test for impropriety of complex signals. *IEEE Sig. Proc. Lett.*, **13**, 433–6.

Shaffer, J. P. (1991). The Gauss-Markov theorem and random regressors. *Am. Stat.*, **45**, 269–73.

Simpson, J., A. Olsen, & J. C. Eden (1975). A Bayesian analysis of a multiplicative treatment effect in weather modifications. *Technometrics*, **17**, 161–6.

Siotani, M., K. Yoshida, H. Kawakami, K. Nojiro, K. Kawashima, et al. (1963). Statistical research on the taste judgement: analysis of the preliminary experiment on sensory and chemical characters of Seishu. *Proc. Inst. Stat. Math.*, **10**, 99–118.

Smirnov, N. V. (1939). On the estimation of the discrepancy between empirical curves of distribution for two independent samples. *Bull. Math. Univ. Moscow. Serie Int.*, **2**, 3–16.

Spearman, C. E. (1904). The proof and measurement of association between two things. *Am. J. Psych.*, **15**, 72–101.

Stephens, M. A. (1970). Use of the Kolmogorov-Smirnov, Cramer-Von Mises and related statistics without extensive tables. *J. R. Stat. Soc.*, **B32**, 115–22.

Stephens, M. A. (1976). Asymptotic results for goodness-of-fit statistics with unknown parameters. *Ann. Stat.*, **4**, 357–69.

Stigler, S. (1977). Do robust estimators work with *real* data? *Ann. Stat.*, **5**, 1055–98.

Stuart, A., & J. K. Ord (1994). *Kendall's Advanced Theory of Statistics,* vol. 1: *Distribution Theory*, 6th edn. London: Edward Arnold.

Stuart, A., J. K. Ord, & S. Arnold (1999). *Kendall's Advanced Theory of Statistics,* vol. 2: *Classical Inference and the Linear Model*, 6th edn. London: Edward Arnold.

Thomson, A., & R. Randall-Maciver (1905). *Ancient Races of the Thebaid.* Oxford University Press.

Thomson, D. J. (1977). Spectrum estimation techniques for characterization and development of WT4 waveguide, *I. Bell. Syst. Tech. J.*, **56**, 1769–815.

Thomson, D. J., & A. D. Chave (1991). Jackknifed error estimates for spectra, coherences and transfer functions. In *Advances in Spectrum Analysis and Array Processing*, vol. 1, ed. S. Haykin. Englewood Cliffs, NJ: Prentice-Hall, pp. 58–113.

Thompson, W. R. (1936). On confidence ranges for the median and other expectation distributions for populations of unknown distribution form. *Ann. Math. Stat.*, **7**, 122–8.

Tibshirani. R. (1996). Regression shrinkage and selection via the lasso. *J. R. Stat. Soc.*, **B58**, 267–88.

Uchaikin, V. V., & V. M. Zolotarev (1999). *Chance and Stability.* Waterbury, VT: VSP Press.

Van Den Bos, A. (1995). A multivariate complex normal distribution: a generalization. *IEEE Trans. Inform. Theory*, **41**, 537–9.

Von Bortkiewicz, L. J. (1898). *Das Gesetz der kleinen Zahlen*. Leipzig: B.G. Teubner.

Von Luxburg, U., & V. H. Franz (2009). A geometric approach to confidence sets for ratios: Fieller's theorem and the general linear model. *Stat. Sinica*, **29**, 1095–117.

Von Mises, R. (1918). Über die "Ganszzahligkeit" der Atomgewichte und verwandte Fragen. *Phys. Z.*, **19**, 490–500.

Von Neumann, J. (1951). Various techniques used in connection with random digits: Monte Carlo methods. *App. Math. Ser. Nat. Bur. Stand.*, **12**, 36–8.

Wald, A. (1940). The fitting of straight lines if both variables are subject to error. *Ann. Math. Stat.*, **11**, 284–300.

Wald, A. (1943). Tests of statistical hypotheses concerning several parameters when the number of observations is large. *Trans. Am. Math. Soc.*, **54**, 426–82.

Walden, A. T., & P. Rubin-Delanchy (2009). On testing for impropriety of complex-valued Gaussian vectors. *IEEE Trans. Sig. Proc.*, **57**, 825–34.

Wasserman, L. (2004). *All of Statistics: A Concise Course in Statistical Inference*. New York: Springer.

Wasserstein, R. L., & N. A. Lazar (2016). The ASA's statement on *p*-values: context, process and purpose. *Am. Stat.*, **70**, 129–33.

Wilcoxon, F. (1945). Individual comparisons by ranking methods. *Biometrics Bull.*, **1**, 80–3.

Wilcoxon, F. (1946). Individual comparisons of grouped data by ranking methods. *J. Econ. Entomol.*, **39**, 269–70.

Wilks, S. S. (1932). Certain generalizations in the analysis of variance. *Biometrika*, **24**, 471–94.

Wilks, S. S. (1938). The large-sample distribution of the likelihood ratio for testing composite hypotheses. *Ann. Math. Stat.*, **9**, 60–2.

Wishart, J. (1928). The generalized product moment distribution in samples from a normal multivariate population. *Biometrika*, **20**, 32–52.

Yohai, V. J. (1987). High breakdown-point and high efficiency robust estimates for regression. *Ann. Stat.*, **15**, 642–56.

Index

additive log ratio transformation 401
additivity property 52, 124
Aitchison distance 398
Aitchison inner product 397
Aitchison norm 398
analysis of variance 192
  and linear regression 289
  error variance 192
  hypothesis variance 192
  Kruskal-Wallis nonparametric 243
Anderson-Darling goodness-of-fit test 228
  bias correction 265
Ansari-Bradley dispersion test 214
arithmetic mean 36

Bartlett's *M* test 189
barycenter 394
Bayes' theorem 17
Bayesian approach 4
beta distribution 76
bias 143–4
binomial distribution 50, 59
biplot 413
bivariate Gaussian distribution 81
Bonferroni method 163, 210, 293, 352–3
Boole's inequality 9
bootstrap
  and linear regression 293, 308
  bias-corrected and accelerated confidence interval 256–7
  bias correction 251
  confidence interval 255
  distribution 247, 309–10
  double bootstrap 293, 309
  nonparametric 250
  one-sample location test 259
  parameter estimation 251
  parametric 250
  percentile confidence interval 256
  p-value 259
  smoothed 254
  t confidence interval 255
  test statistic 259
  two-sample dispersion test 262
  two-sample location test 261
bounded influence estimator 327
  breakdown point 327
  efficiency 328
  weighted least squares 328

canonical correlation 371
Cauchy distribution 66, 102, 125, 147
censoring 119
centered log ratio transformation 401
central limit theorem 124
  classic 51, 102–3
  convergence from Berry-Esseen theorem 104
  directional rvs 105
  generalized 105
  Lindeberg-Lévy 102
  Lyapunov 103
  multivariate 348
  order statistic 102
characteristic function 25, 37, 41, 65
Chebyshev inequality 45, 65, 102
chi square distribution 122
  noncentral 123
chi square goodness-of-fit test 219
chi square test 186
circular variance 43
closure 393
Cochran's theorem 140
combination 12
complement 5
complex multivariate Gaussian distribution 349
  augmented covariance matrix 350
  augmented variable 350
compositional data 38, 378, 391
  additive log ratio 401
  Aitchison distance 398
  Aitchison geometry on the simplex 395
  Aitchison inner product 397
  Aitchison norm 398
  and correlation coefficient 391
  balance 403
  barycenter 394–5
  biplot 413, 421
  center 408
  centered log ratio 401
  closure 393

compositional line 398
contrast matrix 403
correlation coefficient 406
definition 392
empirical orthogonal function 412
inverse 403
isometric log ratio 402
orthonormal basis 402
permutation invariance 395
perturbation 395
powering 396
scale invariance 394
sequential binary partition 403, 418
simplex as sample space 393
subcomposition 393
subcompositional coherence 395
ternary diagram 393, 410
variation matrix 408
conditional expected value 44
conditional variance 45
confidence interval 160
and hypothesis testing 196
and linear regression 292
and order statistics 163
and Student *t* 161
Bonferroni 163
bootstrap 255
central 161
conservative 162
delta method 167
Fieller 166
confidence interval
jackknife 279
consistent estimator 142
convergence
almost surely 46
in distribution 46
in probability 46
correlation 44
correlation coefficient 130, 295
Spearman 241
covariance 44
covariance matrix
population 345
sample 344
Cramér-Rao bound 145–7
cumulative distribution function 23

data sets
battery life 186
Cavendish density of Earth 90, 92, 94
cloud seeding 184, 239, 261, 272
earthquake interval 217, 221, 224, 226, 257, 262, 265
Egyptian skulls 361, 363
Gossett wheat yield 181, 235, 237
household spending 190, 242–3
Hubble recession velocity 298, 324
iris flower 341, 343
komatiite oxides 422
leukemia remission 234, 237, 274
Longley 388
MAR oxides 406, 408–10, 412, 414, 418
medical tablet 193, 352
Michelson speed of light 182, 188–9, 208, 238, 240, 271, 275
oceanographer days at sea 179, 195, 233, 236, 260, 269
Old Faithful geyser 159, 163, 248, 251–2, 254, 256–7, 279
paleocurrent 97
radioactive decay 216, 220
sake 367, 372, 387
star data 301, 326, 331
temperature-humidity-evaporation 369, 372
tornados 206
V-1 flying bomb 217, 221
wheat reflectance 312, 333, 338
wind direction 203, 209
zonally averaged temperature 357, 365, 378
De Moivre-Laplace Theorem 51
decision theory 152
delta method 105, 167
DeMoivre-Laplace Theorem 57
direction parameter 42
disjoint 8
dispersion parameter 39
distribution
Bernoulli 50
beta 76
binomial 50, 59
bivariate 27
bivariate Gaussian 81
Cauchy 35, 66, 103, 198
chi square 122
complex multivariate Gaussian 349
conditional 30
continuous 20, 24
correlation coefficient 130
discrete 18, 23
exponential 71
extreme value 74
F 128
Fisher 85
gamma 70
Gaussian 62, 66
generalized extreme value 78
geometric 54
Hotelling's T^2 *See* Hotelling's T^2 distribution
hypergeometric 56
interquartile range 138
inverted 121
joint pdf of pair of order statistics 138
largest-order statistic 33

distribution (cont.)
  Lévy 66
  lognormal 69
  mixed 22
  multinomial 54
  multivariate Gaussian 346
  negative binomial 53
  noncentral chi square 123
  noncentral *F* 128
  noncentral Student *t* 125
  Poisson 59
  product 35, 121
  quotient 34, 120
  Rayleigh 67
  sampling 122
  single-order statistic 135
  smallest-order statistic 33
  stable 65
  Student *t* 125
  truncated 120
  von Mises 82
  Weibull 74
  Wilks' Λ *See* Wilks' Λ statistic
  Wishart 348
Durbin-Watson test 295

efficiency 146, 317
  relative 146
efficient estimator 146
empirical cumulative distribution function 108–9
  confidence interval 226
empirical orthogonal function 412
empirical orthogonal function analysis 374
  biplot 413
  choosing the number of eofs 377
  empirical orthogonal function 375
  empirical orthogonal function amplitude 375
  eof regression 385
  from correlation matrix 381
  from covariance matrix 379
  from sample correlation matrix 376
  from sample covariance matrix 375
  from singular-value decomposition 376
  Rayleigh's principle 375
  rotation 378
empty set 5
estimate 87
estimator 86
  admissible 153
  bootstrap 251
  consistent 142
  efficient 146
  empirical cumulative distribution function 108
  geometric mean 89
  harmonic mean 88
  histogram 108
  interquartile range 92
  jackknife 277
  jackknife variance 278
  kernel density 111
  kurtosis 94
  least squares *See* least squares estimator
  maximum likelihood 155
  minimax 153
  minimum variance unbiased 146–7
  moments 154
  range 92
  ratio 166
  resultant 96
  risk 152
  sample circular variance 96
  sample covariance matrix *See* covariance matrix
  sample mean 87
  sample mean direction 96
  sample median *See* median
  sample variance 91
  skewness 93
  standard deviation 91
  unbiased 143
event 1
exchangeability 29, 268
expectation *See* expected value
expected value 2, 36–7
  order statistic 136
exponential distribution 71
extreme value distribution 74

F distribution 128
  noncentral 128
F test 188
finite sample breakdown point 148
Fisher distribution 85
Fisher information 145
  matrix 145, 207
frequentist approach 5, 86

gamma distribution 70
Gauss hypergeometric function 130
Gaussian distribution 62
general addition law 9, 13, 16
generalized extreme value distribution 78
generalized linear model 342
generalized sample variance 344
geometric distribution 54
geometric mean 38, 89
Glivenko-Cantelli theorem 110, 222
goodness-of-fit test
  Anderson-Darling 228
  chi square 219
  Cramér–von Mises 228
  Jarque-Bera 230
  Kolmogorov-Smirnov 222
  Lilliefors 223
  multinomial likelihood ratio 215

harmonic mean 38, 88
hat matrix 285, 328
  diagonal 317
  distribution of diagonal elements 285, 297
  properties 285
hazard function 72, 75
histogram 108
Hoeffding's inequality 46, 111
Hotelling's T^2 distribution 351, 353
  confidence interval 352
hypergeometric distribution 56
hypothesis test
  analysis of variance 192
  Ansari-Bradley 214
  Bartlett's M test 189
  bootstrap 259
  bootstrap one-sample location 259
  bootstrap two-sample dispersion 262
  bootstrap two-sample location 261
  chi square 186
  comparing covariance matrices 364
  Durbin-Watson 295
  F test 188
  for independence 366
  for multivariate regression 367
  for propriety of complex Gaussian variables 366
  Kendall τ zero correlation 242
  Mann-Whitney 237
  on covariance matrix 362
  one-sample t test 179
  one-sample location parameter permutation 269
  paired-sample t test 180
  permutation test 267
  Poisson dispersion test 206
  power 178, 183
  p-value 176
  rank sum 237
  Rayleigh test for uniformity 208
  runs test 294
  score test 208
  sign test 232
  signed rank 235
  Spearman zero correlation 241
  sphericity test 363
  two-sample dispersion permutation 275
  two-sample location parameter permutation 270
  two-sample paired data permutation 274
  two-sample Student t 181
  Wald test 207
  z test 177
  zero correlation 190
hypothesis testing 169
  alternate hypothesis 169
  and confidence interval 196
  and sufficient statistics 199
  composite hypothesis 170
  exact test 174
  lower-tail test 171
  Neyman-Pearson lemma 169–70, 196–8
  null hypothesis 169
  power 173, 179, 187–8, 205
  power and sample size 194
  p-value 176
  simple hypothesis 170
  two-tail test 171
  Type 1 error 172
  Type 2 error 172
  uniformly most powerful 174, 198
  upper-tail test 171

improper complex number 349
independence 15, 28
independent and identically distributed 29
indicator function 24
infinite set
  countably 5
  uncountably 5
influence function 318
influential data 295
instrumental variables 335
integral
  Riemann 22
  Riemann-Stieltjes 22
interquartile range 40, 92
  distribution 138
intersection 5, 7
inverse transform sampling 106
isometric log ratio transformation 402

jackknife 277
  and non-Gaussian statistics 279
  confidence interval 279
Jarque-Bera goodness-of-fit test 230

Kendall τ zero correlation test 242
kernel density estimator 111
Kolmogorov axioms 8, 20
Kolmogorov-Smirnov goodness-of-fit test 222
  bias correction 265
Kruskal-Wallis nonparametric analysis of variance 243
kurtosis 41, 94

lasso 337
law of large numbers 101, 110
law of total covariance 45, 285
law of total expectation 45
law of total probability 9, 16, 74
law of total variance 45
least squares estimator 284
  analysis of variance 289
  and bootstrap 293, 308
  and Durbin-Watson test 295
  and jackknife 293
  and maximum likelihood estimate 286

least squares estimator (cont.)
  and sufficiency 286
  backward stepwise selection 316, 336
  best linear unbiased estimator 286
  bounded influence estimator *See* bounded influence estimator
  confidence interval 292
  constrained 289
  correlation coefficient 290
  covariance matrix 284
  distributional assumptions 286
  eof regression 385
  expected value 284
  generalized linear model 342
  hat matrix *See* hat matrix
  instrumental variables 335
  linear model 283
  logistic regression 340–1
  model assessment 296
  multicollinearity 313
  multivariate *See* multivariate regression
  normal equations 282, 286
  predicted values 285
  predictor variables 283
  qr decomposition 288
  random errors 283
  residuals 285
  response variable 283
  ridge estimator 313
  robust estimator *See* robust estimator
  runs test 294
  shrinkage estimator *See* shrinkage estimator
  singular-value decomposition *See* singular-value decomposition
  statistical inference 289
  Student *t* test 291
  two-stage least squares 335
leptokurtic 41
leverage point 285, 295, 297, 317, 327
likelihood function 145
likelihood ratio test
  comparing covariance matrices 364
  composite hypothesis 200
  for independence 366
  multinomial goodness-of-fit 215
  simple hypothesis 197
  sphericity test 363
  Wilks' theorem 202
  Wilks' Λ 355
Lilliefors goodness-of-fit test 223
linear regression 282
  and conditional covariance 284
  and conditional expectation 284
location parameter 36
logistic regression 340
lognormal distribution 69
loss function 152, 318
MAD 40
  sample 92
Mahalanobis distance 345–6, 351, 363
Mann-Whitney *U* test 237
Markov inequality 45
MATLAB
  ADistance 398
  adtest 229
  AInnerProduct 397
  ANorm 398
  anova1 193
  Anovar 290
  ansaribradley 229
  Benjamini 212
  betacdf 164
  BI 329
  biplot 378
  bootci 257
  bootstrp 252
  boxplot 193
  canoncorr 371
  Cart_to_Comp 429
  cdf 49
  chi2cdf 125
  chi2gof 221
  chi2inv 125
  chi2pdf 125
  chi2rnd 125
  Close 393
  Clr 401
  ClrI 401
  Comp_to_Cart 429
  corr 241, 243
  corrcoef 100
  corrcov 100
  cov 100
  dendrogram 411
  disttool 19
  dwtest 295
  ecdf 108
  ecdfhist 108
  fcdf 129
  finv 129
  fitdist 158
  fpdf 129
  frnd 129
  fzero 77, 99
  geomean 89
  geopdf 54
  glmfit 342
  harmmean 89
  histogram 2, 108
  Huber 321
  hygecdf 58
  hygepdf 58
  Hypergeometric2f1 130
  icdf 49

Ilr 405
IlrI 405
integral 137
jackknife 280
jbtest 230
kruskalwallis 244
ksdensity 111
kstest 223
kstest2 223
kurtosis 94
lasso 337
lassoPlot 337
lillietest 223
linkage 411
lsqlin 289
lsqnonneg 289
mad 92
makedist 49
manova1 357
manovacluster 360
mean 88
median 89
mldivide 287
mle 158
mlecov 158
mnpdf 55
mode 89
mvncdf 81
mvnpdf 81, 347
mvnrnd 81
mvregress 370
nancov 100
nanmean 88
nanstd 92
nanvar 92
ncfcdf 129
ncfinv 129
ncfpdf 129
ncfrnd 129
nchoosek 12
nctcdf 127
nctinv 127
nctpdf 127
nctrnd 127
ncx2cdf 125
ncx2inv 125
ncx2pdf 125
ncx2rnd 125
normcdf 137
normpdf 137
normrnd 120
NVariation 409
Padesum 130
pca 377
pcacov 377
pdf 49
Perturbation 396
Pochhammer 130
Powering 397
qqplot 118
qr 288
random 49
randperm 269
range 92
ranksum 238
raylrnd 120
regress 288
ridge 337
rng default 251
robustfit 321
rose 98
rotatefactors 378
runstest 295
sampsizepwr 194–5
signrank 236
signtest 233
skewness 94
std 92
tcdf 127
Ternary 429
TernaryLabel 429
TernaryLine 429
TernaryPlot 433
tiedranks 232
tpdf 127
Treg 292
trimmean 88
trnd 127
ttest 180
ttest2 183
unidpdf 19
unidrnd 1
unifpdf 21
var 91
vartest 188
vartest2 189
vartestn 189
WilksLambda 356
wishrnd 349
zscore 92
ztest 178
maximum likelihood estimator 155
and Cramér-Rao bound 158
and equivariance 157
and sufficient statistic 157
for Gaussian distribution 155
for Poisson distribution 155
for von Mises distribution 156
multivariate Gaussian 348
mean difference 40
mean direction 42
mean squared error 144, 336, 338
median 38, 42
memory-less property 54, 73

mesokurtic 41
meta-analysis 245
  inverse chi square method 245
  probability integral transformation method 245
mode 39
Monte Carlo test 259
  bias correction of goodness-of-fit test 265
multicollinearity 385
multinomial distribution 54
multinomial likelihood ratio goodness-of-fit test 215
multiple hypothesis testing 210
  Benjamini-Hochberg method 211
  comparison-wise error rate 210
  false discovery rate 211
  family-wide error rate 210
multivariate analysis of variance 354
  error matrix 355
  hypothesis matrix 355
  multivariate regression 367
multivariate Gaussian distribution 346
multivariate regression 366
  adequate set of predictors 368
  backward elimination 368
mutually exclusive 8

negative binomial distribution 53
noncentrality parameter 123

order statistic 34, 89, 135
  expected value 136
outlier 295, 297, 317
overdispersed 54, 206

Padé approximant 130
percent-percent plot 115
  confidence interval 227
  stabilized 115–16
permutation 11, 29
permutation distribution 268
permutation test 267
  exchangeability 268
  one-sample location parameter 269
  p-value 268
  sampling without replacement 268
  symmetry of permutation distribution 269
  two-sample dispersion 275
  two-sample location parameter 270
  two-sample paired 274
perturbation 395
pivot 166
platykurtic 41
p-norm
  and robustness 147
Poisson distribution 59
Poisson process 60
power transformation *See* powering
powering 396
principal components *See* empirical orthogonal function analysis
probability
  conditional 14
  joint 15
  marginal 15
probability density function 19–20
probability inequality 45
probability integral transform 106
probability measure 2
proper complex number 349
p-value 176
  cautions about 176

quantile function 24
quantile-quantile plot 117–18, 185, 297, 300, 303, 305–6, 314, 325–6, 332
  confidence interval 227
quantiles 39

random variable 2
randomization distribution 274
range 40, 92
ranks 231
ratio estimator 166
  bias of 166
  delta method 167
  Fieller's method 166
Rayleigh distribution 67
reproductive property 71
resultant 42, 96
ridge regression 313, 336
robust estimator 317
  finite sample breakdown point 317
  Huber function 318
  implementation 320
  influence function 318
  loss function 318
  M-estimator 317
  weighted least squares 320
robustness 147
runs test 294

sample circular variance 96
sample covariance matrix
  biased 100
  unbiased 100
sample mean 87
  independent of the sample variance 140
  variance 87
  vector 344
sample mean direction 96
sample median 89
  expected value 137
  pdf 137
  variance 137, 144

sample space 1, 5, 10
  simple 10
sample variance 91
  biased 91
  generalized 344
  independent of the sample mean 140
  mean squared error 144
  total 344
  unbiased 91
sampling distribution 86, 122
sampling with replacement 50, 247, 277
sampling without replacement 56, 268
score function 145
sequential binary partition 403, 418
set theory 5, 171
shape parameter 41
shrinkage estimator 336
  lasso 337
  ridge estimator 336
sign test 232
simplex 393
simulation 120
singular-value decomposition 288
skewness 41, 93
sphericity test 363
stable distribution 65
standard deviation 40, 91
Student *t* distribution 125
  noncentral 125
Student *t* test 178
sufficiency 148
  and data 149
  and exponential family 150
  and factorization of the pdf 149
  and order statistics 150
  Fisher-Neyman factorization theorem 149
  Rao-Blackwell theorem 151
sufficient statistic
  and ancillary statistic 151
  complete 150
  joint 150
  minimal 150
supremum 110, 153
survival function 72, 75

ternary diagram 393, 410, 416, 429
tolerance interval 164
trigonometric moments 42
two-stage least squares 335

unbiased estimator 143
union 5

variance 39
Venn diagram 5
von Mises distribution 82

Weibull distribution 74
Wilcoxon rank sum test 237
Wilcoxon signed rank test 235
Wilks' Λ statistic 355, 358, 364, 366, 387
  for multivariate regression 367
  partial 368
Wishart distribution 348

zero correlation test 190